Y. Xu, D. Ruan, K. Qin, J. Liu

Lattice-Valued Logic

Springer

Berlin
Heidelberg
New York
Hong Kong
London
Milano
Paris
Tokyo

Studies in Fuzziness and Soft Computing, Volume 132
http://www.springer.de/cgi-bin/search_book.pl?series=2941

Editor-in-chief
Prof. Janusz Kacprzyk
Systems Research Institute
Polish Academy of Sciences
ul. Newelska 6
01-447 Warsaw
Poland
E-mail: kacprzyk@ibspan.waw.pl

Further volumes of this series can be found on our homepage

Vol. 113. A. Abraham, L.C. Jain and J. Kacprzyk (Eds.)
Recent Advances in Intelligent Paradigms and Applications", 2003
ISBN 3-7908-1538-1

Vol. 114. M. Fitting and E. Orowska (Eds.)
Beyond Two: Theory and Applications of Multiple Valued Logic, 2003
ISBN 3-7908-1541-1

Vol. 115. J.J. Buckley
Fuzzy Probabilities, 2003
ISBN 3-7908-1542-X

Vol. 116. C. Zhou, D. Maravall and D. Ruan (Eds.)
Autonomous Robotic Systems, 2003
ISBN 3-7908-1546-2

Vol 117. O. Castillo, P. Melin
Soft Computing and Fractal Theory for Intelligent Manufacturing, 2003
ISBN 3-7908-1547-0

Vol. 118. M. Wygralak
Cardinalities of Fuzzy Sets, 2003
ISBN 3-540-00337-1

Vol. 119. Karmeshu (Ed.)
Entropy Measures, Maximum Entropy Principle and Emerging Applications, 2003
ISBN 3-540-00242-1

Vol. 120. H.M. Cartwright, L.M. Sztandera (Eds.)
Soft Computing Approaches in Chemistry, 2003
ISBN 3-540-00245-6

Vol. 121. J. Lee (Ed.)
Software Engineering with Computational Intelligence, 2003
ISBN 3-540-00472-6

Vol. 122. M. Nachtegael, D. Van der Weken, D. Van de Ville and E.E. Kerre (Eds.)
Fuzzy Filters for Image Processing, 2003
ISBN 3-540-00465-3

Vol. 123. V. Torra (Ed.)
Information Fusion in Data Mining, 2003
ISBN 3-540-00676-1

Vol. 124. X. Yu, J. Kacprzyk (Eds.)
Applied Decision Support with Soft Computing, 2003
ISBN 3-540-02491-3

Vol. 125. M. Inuiguchi, S. Hirano and S. Tsumoto (Eds.)
Rough Set Theory and Granular Computing, 2003
ISBN 3-540-00574-9

Vol. 126. J.-L. Verdegay (Ed.)
Fuzzy Sets Based Heuristics for Optimization, 2003
ISBN 3-540-00551-X

Vol 127. L. Reznik, V. Kreinovich (Eds.)
Soft Computing in Measurement and Information Acquisition, 2003
ISBN 3-540-00246-4

Vol 128. J. Casillas, O. Cordón, F. Herrera, L. Magdalena (Eds.)
Interpretability Issues in Fuzzy Modeling, 2003
ISBN 3-540-02932-X

Vol 129. J. Casillas, O. Cordón, F. Herrera, L. Magdalena (Eds.)
Accuracy Improvements in Linguistic Fuzzy Modeling, 2003
ISBN 3-540-02933-8

Vol 130. P.S. Nair
Uncertainty in Multi-Source Databases, 2003
ISBN 3-540-03242-8

Vol 131. J.N. Mordeson, D.S. Malik, N. Kuroki
Fuzzy Semigroups, 2003
ISBN 3-540-03243-6

Yang Xu
Da Ruan
Keyun Qin
Jun Liu

Lattice-Valued Logic

An Alternative Approach to Treat Fuzziness and Incomparability

Springer

Professor Dr. Yang Xu
E-mail: yxu@home.swjtu.edu.cn
Professor Dr. Keyun Qin
E-mail: keyunqin@263.net
Southwest Jiaotong University
Department of Applied Mathematics
Chengdu
610031 Sichuan
PR China

Professor Dr. Da Ruan
E-mail: druan@sckcen.be
Belgian Nuclear Research Center (SCK*CEN)
Boeretang 200
2400 Mol
Belgium

Dr. Jun Liu
The University of Manchester Institute
of Science and Technology
Department of Mathematics
PO Box 88
Manchester M60 1 QD, UK

ISSN 1434-9922
ISBN 3-540-40175-X Springer-Verlag Berlin Heidelberg New York

Library of Congress Cataloging-in-Publication-Data applied for

A catalog record for this book is available from the Library of Congress.

Bibliographic information published by Die Deutsche Bibliothek
Die Deutsche Bibliothek lists this publication in the Deutsche Nationalbibliographie; detailed bibliographic data is available in the internet at <http://dnb.ddb.de>.

Springer-Verlag Berlin Heidelberg New York
a member of BertelsmannSpringer Science+Business Media GmbH
http://www.springer.de

Typesetting: Data delivered by authors
Cover design: E. Kirchner, Springer-Verlag, Heidelberg
Printed on acid free paper 62/3020/M - 5 4 3 2 1 0

Preface

One of the fundamental goals of artificial intelligence (AI) is to build artificially computer-based systems which make computer simulate, extend and expand human's intelligence and empower computers to perform tasks which are routinely performed by human beings. Due to the fact that human intelligence actions are always involved with uncertainty information processing, one important task of AI is to study how to make the computer simulate human being to deal with uncertainty information. Among major ways in which human being deal with uncertainty information, the uncertainty inference becomes an essential mechanism in AI. From the viewpoint of symbolism, it is highly necessary to study and establish the logical foundation for uncertainty inference. Note that classical logic has already been the foundation of certainty inference, logical foundation for uncertainty inference should be the extension and development of classical logic, which are often generally called non-classical logics, where many-valued logic has been one kind of important non-classical logics.

Fuzziness and incomparability are two kinds of uncertainty often associated with human's intelligent activities in the real world, and they exist not only in the processed object itself, but also in the course of the object being dealt with. Therefore, it is necessary to investigate the logical foundation and the corresponding uncertainty inference theory and methods for characterizing and dealing with not only the fuzziness and incomparability associated with the object itself, but also the uncertainty involved within the course of the object being processed as well. These have become the motivation of our research work. Accordingly, we started the corresponding research work from 1980's. We proposed and studied a many-valued logic–lattice-valued logic based on lattice implication algebra and further studied the theories and methods of uncertainty inference and automated reasoning. Up to now, there has been a lot of work being done. In this present book, we systematically summarize our research work over last decades mainly by our research group, but also include some other researchers' work in this subject.

We expect the book to be useful for AI research and as a reference book for logicians, mathematicians and computer scientists in uncertainty information processing. This book may be used as a text book for graduate students in the relevant areas. The book is divided into three parts.

Part I (Chapter 1) provides the academic background of this book and the major methodologies of AI, where the related research work outline of many-valued logic, especially lattice-valued logic as well as the related uncertainty inference and automated reasoning based on many-valued logic are reviewed.

Part II (Chapters 2–8) introduces the logic algebra– lattice implication algebra and studies its properties. Concretely, Chapter 2 proposes the concept of lattice implication algebra by combining lattice and implication algebra, and gives some typical examples of lattice implication algebra, *e.g.*, Boolean algebra, Łukasiewicz implication algebra. Furthermore, it discusses some basic properties of lattice implication algebras and lattice H implication algebras as well as homomorphisms between lattice implication algebras. The fact that all lattice implication algebras form a proper class is proved. Chapter 3 discusses a special kind of substructures in lattice implication algebra–filters, and gives various kinds of filters and their properties. It is shown that filter is a kind of abstract of MP rule. Chapter 4 discusses the dual structure of filter, *i.e.*, LI-ideal, and introduces some special LI-ideal structures in lattice implication algebra and their properties. Chapter 5 devotes to the homomorphism between lattice implication algebras, discusses the lattice implication quotient algebras based on the congruence relations induced by filters, fuzzy filters, LI-ideals and fuzzy LI-ideals, and gives the corresponding homomorphism theorem and the isomorphism theorem of lattice implication algebras. Moreover it discusses the characteristics of proper lattice implication algebra. Chapter 6 studies lattice implication algebra from the viewpoint of topology, and discusses the filter spaces for lattice implication algebras and their topological properties, such as countability, separability, compactness and connectedness. It further investigates prime topological space based on prime filters. Chapter 7 describes the relations between lattice implication algebra and BCK-algebra, MV-algebra, FI-algebra, R_0-algebra, (2, 2, 2, 1, 0, 0)-type algebra, (2, 1, 0, 0)-type algebra, (2, 0, 0)-type algebra, respectively, and gives several equivalent definitions of lattice implication algebras. Chapter 8 treats the category theories of lattice implication algebras, fuzzy lattice implication algebras, and fuzzy power set theory based on lattice implication algebra. It discusses the adjoint semigroups induced by lattice implication algebras, and formalizes the lattice implication algebra theory in the first-order language with identity.

Part III (Chapters 9–11) introduces the algebraic logic – lattice-valued logics based on lattice implication algebra and their applications in uncertainty reasoning and automated reasoning. Concretely, Chapter 9 establishes the lattice-valued propositional logic $LP(X)$ and the gradational lattice-valued propositional logic L_{vpl} based on lattice implication algebra and discusses their properties. The main results include the Soundness Theorem, Deduction Theorem and Consistent Theorem of $LP(X)$, the (α, β)-i type Completeness and Soundness Theorem, (α, β)-i type Consistent Theorem, (α, β, θ)-i type Deduction Theorem of L_{vpl}. It also discusses the compactness in L_{vpl}, and gives the gradational Łukasiewicz propositional logic L_u and F-valued Łukasiewicz propositional logic FL_u as the special cases. Chapter 10 further introduces the corresponding lattice-valued first-order logics based on lattice implication algebra, *i.e.*, lattice-valued first-order logic $LF(X)$ and the gradational lattice-valued first-order logic L_{vfl}, and discusses the analogous properties in Chapter 9 accordingly. Also the gradational Łukasiewicz first-order logic L_{uf} and F-valued Łukasiewicz first-order logic FL_{uf} are introduced as the special cases. Chapter 11 starts with the uncertainty reasoning approaches based on lattice-valued logic $LP(X)$ corresponding to four kinds of uncertainty reasoning models, then introduces the uncertainty reasoning theory and approach based on L_{vpl}, where the reasoning scheme may be implemented according to semantic interpretation as well as syntactic deduction in L_{vpl}. Finally, it establishes the α-resolution principles based on lattice-valued logic $LP(X)$ and $LF(X)$.

The outcome of this book involves many people's efforts. Our thanks are due to Dr. Jun Ma for his excellent typesetting and editing of the manuscript with some help from Wei Wang and Xuefang Wang; to Xiaodong Guan, Dr. Zheng Pei, Dr. Tianrui Li, Dr. Wenjiang Li, Shuwei Chen, Dan Meng, Yongchuan Tang, Xiaoping Qiu, and Xiaowei Yang for their assistance; to Prof. Young Bae Jun at Gyeongsang National University, Korea and Prof. Yiquan Zhu at Zhao Qing Normal College, China for providing us with some research materials in line with the subject of this book. Furthermore, we acknowledge gratefully for great support on our research by National Natural Science Foundation Committee of China (No. 60074014), Science & Technology Department of Sichuan Province, China, and the Flanders-China Cooperation Project, Belgium. Special thanks are also due to all the support in the preparation of the book by Southwest Jiaotong University of China, the Belgian Nuclear Research Centre (SCK•CEN), and UMIST in the UK. In addition, we acknowledge all the authors whose research papers and books are cited in this book. We do realize that many scholars' research results on the subject are not yet fully cited in the book, but will be part of our future research tasks. Last but not least, our thanks go to Prof. Janusz Kacprzyk (Editor-in-chief), to Katharina Wetzel-Vandai and Dr. Thomas Ditzinger (Editors, Springer-Verlag) for their kind consideration and advice to include this book into their "Studies in Fuzziness and Soft Computing" series.

Chengdu • Mol • Manchester *Yang Xu, Da Ruan, Keyun Qin, Jun Liu*
March 2003

List of symbols

$/\!/$ incomparable
$\pitchfork$ disjoint
$\emptyset$ empty set
$\leqslant$ partial order relation
$'$ complementary, or negation
$\rightarrow$ implication
$\otimes$ residuated operation
$\vee$ supremum, or disjunction
$\wedge$ infimum, or conjunction
$[a)$ generated filter by the element a
$\langle a\rangle$ generated ideal by a
$ord(a)$ order of the element a
a° inner point of a
$\bar{a}$ the closure of a
$[A)$ generated filter by the set A
$\langle A\rangle$ generated ideal by the set A
$c(A)$ the closure of A
$d(A)$ the derived sets of A
$i(A)$ inner of A
$ob(A)$ the set of all objects of A
$P(A)$ power set of the set A
L Lattice; short for lattice implication algebra
$C(L)$ the set of completely normal fuzzy LI-ideal
$Dat(L)$ the set of all dual atoms of L
$IF(L)$ the set of all implicative filters of lattice implication algebra L
$N(L)$ the set of normal fuzzy LI-ideal of L
$PF(L)$ the set of all prime filters of lattice implication algebra L
$\boldsymbol{LC}$ the category of lattice implication algebra

FLC the category of fuzzy lattice implication algebra

F(LC) the forgetful category of LC

I the greatest element of lattice implication algebra; the greatest element of lattice

O the smallest element of lattice implication algebra; the smallest element of lattice

$[a, b\ ]$ the interval in lattice

$\cong_A$ congruence relation induced by the fuzzy LI-ideal A

$\equiv_f$ congruence induced by the mapping of f

D- ker(f) dual kernel of mapping f

ker(f) kernel of mapping f

χ_A the characteristic function of the set A

$\mathscr{F}_L(U)$ the set of all L-fuzzy subsets on U

$\mathscr{F}(X)$ the set of all fuzzy subsets on X

L/F lattice implication quotient algebra induced by filter F

hom(L_1, L_2) the set of all lattice implication homomorphisms from L_1 to L_2

max maximum

min minimum

$\mathcal{T}_{F(L)}$ generated topology by $F(L)$

$\mathcal{U}_x$ the neighborhood base of x

$(L, \mathcal{A})$ topology lattice

v valuation

$\models, \|\!\models$ semantical implies

$\vdash, \|\!\vdash$ syntactical implies

$\models_\alpha, \|\!\models_\alpha$ α semantically implies

$\vdash_\alpha, \|\!\vdash_\alpha$ α syntactical implies

$\models_{\underset{\cdot}{\alpha}}$ strong α syntactical implies

$LP(X)$ the lattice-valued propositional logic

$LF(X)$ the lattice-valued first-ordered logic

L_{vpl} the gradational lattice-valued propositional logic

L_{vfl} the gradational lattice-valued first-ordered logic

$\mathcal{F}$ the set of formulae for propositional logic $LP(X)$

$\mathscr{F}$ the set of formulae for first-ordered logic $LF(X)$

$\mathscr{F}_p$ the set of formulae for gradational lattice-valued propositional logic L_{vpl}

$\mathcal{F}_f$ the set of formulae for gradational lattice-valued first-ordered logic L_{vfl}

Con the set of consequence

Ded the set of deduction

$\mathcal{D}$ interpretation

$\mathscr{I}$ interpretation

(r_n, t_n) n-ary inference rule

$\mathscr{R}$ the set of inference rules

$\mathrm{val}(\omega)$ the value the proof ω

$l(\omega)$ the length of proof ω

$R_\alpha(D_j, D_k)$ α-resolvent of D_j and D_k

α-$\circledcirc$ α-empty clause

$\aleph_0$ denumerable infinity

$\triangleq$ be defined as

$\mathbb{N}$ the set of $\{0, 1, \cdots\}$

$\mathbb{N}^+$ the set of $\{1, 2, \cdots\}$

List of Figures

List of Tables

Table of Contents

Part III Lattice-Valued Logic Systems

Part I

Introduction

1 Introduction

1.1 Major Methodologies in Artificial Intelligence

Artificial intelligence (AI) has been one of top research fields such as biology engineering, space technology, and material engineering, *etc.* It is a comprehensive discipline developed on the basis of multidisciplinary studies, including computer science and technology, information science and technology, control theory, psychology, physiology, philosophy, mathematics, and linguistics, *etc.* The key idea of AI is to let machines simulate, extend and expand human's intelligence by artificial methods and technologies, in order to realize some "machinery thinking," and endow machines with the ability of simulating human's mental faculties to solve the complex real problems, *e.g.*, learning, reasoning, judgment, and decision-making, *etc.* Accordingly, in the framework of the different levels of human's mental faculties, three kinds of simulation methodologies have been gradually established, *i.e.*, *symbolism* which simulates human's abstract mental faculties, *connectionism* which simulates human's visualized mental faculties, and *behaviorism* which simulates human's apperception mental faculties [430].

Symbolism [514] — based on logic, aims to realize human brain's ability such as reasoning, judgment, and decision-making, *etc*, by symbol computing. Symbolism was originated from 1956, while Newell and Simon developed the Heuristic Program Logic Theorist, proved 38 theorems of the book "Mathematic Principle" and studied GPS (General Problem Solver), then it was applied to simulate the human's mental faculties to solve problems. Thereafter, symbol systems were established gradually. Symbolism regarded that the aim of AI is to realize mechanic intelligence. The computer itself possesses the ability to deal with problems by symbols, which implicates the capability of deductive reasoning. So we can attain the purpose of simulating human intelligent activities by running the corresponding programs to embody some kinds of intelligent actions of logical thinking. Currently, most research achievements of AI are based on this kind of methods. The major features of symbolism in information processing can be summarized as follows:

- Based on logic operations and symbol computing, applied to simulate human's logical thinking, can solve the complex problem which needs logic reasoning;
- Knowledge can be represented by definite symbols. In the case of knowing basic rules, you need not to provide a large amount of detailed knowledge;
- Easy to be modularized and revised upon any change in individual facts arises;
- It can connect with traditional databases;
- It can explain the reasoning conclusions so that it is convenient for making a choice from possible conclusions.

Connectionism [514]— adopts the view that the basic function unit of the brain is neuron. It is a kind of simulation as brain processing information using neurons and the network formed among neurons. Moreover, it implements the simulation of human intelligence through the parallel associated ability among neurons, *i.e.*, neural computing. One major model of neural computing is neural network model. The major features of connectionism in information processing can be summarized as follows:

- Processing information by the parallel associated function among neurons, where the process is parallel, dynamic, and global;
- The information is stored in the distributive physical association among neurons, and the association function can be implemented;
- Implementing the simulation of human's learning, classification and recognition by dynamically adjusting the weights of connections among neurons;
- Being suitable for simulating human's visualized mental process;
- Can obtain the approximate solutions of the problem relatively quick.

Behaviorism – proposed by Brooks (1991) [38] is a new artificial intelligence methodology which does not need knowledge representation and reasoning. Brooks takes the point of view that the intelligence is reflected only in the interaction with environment.

Roughly, Table 1.1 shows the major characteristics of these three methodologies in AI [430].

Table 1.1. Comparison of the Characteristics of Symbolism, Connectionism and Behaviorism.

	Symbolism	Connectionism	Behaviorism
Cognition level	discrete	continuous	continuous
Representation level	symbol	connection	activity
Solving level	from top to bottom	from bottom to top	from bottom to top
Process level	serial	parallel	parallel
Operating level	reasoning	mapping	interaction
System level	local	distributive	distributive
Basis level	logic	simulation	judging by intuition

Due to the fact that usually the process human deals with information is the synthesis of multiple mental faculties, we state that AI research need the synthesis of symbolism, connectionism and behaviorism in order to endow machines with much richer intelligence.

Here we mention that the contents of this book are only restricted within the category of symbolism. Furthermore, the focus is on describing and dealing with uncertainty based on symbol systems.

1.2 Basic Academic Ideas

Chronically, humans have made virtual investigations and great achievements in certainty information processing. However, not only does there exist various kinds and large portions of uncertainty in the real world but also the uncertainty is naturally generalized in the course of outer information being reflected to human brain. Such uncertainty includes fuzziness, non-partitionality, incomparability, incompleteness, unreliability, and randomicity, *etc.*

While human dealing with uncertainty information, it is often associated with two kinds of uncertainty: one exists in the processed object itself and the other exists in the course of information processing. The latter results from the former and is more complex than the former. In order to handle the latter, the uncertainty in the course of information processing is often partitioned into different levels, therefore, it can be handled within one or several levels according to the demand. Hitherto, the research work in dealing with these two kinds of uncertainty simultaneously seems still weak.

As it is well known, certainty information processing, especially reasoning based on certainty information, is on the basis of classical two-valued logic. Due to its strict and complete logical foundation (classical logic), inference for certainty information has reached high confidence. Accordingly, it is natural and necessary to establish some reasonable logic systems as the logical foundation for uncertainty information processing. Certainly, this kind of logics is not two-valued logic but the development and extension of two-valued logic. Various kinds of non-classical logic systems hitherto have been extensively studied in the context of finding natural and efficient inference systems to deal with uncertainty.

Many-valued logic, an important non-classical logic, on the one hand is a great extension and development of classical logic and aims at establishing the logical foundation for uncertainty information process, on the other hand, it has always been a crucial direction in non-classical logic.

Especially, in order to establish the theories and methods to simultaneously deal with fuzziness and incomparability of processed object itself and uncertainty in the course of information processing, it is necessary and important to investigate many-valued logical systems with the corresponding information features.

According to the above academic ideas, we consider it necessary to establish a logic system with truth value in a lattice, *i.e.*, a kind of lattice-valued logic which is one kind of many-valued logic and extends the chain-type truth value field to lattice. Since 1980s, we have made extensive study on lattice-valued logic system on the base of former investigations on many-valued logic, especially Pavelka [351] and Novak's [328,330] works. We proposed the lattice-valued logic systems based on lattice implication algebra and studied the corresponding reasoning theories and methods The purpose of this research attempts to enrich and develop the many-valued logic systems and provide a kind of suitable logical foundation for intelligent information processing.

The basic framework of our research work on lattice implication algebras, lattice-valued logical systems and their applications are illustrated in Figure 1.1:

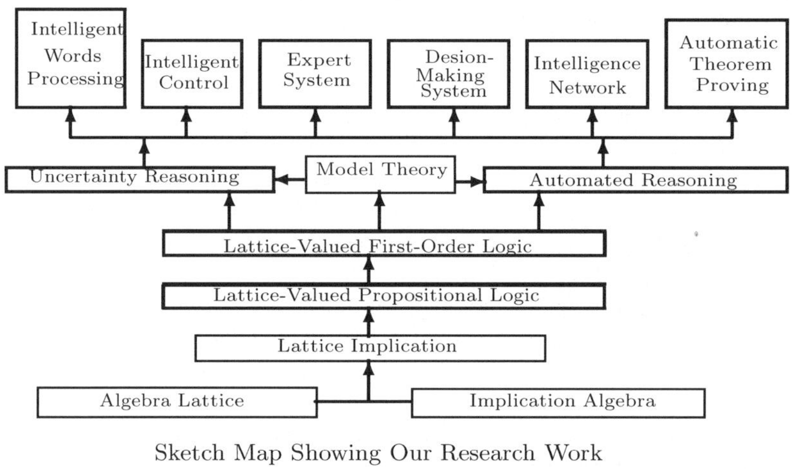

Fig. 1.1. Research Structure Graph

The main contents of the research work introduced in this book include three aspects as follows:

– The lattice-valued logical algebra.
– The lattice-valued logic.
– Uncertainty reasoning and automated reasoning based on logic.

1.3 Some Related Concepts

Here we first introduce some basic concepts about lattice theory, it may be convenience for the reader going to subsequent sections.

Definition 1.3.1 *A partial ordered set (poset) is a set in which a binary relation $\leqslant$ is defined, which satisfies the following conditions: for any x, y, z,*

P1. $x \leqslant x$, *for any* x *(Reflexive),*
P2. $x \leqslant y$ *and* $y \leqslant x$ *implies* $x = y$ *(Antisymmetry),*
P3. $x \leqslant y$ *and* $y \leqslant z$ *implies* $x \leqslant z$ *(Transitivity).*

An upper bound of a subset X of a poset P is an element $a \in P$ containing every $x \in X$. The least upper bound (l.u.b.) is an upper bound contained in every other upper bound. The notions lower bound of X and greatest lower bound (g.l.b.) of X are defined dually. A lattice is one particular type of partly ordered set as the following definition implies.

Definition 1.3.2 *A lattice is a poset any two of whose elements have a greatest lower bound (g.l.b.), denoted by $x \wedge y$, and a least upper bound (l.u.b.), denoted by $x \vee y$. A lattice L is complete when each of its subsets X has a l.u.b. and a g.l.b. in L.*

Here two new operations are introduced on L:

$$a \wedge b = \inf\{a, b\}, a \vee b = \sup\{a, b\},$$

calling $a \wedge b$ the meet and $a \vee b$ the join of a and b.

Here is an equivalent definition of lattice using the operations of meet and join:

Definition 1.3.3 *let L be an arbitrary set, and let there be given two binary operations on L, denoted by $\wedge$ and $\vee$. Then the structure $(L, \wedge, \vee)$ is an algebraic structure with two binary operations. We call the structure $(L, \wedge, \vee)$ a lattice provided that it satisfies the following properties:*

(1). For any $x, y, z \in L$, $x \wedge (y \wedge z) = (x \wedge y) \wedge z$, *and* $x \vee (y \vee z) = (x \vee y) \vee z$.
(2). For any $x, y \in L$, $x \wedge y = y \wedge x$, *and* $x \vee y = y \vee x$.
(3). For any $x \in L$, $x \wedge x = x$, *and* $x \vee x = x$.
(4). For any $x, y \in L$, $x \wedge (x \vee y) = x$, *and* $x \vee (x \wedge y) = x$.

Definition 1.3.4 *A poset $\langle A, \leqslant\rangle$ is a join semilattice if $\sup\{a, b\}$ exists for all $a, b \in A$. A poset $\langle A, \leqslant\rangle$ is a meet semilattice if $\inf\{a, b\}$ exists for all $a, b \in A$.*

Definition 1.3.5 *If a lattice has a smallest element, denoted by O, and a greatest element, denoted by I, then it is called a bounded lattice.*

Definition 1.3.6 *We say that the complement of x is x' if the following conditions are satisfied.*

(1). $(x')' = x$,
(2). $x \wedge x' = O$,
(3). $x \vee x' = I$,
(4). $x \geqslant y$ *implies* $x' \leqslant y'$.

Definition 1.3.7 *A distributive lattice* L *is a lattice which satisfies either of the distributive laws for any* $x, y, z \in L$,

$$z \vee (x \wedge y) = (x \vee z) \wedge (y \vee z),$$
$$x \wedge (y \vee z) = (x \wedge y) \vee (x \wedge z).$$

Definition 1.3.8 *A boolean lattice is a uniquely complemented distributive lattice.*

Definition 1.3.9 *A lattice* L *is called modular if for any* $x, y, z \in L$ *we have*

$$x \leqslant y \text{ implies } x \vee (y \wedge z) = y \wedge (x \vee z).$$

This is called the modular law. Every boolean lattice is a modular lattice.

Definition 1.3.10 (Orthomodular lattices) *[217] A lattice* L *is called orthomodular if for any* $x, y, z \in L$ *we have* $x \leqslant y$ *implies* $x \vee (x' \wedge y) = y$.

Definition 1.3.11 *[150] A bilattice is a structure* $\langle B, \leqslant_t, \leqslant_k, \neg \rangle$ *consisting of a non-empty set* B*, two partial orders of it,* $\leqslant_t$ *and* $\leqslant_k$*, and a mapping from it to itself,* $\neg$*, such that:*

(1). *each of* $\leqslant_t$*, and* $\leqslant_k$ *makes* B *a complete lattice;*
(2). $x \leqslant_t y$ *implies* $\neg y \leqslant_t \neg x$;
(3). $x \leqslant_k y$ *implies* $\neg x \leqslant_k \neg y$;
(4). $\neg\neg x = x$.

Definition 1.3.12 (Distributive p-lattice) *[441] A distributive p-lattice is an algebra* $(A, \wedge, \vee, *, 0, 1)$ *of type (2, 2, 1, 0, 0) such that* $(A, \wedge, \vee, 0, 1)$ *is a distributive lattice with 0 and 1, and* $*$ *is an operation of pseudocomplementation, i.e.,* $a^* = \max\{b \in A \mid a \wedge b = 0\}$.

One extension from classical logic to many-valued logic is the extension of two-valued logic operation to many-valued logic. Among them, two important cases of lattices are of interest and often being used: when lattice L is a finite simple ordered set; and when lattice L is the unit interval $[0, 1]$. More general, L should be a lattice with suitable operations like $\vee, \wedge, \rightarrow, \prime$. The question of the appropriate operation and lattice structure has generated much literature. Goguen [152] extended the concept of fuzzy sets to L-fuzzy sets in which membership grades form a partial ordered set instead of a linear ordered set as $[0, 1]$. Moreover, according to such an extension, Goguen [153] established L-fuzzy logic of which truth value set is a complete lattice-ordered monoid, which is also called a complete residuated lattice in Pavelka and Novak's L-fuzzy logic [328, 351]. Since this algebraic structure is quite general, many commonly used algebraic structures may be considered as the residuated lattice-like structures [485].

Definition 1.3.13 *[351] A residuated lattice is an algebraic structure* $\langle L, \vee, \wedge, \otimes, \rightarrow, O, I\rangle$ *of type* $(2, 2, 2, 2, 0, 0)$*, where*

(1). $L = \langle L, \vee, \wedge, \otimes, \rightarrow, O, I\rangle$ *is a bounded lattice with the smallest element* O *and the greatest element* I*.*
(2). $\langle \otimes, \rightarrow\rangle$ *is an adjoint couple on* L*, i.e.,*
 a) $\otimes$ *is isotone (order-preserving) on* $L \times L$*;*
 b) $\rightarrow$ *is antitone (order-reversing) in the first and isotone in the second variable on* $L \times L$*;*
 c) for all $x, y, z \in L$ *hold the adjointness condition or Galois correspondence:* $x \otimes y \leqslant c$ *if and only if* $x \leqslant y \rightarrow z$*.*
(3). $\langle L, \otimes, I\rangle$ *is a commutative monoid.*

The operation $\otimes$ is called multiplication and $\rightarrow$ is called residuation. A residuated lattice is said to be complete, if L is a complete lattice. The algebraic structure in Definition 1.3.13 is also known by several other names. Birkhoff [26] uses the term integral commutative residuated monoid, Blyth and Janowitz [29] would call it a residuated Abelian semigroup (with an additional specification of the unit), Goguen [153] calls it complete lattice ordered semigroup (and later also cl_∞-monoid). The original terminology residuated lattice was introduced in [81]. For convenience, in the following, we always refer to enriched residuation lattice. The Gödel structure $([0, 1], \min, \rightarrow)$, where $x \rightarrow y$ is 1 for $x \leqslant y$ and y elsewhere, is an example of a complete residuated lattice. The Gaines structure $([0, 1], \cdot, \rightarrow)$, where $\cdot$ is the usual multiplication of reals and $x \rightarrow y$ is 1 for $x \leqslant y$ and b/a elsewhere, is also an example of a complete residuated lattice.

The following algebra are considered as a residuated lattice-like structures [485]:

Definition 1.3.14 *A De Morgan Algebra, also called soft algebra* $L = (L, \vee, \wedge, \neg)$ *is a distributive lattice* $(A, \wedge, \vee)$ *with a unary operation* $\neg$ *such that the De Morgan laws hold.*

Fuzzy logic systems based on soft algebras have been studied by Di Nola and Ventre [83], Preparata and Yeh [364], Lee and Chang [234], Kaufmann [219], Esteva [118,119], Garcia and Esteva [136], and Ying [577] among others.

Definition 1.3.15 *A pseudo-Boolean algebra (also called Heyting algebra, Brouwerian algebra, Godelian algebra), is an algebra* $L = (L, \wedge, \vee, \rightarrow, \neg)$*, where* $\wedge$ *and* $\vee$ *are the lattice infimum and supremum respectively,* $\neg x = x \rightarrow O$ *and* $\rightarrow$ *is determined by the following condition:* $x \wedge y \leqslant z$ *if and only if* $x \leqslant y \rightarrow z$*.*

Fuzzy-valued logics and many-valued logics in pseudo-Boolean algebras have been studied by Horn [185, 298, 298], Takeuti and Titani [471], among others. Rasiowa [394] and Di Zenzo's [86] studies are pseudo-Boolean algebras-like, too, since there the value lattice is a Post algebra, *i.e.*, a pseudo-Boolean algebra equipped with some additional operations.

Definition 1.3.16 *[351] Let $L = [0, 1]$. If for any $a, b \in L$, put*

$$a \otimes b = \max\{0, a + b - 1\}, a \rightarrow b = \min\{1, 1 - a + b\}.$$

Then $\langle L, \otimes, \rightarrow \rangle$ is called a Łukasiewicz algebra. Here $\rightarrow$ is called the Łukasiewicz implication, $\otimes$ is called the Łukasiewicz product and another operator $\oplus$, $a \oplus b = \min\{1, a + b\}$, is called the Łukasiewicz sum. Note that $a' = 1 - a = a \rightarrow 0$, $a \otimes b = \max\{0, a + b - 1\} = (a \rightarrow b')'$, and $a \oplus b = a' \rightarrow b$. Similarly,

Pavelka [351] and Novak [328,330] have studied fuzzy logic on Łukasiewicz lattice where L is either the unit interval or a finite chain.

There are different Łukasiewicz systems according to truth values allowed. There are finite-valued system L_n $(n = 2, 3, 4, \cdots)$ with the set of value $\{i/n \mid i = 0, 1, \cdots, (n - 1)\}$, *i.e.*, Let $L_{m+1} = \{0 = a_0 < \cdots < a_m = 1\}$ be the $(m + 1)$-element chain, $m \geqslant 1$. We set $a_k \otimes b_p = a_{\max\{0,k+p-m\}}, a_k \rightarrow b_p = a_{\min\{m,m-k+p\}}$ for $0 \leqslant k,\ p \leqslant m$. Then $\langle L, \otimes, \rightarrow \rangle$ is also a residuated lattice, is called as $(m + 1)$-element Łukasiewicz chain, denoted as L_{m+1}.

The countably infinite valued system L_w has all rationals from $[0, 1]$ for its truth values and L_ω has the w whole continuum [0,1] of values [161, 406].

One of the essential content of the brilliant work of Goguen [153] is that the algebra of inexact concept is residuated lattice-like [485]. Turunen [485] showed that residuated lattice possesses the common features in various fuzzy logical systems based on the different particular algebraic structures. He proved that Pseudo-Boolean algebra, MV-algebra, Łukasiewicz algebra, Gaines algebra, and t-norm algebra all define a residuated lattice (for more details about all these algebras, see the references in [485]).

Definition 1.3.17 *[127] An Wajsberg algebra is an algebra $(L, \rightarrow, *, 1)$ of type $(2, 1, 0)$ such that*

(1). $1 \rightarrow x = x;$
(2). $(x \rightarrow y) \rightarrow ((y \rightarrow z) \rightarrow (x \rightarrow z)) = 1;$
(3). $(x \rightarrow y) \rightarrow y = (y \rightarrow x) \rightarrow x;$
(4). $(x^* \rightarrow y^*) \rightarrow (y \rightarrow x) = 1.$

There is a one-to-one corresponding between Wajsberg algebras and MV-algebras [182, 183].

Definition 1.3.18 (Lindenbaum Algebra) *[392, 465] Let L be a first-order language. We define the equivalence relation $\sim$ over formula of L by $\varphi \sim \psi$ if and only if $\vdash \varphi \leftrightarrow \psi$. The equivalence class of a formula φ is denoted* *by* $[\varphi]$. *Let* $B = L/\sim$ *be the collection of all equivalence classes under this equivalence relation. We define the operations* $\oplus$ *and* $\cdot$ *and complementation, denoted* $\overline{[\varphi]}$ *by:*

$$\begin{aligned}[j] \oplus [y] &= [j \vee y], \\ [j] \cdot [y] &= [j \wedge y], \\ \overline{[\varphi]} &= [j'].\end{aligned}$$

We let $0 = [j \wedge j']$, *and and* $1 = [j \vee j']$. *Then the structure* $(B, \oplus, \cdot, \bar{}, 0, 1)$ *is a Boolean algebra, called the Lindenbaum algebra.*

Hájek [174, 175] has proposed BL-algebras in order to study the "Basic Logic" arising from these applications, in other words he proved that the "Basic Logic" (here understood in "narrow sense") is that logic which generalizes the three logics, more commonly used in fuzzy set theory: Łukasiewicz logic, Product logic, and Gödel logic. The related Lindebaum algebras are the MV-algebras [52], the Product algebras [173] and the Heyting algebras satisfying the additional condition of prelinearity [173].

Definition 1.3.19 *A* BL*-algebra is an algebra* $L = (L, \wedge, \vee, \otimes, \to, 0, 1)$ *with four binary operations* $\wedge, \vee, \otimes, \to$ *and two different constants 0, 1 such that for each* $x, y, z \in L$, *by setting* $x \leqslant y$ *if and only if* $x \wedge y = x$ *hold:*

(1). $(L, \wedge, \vee, 0, 1)$ *is a distributive lattice with universal bounds 0 and 1,*
(2). $\otimes$ *is an associative, commutative and isotone operation having 1 as neutral element, i.e.,* $x \otimes 1 = x$,
(3). $x \wedge y = x \otimes (x \to y)$ *(divisibility),*
(4). $x \otimes y \leqslant z$ *if and only if* $x \leqslant y \to z$, *i.e., by (4),* $y \to z = \vee\{x \in L; x \otimes y \leqslant z\}$ *(residuation),*
(5). $(x \to y) \vee (y \to x) = 1$ *(pertinacity).*

Definition 1.3.20 (Gaines algebra) *[131]. Let* $L = [0, 1]$. *If for any* $a, b \in L$,

$$a \to b = \begin{cases} 1, & a \leqslant b, \\ 0, & a > b. \end{cases}$$

Definition 1.3.21 *A t-norm* $T\colon [0,1] \longrightarrow [0,1]$ *is a function satisfying the following conditions for any* $a, b, c \in [0,1]$:

T1 boundary conditions: $T(a, 1) = T(1, a) = a$, $T(a, 0) = T(0, a) = 0$;
T2 monotonicity: $a \leqslant b$ *implies* $T(a, c) \leqslant T(b, c)$, $T(c, a) \leqslant T(c, b)$;
T3 associativity: $T(a, T(b, c)) = T(T(a, b), c)$;
T4 commutativity: $T(a, b) = T(b, a)$.

Important examples of BL-algebras are the so-called t-norm algebra [64]:

Definition 1.3.22 *t-norm algebra [174] is an algebra:*

$$L = ([0,1], \wedge, \vee, \otimes, \rightarrow, 0, 1).$$

Where $L = ([0,1], \wedge, \vee, 0, 1)$ is the usual lattice on the real unit interval, $\otimes$ is a continuous t-norm and $\rightarrow$ is the corresponding residuum [64, 174].

The most used t-norm algebras in fuzzy set theory are the following [174, 175]

Gödel algebra:

$$x \otimes y = \min\{x, y\},$$
$$x \rightarrow y = \begin{cases} 1, \, a \leqslant b, \\ y, \, a > b. \end{cases}$$

Goguen algebra:

$$x \otimes y = x \cdot y,$$
$$x \rightarrow y = \begin{cases} 1, & a \leqslant b, \\ y/x, & a > b. \end{cases}$$

Łukasiewicz algebra (also see Definition 1.3.16)

$$x \otimes y = \max\{o, x + y - 1\}, x \rightarrow y = \min\{1, 1 - x + y\}.$$

MV-algebra and BL algebra are closely related, a BL-algebra L is an MV-algebra if and only if $x^{**} = x$ for all $x \in L$ [174].

MV-algebra, BCK-algebra, FI-algebra, R_0-algebra and Lattice implication algebra are refered to Sections 7.2, 7.1, 7.2, 7.2, 2.1, respectively.

1.4 Many-Valued Logic and Lattice-Valued Logic

Many-valued logics provide an interesting alternative to the classical logic for modelling and reasoning about systems. By allowing additional truth values, they support the explicit modelling of uncertainty and disagreement. For these reasons, they have been explored for a variety of applications in databases [133], knowledge representation [148, 150], machine learning [297], circuit design [178], and many other fields.

The first step in many-valued logic was taken by Polish logician Łukasiewicz [269] and American logician Post [363]. In 1920, Łukasiewicz introduced a 3-valued logic system, where the third truth value can represent "neutral," "undefined" or certain intermediate state. It is the first formal system in many-valued logic. Shortly after then, Post proposed a complete n-valued logic system independently different from the Łukasiewicz logic system, which does not pay attention to the philosophical problem specially, but to a purely formalized many-valued logic system. Afterwards, various many-valued logic systems with different philosophical background and different application fields have been proposed and studied. Some representative works are (see [31] for more details: 3-valued logic systems introduced by Heyting (1930), Sobocinski (1936), Webb (1936), Bochvar (1939), Reichenbach (1946), Slupecki (1946), Hallden (1949), Kleene (1952), Aqvist (1962), Segerberg (1965), Finn (1972), Pirrog-Rzepecka (1977), Zhu and Xiao [603–607], and Douglas (1993) [89] *et al.*, respectively; n-valued logic systems (see also [31] for details) mainly introduced by Gödel (1930), Kalmar (1934), Sobocinski (1936), Slupecki (1938), Chang (1958), Rasiowa (1972), Cignoli (1982), Mangani (1973), Komori (1978), Di Zenzo (1986), De Glas (1989), respectively; logic systems with truth values in $[0,1]$ mainly introduced by Zadeh [588], Lee and Chang [234], Rasiowa [393], Gaines [131], Baldwin [11], Gottwald [160], Todt [476], Wu [524], and Wang [497, 498]*etc.*, respectively.

In 1965, Zadeh [588] introduced the concept "fuzzy sets" which is of momentous significance on the study of non-classical logic. And in 1967, Goguen [152] extended the concept of fuzzy sets to L-fuzzy sets in which membership degrees form a partially ordered set instead of a linearly ordered set as $[0,1]$.

Goguen studied lattice-valued logic and further proposed the first lattice-valued logic formal system based on complete lattice-ordered semigroups which provided a new instrument and approach to study the lattice-valued logic. Thereafter, Rosser, Turquette and Pavelka, *etc*, made some improvements and developments on Goguen's logic system, where Pavelka's work is the most typical one. In 1979, Pavelka incorporated internal truth value in the language, established a fuzzy propositional logic system whose truth value set is an enriched residuated lattice and proved a lot of important results about its axiomatizability. The followings are most significant [351]:

(1). Any propositional logic with $L = [0,1]$ and a continuous residuation operation (in particular, Łukasiewicz's implication) is axiomatizable.
(2). Any propositional logic in which L is a finite chain, is axiomatizable.

Pavelka's work is concerned mainly with propositional fuzzy logic. Pavelka [351] showed that the only natural way of formalizing fuzzy logic for truth values in the unit interval $[0,1]$ is by using the Łukasiewicz's implication operator or some isomorphic forms of it; Novak [328, 330] extended it to the first-order fuzzy logic. Pavelka and Novak's work provided a relative general frame for lattice-valued logic system. There is no doubt that Pavelka's works [351] and Novak's works [328, 330] are of crucial importance for fuzzy logic. After that, many authors still devoted themselves into fuzzy logic in the light of Pavelka and Novak's works, (*i.e.*, many-valued logic based on residuated lattice, such as Turunen, Esteva, Godo, Hajek, and Ying *et al.* [120, 121, 172, 173, 175, 337, 483, 484, 487, 581, 584], where many of those important and remarkable results reflected some fundamental characteristics of fuzzy logic. For more details, *cf.* [175, 337].

We mention also an extensive literature on triangular norms and conorms as possible truth-function of the conjunction and disjunction, and the induced truth-function for implication. Triangular norms were introduced in the framework of probabilistic metric spaces [419–421], and they are applied in several fields, *e.g.*, fuzzy sets [588], fuzzy logics [44, 173, 351] and their applications.

Basically, residuated many-valued logics [174] are related to continuous t-norms which are used as truth functions for the conjunction connective, and their residua as truth functions for the implication. Main examples are Łukasiewicz, Gödel and Product logics, related to Łukasiewicz t-norm ($t(x, y) = \max\{0, x+y-1\}$), Gödel t-norm ($t(x, y) = \min\{x, y\}$) and product t-norm ($t(x, y) = x \cdot y$), respectively. Rose and Rosser [406] provided completeness results for Łukasiewicz logic, and Dummet [109] for Gödel logic, and recently three of the authors [173] axiomatized product logic. More recently, Hajek (1998) has proposed the axiomatic system BL corresponding to a generic continuous t-norm and taking Łukasiewicz, Gödel and Product logics as special cases. Chang [52,53], Mundici [314,315,317], Belluce [16], Turunen [488], Hajek [175] among others have studied many-valued logic based on MV-algebras.

From the logical viewpoint, fuzzy logic can be regarded as a kind of many-valued or infinite-valued logic. A many-valued propositional logic with truth values modelled by the unit interval $[0, 1]$ is quite often called a fuzzy logic (in a narrow sense). In such logic, the conjunction is usually interpreted by a triangular norm. There are different ways of interpretation of the implication; this, together with the choice of the triangular norm, leads to a large collection of fuzzy logics with different semantics. For details about the triangular-norm-based fuzzy logic, see [44].

Dubois and Prade [95,100] also investigated the possibilitic logic, and Liau *et al.* [243,244] introduced a possibilitic residuated implication logics with applications for reasoning, where the semantics of the logics is uniformly based on possibility theory. Each logic in the class is parameterized by a t-norm operation on $[0, 1]$, and the degree of implication between the possibilities of two formulas explicitly by using residuated implication with respect to the t-norm.

Another kind of important lattice-valued logic, *i.e.*, quantum logic, was born of the attempts of Jordan, *et al.* [204], and Birkhoff [25], and more fully in [468], aim to provide an axiomatic foundation for quantum mechanics [391]. Quantum logic based on the orthomodular lattices [217], for example, as lattices of closed convex subsets or closed linear subspaces in a real or complex Hilbert space. These lattices in either case are non-Boolean [217]. Quantum logic can be seen as being more general than Boolean logic since it does not require lattice distributivity. In fact, quantum logic does not even require modularity. The uncountably infinite orthonormal bases of a non-separable complex Hilbert space (H) can represent all conceivable experiments specifying a quantum mechanical system. The relations among classical logic, fuzzy logic and quantum logic may be seen in [347].

Boolean logic, fuzzy logic and quantum logic are typically modelled mathematically using lattice theory. In particular, Boolean lattices are typically used to model Boolean logic systems, the narrow version of fuzzy logic developed by Zadeh (1965) is typically modelled using the interval $[0, 1]$ in the real number line, equipped with its associated distributive lattice structure, and the general form of fuzzy logic allows the truth values to be taken in an arbitrary universal algebra. Quantum logics are typically constructed so that the truth values of propositions are taken in an ortholattice.

Among existing methods exploiting the structure of the lattice-valued algebra of truth values, we further mention [305,341], which gave theorem proving systems for m-valued Post logics and for algorithmic logics. Belnap [17] proposed a 4-valued logic that also introduces the value 'both' (*i.e.*, "true and false"), to handle inconsistent assertions in database systems. Salzer [413] studied operators and distribution quantifiers in finite-valued logics based on semi-lattices, and Hähnle [171] who derived tableau-style axiomatizations of distribution quantifiers by using Birkhoff's representation theorem for finite distributive lattices.

Another kind of particular lattice-value logics is the logic based on bilattice. It is a natural generalization of classical two-valued logic, which is introduced by Ginsberg in [148], and more fully in [123–125, 150]. The logics based on bilattice constitute an elegant family of many-valued logics. Those meeting certain natural conditions have provided the basis for providing a uniform semantics for a family of logic programming languages [123, 124]. Essentially, a bilattice is a space of generalized truth values with two lattices orderings. One ordering records degree of truth, whatever that means, while the other ordering records degree of information or knowledge [123, 150].

Sofronie-Stokkermans [439, 440] discussed the finite-valued logic having as an algebras of true values distributive lattices with certain type of operators, and their automated theorem proving approach, where many results are based on the Priestley representation theorem [72] of distributive lattice. In many cases the algebras of truth values associated to non-classical logic are of this type, such as SH_n-logics [197, 199, 439] (which are a series of propositional finite-valued logics based on symmetric Heyting algebras of order n). Iturrior (1995) showed that SH_n-logics are instances of L_{tf}-logics, which were introduced by Rasiowa (see *e.g.*, [393]) to formalize the reasoning of poset of intelligence (for further details see also [393, 395]). Logics based on p lattices was also investigated in [439–441]. They focus on non-classical logics having as algebraic models bounded distributive lattices with certain types of operators. Automated theorem proving method for this kind of logic were also discussed. The idea is to use a Priestley-style representation for distributive lattices with operators in order to define a class of Kripke-style models with respect to which the logic is sound and complete. If this class of Kripke-style models is elementary, it can then be used for a translation to clause form; satisfiability of the resulting clauses can be checked by resolution [441, 442, 446].

Zadeh [589] proposed a logic system based on a linguistic truth values lattice. Liu [265] proposed a logic system based on a complete lattice with the dividing element. The implication connective in Liu's logic system is Kleene's implication. Also its truth values lattice contains a special element, *i.e.*, the dividing element, and some important results are given for $[0, 1]$. Wang [497,498] investigated the logic system based on certain class of algebras and also obtained some important results.

In order to provide a logical foundation for uncertain information processing theory, especially for the fuzziness, the incomparability in uncertain information in the reasoning, Xu [534] established lattice implication algebra by combining lattice and implication algebra. Based on lattice implication algebra, we established lattice-valued propositional logic $LP(X)$, lattice-valued first order logic $LF(X)$, as well as fuzzy propositional logic $FP(X)$. In order to reflect the difference of degree among attributes of each intermediate-variation state of objects and the incomparability of objects, based on Goguen, Pavelka and Novak's works as well as on ours, we established gradational L-type lattice-valued propositional logic system L_{vpl} and first order logic system L_{vfl}. Moreover, we discussed the uncertainty reasoning methods and the resolution-based automated reasoning based on these logic systems.

The above important works for many-valued logics and lattice-valued logics have become an important foundation for investigating more deeply about many-valued logics and lattice-valued logics.

1.5 Uncertainty Inference

Methods for dealing with uncertainty in many areas of AI have received considerable attention for more than a decade. Several numerical and symbolic methods have been proposed for handling uncertainty information [65, 229, 230].

Three of the most common methods of representing and reasoning with uncertain knowledge are probability-based uncertainty reasoning, uncertainty reasoning based on fuzzy set theory, and uncertainty reasoning based on nonmonotonic logic. It is hard to rank the approaches in general, because each of them is usually aimed at a special application environment. In the subsequent subsection, an overview of major uncertainty reasoning methods are given.

1.5.1 Probability-Based Uncertainty Reasoning

The classical approach to address uncertainty is the Bayesian theory of probability. Probability theory has been used to model precisely described, repetitive experiments with observable but uncertain outcomes. In this approach, uncertain variables are assumed to be described by statistical parameters, which define the probability of the variable having a given value. The two main schools of thought in this field are the frequentist and the Bayesian. According to Pat-Cornell [350] the frequentist school (including classical statisticians) defines probability as a limiting frequency, which applies only if one can identify a sample of independent, identically distributed observations of the phenomenon of interest. The Bayesian side, on the other hand, looks upon the concept of probability as a degree of belief. The Bayesian framework also provides methods for updating your probabilities when new data are introduced based on Bay's rule. In theory, when there is sufficient data for the frequentist to estimate a probability, the subjectivist's assessment of his or her probability will converge to the frequentist's estimate of the probability of the event. In other words, they will tend to agree as information becomes available.

A basis for the application of probabilistic-oriented methods is the validity of statistical laws for stochastic input parameters. The heavy reliance on the probability theory as the only effective and reliable methodology to deal with uncertainty has historical roots. Probability theory has well established and sound scientific foundations and has been widely used for centuries. There have been several adaptations of probability theory within the literature of artificial intelligence including the odds-likelihood formulation used in Prospector [108], the certainty factors approach [40,145,431], the cautious approach adopted by Inferno [385], diagnosis system application [474], and the probabilistic logic approach by Nillson [327], Pearl's probabilistic reasoning approach in Bayesian networks [352].

It is recognized that in practical applications there are many subjective uncertainties, which are due to the lack of knowledge. To address these, subjective probability (Bayesian probability) is often applied. This approach uses the common probability approach, however expert judgment is used to generate the probability or probability distribution functions representing the resulting state of knowledge. Subjective probability theory matches quite well to our knowledge base of information from a human expert in addition to his or her subjective beliefs about the accuracy of that information [354]. The Bayesian approach for uncertainty reasoning is characterized by the following features [354]:

1) Probability is interpreted as degree of belief, based on available evidence; 2) Current knowledge is represented by a (real-valued) probability distribution on a proposition space; 3) New knowledge is learned by conditionalization.

An interesting important approach to using probability theory in computing is Bayesian networks, which is firstly explored by [352], also called probabilistic network, causal networks [179, 323, 353, 354], which are rooted in traditional subjective probability theory.

A Bayesian belief network consists of a graphical structure that is augmented by a set of probabilities. The graphical structure is a directed, acyclic graph in which nodes represent domain variables which have a finite number of possible values. Prior probabilities are assigned to source nodes, and conditional probabilities are associated with arcs.

Bayesian belief network is capable of representing the probabilities over any discrete sample space, such that the probability of any sample point in that space can be computed from the probabilities in the Bayesian belief network. The key feature of Bayesian belief network is their explicit representation of the conditional independence among events.

Today, Bayesian networks represent a culmination of Bayesian probability theory and causal graphical representations for modeling causal and probabilistic applications [57, 179, 402].

Some more research and applications of Bayesian approach will be found in [162, 216, 249, 321, 478, 566]

More thorough introductions to Bayesian analysis and its applications can be found in [20,138,389]. Bayesian networks have been applied in diverse fields like medical diagnosis [456], image recognition [33], search algorithms [177], and many others. A detailed list of current applications of BN is provided by Welman and Heckerman [519]. One of the most important features of BN is the fact that this approach provides an elegant mathematical structure for modeling complicated relationships among random variables while keeping a relatively simple visualization of these relationships. However, here we are not intended to provide an explanation of BN. For more information on this subject, see Pearl [354], Neapolitan [322], Oliver *et al.* [340], Charniak [57].

The Dempster-Shafer theory of evidence (DS theory) (sometimes called evidential reasoning or belief function theory) is formalized by Shafer [438] for representing and reasoning with uncertainty, imprecise and incomplete information. It is based on Dempster's original work [78] on the modelling of uncertainty in terms of upper and lower probabilities that are induced by a multi-valued mapping rather than as a single probability value. The Dempster-Shafer theory (DST) of evidence recognizes the distinction between uncertainty and ignorance by introducing belief functions that satisfy axioms that are weaker than those of probability functions [80,356]. Thus, probability functions are a subclass of belief functions, and the theory of evidence reduces to probability theory when the probability values are known. Roughly speaking, the belief functions allow us to use our knowledge to put constraints or bounds on the assignment of probabilities to events without having to specify the probabilities themselves [564]. In addition, the theory of evidence provides appropriate methods for computing belief functions for combinations of evidence. DS theory has been popular since early 1980s when AI researchers were searching for different mechanisms to cope with those situations where Bayesian probability is powerless, with particular emphasis on combining evidence from different sources. Its relationships with related theories have been intensively discussed [564].

A mathematical description of the condition of applying Dempster's combination rule is formalized in Voorbraak [527]. Some discussions and criticisms of the combination rule have appeared in the literature [267, 527].

Although D-S theory has been applied with some success to such topics as face recognition [193], statistical classification [79] and target identification [41]. Additional applications centered around multi-source information, including medical diagnosis [574] and plan recognition [14], and in the MCDM area [569–572], safety analysis and synthesis [501–503] *etc.* Some more works can be found in [176, 190, 250, 412, 438, 515, 516, 560] *etc.*

1.5.2 Fuzzy Set Based Uncertainty Reasoning

Since Zadeh proposed compositional rule of reasoning (CRI) in 1975 [589], fuzzy reasoning has attached considerable attention. The main works are: Zadeh's CRI method (1975); Baldwin's (1979) truth value restrictive method by combining CRI method with linguistic truth value implication; Mizumoto's [301] method by combining the extension principle with linguistic truth value implication; Tsukamoto's [480] method by combining the extension principle with linguistic truth value Łukasiewicz implication and Yager's [560] reasoning method based on a similarity measure. Zhang proposed a numerical-value reasoning model [599], and Xu proposed a reasoning method based on attribute description and a reasoning method based on set mapping [539, 610].

Bien *et al.* (1994) presented a reasoning network for bi-directional uncertainty reasoning based on fuzzy sets. The uncertainty reasoning scheme based on interval fuzzy sets has been discussed by many researchers [59, 61, 157, 187, 482, 587], Chen and Hsiao [62] proposed the approach dealing with bi-directional uncertainty reasoning of rule-based systems. [115] proposed a unified parameterized formulation of reasoning by combining Mamdani's approximation and formal logical reasoning approaches. Similarity-based fuzzy reasoning method has been studied [325, 575].

Yager [567] provided mechanisms to enable the fuzzy set based uncertainty reasoning system to deal with conflicts in knowledge; Raha and Ray [386] investigated the uncertainty reasoning with time providing a representation and manipulation of some vague concepts in the temporally changing world. Raha and Ray [387] proposed a theoretical study of an approach to default reasoning with vague default based on fuzzy logic, and also studied a formal uncertainty reasoning technique for representing and manipulating partially true vague knowledge [388]; Xu [555] proposed an uncertainty reasoning method based on lattice-valued propositional logic L_{vpl}, and also proposed fuzzy reasoning approach based on the extension principle [558]; Dubois and Prade [103] proposed a fuzzy reasoning method using fuzzy relation equations. Takagi *et al.* [469] proposed a multi-layered fuzzy reasoning method by means of multi-layered conceptual fuzzy sets. Nguyen *et al.* [324] constructed algebraic structures to model linguistic domains, and developed a method of linguistic reasoning. Kuhu *et al.* [232] proposed a neural network architecture for fuzzy reasonings. Wang [499] proposed the full implication 3-I fuzzy reasoning scheme using implication operator R_0, and Song *et al.* proposed backword 3-I fuzzy reasoning scheme [449]. Some other new methods may be, among others [60, 68, 191, 481, 575].

General introductions to fuzzy sets and fuzzy reasoning can be found in [23, 91, 95, 96, 101, 106, 107, 227, 231, 587]. An anthology of basic papers in fuzzy logic is in Dubois *et al.* [101]. Two anthologies of Zadeh's personal contributions have been published by Yager *et al.* [560] and [227, 587]. An overview of applications of fuzzy set theory is proposed under the form of an anthology by Dubois *et al.* [106].

Besides these methods, there are still many kinds of reasoning methods being proposed. One important and promising direction for fuzzy reasoning research is to establish fuzzy reasoning theories and methods based on strict and complete logic system so that these fuzzy reasoning methods may be more reasonable, suitable universally, sound and complete. Otherwise, it is impossible to make a great breakthrough.

1.5.3 Non-Monotonic Logic Based Uncertainty Reasoning

One of the most significant developments both in logic and AI is the emergence of a number of non-monotonic formalisms, devised expressly for the purpose of capturing defensible reasoning in a mathematically precise manner. Non-monotonic reasoning means that our intermediate belief may be changed according to additional information. This kind of reasoning of everyday life in which reasoners often draw conclusions that have to be withdrawn when further information is obtained. The set of conclusions thus does not grow monotonically with the given information. It is this phenomenon that non-monotonic reasoning methods try to formalize. This is in contrast to classical (first-order) logic, whose reasoning, being deductively valid, can never be "undone" by new information. Hence, non-monotonic reasoning is a subfield of AI trying to find more realistic formal models of reasoning than classical logic. Non-monotonic reasoning is suitable for reasoning in the cases of incomplete information, incomplete information knowledge and inconsistent information. Pioneering work in the field of non-monotonic logics began with the realization that ordinary first-order logic is inadequate for the representation of defensible reasoning accompanied by the effort to reproduce the success of classical first-order logic in the representation of mathematical, or formal, reasoning. Among the pioneers of the field in the late 1970's were (among others) McCarthy, McDermott & Doyle, and Reiter (see [148] for a collection of early papers in the field and [130] for a more recent collection of excellent survey papers).

In 1978, Reiter [398] firstly studied the problem on reasoning by default. Afterwards, he proposed a logic system for default reasoning [399]. At the same time, McDermott [289] put forward a non-monotonic logic, McCarthy [285, 287] gave a kind of Circumscription. Among these, Reiter's default theory, McDermott's non-monotonic logic and McCarthy's Circumscription are similar in many aspects. At first, they all hope to validate all the fixed points (or belief extension, or minimal model) have properties needed, in order that we have the same viewpoint and understanding to the world. Secondly, these three theories are all limited to deal with default reasoning, namely reasoning on condition that we do not know the contrary evidence. Furthermore, they all tried to put up a full formalization theory, but this is impossible. In spite of this, these four kinds of theories represented the main work of non-monotonic reasoning. A lot of research work had begun after them. For example, Poole [362] discussed the problem of logical framework for default reasoning and so on. Moreover, in 1982, McDermott [290] ulteriorly studied non-monotonic logic, proposed non-monotonic model theory. Later, McCarthy [287] proposed the formalized method on circumscription reasoning in 1986. After that, Brown *et al.* [39] did some work on algebraic foundation for truth maintenance, McDermott [291] proposed a general framework for reason maintenance, Brewka [37], Dix [84] , Georg, Zhang [159] gave many valuable results on commutativity of default logic, Przymusinski [367], Besnard and Schaub [22], Eiter *et al.* [114] studied the semantics of non-monotonic logic, Lifschitz [248], Marek and Truszczynski [282], Thirunarayan [475] discussed the problem of the relation between non-monotonic logic and other logic. Besides these, Papadimitriou [349], Zhang [597] made a research on extensions of default theories. Especially, Eiter and Gottlob [113] studied the completeness of propositional ciucumscription and extended closed world reasoning in 1993. Moreover, Antoniou [2] did much work on the decomposition of default theories. It is worthy of being concerned that Stillman [462] and Ben-David *et al.* [18] studied propositional default logic and modal logic for subjective default reasoning, respectively. With the development of non-monotonic logic, a lot of outcomes had been attained on non-monotonic reasoning and its applications. Later, Goodwin [156] made a further research on non-monotonic reasoning. In 1991, Rayner [396] got some results on non-monotonic formal reasoning in continuous time. At the same time, Baker [10] applied non-monotonic reasoning to situation calculus and attained some conclusions. Ben-Eliyahu [19] studied the problem of default reasoning using classical logic, and Moinard [302] and Roos [405] made a discussion on reasoning by cases in default logic, respectively. During recent years, El-Azhary *et al.* [112] put up a diagnostic expert system using non-monotonic reasoning. Their work accelerated the course of non-monotonic reasoning's conversion from theory research to applying. Some other different non-monotonic reasoning were also developed namely: autoepistemic reasoning [304], abduction [397, 423], reasoning in the presence of inconsistent information [404], counterfactual reasoning [147, 242], priority reasoning [585], and negative reasoning [357].

Some more recent works on non-monotonic logic and non-monotonic reasoning have been done, *e.g.*, among others [2,18,75,137,139,155,158,186,228, 240,271,277,283,355,362,416,427], and some applications of non-monotonic reasoning (*e.g.*, [85, 112, 308, 309]).

We refer the reader to the references [2,21,45,88,130,148,271] for extensive surveys on non-monotonic reasoning.

The typical non-monotonic reasoning methods are all based on the corresponding non-monotone logic system. Up to now, non-monotone logic system has been a promising direction and has attached considerable attention in AI. In the non-monotonic reasoning, different non-monotonic reasoning was developed.

1.6 Automated Reasoning in Many-Valued Logic

As the use of non-classical logics becomes increasingly important in computer science, information science and AI, the development of efficient automated theorem proving based on non-classical logic is currently an active area of research, especially for many-valued logics.

In classical logic, one approach of automated theorem proving is resolution and its variants. Since the introduction of resolution principle in 1965 by Robinson [401], resolution-based automated reasoning has been extensively studied [241, 280, 523] in the context of finding natural and efficient proof systems to support a wide spectrum of computational tasks. A number of important applications of such systems have been found in areas such as AI, logic programming, problem solving and question answering systems, database theory, and so on. Because the real world is dealing with uncertainty, it is difficult to design any intelligent system based on traditional logic. With the development of Zadeh's fuzzy logic [588], expert systems and knowledge engineering, especially since non-classical logics became a considerable formal tool for computer science, information science and AI, the area of automated reasoning based on non-classical logic (especially multi-valued logic and fuzzy logic) has drawn many researchers' attention.

The first step in automated deduction in fuzzy logic was made by Lee [235]. Many subsequent works were based on Lee's definitions and results. They proved that fuzzy logical inference by the resolution is significant when the truth values of all variables are taken in the half-open, half-closed interval (0.5, 1]. Mukaidono [310, 311] generalized Lee's results by introducing a fuzzy resolvent, allowing the truth values of all variables to be taken in the closed interval [0, 1] and investigated some fuzzy Prolog system. Yager [559] independently proposed a multi-valued propositional logic system similar to Lee's fuzzy logic system, with a notion of logical consequence which is stronger than Lee's. He generalized resolution in the same spirit as Mukaidono's. Lee's fuzzy formulae are syntactically defined exactly as classical first-order formulae, the only difference between them being the value of the atom, *i.e.*, they differ semantically. Therefore, these fuzzy logics cannot describe fuzzy reasoning completely. In 1985, Liu and Xiao [262] proposed an operator fuzzy logic system, which is able to represent the fuzzy-degree of a fuzzy proposition [265]. These fuzzy logic systems have a common feature, *i.e.*, a clause set S is unsatisfiable in fuzzy logic if and only if S is unsatisfiable in the classical logic. Orlowska and Wierzchon [344] introduced a deduction system based on decomposition rules, for a fuzzy logic with the classical language, Gödel's implication and an intuitionistic negation. Dubois and Prade proposed a resolution principle in possibilistic logic [94, 98, 99].

In the framework of many-valued logic, perhaps the earliest reference to a many-valued resolution-based deduction system is the one proposed by Morgan [305]. Afterwards, Orlowska [341, 343, 344] proposed an automated reasoning theory and an application in ω^+-valued logic and in Post's finitely-valued logic. Schmitt [417] proposed a resolution-based proving system for a certain 3-valued logic. Recently, a general treatment of resolution for many-valued logic based on signed clauses has been published. One feature common to much of these works is the use of signs (subsets of the set of truth values). This method was independently proposed and studied by Hahnle [164], Murray and Rosenthal [319]. Signed logics are characterized by having capability for semantic reflection: elements of their semantic domain can be directly expressed by syntactical entities: signs. Semantic reflection does, in principle, not depend on the presence of signs [164]. Signed logics are used in various deductive approaches to reasoning in non-classical logics, in particular, many-valued logics [4, 168, 409, 410]. In general, one takes truth value sets as signs, because this can speed up proof search exponentially [168]. Other authors, such as Blair, da Costa [69], Blair and Subrahamnian [27], Subrahamnian [464], Kifer and Subrahamnian [223], Kifer and Lozinski [221, 222], Lu *et al.* [272, 273] *et al.*, studied the annotated logic systems. It has been proved that the annotated logic and operator fuzzy logic are special cases of the signed logic [273]. Baaz and Fermuller investigated many-valued resolution method [5, 6], which is also a general many-valued resolution method for automated theorem proving in finitely-valued logics. They extended the resolution procedure to arbitrary finitely-valued logics, and described methods for translation to clause form, formulated a sound and complete many-valued resolution calculus, and showed that the completeness of the calculus is preserved when applying simplification rules such as subsumption and deletion of certain types of tautologies. Many-valued literals are atomic formulae superscripted by truth values; many-valued clauses are disjunctions of many-valued literals. Many-valued logic may lack the expressive power to allow normal form. Not even a disjunctive operator needs to be present. One way to deal with this situation is to embody the multi-valued version of non-classical resolution system as it has been presented for classical logic [318]. Concerning non-clausal method in many-valued logics, we may see the tableaux methods such as the method of Surma (1984), further developed by Carnielli [47], see also [48], and the method of Hahnle who defined a general notion of signs in tableaux that makes it possible to speak concisely about the truth values a formula can take at a certain stage during the construction of the tableau [164, 165]. Ideas from tableaux-based theorem proving, such as the use of sets as signs, turned out to be also useful in theorem proving by resolution [167, 169]. An example of a multi-valued logic system, which can deal with this issue is Stachniak's resolution-based proof system [338, 457–459]. To summarize, Stachniak's resolution-based approach to theorem proving in multi-valued logic is of considerable flexibility and covers a wide scope of logic. The chaining techniques [7, 9] to automated theorem proving in many-valued (lattice-valued) logics were presented in [135]. Up to now, the study of proving systems and automated reasoning for Pavelka type fuzzy logic has not been extensively reported. In particular, the widespread use of resolution in automated theorem proving and its relatives (*e.g.*, Prolog and Logic programming), as well as the central role that the Łukasiewicz implication connective plays in all the developments and applications of fuzzy logic, suggest that the study of the resolution procedure containing a more general implication, especially the Łukasiewicz implication, can provide a useful framework for the development of uncertainty reasoning. The important investigations can be found in [236, 239].

Particularly, the work of Hähnle and others [168, 442–446] has led to the development of several theorem-provers for first-order multi-valued logics.

The major textbooks available on the subject so far include Rosser and Turquette [408], Rescher [400], Rasiowa [393], among others, more recent, books by Hahnle [167], Hohle and Klement [182], Novak, Perfilieva and Mojckojr [337], and Hajek [175].

Part II

Lattice Implication Algebras

2 Concepts and Properties

In this chapter, we establish the concept of lattice implication algebra by combining lattice and implication algebra, which is a new logical algebraic system, and discuss its algebraic properties. This chapter serves as a foundation of the rest chapters of the book.

2.1 Lattice Implication Algebras

2.1.1 Concepts and Examples

Definition 2.1.1 (Quasi-Lattice Implication Algebra) *Let* $(L, \vee, \wedge, O, I)$ *be a bounded lattice with an order-reversing involution* $'$*,* I *and* O *the greatest and the smallest element of* L *respectively, and*

$$\rightarrow: \ L \times L \longrightarrow L$$

be a mapping. $(L, \vee, \wedge, ', \rightarrow, O, I)$ *is called a quasi-lattice implication algebra if the following conditions hold for any* $x, y, z \in L$*:*

$$(I_1)\ x \rightarrow (y \rightarrow z) = y \rightarrow (x \rightarrow z);$$
$$(I_2)\ x \rightarrow x = I;$$
$$(I_3)\ x \rightarrow y = y' \rightarrow x';$$
$$(I_4)\ x \rightarrow y = y \rightarrow x = I \ \textit{implies}\ \ x = y;$$
$$(I_5)\ (x \rightarrow y) \rightarrow y = (y \rightarrow x) \rightarrow x.$$

Remark 1. Since in a bounded lattice $(L, \vee, \wedge)$ with an order-reversing involution $'$,

$$(x \vee y)' = x' \wedge y' \tag{2.1}$$

holds for any $x, y \in L$ if and only if

$$(x \wedge y)' = x' \vee y' \tag{2.2}$$

holds for any $x, y \in L$, it is obvious that Eqs. (2.1) and (2.2) hold in a quasi-lattice implication algebra.

Theorem 2.1.1 *Let* $(L, \vee, \wedge, ', \rightarrow, O, I)$ *be a quasi-lattice implication algebra, then*

(1). (l_2) *holds if and only if* (l_3) *holds;*
(2). (l_1) *holds if and only if* (l_4) *holds,*

where

$$(l_1)\ (x \vee y) \rightarrow z = (x \rightarrow z) \wedge (y \rightarrow z);$$
$$(l_2)\ (x \wedge y) \rightarrow z = (x \rightarrow z) \vee (y \rightarrow z);$$
$$(l_3)\ x \rightarrow (y \vee z) = (x \rightarrow y) \vee (x \rightarrow z);$$
$$(l_4)\ x \rightarrow (y \wedge z) = (x \rightarrow y) \wedge (x \rightarrow z).$$

Proof. (1). For any $x, y, z \in L$, if (l_2) holds, then

$$\begin{aligned} x \rightarrow (y \vee z) &= (y \vee z)' \rightarrow x' \\ &= (y' \wedge z') \rightarrow x' \\ &= (y' \rightarrow x') \vee (z' \rightarrow x') \\ &= (x \rightarrow y) \vee (x \rightarrow z), \end{aligned}$$

i.e., (l_3) holds. Conversely, if (l_3) holds, then

$$\begin{aligned} (x \wedge y) \rightarrow z &= z' \rightarrow (x \wedge y)' \\ &= z' \rightarrow (x' \vee y') \\ &= (z' \rightarrow x') \vee (z' \rightarrow y') \\ &= (x \rightarrow z) \vee (y \rightarrow z), \end{aligned}$$

i.e., (l_2) holds.
(2). It can be proved similarly. □

Definition 2.1.2 (Lattice Implication Algebra) *A quasi-lattice implication algebra is called a lattice implication algebra, if* (l_1) *and* (l_2) *hold.*

Now, we give some examples of lattice implication algebra. From these examples, we know that the concept of lattice implication algebra is a generalization of Boolean algebra and Łukasiewicz implication algebra.

Example 2.1.1 (Boolean Algebra) *Let* $(L, \vee, \wedge, ')$ *be a Boolean lattice, for any* $x, y \in L$, *define*

$$x \rightarrow y = x' \vee y,$$

then $(L, \vee, \wedge, ', \rightarrow)$ *is a lattice implication algebra.*

From Example 2.1.1, we can define a lattice implication algebra on any Boolean algebra. This implies that the class of lattice implication algebras is a proper class.

Example 2.1.2 (Łukasiewicz implication algebra on [0,1]) *If the operations on* $[0,1]$ *are defined respectively as follows:*

$$\begin{aligned} x \vee y &= \max\{x, y\}, \\ x \wedge y &= \min\{x, y\}, \\ x' &= 1 - x, \\ x \to y &= \min\{1, 1 - x + y\}, \end{aligned}$$

then $([0,1], \vee, \wedge, ', \to)$ *is a lattice implication algebra.*

Example 2.1.3 (Łukasiewicz implication algebra on finite chains) *Consider the set* $L = \{a_i \mid i = 1, 2, \cdots, n\}$. *For any* $1 \leqslant j, k \leqslant n$, *define*

$$\begin{aligned} a_j \vee a_k &= a_{\max\{j,k\}}, \\ a_j \wedge a_k &= a_{\min\{j,k\}}, \\ (a_j)' &= a_{n-j+1}, \\ a_j \to a_k &= a_{\min\{n-j+k,n\}}, \end{aligned}$$

then $(L, \vee, \wedge, ', \to)$ *is a lattice implication algebra. Specially, let*

$$L = \{O, a, b, c, I\},$$

the partial order of L *is given by* $O < a < b < c < I$, $'$ *is defined as*

$$O' = I, a' = c, b' = b, c' = a, I' = O;$$

and $\to$ *is defined as Table 2.1. We call such Łukasiewicz implication algebra as* C_5.

Table 2.1. Implication Operator of $L = \{O, a, b, c, I\}$.

$\to$	O	a	b	c	I
O	I	I	I	I	I
a	c	I	I	I	I
b	b	b	I	I	I
c	a	b	b	I	I
I	O	a	b	c	I

Example 2.1.4 *Let* $L = \{O, a, b, c, d, I\}$, $O' = I$, $a' = c$, $b' = d$, $c' = a$, $d' = b$, $I' = O$, *the Hasse diagram of* L *be defined as Fig. 2.1 and its implication operator be defined as Table 2.2, then* $(L, \vee, \wedge, ', \to)$ *is a lattice implication algebra.*

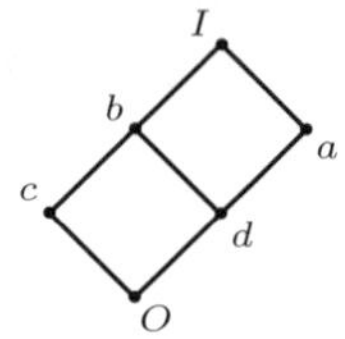

Fig. 2.1. Hasse Diagram of $L = \{O, a, b, c, d, I\}$

Table 2.2. Implication Operator of $L = \{O, a, b, c, d, I\}$

$\rightarrow$	O	a	b	c	d	I
O	I	I	I	I	I	I
a	c	I	b	c	b	I
b	d	a	I	b	a	I
c	a	a	I	I	a	I
d	b	I	I	b	I	I
I	O	a	b	c	d	I

Example 2.1.5 *Let $L = \{O, a, b, I\}$, the partial order on L be defined as*

$$O < a < b < I,$$

and $O' = I$, $a' = b$, $b' = a$, $I' = O$, the implication operator $\rightarrow$ be defined as Table 2.3. It is routine to verify that $(L, \vee, \wedge, ', \rightarrow)$ is a quasi-lattice implication algebra instead of a lattice implication algebra because $a \rightarrow (a \vee b) = a \rightarrow b = b$ and $a \rightarrow (a \vee b) = (a \rightarrow a) \vee (a \rightarrow b) = I$.

Table 2.3. Implication Operator of $L = \{O, a, b, I\}$

$\rightarrow$	O	a	b	I
O	I	I	I	I
a	b	I	b	I
b	a	a	I	I
I	O	a	b	I

By Example 2.1.5, a quasi-lattice implication algebra may not be a lattice implication algebra in general.

Example 2.1.6 *Let $L = \{O, a, b, I\}$ and $O' = I$, $a' = b$, $b' = a$, $I' = O$, the Hasse diagram of L be defined as Fig. 2.2 and operators of L be defined in Table 2.4, then $(L, \vee, \wedge, ', \rightarrow)$ is a lattice implication algebra.*

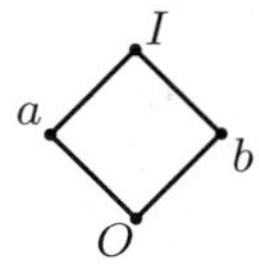

Fig. 2.2. Hasse Diagram of $L = \{O, a, b, I\}$

Table 2.4. Operators of $L = \{O, a, b, I\}$

$\vee$	O	a	b	I
O	O	a	b	I
a	a	a	I	I
b	b	I	b	I
I	I	I	I	I

$\wedge$	O	a	b	I
O	O	O	O	O
a	O	a	O	a
b	O	O	b	b
I	O	a	b	I

$\rightarrow$	O	a	b	I
O	I	I	I	I
a	b	I	b	I
b	a	a	I	I
I	O	a	b	I

$'$	O	a	b	I
	I	b	a	O

In L, a and b are incomparable.

Remark 2. In Example 2.1.6, the implication operator $\rightarrow$ can be written as

$$a \rightarrow b = a' \vee b.$$

Hence, it is the Kleene's implication operator.

Example 2.1.7 *Let U be any non-empty set and $(L, \vee, \wedge, ', \rightarrow)$ a lattice implication algebra. Denote the set of all the L-fuzzy subsets on U as $\mathscr{F}_L(U)$. Now we define four operators of $\mathscr{F}_L(U)$ as follows: for any $A, B \in \mathscr{F}_L(U)$ and $x \in U$,*

$$\begin{aligned}(A \vee B)(x) &= A(x) \vee B(x),\\ (A \wedge B)(x) &= A(x) \wedge B(x),\\ (A \rightarrow B)(x) &= A(x) \rightarrow B(x),\\ A'(x) &= (A(x))'.\end{aligned}$$

It is routine to verify that $(\mathscr{F}_L(U), \vee, \wedge, ', \rightarrow)$ is a lattice implication algebra. Specially, if $U = [0, 1]$ and $L = ([0, 1], \vee, \wedge, ', \rightarrow)$ is the Łukasiewicz implication algebra defined in Example 2.1.2, then $(\mathscr{F}([0, 1]), \vee, \wedge, ', \rightarrow)$ is a lattice implication algebra, where $\vee$, $\wedge$, $'$, $\rightarrow$ are defined as follows respectively:

$$\begin{aligned}(f \vee g)(x) &= \max\{f(x), g(x)\},\\ (f \wedge g)(x) &= \min\{f(x), g(x)\},\\ f'(x) &= 1 - f(x),\\ (f \rightarrow g)(x) &= \min\{1, 1 - f(x) + g(x)\}.\end{aligned}$$

Remark 3. L-fuzzy subsets defined in Example 2.1.7 can be used for studying truth-value operation in fuzzy subsets.

2.1.2 Basic Properties

In this section, we discuss basic properties of quasi-lattice implication algebras and lattice implication algebras. In what follows, the lattice implication algebra $(L, \vee, \wedge, ', \rightarrow, O, I)$ is denoted by L simply.

Theorem 2.1.2 *Let L be a quasi-lattice implication algebra, then for any $x, y, z \in L$,*

(1). If $I \rightarrow x = I$, then $x = I$;
(2). $I \rightarrow x = x$ and $x \rightarrow O = x'$;
(3). $O \rightarrow x = I$ and $x \rightarrow I = I$;
(4). $(x \rightarrow y) \rightarrow ((y \rightarrow z) \rightarrow (x \rightarrow z)) = I$;
(5). $(x \rightarrow y) \rightarrow x' = (y \rightarrow x) \rightarrow y'$.

Proof. (1). Suppose that $I \rightarrow x = I$, by

$$\begin{aligned}x \rightarrow I &= x \rightarrow (I \rightarrow x)\\ &= I \rightarrow (x \rightarrow x)\\ &= I \rightarrow I\\ &= I,\end{aligned}$$

it follows that $x = I$.

(2). By (1) and

$$I \rightarrow ((I \rightarrow x) \rightarrow x) = (I \rightarrow x) \rightarrow (I \rightarrow x) = I,$$

it follows that

$$(I \rightarrow x) \rightarrow x = I.$$

On the other hand,

$$\begin{aligned}x \rightarrow (I \rightarrow x) &= I \rightarrow (x \rightarrow x)\\ &= I \rightarrow I\\ &= I,\end{aligned}$$

that is to say $I \rightarrow x = x$ and hence

$$\begin{aligned} x \to O &= O' \to x' \\ &= I \to x' \\ &= x'. \end{aligned}$$

(3). By (2) it follows that

$$\begin{aligned} x \to I &= x \to (O \to O) \\ &= O \to (x \to O) \\ &= O \to ((x' \to O) \to O) \\ &= O \to ((O \to x') \to x') \\ &= (O \to x') \to (O \to x') \\ &= I, \end{aligned}$$

and

$$O \to x = x' \to I = I.$$

(4).

$$\begin{aligned} (x \to y) \to ((y \to z) \to (x \to z)) &= (x \to y) \to (x \to ((y \to z) \to z)) \\ &= (x \to y) \to (x \to ((z \to y) \to y)) \\ &= (x \to y) \to ((z \to y) \to (x \to y)) \\ &= (z \to y) \to ((x \to y) \to (x \to y)) \\ &= (z \to y) \to I \\ &= I. \end{aligned}$$

(5).

$$\begin{aligned} (x \to y) \to x' &= (y' \to x') \to x' \\ &= (x' \to y') \to y' \\ &= (y \to x) \to y'. \ \square \end{aligned}$$

Theorem 2.1.3 *Let L be a quasi-lattice implication algebra, then for any $x, y \in L$, the following statements are equivalent:*

(1). $x \to y = I$;
(2). for any $z \in L$, $(y \to z) \to (x \to z) = I$;
(3). for any $z \in L$, $(z \to x) \to (z \to y) = I$;
(4). for any $z \in L$, $z \to y = ((z \to x) \to x) \to y$;
(5). for any $z \in L$, $x \to z = x \to (z \to (z \to y)')'$.

Proof. First, we prove (1) and (2) are equivalent. Assume (1), *i.e.*, $x \to y = I$, then

$$\begin{aligned}(y \to z) \to (x \to z) &= x \to ((y \to z) \to z)\\ &= x \to ((z \to y) \to y)\\ &= (z \to y) \to (x \to y)\\ &= (z \to y) \to I\\ &= I,\end{aligned}$$

i.e., (2) holds. Conversely, assume (2) holds. Let $z = O$ in (2), it follows that

$$\begin{aligned}x \to y &= y' \to x'\\ &= (y \to O) \to (x \to O)\\ &= I.\end{aligned}$$

i.e., (1) holds. Hence, (1) and (2) are equivalent.

The equivalence of (1) and (3) can be proved similarly.

Now we prove (1) and (4) are equivalent. Assume (1), *i.e.*, $x \to y = I$, by $z \to ((z \to x) \to x) = I$, it follows that

$$\begin{aligned}I &= (z \to ((z \to x) \to x)) \to ((((z \to x) \to x) \to y) \to (z \to y))\\ &= I \to ((((z \to x) \to x) \to y) \to (z \to y))\\ &= (((z \to x) \to x) \to y) \to (z \to y).\end{aligned}$$

On the other hand, by $x \to y = I$, it follows that

$$(y \to z) \to (x \to z) = (x \to y) \to ((y \to z) \to (x \to z)) = I$$

and hence

$$\begin{aligned}(z \to y) \to (((z \to x) \to x) \to y) &= ((z \to x) \to x) \to ((z \to y) \to y)\\ &= ((x \to z) \to z) \to ((y \to z) \to z)\\ &= (y \to z) \to (((x \to z) \to z) \to z)\\ &= (y \to z) \to ((z \to (x \to z)) \to (x \to z))\\ &= (y \to z) \to (I \to (x \to z))\\ &= (y \to z) \to (x \to z)\\ &= I.\end{aligned}$$

It follows that (4) holds. Conversely, assume (4) holds. Let $z = O$, then

$$\begin{aligned}I &= O \to y\\ &= ((O \to x) \to x) \to y\\ &= (I \to x) \to y\\ &= x \to y.\end{aligned}$$

This shows that (1) holds. Hence, (1) and (4) are equivalent.

Finally, we prove (1) and (5) are equivalent. Assume (1), by

$$\begin{aligned}
&(z \to (z \to y)')' \to z \\
&= z' \to (z \to (z \to y)') \\
&= z' \to ((z \to y) \to z') \\
&= (z \to y) \to (z' \to z') \\
&= (z \to y) \to I \\
&= I,
\end{aligned}$$

it follows that

$$\begin{aligned}
&(x \to (z \to (z \to y)')') \to (x \to z) \\
&= ((z \to (z \to y)') \to x') \to (z' \to x') \\
&= z' \to ((x' \to (z \to (z \to y)')) \to (z \to (z \to y)')) \\
&= (x' \to (z \to (z \to y)')) \to (z' \to (z \to (z \to y)')) \\
&= (x' \to (z \to (z \to y)')) \to I \\
&= I.
\end{aligned}$$

On the other hand, by $x \to y = I$, it follows that

$$(z \to x) \to (z \to y) = (x \to y) \to ((z \to x) \to (z \to y)) = I$$

and hence

$$\begin{aligned}
&(x \to z) \to (x \to (z \to (z \to y)')') \\
&= (x \to z) \to ((z \to (z \to y)') \to x') \\
&= (z \to (z \to y)') \to ((x \to z) \to x') \\
&= (z \to (z \to y)') \to (z \to (z \to x)') \\
&= ((z \to y) \to z') \to ((z \to x) \to z') \\
&= (z \to x) \to ((z' \to (z \to y)) \to (z \to y)) \\
&= (z' \to (z \to y)) \to ((z \to x) \to (z \to y)) \\
&= (z' \to (z \to y)) \to I \\
&= I,
\end{aligned}$$

i.e., (5) holds. Conversely, assume (5) holds. Let $z = I$, then

$$\begin{aligned}
I &= x \to I \\
&= x \to (I \to (I \to y)')' \\
&= x \to y.
\end{aligned}$$

i.e., (1) holds. □

Theorem 2.1.4 *Let L be a quasi-lattice implication algebra, then for any $x, y \in L$,*

$$((x \to y) \to y) \to y = x \to y.$$

Proof. For any $x, y \in L$,

$$\begin{aligned}
&(((x \to y) \to y) \to y) \to (x \to y) \\
&= x \to ((y \to ((x \to y) \to y)) \to ((x \to y) \to y)) \\
&= x \to (I \to ((x \to y) \to y)) \\
&= x \to ((x \to y) \to y) \\
&= I, \\
&(x \to y) \to (((x \to y) \to y) \to y) \\
&= ((x \to y) \to y) \to ((x \to y) \to y) \\
&= I,
\end{aligned}$$

i.e., $((x \to y) \to y) \to y = x \to y$. □

Theorem 2.1.5 *Let L be a quasi-lattice implication algebra, then for any $x, y \in L$, the following statements are equivalent:*

(1). $x \vee y = (x \to y) \to y$;
(2). $x \wedge y = ((x \to y) \to x')'$;
(3). $x \leqslant y$ *if and only if* $x \to y = I$.

Proof. ((1) $\Longrightarrow$ (2)).

$$\begin{aligned}
x \wedge y &= (x' \vee y')' \\
&= ((y' \to x') \to x')' \\
&= ((x \to y) \to x')'.
\end{aligned}$$

((2) $\Longrightarrow$ (1)).

$$\begin{aligned}
x \vee y &= (x' \wedge y')' \\
&= (y' \to x') \to y \\
&= (x \to y) \to y.
\end{aligned}$$

((1) $\Longrightarrow$ (3)). If $x \leqslant y$, then $x \vee y = y$ and hence

$$(x \to y) \to y = y,$$

it follows that

$$\begin{aligned}
x \to y &= x \to ((x \to y) \to y) \\
&= (x \to y) \to (x \to y) \\
&= I.
\end{aligned}$$

Conversely, suppose that $x \to y = I$, then

$$\begin{aligned} x \vee y &= (x \to y) \to y \\ &= I \to y \\ &= y, \end{aligned}$$

it follows that $x \leqslant y$.

((3) $\Longrightarrow$ (1)). By

$$x \to ((x \to y) \to y) = (x \to y) \to (x \to y) = I,$$

and

$$\begin{aligned} y \to ((x \to y) \to y) &= (x \to y) \to (y \to y) \\ &= (x \to y) \to I \\ &= I, \end{aligned}$$

it follows that $x \leqslant (x \to y) \to y$ and $y \leqslant (x \to y) \to y$, *i.e.*,

$$x \vee y \leqslant (x \to y) \to y.$$

On the other hand, for any $a \in L$, if $x \leqslant a, y \leqslant a$, then $x \to a = I, y \to a = I$, and

$$\begin{aligned} &((x \to y) \to y) \to a \\ =\ &((x \to y) \to y) \to (I \to a) \\ =\ &((x \to y) \to y) \to ((y \to a) \to a) \\ =\ &((x \to y) \to y) \to ((a \to y) \to y) \\ =\ &(a \to y) \to (((x \to y) \to y) \to y) \\ =\ &(a \to y) \to (x \to y) \\ =\ &I, \end{aligned}$$

i.e., $(x \to y) \to y \leqslant a$, it follows that $(x \to y) \to y = x \vee y$. □

Corollary 2.1.1 *In a quasi-lattice implication algebra L, for any $x, y, z \in L$, if $x \leqslant y$, then $x \to z \geqslant y \to z$ and $z \to x \leqslant z \to y$.* □

Theorem 2.1.6 *Let L be a quasi-lattice implication algebra and $x \vee y = (x \to y) \to y$ for any $x, y, z \in L$, then*

(1). $((x \vee y) \to x) \vee ((x \vee y) \to y) = I$;
(2). *for* $x \leqslant z$ *and* $y \leqslant z$, $x \vee y = (z \to ((z \to x)' \wedge (z \to y)'))'$;
(3). $z \to (x \wedge y) \leqslant (z \to x) \wedge (z \to y)$;
(4). $(x \vee y) \to z \leqslant (x \to z) \wedge (y \to z)$;
(5). $(x \to z) \vee (y \to z) \leqslant (x \wedge y) \to z$;
(6). $(z \to x) \vee (z \to y) \leqslant z \to (x \vee y)$.

Proof. (1). Let $z = x \vee y$, then

$$\begin{aligned} z \to x &= (x \vee y) \to x \\ &= ((y \to x) \to x) \to x \\ &= y \to x \end{aligned}$$

and

$$\begin{aligned} &(z \to y) \to (z \to x) \\ &= z \to ((z \to y) \to x) \\ &= z \to (x' \to (z \to y)') \\ &= x' \to (z \to (z \to y)') \\ &= (z \to (z \to y)')' \to x \\ &= ((z \to y) \to z')' \to x, \end{aligned}$$

it follows that

$$\begin{aligned} &(z \to x) \vee (z \to y) \\ &= ((z \to y) \to (z \to x)) \to (z \to x) \\ &= (((z \to y) \to z')' \to x) \to (z \to x) \\ &= ((z \wedge y) \to x) \to (z \to x) \\ &= (y \to x) \to (z \to x) = I. \end{aligned}$$

(2). By (1),

$$\begin{aligned} &(z \to x) \vee (z \to y) \\ &= ((z \wedge y) \to x) \to (z \to x) \\ &= (y \to x) \to (z \to x) \\ &= z \to ((y \to x) \to x) \\ &= z \to (x \vee y), \end{aligned}$$

and hence

$$\begin{aligned} &(z \to ((z \to x)' \wedge (z \to y)'))' \\ &= (((z \to x) \vee (z \to y)) \to z')' \\ &= ((z \to (x \vee y)) \to z')' \\ &= z \wedge (x \vee y) \\ &= x \vee y. \end{aligned}$$

(3). By $x \wedge y \leqslant x$ and $x \wedge y \leqslant y$, it follows that $z \to (x \wedge y) \leqslant z \to x$ and $z \to (x \wedge y) \leqslant z \to y$, hence

$$z \to (x \wedge y) \leqslant (z \to x) \wedge (z \to y).$$

(4), (5) and (6) can be proved similarly. □

Theorem 2.1.7 *Let L be a lattice implication algebra, then $x \leqslant y$ if and only if $x \to y = I$ for any $x, y \in L$.*

Proof. If $x \leqslant y$, then

$$\begin{aligned} x \to y &= (x \wedge y) \to y \\ &= (x \to y) \vee (y \to y) \\ &= (x \to y) \vee I \\ &= I. \end{aligned}$$

If $x \to y = I$, then

$$\begin{aligned} (x \vee y) \to y &= (x \to y) \wedge (y \to y) \\ &= I. \end{aligned}$$

On the other hand,

$$\begin{aligned} y \to (x \vee y) &= (y \to x) \vee (y \to y) \\ &= (y \to x) \vee I \\ &= I, \end{aligned}$$

it follows from (I_4) that $x \vee y = y$, *i.e.*, $x \leqslant y$. □

Theorem 2.1.8 *Let L be a quasi-lattice implication algebra, then L is a lattice implication algebra if and only if $x \vee y = (x \to y) \to y$ for any $x, y \in L$.*

Proof. Suppose that L is a lattice implication algebra, by Theorems 2.1.5 and 2.1.7, $x \vee y = (x \to y) \to y$ holds for any $x, y \in L$.

Conversely, suppose that for any $x, y \in L$, $x \vee y = (x \to y) \to y$. For any $x, y, z \in L$,

$$\begin{aligned} &((x \to z) \wedge (y \to z)) \to ((x \vee y) \to z) \\ =\ &(((x \to z) \to (y \to z)) \to (x \to z)')' \to ((x \vee y) \to z) \\ =\ &((y \to (x \vee y)) \to (x \to y)')' \to ((x \vee y) \to z) \\ =\ &((x \vee y) \to z)' \to ((y \to (x \vee z)) \to (x \to y)') \\ =\ &(y \to (x \vee z)) \to ((x \to z) \to ((x \vee y) \to z)) \\ =\ &(y \to (x \vee z)) \to ((x \vee y) \to (x \vee z)) \\ =\ &(x \vee y) \to ((y \to (x \vee z)) \to (x \vee z)) \\ =\ &(x \vee y) \to (y \vee x \vee z) \\ =\ &I, \end{aligned}$$

i.e.,

$$(x \to z) \wedge (y \to z) \leqslant (x \vee y) \to z,$$

and hence (l_1) holds. (l_2) can be proved similarly. □

Theorem 2.1.9 *Let L be a lattice implication algebra, then for any $x, y \in L$,*

(1). $x \to y \geqslant x' \vee y$;
(2). for any $z \in L$, $x \to z = y \to z$ if and only if $x = y$ if and only if for any $z \in L$, $z \to x = z \to y$.

Proof. (1).

$$\begin{aligned}(x' \vee y) \to (x \to y) &= (x' \to (x \to y)) \wedge (y \to (x \to y)) \\ &= (x' \to (y' \to x')) \wedge (x \to (y \to y)) \\ &= y' \to (x' \to x') = I,\end{aligned}$$

i.e., $x' \vee y \leqslant x \to y$.

(2).

$$\begin{aligned} &\text{for any } z \in L, x \to z = y \to z \\ \text{iff } &\text{for any } z \in L, x \to z \leqslant y \to z, x \to z \geqslant y \to z \\ \text{iff } &x \leqslant y, y \leqslant x \\ \text{iff } &x = y.\end{aligned}$$

$x = y$ if and only if that for any $z \in L, z \to x = z \to y$ can be proved similarly. □

Theorem 2.1.10 *Let L be a lattice implication algebra, then for any $x, y \in L$,*

$$(x \to y) \vee (y \to x) = I.$$

Proof. For any $x, y \in L$,

$$\begin{aligned} I &= (x \vee y) \to (x \vee y) \\ &= ((x \vee y) \to x) \vee ((x \vee y) \to y) \\ &= ((x \to x) \wedge (y \to x)) \vee ((x \to y) \wedge (y \to y)) \\ &= (x \to y) \vee (y \to x). \ \square\end{aligned}$$

Theorem 2.1.11 *Let L be a lattice implication algebra, then for any $x, y, z \in L$,*

(1). $(x \to z) \to (y \to z) = y \to (x \vee z) = (z \to x) \to (y \to x)$;
(2). $(z \to x) \to (z \to y) = (x \wedge z) \to y = (x \to z) \to (x \to y)$.

Proof. (1).

$$(x \to z) \to (y \to z) = y \to ((x \to z) \to z) = y \to (x \vee z);$$

$$(z \to x) \to (y \to x) = y \to ((z \to x) \to x) = y \to (x \vee z).$$

(2). It can be proved similarly. □

Theorem 2.1.12 *Let L be a lattice implication algebra, then $(L, \vee, \wedge)$ is a distributive lattice.*

Proof. We need only to prove that for any $x, y, z \in L$,

$$x \vee (y \wedge z) = (x \vee y) \wedge (x \vee z).$$

In fact,

$$\begin{aligned} x \vee (y \wedge z) &= ((y \wedge z) \to x) \to x \\ &= ((y \to x) \vee (z \to x)) \to x \\ &= ((y \to x) \to x) \wedge ((z \to x) \to x) \\ &= (x \vee y) \wedge (x \vee z). \ \Box \end{aligned}$$

Theorem 2.1.13 *Let L be a lattice implication algebra, then for any $x, y \in L$,*

(1). $x \vee y = I$ if and only if $x \to y = y$;

(2). $x \wedge y = O$ if and only if $x \to y = x'$.

Proof. (1).

$$\begin{aligned} x \vee y = I \quad &\text{iff } (x \to y) \to y = I \\ &\text{iff } x \to y \leqslant y \\ &\text{iff } x \to y = y. \end{aligned}$$

(2).

$$\begin{aligned} x \wedge y = O \quad &\text{iff } x' \vee y' = I \\ &\text{iff } y' \to x' = x' \\ &\text{iff } x \to y = x'. \ \Box \end{aligned}$$

Theorem 2.1.14 *Let L be a lattice implication algebra. For any $x, y, z \in L$,*

(1). if $x \to y = x \to z$, then $x \to (y \vee z) = x \to (y \wedge z) = x \to y$;
(2). if $y \to x = z \to x$, then $(y \wedge z) \to x = (y \vee z) \to x = y \to x$;
(3). $x \to (y \to z) = (x \vee y) \to z$ if and only if $x \to (y \to z) = x \to z = y \to z$.

Proof. (1). If $x \to y = x \to z$, then

$$\begin{aligned} x \to (y \vee z) &= (x \to y) \vee (x \to z) \\ &= (x \to y) \vee (x \to y) \\ &= x \to y, \end{aligned}$$

and

$$\begin{aligned} x \to (y \wedge z) &= (x \to y) \wedge (x \to z) \\ &= (x \to y) \wedge (x \to y) \\ &= x \to y. \end{aligned}$$

The proof of (2) is similar to that of (1).

(3). If $x \to (y \to z) = (x \vee y) \to z$, then

$$\begin{aligned} y \to z &\leqslant x \to (y \to z) \\ &= (x \vee y) \to z \\ &= (x \to z) \wedge (y \to z) \\ &\leqslant y \to z, \end{aligned}$$

i.e., $x \to (y \to z) = y \to z$. $x \to (y \to z) = x \to z$ can be proved similarly. Conversely, *i.e.*, $x \to (y \to z) = x \to z = y \to z$ holds, then

$$\begin{aligned} &(x \vee y) \to z \\ =\ &(x \to z) \wedge (y \to z) \\ =\ &x \to (y \to z). \ \square \end{aligned}$$

Corollary 2.1.2 *Let L be a lattice implication algebra, then for any $x, y, z \in L$,*

(1). if $x \to y = x \to z$, then for any $t \in [y \wedge z, y \vee z]$, $x \to t = x \to y$;
(2). if $y \to x = z \to x$, then for any $t \in [y \wedge z, y \vee z]$, $t \to x = y \to x$. $\square$

Theorem 2.1.15 *Let L be a lattice implication algebra, then for any $x, y, z \in L$,*

(1). $(x \wedge z) \to (y \wedge z) = (x \wedge z) \to y$;
(2). $(x \to y) \wedge z \leqslant (x \to (y \wedge z)) \wedge ((x \vee z') \to y)$;
(3). $((x \to y) \wedge (y \to z)) \to (x \to z) = (x \to y) \vee (y \to z) \vee (x \to z)$.

Proof. (1).

$$\begin{aligned} &(x \wedge z) \to (y \wedge z) \\ =\ &((x \wedge z) \to y) \wedge ((x \wedge z) \to z) \\ =\ &((x \wedge z) \to y) \wedge I \\ =\ &(x \wedge z) \to y. \end{aligned}$$

(2). By $x \to z \geqslant z$ and $z' \to y \geqslant z$, it follows that

$$x \to (y \wedge z) = (x \to y) \wedge (x \to z) \geqslant (x \to y) \wedge z,$$

$$(x \vee z') \to y = (x \to y) \wedge (z' \to y) \geqslant (x \to y) \wedge z,$$

and hence $(x \to y) \wedge z \leqslant (x \to (y \wedge z)) \wedge ((x \vee z') \to y)$.

(3).

$$\begin{aligned} &((x \to y) \wedge (y \to z)) \to (x \to z) \\ =\ &((x \to y) \to (x \to z)) \vee ((y \to z) \to (x \to z)) \\ =\ &((x \wedge y) \to z) \vee (x \to (y \vee z)) \\ =\ &(x \to z) \vee (y \to z) \vee (x \to y). \ \square \end{aligned}$$

Theorem 2.1.16 *Let L be a lattice implication algebra, $a, b \in L$. For any $x, y \in L$,*

$$(a \to b)' \leqslant (x \to a) \wedge (b \to y).$$

Proof.

$$\begin{aligned} &(a \to b)' \to ((x \to a) \wedge (b \to y)) \\ = &((a \to b)' \to (x \to a)) \wedge ((a \to b)' \to (b \to y)) \\ = &(x \to ((a \to b)' \to a)) \wedge (y' \to ((a \to b)' \to b')) \\ = &I \wedge I \\ = &I, \end{aligned}$$

i.e., $(a \to b)' \leqslant (x \to a) \wedge (b \to y)$. □

Theorem 2.1.17 *Let L be a lattice implication algebra, $a, b \in L$, then $a \to b = O$ if and only if $a = I$ and $b = O$.* □

Theorem 2.1.18 *Let L be a lattice implication algebra, then for any $x, y, z \in L$,*

(1). $z \to (y \to x) \geqslant (z \to y) \to (z \to x)$;
(2). $z \leqslant y \to x$ if and only if $y \leqslant z \to x$.

Proof. (1). $z \to (y \to x) = y \to (z \to x) \geqslant (z \to y) \to (z \to x)$.
(2).

$$\begin{aligned} z \leqslant y \to x \quad &\text{iff } z \to (y \to x) = I \\ &\text{iff } y \to (z \to x) = I \\ &\text{iff } y \leqslant z \to x. \ \square \end{aligned}$$

Theorem 2.1.19 *Let L be a lattice implication algebra, then for any $x, y, z \in L$,*

$$(x \to (y \to z)) \to ((x \to y) \to (x \to z)) = x \vee y \vee ((x \wedge y) \to z).$$

Proof. For any $x, y, z \in L$,

$$\begin{aligned} &(x \to (y \to z)) \to ((x \to y) \to (x \to z)) \\ = &(x \to (y \to z)) \to ((y \to z) \vee (x \to z)) \\ = &((x \to (y \to z)) \to (y \to z)) \vee ((x \to (y \to z)) \to (x \to z)) \\ = &x \vee (y \to z) \vee y \vee (x \to z) \\ = &x \vee y \vee ((x \wedge y) \to z). \ \square \end{aligned}$$

Theorem 2.1.20 *Let L be a lattice implication algebra, then the following statements are equivalent:*

(1). for any $x, y, z \in L$, $x \to (y \to z) = (x \wedge y) \to z$;
(2). for any $x, y, z \in L$, $(x \to (y \to z)) \to ((x \to y) \to (x \to z)) = I$;
(3). for any $x, y \in L$, $x \to (x \to y) = x \to y$.

Proof. Assume (1) holds, we prove (2). In fact, for any $x, y, z \in L$,

$$\begin{aligned} &(x \to (y \to z)) \to ((x \to y) \to (x \to z)) \\ =& ((x \wedge y) \to z) \to ((x \to y) \to (x \to z)) \\ =& ((x \to y) \to (x \to z)) \to ((x \to y) \to (x \to z)) \\ =& I. \end{aligned}$$

Now we assume (2) holds and prove (3). For any $x, y \in L$,

$$\begin{aligned} &(x \to (x \to y)) \to (x \to y) \\ =& (x \to (x \to y)) \to ((x \to x) \to (x \to y)) \\ =& I, \end{aligned}$$

i.e., $x \to (x \to y) \leqslant x \to y$; $x \to y \leqslant x \to (x \to y)$ is trivial.

Finally, assume (3) holds, we prove (1). For any $x, y, z \in L$,

$$\begin{aligned} &(x \wedge y) \to z \\ =& (x \wedge y) \to ((x \wedge y) \to z) \\ =& (x \to (x \to z)) \vee (x \to (y \to z)) \vee (y \to (x \to z)) \vee (y \to (y \to z)) \\ =& (x \to z) \vee (x \to (y \to z)) \vee (y \to (x \to z)) \vee (y \to z) \\ =& (x \to (y \to z)) \vee (y \to (x \to z)) \\ =& x \to (y \to z). \ \square \end{aligned}$$

In a lattice implication algebra L, we define binary operations $\otimes$ and $\oplus$ as follows: for any $x, y \in L$,

$$\begin{aligned} x \otimes y &= (x \to y')'; \\ x \oplus y &= x' \to y. \end{aligned}$$

Theorem 2.1.21 *In a lattice implication algebra L,*

(1). $x \otimes y = y \otimes x, \quad x \oplus y = y \oplus x$;
(2). $(x \otimes y) \otimes z = x \otimes (y \otimes z), \quad (x \oplus y) \oplus z = x \oplus (y \oplus z)$;
(3). $(x \otimes y)' = x' \oplus y'$, $(x \oplus y)' = x' \otimes y'$;
(4). $x \otimes y \leqslant x \leqslant x \oplus y, \quad x \otimes y \leqslant y \leqslant x \oplus y$;
(5). $O \otimes x = O$, $I \otimes x = x$, $x \otimes x' = O$, $O \oplus x = x$, $I \oplus x = I$, $x \oplus x' = I$;
(6). $x \to (x \otimes y) = x' \vee y = (x \oplus y) \to y$;
(7). $(x \to y) \otimes x = x \wedge y$;
(8). *if* $x \leqslant a$ *and* $y \leqslant b$, *then* $x \otimes y \leqslant a \otimes b$ *and* $x \oplus y \leqslant a \oplus b$;
(9). $(x \otimes y)' = x' \oplus y'$, $(x \oplus y)' = x' \otimes y'$;
(10). $x \to (y \to z) = (x \otimes y) \to z$;
(11). $x \leqslant y \to z$ *if and only if* $x \otimes y \leqslant z$;
(12). $a \otimes (b \vee c) = (a \otimes b) \vee (a \otimes c), \quad a \otimes (b \wedge c) = (a \otimes b) \wedge (a \otimes c)$;
(13). $x \otimes (y \to z) \leqslant y \to (x \otimes z), \quad x \oplus (y \to z) = y \to (x \oplus z)$;
(14). $a \oplus (b \vee c) = (a \oplus b) \vee (a \oplus c), \quad a \oplus (b \wedge c) = (a \oplus b) \wedge (a \oplus c)$;
(15). $a \otimes b = (a \vee b) \otimes (a \wedge b), \quad a \oplus b = (a \vee b) \oplus (a \wedge b)$,

hold for any $x, y, z, a, b, c \in L$. $\square$

Corollary 2.1.3 *If $(L, \vee, \wedge, ', \rightarrow, O, I)$ is a lattice implication algebra, then $(L, \otimes, \rightarrow, I)$ is a residuated lattice.* $\square$

Theorem 2.1.22 *Let L be a lattice implication algebra. Then $(L, \leqslant, \oplus)$ and $(L, \leqslant, \otimes)$ are all lattice-ordered semigroups.*

Proof. For any $a, b, c \in L$, if $a \leqslant b$, then

$$a \oplus c = a' \rightarrow c \leqslant b' \rightarrow c = b \oplus c;$$

$$a \otimes c = (a \rightarrow c')' \leqslant (b \rightarrow c')' = b \otimes c.$$

By the commutativity of $(L, \oplus)$ and $(L, \otimes)$, $c \oplus a \leqslant c \oplus b$, $c \otimes a \leqslant c \otimes b$. $\square$

In a lattice implication algebra, $x \oplus y = O$ if and only if $x = y = O$; $x \otimes y = I$ if and only if $x = y = I$. So, we have the following theorem.

Theorem 2.1.23 *Let L be a lattice implication algebra.*

(1). In semigroup $(L, \oplus)$, the unique element with inverse is O.
(2). In semigroup $(L, \otimes)$, the unique element with inverse is I. $\square$

Corollary 2.1.4 *Suppose that L is a lattice implication algebra and $|L| \geqslant 2$, then neither $(L, \oplus)$ nor $(L, \otimes)$ is a lattice-ordered group.* $\square$

Theorem 2.1.24 *Let L be a lattice implication algebra, and $a, b \in L$, then*

$$a \otimes b = \bigwedge \{x \mid a \leqslant b \rightarrow x\}.$$

Proof. By

$$\begin{aligned} & a \rightarrow (b \rightarrow (a \otimes b)) \\ = & a \rightarrow (b \rightarrow (a \rightarrow b')') \\ = & (a \rightarrow b') \rightarrow (a \rightarrow b') \\ = & I, \end{aligned}$$

it follows that $a \leqslant b \rightarrow (a \otimes b)$ and hence

$$a \otimes b \in \{x \mid a \leqslant b \rightarrow x\}.$$

On the other hand, if $y \in \{x \mid a \leqslant b \rightarrow x\}$, then $a \leqslant b \rightarrow y$, and hence, $a \otimes b \leqslant y$, it follows that

$$a \otimes b = \bigwedge \{x \mid a \leqslant b \rightarrow x\}. \ \square$$

In a lattice implication algebra L, for any $a \in L$ and $n \in \mathbb{N}$, we denote that

$$a^0 = I, \quad a^1 = a, \quad a^{n+1} = (a^n) \otimes a;$$

$$0a = O, \quad 1a = a, \quad (n+1)a = (na) \oplus a.$$

Theorem 2.1.25 *Let L be a lattice implication algebra, then for any $a, b \in L$ and $m, n \in \mathbb{N}^+(\mathbb{N}^+ = \mathbb{N} \setminus \{0\})$,*

(1). $(a \vee b)^n = a^n \vee b^n$;
(2). $(a \wedge b)^n = a^n \wedge b^n$;
(3). $n(a \vee b) = (na) \vee (nb)$;
(4). $n(a \wedge b) = (na) \wedge (nb)$;
(5). if $m \leqslant n$, then $a^n \leqslant a^m$, $ma \leqslant na$.

Proof. (1). We use the induction over n to prove the conclusion. If $n = 1$, the conclusion is trivial. Now suppose $n > 1$, and the conclusion holds for n, then by Theorem 2.1.21, we have

$$\begin{aligned}&(a \vee b)^{n+1}\\ &= (a \vee b)^n \otimes (a \vee b)\\ &= (a^n \vee b^n) \otimes (a \vee b)\\ &= (a^n \otimes (a \vee b)) \vee (b^n \otimes (a \vee b))\\ &= a^{n+1} \vee (a^n \otimes b) \vee (b^n \otimes a) \vee b^{n+1}\end{aligned}$$

and

$$\begin{aligned}&(a \otimes b^n) \vee (a^n \otimes b)\\ &= (a \otimes b) \otimes (a^{n-1} \vee b^{n-1})\\ &= a \otimes b \otimes (a \vee b)^{n-1}\\ &\leqslant (a \vee b) \otimes (a \wedge b) \otimes (a \vee b)^{n-1}\\ &= (a \wedge b) \otimes (a \vee b)^n\\ &= (a^n \otimes (a \wedge b)) \vee (b^n \otimes (a \wedge b))\\ &\leqslant a^{n+1} \vee b^{n+1},\end{aligned}$$

then $(a \vee b)^{n+1} = a^{n+1} \vee b^{n+1}$, *i.e.*, the conclusion holds for $n + 1$. Hence, by the induction, (1) holds.

(5). If $m \leqslant n$, then it follows from Theorem 2.1.21 that

$$a^n = a^m \otimes a^{n-m} \leqslant a^m;$$

$$na = ma \oplus (n - m)a \geqslant ma.$$

(2), (3) and (4) can be proved similarly. □

Theorem 2.1.26 *Let L be a lattice implication algebra and $a, b, x \in L$, if there exist $n, k \in \mathbb{N}^+$ such that $a^n \to x = I$ and $b^k \to x = I$, then there exists $m \in \mathbb{N}^+$ such that $(a \vee b)^m \to x = I$.*

Proof. Without losing generality, we suppose that $k \leqslant n$. It follows that $b^n \to x = I$. Taking $m = n$, then by Theorem 2.1.25, we have

$$\begin{aligned}(a \vee b)^n \to x &= (a^n \vee b^n) \to x\\ &= (a^n \to x) \wedge (b^n \to x)\\ &= I. \square\end{aligned}$$

Theorem 2.1.27 *Let L be a lattice implication algebra, $x, y, z \in L$ and $m, n \in \mathbb{N}$, then*

$$(x^m \to (y \to z)) \to ((x^n \to y) \to (x^{m+n} \to z)) = I.$$

Proof. By

$$\begin{aligned}&(y \to z) \to ((x^n \to y) \to (x^n \to z)) \\ =\ &(x^n \to y) \to ((y \to z) \to (x^n \to z)) = I,\end{aligned}$$

it follows that $y \to z \leqslant (x^n \to y) \to (x^n \to z)$ and hence

$$x^m \to (y \to z) \leqslant x^m \to ((x^n \to y) \to (x^n \to z)),$$

which implies that

$$\begin{aligned}&(x^m \to (y \to z)) \to ((x^n \to y) \to (x^{m+n} \to z)) \\ =\ &(x^m \to (y \to z)) \to ((x^n \to y) \to (x^m \to (x^n \to z))) \\ =\ &(x^m \to (y \to z)) \to (x^m \to ((x^n \to y) \to (x^n \to z))) \\ =\ &I. \quad \square\end{aligned}$$

2.2 Lattice *H* Implication Algebras

In classical logic, the formulae with the form

$$(p \to (q \to r)) \to ((p \to q) \to (p \to r))$$

are tautologies. But in non-classical logic, these formulae may not be tautologies in general. This is one of the essential differences between classical logic and non-classical logic. The reflection of this difference on logical algebra is that

$$(x \to (y \to z)) \to ((x \to y) \to (x \to z)) = I$$

holds in Boolean algebra but may not hold in non-classical logical algebra in general. In this section, we introduce the concept of lattice H implication algebra and study the conditions of a lattice implication algebra being a lattice H implication algebra. We prove that the concepts of lattice H implication algebras and Boolean algebras coincide.

Definition 2.2.1 *A lattice implication algebra L is called a lattice H implication algebra, if for any $x, y, z \in L$,*

$$x \vee y \vee ((x \wedge y) \to z) = I.$$

It is routine to check that the lattice implication algebras defined in Examples 2.1.1 and 2.1.6 are both lattice H implication algebras.

Theorem 2.2.1 *Let L be a lattice H implication algebra, then for any $x, y, z \in L$, $x \to (y \to z) = (x \to y) \to (x \to z)$.*

Proof. By Definition 2.2.1 and Theorem 2.1.19,

$$x \to (y \to z) \leqslant (x \to y) \to (x \to z).$$

On the other hand,

$$\begin{aligned}
&((x \to y) \to (x \to z)) \to (x \to (y \to z)) \\
&= ((x \to z) \vee (y \to z)) \to (x \to (y \to z)) \\
&= ((x \to z) \to (x \to (y \to z))) \wedge ((y \to z) \to (x \to (y \to z))) \\
&= (y \to ((x \to z) \to (x \to z))) \wedge (x \to ((y \to z) \to (y \to z))) \\
&= I,
\end{aligned}$$

i.e., $(x \to y) \to (x \to z) \leqslant x \to (y \to z)$. □

By Theorem 2.1.20, the following corollary can be obtained.

Corollary 2.2.1 *Let L be a lattice implication algebra, then the following statements are equivalent:*

(1). L is a lattice H implication algebra;
(2). for any $x, y \in L$, $x \to (x \to y) = x \to y$;
(3). for any $x, y, z \in L$, $x \to (y \to z) = (x \to y) \to (x \to z)$;
(4). for any $x, y, z \in L$, $x \to (y \to z) = (x \wedge y) \to z$;
(5). for any $x, y, z \in L$, $(x \to (y \to z)) \to ((x \to y) \to (x \to z)) = I$. □

Theorem 2.2.2 *Let L be a lattice implication algebra. The following statements are equivalent:*

(1). L is a lattice H implication algebra;
(2). for any $x, y \in L$ and $z \in [x, I]$, $z \to (x \to y) = x \to y$;
(3). for any $x, y \in L$, $(x \to y) \to x = x$;
(4). for any $x \in L$, $x \vee x' = I$.

Proof. Suppose (1) holds, *i.e.*, L is a lattice H implication algebra, then for any $x, y \in L$ and $z \in [x, I]$,

$$x \to y \leqslant z \to (x \to y) \leqslant x \to (x \to y) = x \to y,$$

i.e., $z \to (x \to y) = x \to y$, (2) holds. It is trivial that (2) implies (1). Hence, (1) and (2) are equivalent.

Now we prove (1) implies (3). In fact, for any $x, y \in L$,

$$\begin{aligned}
((x \to y) \to x) \to x &= (x \to (x \to y)) \to (x \to y) \\
&= (x \to y) \to (x \to y) \\
&= I,
\end{aligned}$$

i.e., $(x \to y) \to x \leqslant x$, $x \leqslant (x \to y) \to x$ is trivial, *i.e.*, (3) holds. Conversely, assume (3) holds, we prove (1). For any $x, y \in L$,

$$\begin{aligned}(x \to (x \to y)) \to (x \to y) &= ((x \to y) \to x) \to x \\ &= x \to x \\ &= I,\end{aligned}$$

it follows that $x \to (x \to y) \leqslant x \to y$.

By $x \to y \leqslant x \to (x \to y)$,

$$x \to y = x \to (x \to y)$$

followed and hence L is a lattice H implication algebra.

Finally, we prove (3) and (4) are equivalent. Assume (3) holds, then for any $x \in L$,

$$x' = (x' \to O) \to x' = x \to x',$$

it follows that

$$\begin{aligned}&x \vee x' \\ =\ &x \vee (x \to x') \\ =\ &(x \to (x \to x')) \to (x \to x') \\ =\ &(x \to x') \to (x \to x') \\ =\ &I,\end{aligned}$$

i.e., (4) holds. Now assume (4), for any $x, y \in L$,

$$\begin{aligned}&(x \to y) \to (x' \vee y) \\ =\ &((x \to y) \to x') \vee ((x \to y) \to y) \\ =\ &y' \vee x' \vee x \vee y = I,\end{aligned}$$

i.e., $x \to y \leqslant x' \vee y$ and hence $x \to y = x' \vee y$. It follows that

$$\begin{aligned}(x \to y) \to x &= (x' \vee y) \to x \\ &= (x' \vee y)' \vee x \\ &= (x \wedge y') \vee x \\ &= x,\end{aligned}$$

i.e., (3) holds. □

Corollary 2.2.2 *$(L, \vee, \wedge, ', \to)$ is a lattice H implication algebra if and only if $(L, \vee, \wedge, ')$ is a Boolean lattice, x' is the complement of x and $x \to y = x' \vee y$ for any $x, y \in L$.* □

Lemma 2.2.1 *Let L be a lattice implication algebra. Then it is a lattice H implication algebra if and only if $x \otimes x = x$ holds for any $x \in L$.*

Proof. By Theorem 2.2.2, we know that L is a lattice H implication algebra if and only if $x \vee x' = I$ holds for any $x \in L$. And since

$$\begin{aligned} x \vee x' = I \text{ iff } & (x \rightarrow x') \rightarrow x' = I \\ \text{iff } & x \rightarrow x' = x' \\ \text{iff } & (x \rightarrow x')' = x \\ \text{iff } & x \otimes x = x. \ \Box \end{aligned}$$

Theorem 2.2.3 *If L is a lattice H implication algebra, then for any $x \in L$ and $n \in \mathbb{N}^+$, $x^n = x$.* $\Box$

2.3 Lattice Properties

Theorem 2.3.1 *Let L be a lattice implication algebra, $x_0 \in L$, define θ_l, θ_r: $L \longrightarrow L$ as follows: for any $x \in L$,*

$$\theta_l(x) = x_0 \rightarrow x, \theta_r(x) = x \rightarrow x_0,$$

then θ_l (named a left-mapping) is a lattice homomorphism and θ_r (named a right-mapping) is a dual lattice homomorphism.

Remark 4. In Theorem 2.3.1, θ_l and θ_r are related to x_0. For simplicity, we omit x_0 in them.

Proof. By Definition 2.1.2, for any $x, y \in L$,

$$\begin{aligned} \theta_l(x \vee y) &= x_0 \rightarrow (x \vee y) \\ &= (x_0 \rightarrow x) \vee (x_0 \rightarrow y) \\ &= \theta_l(x) \vee \theta_l(y), \\ \theta_l(x \wedge y) &= x_0 \rightarrow (x \wedge y) \\ &= (x_0 \rightarrow x) \wedge (x_0 \rightarrow y) \\ &= \theta_l(x) \wedge \theta_l(y), \\ \theta_r(x \vee y) &= (x \vee y) \rightarrow x_0 \\ &= (x \rightarrow x_0) \wedge (y \rightarrow x_0) \\ &= \theta_r(x) \wedge \theta_r(y), \\ \theta_r(x \wedge y) &= (x \wedge y) \rightarrow x_0 \\ &= (x \rightarrow x_0) \vee (y \rightarrow x_0) \\ &= \theta_r(x) \vee \theta_r(y). \ \Box \end{aligned}$$

Theorem 2.3.2 *Let L be a lattice implication algebra, $J \subseteq L$.*

(1). If J is a dual ideal of lattice $(L, \vee, \wedge)$, then $x \rightarrow y \in J$ for any $x, y \in J$.

(2). Let $J' = \{x' \mid x \in J\}$. J' is an ideal of $(L, \vee, \wedge)$ if and only if J is a dual ideal of $(L, \vee, \wedge)$. □

Theorem 2.3.3 *Let L be a lattice implication algebra, $a, b \in L$,*

$$\begin{aligned} A_{a,b} &= \{x \mid a \to x \geqslant b\}, \\ B_{a,b} &= \{x \mid a \to x \leqslant b\}, \\ C_{a,b} &= \{x \mid x \to a \geqslant b\}, \\ D_{a,b} &= \{x \mid x \to a \leqslant b\}, \end{aligned}$$

then

(1). $A_{a,b}, B_{a,b}, C_{a,b}$ and $D_{a,b}$ are convex sublattices of $(L, \vee, \wedge)$;
(2). $A_{a,b}$ and $D_{a,b}$ are dual ideals of $(L, \vee, \wedge)$, $B_{a,b}$ and $C_{a,b}$ are ideals of $(L, \vee, \wedge)$.

Proof. (1). For any $x, y \in A_{a,b}$, $a \to x \geqslant b, a \to y \geqslant b$ and hence

$$\begin{aligned} a \to (x \vee y) = (a \to x) \vee (a \to y) \geqslant b, \\ a \to (x \wedge y) = (a \to x) \wedge (a \to y) \geqslant b, \end{aligned}$$

i.e., $x \vee y$, $x \wedge y \in A_{a,b}$ and $A_{a,b}$ is a sublattice of $(L, \vee, \wedge)$.

If $z \in [x \wedge y, x \vee y]$, then

$$a \to z \geqslant a \to (x \wedge y) \geqslant b,$$

it follows that $A_{a,b}$ is a convex sublattice of $(L, \vee, \wedge)$. Similarly, we can prove that $B_{a,b}, C_{a,b}$ and $D_{a,b}$ are all convex sublattices of $(L, \vee, \wedge)$.

(2). For any $x, y \in L$, by (1), $x, y \in A_{a,b}$ implies that $x \wedge y \in A_{a,b}$. If $x \in A_{a,b}, x \leqslant y$, then

$$a \to y \geqslant a \to x \geqslant b.$$

It follows that $y \in A_{a,b}$ and hence $A_{a,b}$ is a dual ideal of $(L, \vee, \wedge)$. The remained conclusions can be proved similarly. □

Corollary 2.3.1 *Let L be a lattice implication algebra, $a, b \in L$,*

$$\begin{aligned} A_{(a,b)} &= \{x \mid a \to x = b\}, \\ B_{(a,b)} &= \{x \mid x \to a = b\}, \end{aligned}$$

then both $A_{(a,b)}$ and $B_{(a,b)}$ are convex sublattices of $(L, \vee, \wedge)$. □

Corollary 2.3.2 *Let L be a lattice implication algebra, $a, b \in L$,*

$$\begin{aligned} A_{(a,a\to b)} &= \{x \mid a \to x = a \to b\}, \\ B_{(a,b\to a)} &= \{x \mid x \to a = b \to a\}, \\ A_{(a,b\to a)} &= \{x \mid a \to x = b \to a\}, \\ B_{(a,a\to b)} &= \{x \mid x \to a = a \to b\}, \end{aligned}$$

then $A_{(a,a\to b)}$, $B_{(a,b\to a)}$, $A_{(a,b\to a)}$ and $B_{(a,a\to b)}$ are all convex sublattices of $(L, \vee, \wedge)$. □

Theorem 2.3.4 *Let L be a lattice implication algebra, $a, b, c \in L$,*

$$A_{a,b,c} = \{x \mid (a \to x) \wedge (x \to b) \leqslant c\},$$
$$B_{a,b,c} = \{x \mid (a \to x) \wedge (x \to b) \geqslant c\},$$
$$C_{a,b,c} = \{x \mid (a \to x) \vee (x \to b) \leqslant c\},$$
$$D_{a,b,c} = \{x \mid (a \to x) \vee (x \to b) \geqslant c\},$$

then $A_{a,b,c}, B_{a,b,c}, C_{a,b,c}$ and $D_{a,b,c}$ are all sublattices of $(L, \vee, \wedge)$.

Proof. For any $x, y \in A_{a,b,c}$, $(a \to x) \wedge (x \to b) \leqslant c$, $(a \to y) \wedge (y \to b) \leqslant c$, it follows that

$$\begin{aligned}
&(a \to (x \wedge y)) \wedge ((x \wedge y) \to b) \\
&= (a \to x) \wedge (a \to y) \wedge ((x \to b) \vee (y \to b)) \\
&= ((a \to x) \wedge (a \to y) \wedge (x \to b)) \vee ((a \to x) \wedge (a \to y) \wedge (y \to b)) \\
&\leqslant ((a \to x) \wedge (x \to b)) \vee ((a \to y) \wedge (y \to b)) \\
&\leqslant c
\end{aligned}$$

and

$$\begin{aligned}
&(a \to (x \vee y)) \wedge ((x \vee y) \to b) \\
&= ((a \to x) \vee (a \to y)) \wedge (x \to b) \wedge (y \to b) \\
&= ((a \to x) \wedge (x \to b) \wedge (y \to b)) \vee ((a \to y) \wedge (x \to b) \wedge (y \to b)) \\
&\leqslant ((a \to x) \wedge (x \to b)) \vee ((a \to y) \wedge (y \to b)) \\
&\leqslant c,
\end{aligned}$$

i.e., $x \wedge y, x \vee y \in A_{a,b,c}$ and hence $A_{a,b,c}$ is a sublattice of $(L, \vee, \wedge)$. The remained conclusions can be proved similarly. □

Corollary 2.3.3 *Let L be a lattice implication algebra, $a, b \in L$. $A_{a,b,a \to b}$ is a sublattice of $(L, \vee, \wedge)$ and $O, I \in A_{a,b,a \to b}$.* □

Corollary 2.3.4 *Let L be a lattice implication algebra, $a, b, c \in L$. Both $\{x \mid (a \to x) \wedge (x \to b) = c\}$ and $\{x \mid (a \to x) \vee (x \to b) = c\}$ are sublattices of $(L, \vee, \wedge)$.* □

Definition 2.3.1 *Let L be a lattice implication algebra, $S \subseteq L$ is called a lattice implication subalgebra of L, if the following conditions hold:*

(1). $(S, \vee, \wedge, ')$ is a bounded sublattice of $(L, \vee, \wedge)$ with an order-reversing involution $'$;
(2). if $x, y \in S$, then $x \to y \in S$.

It is clear that lattice implication subalgebra is a lattice implication algebra itself.

Theorem 2.3.5 *Let L be a lattice implication algebra, $S \subseteq L$, if*

(1). $O \in S$;

(2). $x, y \in S$ implies $x \to y \in S$,

then S is a lattice implication subalgebra of L.

Proof. We only need to prove that $(S, \vee, \wedge, ')$ is a sublattice of $(L, \vee, \wedge)$ with an order-reversing involution $'$ and $I \in S$. In fact, for any $x, y \in S$,

(1). $x' = x \to O \in S$;
(2). $x \vee y = (x \to y) \to y \in S$;
(3). $x \wedge y = (x' \vee y')' \in S$;
(4). $I = O \to O \in S$. □

Theorem 2.3.6 *Let L be an atomless lattice implication algebra, then L is densely ordered, i.e., for any $x, y \in L, x < y$, there exists $z \in L$ such that $x < z < y$.*

Proof. Let $x, y \in L$ and $x < y$. Then $y \to x \neq I$. If $y \to x = O$, *i.e.*, $y = I$, $x = O$. So there exists $z \in L$, $O < z < I$.

Now we suppose that $y \to x \neq O$. Since L is atomless, then L is dual-atomless, so there exists u with $y \to x < u < I$. Hence,

$$x = y \otimes (y \to x) \leqslant y \otimes u \leqslant y.$$

Firstly we suppose that $y = y \otimes u$, then

$$\begin{aligned} y = (u \to y')' &\text{ iff } u \to y' = y' \\ &\text{ iff } u \vee y' = I. \end{aligned}$$

Hence $u = I$, by $y' < u$, which is a contradiction, so $y \otimes u < y$. Now let $y \otimes u = x$. Then

$$\begin{aligned} y \to x &= y \to (y \otimes u) \\ &= y' \vee u, \end{aligned}$$

and since

$$u > y \to x \geqslant y',$$

so $y' \vee u = u$. We have $u = y \to x$, which is a contradiction, hence, $x < y \otimes u < y$. □

2.4 Homomorphisms

Definition 2.4.1 *Let L_1 and L_2 be lattice implication algebras, $f: L_1 \longrightarrow L_2$ a mapping from L_1 to L_2, if*

$$f(x \to y) = f(x) \to f(y)$$

holds for any $x, y \in L_1$, *then* f *is called an implication homomorphism from* L_1 *to* L_2. *If an implication homomorphism* f *is a surjection, then it is called an implication epimorphism. If* f *is an implication homomorphism and satisfies*

$$\begin{aligned} f(x \vee y) &= f(x) \vee f(y), \\ f(x \wedge y) &= f(x) \wedge f(y), \\ f(x') &= (f(x))', \end{aligned}$$

then f *is called a lattice implication homomorphism from* L_1 *to* L_2.

A one-to-one and onto lattice implication homomorphism is called a lattice implication isomorphism. If there exists a lattice implication isomorphism from L_1 *to* L_2, L_1 *and* L_2 *are called to be isomorphic and denoted by* $L_1 \cong L_2$.

Let X and Y be lattice implication algebras, $\mathrm{hom}(X, Y)$ the set of all lattice implication homomorphisms from X to Y.

It is obvious that monotonicity holds if f is an implication homomorphism.

Example 2.4.1 *Let* L *be a lattice implication algebra, then* θ_l *is a lattice homomorphism. If* L *is a lattice* H *implication algebra, then* θ_l *is a lattice implication homomorphism.*

Theorem 2.4.1 *Let* L_1 *and* L_2 *be lattice implication algebras,* f *an implication homomorphism from* L_1 *to* L_2, *then*

(1). for any $x, y \in L$, $f(x \vee y) = f(x) \vee f(y)$;
(2). $f(I) = I$;
(3). if $f(O) = O$, *then* f *is a lattice implication homomorphism.*

Proof. (1). For any $x, y \in L$,

$$\begin{aligned} f(x \vee y) &= f((x \to y) \to y) \\ &= (f(x) \to f(y)) \to f(y) \\ &= f(x) \vee f(y). \end{aligned}$$

(2).

$$\begin{aligned} f(I) &= f(I \to I) \\ &= f(I) \to f(I) \\ &= I. \end{aligned}$$

(3). For any $x, y \in L$,

$$\begin{aligned} f(x') &= f(x \to O) \\ &= f(x) \to f(O) \\ &= f(x) \to O \\ &= (f(x))', \\ f(x \wedge y) &= f((x' \vee y')') \\ &= (f(x') \vee f(y'))' \\ &= (f(x'))' \wedge (f(y'))' \\ &= f(x) \wedge f(y). \ \square \end{aligned}$$

Corollary 2.4.1 *An implication homomorphism from L_1 to L_2 is a lattice implication homomorphism if and only if $f(x') = (f(x))'$ for any $x \in L_1$.* $\square$

Corollary 2.4.2 *An onto mapping f from L_1 to L_2 is an implication homomorphism if and only if f is a lattice implication homomorphism.*

Proof. It is obvious that f is an implication homomorphism if it is a lattice implication homomorphism.

Suppose that f is an onto mapping and an implication homomorphism from L_1 to L_2, then for any $y \in L_2$, there exists $x \in L_1$ such that $f(x) = y$, and hence

$$\begin{aligned} y \vee f(O) &= f(x) \vee f(O) \\ &= f(x \vee O) \\ &= f(x) \\ &= y, \end{aligned}$$

it follows that $f(O) = O$. $\square$

Corollary 2.4.3 *Let L_1 and L_2 be lattice implication algebras, f an onto implication homomorphism from L_1 to L_2, then*

(1). if $J \subseteq L_1$ is a sublattice (ideal, dual ideal) of L_1, then $f(J)$ is a sublattice(ideal, dual ideal) of L_2;
(2). if $J \subseteq L_2$ is a sublattice (ideal, dual ideal) of L_2, then $f^{-1}(J)$ is a sublattice (ideal, dual ideal) of L_1. $\square$

Theorem 2.4.2 *Let L_1 and L_2 be lattice implication algebras, f an implication homomorphism from L_1 to L_2, $\ker(f) = \{x \mid f(x) = O, x \in L_1\}$, then*

(1). $\ker(f) \neq \emptyset$ if and only if $O \in \ker(f)$;
(2). $(\ker(f), \vee, \wedge)$ is a convex sublattice and an ideal of $(L, \vee, \wedge)$.

Proof. (1). If $O \in \ker(f)$, obviously, $\ker(f) \neq \emptyset$. Conversely, if $\ker(f) \neq \emptyset$, then there exists $x \in L_1$ such that $f(x) = O$ and hence $f(O) = O$, *i.e.*, $O \in \ker(f)$.

(2). For any $x, y \in \ker(f)$ and for any $t \in [x \wedge y, x \vee y]$,

$$\begin{aligned} f(t) &\leqslant f(x \vee y) \\ &= f(x) \vee f(y) \\ &= O, \end{aligned}$$

i.e., $t \in \ker(f)$ and $\ker(f)$ is a convex sublattice of $(L, \vee, \wedge)$. The remained conclusion can be proved easily. □

Theorem 2.4.3 *Let L_1 and L_2 be lattice implication algebras, f an implication homomorphism from L_1 to L_2, $f(O) = O$. If $S \subseteq L_1$ is a lattice implication subalgebra of L_1, then $f(S)$ is a lattice implication subalgebra of L_2.*

Proof. By $O \in S$, $O = f(O) \in f(S)$ followed. For any $t, p \in f(S)$, there exist $x, y \in S$, such that $f(x) = t$ and $f(y) = p$. By $t \to p = f(x \to y)$ and $x \to y \in S$, it follows that $t \to p \in f(S)$ and hence $f(S)$ is a lattice implication subalgebra of L_2. □

Corollary 2.4.4 *Let L_1 and L_2 be lattice implication algebras, f an onto implication homomorphism from L_1 to L_2.*

(1). If $S \subseteq L_1$ is a lattice implication subalgebra of L_1, then $f(S)$ is a lattice implication subalgebra of L_2.
(2). If $S \subseteq L_2$ is a lattice implication subalgebra of L_2, then $f^{-1}(S)$ is a lattice implication subalgebra of L_1. □

Let L_1 and L_2 be lattice implication algebras and f a lattice implication homomorphism from L_1 to L_2. Define a binary relation $\equiv_f$ on L_1 as follows: for any $x, y \in L_1$,

$$x \equiv_f y \quad \text{iff} \quad f(x) = f(y).$$

It can be verified easily that $\equiv_f$ is an equivalent relation. We denote

$$[x] = \{y \mid y \in L_1, y \equiv_f x\},$$

$$L_1/_{\equiv_f} = \{[x] \mid x \in L_1\}$$

and define operations on $L_1/_{\equiv_f}$ as follows:

$$\begin{aligned} [x] \vee [y] &= [x \vee y], \\ [x] \wedge [y] &= [x \wedge y], \\ [x] \to [y] &= [x \to y], \\ [x]' &= [x']. \end{aligned}$$

It is routine to check that $(L_1/_{\equiv_f}, \vee, \wedge, ', \to)$ is a lattice implication algebra. We call it the quotient lattice implication algebra of L_1 induced by f.

Theorem 2.4.4 *Let L_1 and L_2 be lattice implication algebras and f a lattice implication homomorphism from L_1 to L_2.*

(1). If $\pi: L_1 \longrightarrow L_1/_{\equiv_f}$ is the natural lattice implication homomorphism, then there exists an unique lattice implication homomorphism $f^*: L_1/_{\equiv_f} \longrightarrow L_2$, such that

$$f = f^* \circ \pi.$$

(2). f^* is an injection, and $L_1/_{\equiv_f} \cong f(L_1)$.

(3). f^* is a lattice implication isomorphism if and only if f is a surjection. □

This theorem holds according to homomorphism theorem of universal algebra. It shows that a lattice implication homomorphism can be decomposed as a composition of two lattice implication homomorphisms, where one is a natural homomorphism, and the other is an injective lattice implication homomorphism. We will further discuss properties of homomorphism of lattice implication algebra in Chapter 5.

3 Filters

In classical logic and many-valued logic, Modus Ponens rule (MP rule) is very useful for a formal deduction. In this chapter, we establish the concept of filters, which is an abstraction of MP rule. We also extend the concept of filters, establish the concepts of implicative filters, prime filters, ultra-filters, I-filters, involution filters and so on, and discuss their properties as well.

3.1 Filters and Implicative Filters

Definition 3.1.1 *Let L be a lattice implication algebra, $J \subseteq L$ is said to be a filter of L, if it satisfies the following conditions:*

(1). $I \in J$;
(2). for any $x, y \in L$, if $x \in J$ and $x \to y \in J$, then $y \in J$.

$J \subseteq L$ is said to be an implicative filter of L, if it satisfies the following conditions:

(1). $I \in J$;
(2). for any $x, y, z \in L$, if $x \to (y \to z) \in J$ and $x \to y \in J$, then $x \to z \in J$.

Example 3.1.1 *As for the lattice implication algebra defined in Example 2.1.4, it is routine to check that $\{a, I\}$ is a filter.*

If L is a lattice implication algebra, then $\{I\}$ and L are filters of L. They are called trivial filters.

Example 3.1.2 *Consider the Łukasiewicz implication algebra defined in Example 2.1.2. If $J \subseteq [0,1]$ is a filter, and $J \setminus \{1\} \neq \emptyset$, we can prove that $J = [0,1]$. In fact, it is trivial to prove that J is an interval with the right endpoint 1. If $J = (u, 1]$ for some $u < 1$, we choose x such that*

$$u < x < 1,$$

it follows that

$$x \to u = 1 - x + u > u,$$

i.e., $x \to u \in J$ and hence $u \in J$. This proves that J is a closed interval. Suppose that $J = [u, 1]$ for some $u < 1$. For any $x \in [u, 1)$, if $x + u \geqslant 1$, then

$$u \to (x + u - 1) = 1 - u + (x + u - 1) = x \in J$$

and hence $u + (x - 1) \in J$, *this is a contradiction. So, for any* $x \in [u, 1)$, $x + u < 1$ *and hence* $u = 0$.

From Example 3.1.2, we know that the Łukasiewicz implication algebra has only trivial filters.

Theorem 3.1.1 *If* J *is a filter of a lattice implication algebra* L, *then* J *is a dual ideal of* $(L, \vee, \wedge)$.

Proof. If $x \in J$ and $x \leqslant y$, then $x \to y = I \in J$ and hence $y \in J$. If $x, y \in J$, by

$$x \to (x \wedge y) = x \to y \geqslant y$$

it follows that $x \to (x \wedge y) \in J$ and hence $x \wedge y \in J$. □

The converse of Theorem 3.1.1 is not true. For the Łukasiewicz implication algebra defined in Example 2.1.2, $J = [1/4, 1]$ is a dual ideal but not a filter.

Theorem 3.1.2 *If* J *is a dual ideal of lattice* H *implication algebra* L *and* $J \neq \emptyset$, *then* J *is a filter of* L.

Proof. By $J \neq \emptyset$, we suppose that $a \in J$, it follows that $I \in J$ since $a \leqslant I$. If $x \in J$ and $x \to y \in J$, then $x \wedge (x \to y) \in J$. By

$$\begin{aligned} x \wedge (x \to y) &= x \wedge (x' \vee y) \\ &= (x \wedge x') \vee (x \wedge y) \\ &= x \wedge y \leqslant y, \end{aligned}$$

$y \in J$ followed. □

Theorem 3.1.3 *An implicative filter is a filter in a lattice implication algebra.*

Proof. Let L be a lattice implication algebra, $J \subseteq L$ an implicative filter of L. If $x \in J$ and $x \to y \in J$, then $I \to x \in J$ and $I \to (x \to y) \in J$. It follows that

$$y = I \to y \in J. \ \Box$$

We notice that the converse of this theorem is not true. For example, in C_5, $\{I\}$ is a filter but not an implicative filter since $a \to (b \to O) = a \to b = I \in \{I\}$, and $a \to O = c \notin \{I\}$.

Definition 3.1.2 *Let* L *be a lattice implication algebra,* $a \in L$. *Denote the interval* $[a, I]$

$$[a, I] = \{x \mid x \in L, a \leqslant x\}$$

of L *as* $A(a)$.

Theorem 3.1.4 *Let L be a lattice implication algebra, $a \in L$, then*

(1). $A(a)$ is a dual ideal of L;
(2). $A(a)$ is a filter of L if and only if for any $x, y \in L$, $a \leqslant y \to x$ and $a \leqslant y$ imply $a \leqslant x$;
(3). $\{I\}$ is an implicative filter of L if and only if $A(a)$ is a filter for any $a \in L$.

Proof. We only prove (3). Suppose that $\{I\}$ is an implicative filter of L. For any $a \in L$, $I \in A(a)$ is trivial. If $x \in A(a)$ and $x \to y \in A(a)$, then $a \leqslant x$, $a \leqslant x \to y$, *i.e.*, $a \to x \in \{I\}$ and $a \to (x \to y) \in \{I\}$. It follows that $a \to y \in \{I\}$, $a \leqslant y$, and hence $y \in A(a)$. So $A(a)$ is a filter of L.

On the other hand, suppose that $A(a)$ is a filter for any $a \in L$. For any $x, y, z \in L$, if $x \to (y \to z) \in \{I\}$ and $x \to y \in \{I\}$, then $x \leqslant y \to z$ and $x \leqslant y$, it follows that $x \leqslant z$ because $A(x)$ is a filter, and hence $x \to z = I \in \{I\}$. □

Theorem 3.1.5 *Let L be a lattice implication algebra, $J \subseteq L$. J is a filter of L if and only if for any $x, y \in J$,*

$$A_{x,y} \subseteq J.$$

Proof. We suppose that J is a filter of L, $x, y \in J$. For any $z \in A_{x,y}$,

$$y \leqslant x \to z,$$

by $y \to (x \to z) = I \in J$, it follows that $x \to z \in J$ and hence $z \in J$, *i.e.*, $A_{x,y} \subseteq J$.

On the other hand, we suppose that $A_{x,y} \subseteq J$ for any $x, y \in J$. By $y \leqslant I = x \to I$, it follows that $I \in A_{x,y} \subseteq J$. If $y \in J$, $y \to x \in J$, then $A_{y,y\to x} \subseteq J$ by $y \leqslant x \vee y = (y \to x) \to x$ and $x \in A_{y,y\to x} \subseteq J$. □

Corollary 3.1.1 *$J \subseteq L$ is a filter of L if and only if it satisfies the following conditions:*

(1). $I \in J$;
(2). for any $x, y \in J$ and $z \in L$, $x \to (y \to z) \in J$ implies $z \in J$. □

Theorem 3.1.6 *Let L be a lattice implication algebra, $J \subseteq L$. J is a filter of L if and only if it satisfies the following conditions:*

(1). $I \in J$;
(2). for any $x, y, z \in L$, $x \to y \in J$, $y \to z \in J$ imply $x \to z \in J$.

Proof. Suppose that J is a filter of L, $x \to y \in J$, $y \to z \in J$. By

$$(x \to y) \to ((y \to z) \to (x \to z)) = I \in J,$$

it follows that $x \to z \in J$.

Conversely, if $J \subseteq L$ satisfies (1), (2) and $x, x \to y \in J$, it follows that $I \to x \in J$, $x \to y \in J$, and hence $y = I \to y \in J$. □

Theorem 3.1.7 *Let L be a lattice implication algebra, $J \subseteq L$. J is a filter of L if and only if the following conditions hold:*

(1). $I \in J$;
(2). for any $x, y, z \in L$, $(z \to y) \to x \in J$, $y \in J$ imply $z \to x \in J$.

Proof. Suppose that (1) and (2) hold. If $x \to y \in J$ and $x \in J$, then $(I \to x) \to y \in J$, which implies $y = I \to y \in J$ and J is a filter.

Conversely, if J is a filter, and $(z \to y) \to x \in J$, $y \in J$, then by $x \leqslant z \to x$ and $y \leqslant z \to y$, it follows that

$$\begin{aligned}(z \to y) \to x &\leqslant (z \to y) \to (z \to x) \\ &\leqslant y \to (z \to x)\end{aligned}$$

and hence $y \to (z \to x) \in J$, which implies $z \to x \in J$. □

Theorem 3.1.8 *Let L be a lattice implication algebra, $J \subseteq L$. The following statements are equivalent:*

(1). J is an implicative filter;
(2). J is a filter and for any $x, y \in L$, $x \to (x \to y) \in J$ implies $x \to y \in J$;
(3). J is a filter and for any $x, y, z \in L$, $x \to (y \to z) \in J$ implies

$$(x \to y) \to (x \to z) \in J;$$

(4). $I \in J$ and for any $x, y, z \in L$, $z \to (x \to (x \to y)) \in J$ and $z \in J$ imply $x \to y \in J$.

Proof. First, we prove that (1) implies (2). In fact, for any $x, y \in L$, if $x \to (x \to y) \in J$, then by $x \to x = I \in J$, it follows that $x \to y \in J$.

Assume (2) holds, we prove (3). For any $x, y, z \in L$, we suppose $x \to (y \to z) \in J$, by

$$\begin{aligned}& x \to (x \to ((x \to y) \to z)) \\ &= x \to ((x \to y) \to (x \to z)) \\ &= x \to ((x \wedge y) \to z) \\ &\geqslant x \to (y \to z) \in J,\end{aligned}$$

it follows that

$$x \to ((x \to y) \to z) \in J,$$

i.e., $(x \to y) \to (x \to z) \in J$, (3) holds.

Now we assume (3) holds and prove (4). In fact, $I \in J$ is trivial. For any $x, y, z \in L$, suppose $z \to (x \to (x \to y)) \in J$ and $z \in J$, then

$$x \to (x \to y) \in J.$$

It follows that

$$\begin{aligned} x \to y &= I \to (x \to y) \\ &= (x \to x) \to (x \to y) \\ &\in J. \end{aligned}$$

Finally, we prove (4) implies (1). Suppose $x \in J$ and $x \to y \in J$, then

$$x \to (I \to (I \to y)) = x \to y \in J,$$

it follows that $y = I \to y \in J$ and hence J is a filter. For $x, y, z \in L$, suppose $x \to (y \to z) \in J$ and $x \to y \in J$, notice that

$$\begin{aligned} &(x \to (y \to z)) \to ((x \to y) \to (x \to (x \to z))) \\ = &(x \to y) \to ((x \to (y \to z)) \to (x \to (x \to z))) \\ = &(x \to y) \to ((y \to (x \to z)) \to (x \to (x \to z))) \\ = &(x \to y) \to (x \to (y \vee (x \to z))) \\ = &(x \to y) \to ((x \to y) \vee (x \to (x \to z))) \\ = &I \in J, \end{aligned}$$

it follows that

$$(x \to y) \to (x \to (x \to z)) \in J$$

and hence $x \to z \in J$ by $x \to y \in J$ and (4). □

Theorem 3.1.9 *Let L be a lattice implication algebra, $J \subseteq L$. J is a filter of L if and only if:*

(1). $I \in J$;
(2). if $a \in J$ and $a \leqslant b$, then $b \in J$;
(3). if $a, b \in J$, then $a \otimes b \in J$.

Proof. If J is a filter of L, then (1) and (2) hold. For $a, b \in J$, by

$$a \to (b \to (a \otimes b)) = (a \otimes b) \to (a \otimes b) = I \in J,$$

it follows that $a \otimes b \in J$, *i.e.*, (3) holds.

Conversely, suppose that (1), (2) and (3) hold. If $x \in J$ and $x \to y \in J$, then $x \otimes (x \to y) \in J$. Hence $y \in J$ for $x \otimes (x \to y) = x \wedge y \leqslant y$. □

Theorem 3.1.10 *Let L be a lattice implication algebra, J_1, J_2 filters of L, $J_1 \subseteq J_2$. If J_1 is an implicative filter, so is J_2.*

Proof. Suppose $x \to (x \to y) \in J_2$, we only need to prove $x \to y \in J_2$. By

$$x \to (x \to ((x \to (x \to y)) \to y)) = (x \to (x \to y)) \to (x \to (x \to y)) = I \in J_1,$$

it follows that $x \to ((x \to (x \to y)) \to y) \in J_1 \subseteq J_2$, *i.e.*, $(x \to (x \to y)) \to (x \to y) \in J_2$, and hence $x \to y \in J_2$. □

Theorem 3.1.11 *Let L be a lattice implication algebra, J an implicative filter of L. If $a \in L$, $n \in \mathbb{N}^+$, $na \in J$, then $a \in J$.*

Proof. We use induction over n to prove the statement.

If $n = 1$, then it is trivial.

Suppose the conclusion holds for any $p < n+1$ and $(n+1)a \in J$, then

$$a' \to (a' \to (n-1)a) = a' \to na = (n+1)a \in J,$$

it follows that $na = a' \to (n-1)a \in J$ and hence $a \in J$. □

Theorem 3.1.12 *Let L and L^* be lattice implication algebras, f a lattice implication homomorphism from L to L^*. If $F^* \subseteq L^*$ is a filter of L^*, then $f^{-1}(F^*)$ is a filter of L.*

Proof. Let I and I^* be the greatest elements of L and L^*, respectively. Since

$$f(I) = f(I \to I) = I^* \in F^*,$$

$I \in f^{-1}(F^*)$ followed.

If $x, x \to y \in f^{-1}(F^*)$, then $f(x) \in F^*$ and $f(x) \to f(y) = f(x \to y) \in F^*$. It follows that $f(y) \in F^*$ and hence $y \in f^{-1}(F^*)$. Therefore, $f^{-1}(F^*)$ is a filter of L. □

Theorem 3.1.13 *Let L and L^* be lattice implication algebras, I and I^* the greatest elements of L and L^*, respectively, f an implication homomorphism from L to L^*,*

$$D\text{-}\ker(f) = \{x \mid x \in L, f(x) = I^*\}.$$

Then $D\text{-}\ker(f)$ is a filter of L.

Proof. Because f is an implication homomorphism, then $f(I) = I^*$, and hence $I \in \mathrm{D\text{-}ker}(f)$.

If $x, x \to y \in \mathrm{D\text{-}ker}(f)$, then $f(x) = I^*$, $f(x \to y) = I^*$ and hence

$$\begin{aligned} f(y) &= I^* \to f(y) \\ &= f(x) \to f(y) \\ &= f(x \to y) \\ &= I^*, \end{aligned}$$

which means $y \in \mathrm{D\text{-}ker}(f)$. □

Theorem 3.1.14 *Let f be an implication epimorphism from L to L^*. f is a lattice implication isomorphism if and only if $D\text{-}\ker(f) = \{I\}$, where I is the greatest element of L.* □

3.2 Generated Filters

Let L be a lattice implication algebra, it is easy to verify that the intersection of filters of L is also a filter, and the intersection of implicative filters is an implicative filter, too. Suppose $A \subseteq L$, the least filter containing A is called the filter generated by A and denoted by $[A)$. It is trivial to verify that

$$[A) = \cap\{B \mid A \subseteq B \subseteq L, B \text{ is a filter}\}.$$

In what follows, $[\{a\})$ is denoted by $[a)$.

Theorem 3.2.1 *Let L be a lattice implication algebra, J an implicative filter of L, $a \in L$. Then $[J \cup \{a\}) = \{x \mid x \in L, a \rightarrow x \in J\}$.*

Proof. Suppose

$$B = \{x \mid x \in L, a \rightarrow x \in J\}.$$

Firstly, we prove that B is a filter. In fact, by $a \rightarrow I = I \in J$, $I \in B$ followed. As J is an implicative filter, if $x, x \rightarrow y \in B$, then

$$a \rightarrow x \in J, a \rightarrow (x \rightarrow y) \in J,$$

so that $a \rightarrow y \in J$ and hence $y \in B$. Secondly, we prove that B is the least filter containing $J \cup \{a\}$. By $a \rightarrow a = I \in J$, $a \in B$ followed. For any $x \in J$, $a \rightarrow x \in J$ because J is a filter and $x \leqslant a \rightarrow x$, it follows that $x \in B$ and hence

$$J \cup \{a\} \subseteq B.$$

If $J \cup \{a\} \subseteq C$ and C is a filter, for any $x \in B$,

$$a \rightarrow x \in J \subseteq C$$

and hence $x \in C$ because $a \in C$ and C is a filter. That is to say $B \subseteq C$. Hence $B = [J \cup \{a\})$. □

In what follows,

$$[a_1, a_2, \cdots, a_n, x] \triangleq a_1 \rightarrow (a_2 \rightarrow (\cdots \rightarrow (a_n \rightarrow x)\cdots)).$$

Specially,

$$\begin{aligned}
&[a, x]^0 \triangleq x, \\
&[a, x]^1 \triangleq [a, x] = a \rightarrow x, \\
&[a, x]^n \triangleq \underbrace{a \rightarrow (a \rightarrow (\cdots \rightarrow (a}_{n} \rightarrow x)\cdots))\ (n \geqslant 2).
\end{aligned}$$

Remark 5. It is easy to distinguish $[a, x]$ being an interval or $a \rightarrow x$ from the context.

Theorem 3.2.2 *Let L be a lattice implication algebra, $\emptyset \neq A \subseteq L$. Then*

$$[A) = \{x \mid x \in L, \text{ there exist } a_1, \cdots, a_n \in A, \\ n \in \mathbb{N}^+, \text{ s.t., } [a_1, \cdots, a_n, x] = I\}.$$

Proof. Let

$$B = \{x \mid x \in L, \text{ there exist } a_1, \cdots, a_n \in A, \text{ s.t., } [a_1, \cdots, a_n, x] = I\}.$$

By $A \neq \emptyset$, we suppose $a \in A$. It follows that $a \to I = I$ and hence $I \in B$. If $x \in B$ and $x \to y \in B$, then there exist $a_1, \cdots, a_n, b_1, \cdots, b_m \in A$, such that

$$a_1 \to (\cdots \to (a_n \to (x \to y)) \cdots) = I, \tag{3.1}$$
$$b_1 \to (\cdots \to (b_m \to x) \cdots) = I. \tag{3.2}$$

By Eq.(3.1), $x \to (a_1 \to (\cdots \to (a_n \to y) \cdots)) = I$ and hence $x \leqslant a_1 \to (\cdots \to (a_n \to y) \cdots)$. It follows that $b_m \to x \leqslant b_m \to (a_1 \to (\cdots \to (a_n \to y) \cdots))$ and similarly

$$\begin{aligned} I &= b_1 \to (\cdots \to (b_m \to x) \cdots) \\ &\leqslant b_1 \to (\cdots \to (b_m \to (a_1 \to (\cdots \to (a_n \to y) \cdots))) \cdots), \end{aligned}$$

i.e., $y \in B$ and B is a filter.

The proof of $A \subseteq B$ is trivial.

If $A \subseteq C$ and C is a filter, then for any $x \in B$, there exist $a_1, \cdots, a_n \in A$ such that

$$a_1 \to (\cdots \to (a_n \to x) \cdots) = I.$$

By $a_1, \cdots, a_n \in A \subseteq C$ and C is a filter, it follows that $x \in C$ and hence $B \subseteq C$. □

Corollary 3.2.1 *Let L be a lattice implication algebra, $\emptyset \neq A \subseteq L$. Then*

$$[A) = \{x \mid x \in L, \text{ there exist } a_1, \cdots, a_n \in A, \text{ s.t., } a_1 \otimes \cdots \otimes a_n \leqslant x\}. \ \square$$

Corollary 3.2.2 *Let L be a lattice implication algebra, $a \in L$. Then $[a) = \{x \mid x \in L, \text{ there exists } n \in \mathbb{N}^+, \text{ s.t., } [a, x]^n = I\}$.* □

Definition 3.2.1 *Let L be a lattice implication algebra. $A \subseteq L$ is said to have the finite intersection property if for any $a_1, \cdots, a_n \in A$, $a_1 \otimes \cdots \otimes a_n > O$.*

Theorem 3.2.3 *Let L be a lattice implication algebra, $A \subseteq L$. Then $[A)$ is a proper filter if and only if A has the finite intersection property.* □

Theorem 3.2.4 *Let L be a lattice implication algebra, J a filter of L, $a \in L$. Then $[J \cup \{a\}) = \{x \mid x \in L, \text{ there exists } n \in \mathbb{N}, \text{ s.t., } a^n \to x \in J\}$.*

Proof. Suppose that $x \in [J \cup \{a\})$, by Theorem 3.2.2, there exist $b_1, \cdots, b_m \in J$ and $n \in \mathbb{N}$ such that

$$b_1 \to (\cdots (b_m \to (a^n \to x)) \cdots) = I.$$

By $I \in J$, it follows that $a^n \to x \in J$.

Conversely, if there exists $n \in \mathbb{N}$ such that $a^n \to x \in J$, let $a^n \to x = u$, it follows that $u \to (a^n \to x) = I$ and hence $x \in [J \cup \{a\})$. □

Theorem 3.2.5 *Let L be a lattice implication algebra, $a, b, x \in L$.*

(1). If $a \geqslant b$, then $[a, x]^n \leqslant [b, x]^n$ for any $n \in \mathbb{N}$.
(2). If $n, m \in \mathbb{N}$, $n \geqslant m$, then $[a, x]^n \geqslant [a, x]^m$;
(3). $[a, x]^n \geqslant x$ for any $n \in \mathbb{N}$.

Proof. These conclusions are trivial when $n = 0$ or $m = 0$.

(1). We use induction over n to show $[a, x]^n \leqslant [b, x]^n$. If $n = 1$, then

$$[a, x]^n = a \to x \leqslant b \to x = [b, x]^n.$$

Suppose now $n > 1$, and $[a, x]^m \leqslant [b, x]^m$ for any $m < n$, then

$$[a, x]^n = a \to [a, x]^{n-1} \leqslant a \to [b, x]^{n-1} \leqslant b \to [b, x]^{n-1} = [b, x]^n.$$

(2). Suppose that $n = m + p$, it follows that $p \geqslant 0$. We use induction over p to show $[a, x]^{m+p} \geqslant [a, x]^m$. If $p = 0$, then $[a, x]^{m+p} \geqslant [a, x]^m$ holds. Suppose that $p = 1$, then

$$[a, x]^{m+p} = a \to [a, x]^m \geqslant [a, x]^m.$$

Suppose now $p > 1$, and $[a, x]^{m+q} \geqslant [a, x]^m$ for any $q < p$. It follows that

$$[a, x]^{m+p} = a \to [a, x]^{m+(p-1)} \geqslant a \to [a, x]^m \geqslant [a, x]^m.$$

(3). The proof is similar to that of (2). □

Theorem 3.2.6 *Let L be a lattice implication algebra, $a, b \in L$. Then $[a) \cap [b) = [a \vee b)$.*

Proof. For any $x \in [a \vee b)$, there exists $n \in \mathbb{N}^+$, such that $[a \vee b, x]^n = I$. By (1) of Theorem 3.2.5, $[a, x]^n = I$ and $[b, x]^n = I$, *i.e.*, $x \in [a) \cap [b)$.

Conversely, suppose that $x \in [a) \cap [b)$. Then there exist $m, n \in \mathbb{N}^+$ such that $[a, x]^m = I$ and $[b, x]^n = I$. We prove that $[a \vee b, x]^{m+n} = I$ and hence $x \in [a \vee b)$.

$$\begin{aligned} [a \vee b, x]^{m+n} &= [a, x]^{m+n} \wedge [b, [a, x]^{m+n-1}]^1 \\ &\quad \wedge \cdots \wedge [b, [a, x]^m]^n \\ &\quad \wedge \cdots \wedge [b, [a, x]]^{m+n-1} \wedge [b, x]^{m+n} \\ &= I. \ \square \end{aligned}$$

Corollary 3.2.3 *Let L be a lattice implication algebra, $a, b \in L$.*

(1). If $a \leqslant b$, then $[b) \subseteq [a)$;
(2). $[a) \cup [b) \subseteq [a \wedge b)$;
(3). If $a \vee b = I$, then $[a) \cap [b) = \{I\}$. □

Theorem 3.2.7 *Let L be a lattice implication algebra, $a, b \in L$, J a filter of L. Then $[J \cup \{a\}) \cap [J \cup \{b\}) = [J \cup \{a \vee b\})$.*

Proof. If $x \in [J\cup\{a\})\cap[J\cup\{b\})$, then by Theorem 3.2.4, there exist $m, n \in \mathbb{N}$ such that $a^n \to x \in J$ and $b^m \to x \in J$. Let $u = a^n \to x$ and $v = b^m \to x$, it follows that

$$a^n \to (v \to (u \to x)) = v \to (u \to (a^n \to x)) = v \to (u \to u) = I.$$

Similarly, we have $b^m \to (v \to (u \to x)) = I$ and hence there exists $p \in \mathbb{N}$ such that

$$v \to (u \to ((a \vee b)^p \to x)) = (a \vee b)^p \to (v \to (u \to x)) = I,$$

it follows that $(a \vee b)^p \to x \in J$. By Theorem 3.2.4, $x \in [J \cup \{a \vee b\})$ and hence $[J \cup \{a\}) \cap [J \cup \{b\}) \subseteq [J \cup \{a \vee b\})$.

$[J \cup \{a\}) \cap [J \cup \{b\}) \supseteq [J \cup \{a \vee b\})$ is trivial. □

Corollary 3.2.4 *Let L be a lattice implication algebra, $a, b \in L$, J a filter of L. If $a \vee b \in J$, then $J = J_1 \cap J_2$, where $J_1 = [J \cup \{a\})$, $J_2 = [J \cup \{b\})$.* □

For any subsets A and B of L, let $A \vee B$ denote the set $\{a \vee b \mid a \in A, b \in B\}$.

Theorem 3.2.8 *Let L be a lattice implication algebra, $a, b \in L$, then*

$$A(a) \vee A(b) = A(a \vee b).$$

Proof. Let $x \in A(a \vee b)$, then $a \leqslant a \vee b \leqslant x$ and $b \leqslant a \vee b \leqslant x$, hence $x \in A(a) \vee A(b)$, *i.e.*, $A(a) \vee A(b) \supseteq A(a \vee b)$.

Conversely, suppose that $x \in A(a) \vee A(b)$, then $x = a_1 \vee b_1$ for some $a \leqslant a_1$ and $b \leqslant b_1$, thus $x \in A(a \vee b)$, *i.e.*, $A(a) \vee A(b) \subseteq A(a \vee b)$. □

Theorem 3.2.9 *If L is a lattice H implication algebra, $a \in L$, then $A(a) = [a)$.*

Proof. Let $x, y \in L$ be such that $x \to y \in A(a)$ and $x \in A(a)$, then $a \to (x \to y) = I$ and $a \to x = I$. Since L is a lattice H implication algebra, we get

$$a \to y = (a \to x) \to (a \to y) = a \to (x \to y) = I,$$

i.e., $y \in A(a)$ and hence $A(a)$ is a filter. Let F be a filter containing a. For $x \in A(a)$, we have $a \to x = I \in F$ and hence $x \in F$. Therefore $A(a)$ is the least filter containing a, *i.e.*, $A(a) = [a)$. □

Corollary 3.2.5 *Let L be a lattice H implication algebra, $a, b \in L$, then*

$$A(a) \vee A(b) = [a \vee b). \ \square$$

Let $F(L)$ be the set of all filters of L. We study the structure of $F(L)$.

Theorem 3.2.10 *Let L be a lattice implication algebra. Then $(F(L), \vee, \wedge)$ is a complete distributive lattice and for any $A, B \in F(L)$,*

$$\begin{aligned} A \wedge B &= A \cap B, \\ A \vee B &= [A \cup B). \end{aligned}$$

Proof. It is trivial that $(F(L), \vee, \wedge)$ is a bounded lattice. We need only to prove that the distributive law holds, *i.e.*,

$$F_1 \wedge (F_2 \vee F_3) = (F_1 \wedge F_2) \vee (F_1 \wedge F_3)$$

or

$$F_1 \cap [F_2 \cup F_3) = [(F_1 \cap F_2) \cup (F_1 \cap F_3))$$

hold for any $F_1, F_2, F_3 \in F(L)$.

In fact, by $F_1 \cap [F_2 \cup F_3) \supseteq (F_1 \cap F_2) \cup (F_1 \cap F_3)$ and $F_1 \cap [F_2 \cup F_3)$ is a filter, $F_1 \cap [F_2 \cup F_3) \supseteq [(F_1 \cap F_2) \cup (F_1 \cap F_3))$ followed.

Conversely, suppose that $x \in F_1 \cap [F_2 \cup F_3)$, then there exist $a_1, \cdots, a_n \in F_2 \cup F_3$, such that $[a_1, \cdots, a_n, x] = I$. Let $b_n = a_n \vee x$, $b_{n-1} = a_{n-1} \vee [a_n, x], \cdots$, and $b_1 = a_1 \vee [a_2, \cdots, a_n, x]$. It follows that $b_n \in (F_1 \cap F_2) \cup (F_1 \cap F_3)$ because $b_n \in F_1$ and $a_n \in F_2 \cup F_3$. $b_i \in (F_1 \cap F_2) \cup (F_1 \cap F_3)$ for any $1 \leqslant i \leqslant n-1$ can be proved similarly. In what follows, we need only to proved $[b_1, \cdots, b_n, x] = I$. In fact,

$$\begin{aligned} & [b_1, \cdots, b_n, x] \\ &= b_1 \to (\cdots \to (b_{n-1} \to (b_n \to x)) \cdots) \\ &= b_1 \to (\cdots \to (b_{n-1} \to ((a_n \vee x) \to x)) \cdots) \\ &= b_1 \to (\cdots \to (b_{n-1} \to (a_n \to x)) \cdots) \\ &= b_1 \to (\cdots \to ((a_{n-1} \vee (a_n \to x)) \to (a_n \to x)) \cdots) \\ &= b_1 \to (\cdots \to (a_{n-1} \to (a_n \to x)) \cdots) \\ & \ \vdots \\ &= a_1 \to (\cdots \to (a_n \to x) \cdots) = I. \ \square \end{aligned}$$

3.3 Positive Implicative Filters and Associative Filters

Definition 3.3.1 *Let L be a lattice implication algebra. A subset F of L is called a positive implicative filter of L if it satisfies:*

(1). $I \in F$;
(2). $x \to ((y \to z) \to y) \in F$ and $x \in F$ imply $y \in F$,

for any $x, y, z \in F$.

We give an example of positive implicative filter of a lattice implication algebra. For the lattice implication algebra defined in Example 2.1.4, it is easy to check that $F = \{b, c, I\}$ is a positive implicative filter.

Theorem 3.3.1 *Every positive implicative filter of a lattice implication algebra is a filter.*

Proof. Let L be a lattice implication algebra and F a positive implicative filter of L. If $x \in F$ and $x \to y \in F$, then $x \to ((y \to y) \to y) \in F$, it follows from the definition of positive implicative filters that $y \in F$, hence F is a filter of L. □

Remark 6. In general, the converse of Theorem 3.3.1 is not true. In fact, considering the lattice implication algebra C_5 defined in Example 2.1.3, we know that $\{I\}$ is a filter, but not a positive implicative filter of L, since $c \notin \{I\}$ and $I \to ((c \to a) \to c) = I \in \{I\}$. Also, in the lattice implication algebra defined in Example 2.1.4, $G = \{a, I\}$ is a filter, but not a positive implicative filter, since $b \notin G$ and $I \to ((b \to c) \to b) = I \in G$.

Theorem 3.3.2 *Let L be a lattice implication algebra and F a filter of L. Then F is a positive implicative filter if and only if $(x \to y) \to x \in F$ implies $x \in F$ for any $x, y \in L$.*

Proof. Suppose that F is a positive implicative filter and let $(x \to y) \to x \in F$, then we have $I \to ((x \to y) \to x) = (x \to y) \to x \in F$. Since $I \in F$, it follows that $x \in F$.

Conversely, let $x \in F$ and $x \to ((y \to z) \to y) \in F$, then $(y \to z) \to y \in F$, which implies $y \in F$. Hence F is a positive implicative filter. □

Theorem 3.3.3 *Let L be a lattice implication algebra and $F \subseteq L$. If F is a positive implicative filter, then F is an implicative filter.*

Proof. Let $x \to (y \to z) \in F$ and $x \to y \in F$. Since F is a filter and

$$x \to (y \to z) = y \to (x \to z) \leqslant (x \to y) \to (x \to (x \to z)),$$

it follows that $x \to (x \to z) \in F$. On the other hand, we have

$$\begin{aligned}
&((x \to z) \to z) \to (x \to z) \\
&= x \to (((x \to z) \to z) \to z) \\
&= x \to ((z \to (x \to z)) \to (x \to z)) \\
&= x \to (I \to (x \to z)) \\
&= x \to (x \to z) \in F,
\end{aligned}$$

hence $x \to z \in F$. □

Theorem 3.3.4 *Let L be a lattice implication algebra and F an implicative filter of L, then for any $x, y \in L$, $(x \to y) \to x \in F$ implies $x \in F$.*

Proof. Suppose $x, y \in L$ and $(x \to y) \to x \in F$, since

$$(x \to y) \to x \leqslant (x \to y) \to ((x \to y) \to y),$$

we have $(x \to y) \to ((x \to y) \to y) \in F$, it follows that $(x \to y) \to y \in F$. Moreover, since $(x \to y) \to x \leqslant y \to x$, we get $y \to x \in F$. Thus we have $((x \to y) \to y) \to x \in F$ and hence $x \in F$. □

Corollary 3.3.1 *In a lattice implication algebra, the concepts of implicative filters and positive implicative filters coincide.* □

Definition 3.3.2 *Let L be a lattice implication algebra, $x \in L$. A subset F of L is called an associative filter of L w.r.t. x if it satisfies:*

(1). $I \in F$;
(2). $x \to (y \to z) \in F$ and $x \to y \in F$ imply $z \in F$, for any $y, z \in L$.

F is called an associative filter of L if F is an associative filter w.r.t. x for any $x \neq O$.

Clearly, an associative filter w.r.t. O is L itself. An associative filter w.r.t. I is coincident with a filter.

Now, we give an example of an associative filter of a lattice implication algebra. In the lattice implication algebra defined in Example 2.1.4, $F = \{a, c, I\}$ is an associative filter w.r.t. a and c, but not w.r.t. b or d, since

$$b \to (a \to d) = b \to b = I \in F, b \to a = a \in F, \text{ but } d \notin F,$$

$$d \to (a \to b) = d \to b = I \in F, d \to a = c \in F, \text{ but } b \notin F.$$

Theorem 3.3.5 *Every associative filter w.r.t. x contains x itself.* □

Theorem 3.3.6 *Every associative filter is a filter.*

Proof. Let F be an associative filter of L, $x \to y \in F$ and $x \in F$, then $I \to x = x \in F$ and $I \to (x \to y) \in F$, it follows that $y \in F$. Hence F is a filter of L. □

Remark 7. The converse of Theorem 3.3.6 is not true. In fact, we know that, in the lattice implication algebra C_5, $\{I\}$ is a filter but not an associative filter, since $a \to (b \to c) = a \to I = I \in \{I\}$ and $a \to b = I \in \{I\}$, but $c \notin \{I\}$.

Theorem 3.3.7 *Let L be a lattice implication algebra, F a filter of L. Then the following statements are equivalent.*

(1). F is an associative filter.

(2). For any $x, y, z \in L$,

$$x \to (y \to z) \in F \text{ implies } (x \to y) \to z \in F. \tag{3.3}$$

(3). F satisfies

$$x \to (x \to y) \in F \text{ implies } y \in F \tag{3.4}$$

for any $x, y \in L$.

Proof. ((1)$\Longrightarrow$(2)). If a filter F of L satisfies

$$x \to (y \to z) \in F \text{ implies } (x \to y) \to z \in F$$

for any $x, y, z \in L$, then F is associative.

((2)$\Longrightarrow$(1)). Suppose that F is an associative filter and $x \to (y \to z) \in F$, since

$$\begin{aligned}
&x \to ((y \to z) \to ((x \to y) \to z)) \\
&= (y \to z) \to (x \to ((x \to y) \to z)) \\
&= (y \to z) \to ((x \to y) \to (x \to z)) \\
&= I \in F,
\end{aligned}$$

we get $(x \to y) \to z \in F$.

((2)$\Longrightarrow$(3)). Trivial.

((3)$\Longrightarrow$(2)). Assume Eq. (3.4) holds, if $x \to (y \to z) \in F$, then we have

$$\begin{aligned}
&(x \to (y \to z)) \to (x \to (x \to ((x \to y) \to z))) \\
&= (x \to (y \to z)) \to (x \to ((x \to y) \to (x \to z))) \\
&\geqslant (x \to (y \to z)) \to (x \to (y \to z)) \\
&= I,
\end{aligned}$$

which implies that $(x \to (y \to z)) \to (x \to (x \to ((x \to y) \to z))) = I$, since F is a filter and $x \to (y \to z) \in F$, it follows that $x \to (x \to ((x \to y) \to z)) \in F$. By Eq. (3.4), we conclude that $(x \to y) \to z \in F$. □

3.4 Prime Filters and Ultra-Filters

Definition 3.4.1 *Let L be a lattice implication algebra, P a proper filter of L. P is called a prime filter if $a \vee b \in P$ implies $a \in P$ or $b \in P$ for any $a, b \in L$.*

Theorem 3.4.1 *Let L be a lattice implication algebra, P a proper filter of L. P is a prime filter if and only if $a \to b \in P$ or $b \to a \in P$ for any $a, b \in L$.*

Proof. If P is a prime filter, by Theorem 2.1.10, $(a \to b) \vee (b \to a) \in P$, and hence $a \to b \in P$ or $b \to a \in P$ for any $a, b \in L$.

Conversely, suppose $a \vee b \in P$ and $a \to b \in P$, then by $a \vee b = (a \to b) \to b$, it follows that $b \in P$. If $b \to a \in P$, then $a \in P$. □

Corollary 3.4.1 *If J is a proper filter and P is a prime filter such that $P \subseteq J$, then J is a prime filter.* □

Definition 3.4.2 *Let L be a lattice implication algebra. A proper filter J of L is called irreducible, if $J = J_1 \cap J_2$ implies $J = J_1$ or $J = J_2$ for any filters J_1 and J_2 of L.*

Theorem 3.4.2 *Let L be a lattice implication algebra, J a proper filter of L. The following statements are equivalent:*

(1). J is irreducible;
(2). $J_1 \cap J_2 \subseteq J$ implies $J_1 \subseteq J$ or $J_2 \subseteq J$ for any filters J_1 and J_2 of L;
(3). $[a) \cap [b) \subseteq J$ implies $a \in J$ or $b \in J$ for any $a, b \in L$;
(4). J is prime.

Proof. ((1)$\Longrightarrow$(2)). In fact, by Theorem 3.2.10,

$$J = J \vee (J_1 \wedge J_2) = (J \vee J_1) \wedge (J \vee J_2) = [J \cup J_1) \cap [J \cup J_2),$$

it follows that $J = [J \cup J_1)$ or $J = [J \cup J_2)$, that is $J_1 \subseteq J$ or $J_2 \subseteq J$.

((2)$\Longrightarrow$(3)). It is trivial.

((3)$\Longrightarrow$(1)). Suppose that there exist filters J_1 and J_2 of L such that $J = J_1 \cap J_2$ and $J \neq J_1$, $J \neq J_2$. Then there exist $a \in J_1$ and $b \in J_2$ such that $a \notin J$ and $b \notin J$. By (3), $[a) \cap [b) \not\subseteq J$, this is a contradiction because $[a) \cap [b) \subseteq J_1 \cap J_2$ and $J = J_1 \cap J_2$.

((4)$\Longrightarrow$(2)). Suppose that J is a prime filter, J_1 and J_2 are filters, $J_1 \cap J_2 \subseteq J$, $J_1 \not\subseteq J$ and $J_2 \not\subseteq J$. Then there exist $a \in J_1 \setminus J$ and $b \in J_2 \setminus J$, it follows that $a \vee b \in J_1$, $a \vee b \in J_2$ and hence $a \vee b \in J_1 \cap J_2 \subseteq J$. This is a contradiction to that J is a prime filter.

((2)$\Longrightarrow$(4)). Suppose that $a \vee b \in J$. By Corollary 3.2.4, $J = [J \cup \{a\}) \cap [J \cup \{b\})$, it follows that $[J \cup \{a\}) \subseteq J$ or $[J \cup \{b\}) \subseteq J$, that is $a \in J$ or $b \in J$. □

Let L be a lattice implication algebra. We denote

$$PF(L) = \{J \mid J \text{ is a prime filter of } L\},$$

$$IF(L) = \{J \mid J \text{ is an implicative filter of } L\}.$$

Theorem 3.4.3 *Let L be a lattice implication algebra, J a proper filter of L. Then $J \in PF(L) \cap IF(L)$ if and only if $a \in J$ or $a' \in J$ for any $a \in L$.*

Proof. Suppose $J \in PF(L) \cap IF(L)$ and $a \in L$. It follows that $a \to a' \in J$ or $a' \to a \in J$ by $(a \to a') \vee (a' \to a) = I \in J$, *i.e.*, $2a' \in J$ or $2a \in J$ and hence $a' \in J$ or $a \in J$.

Conversely, suppose $a \in J$ or $a' \in J$ for any $a \in L$. If $a \to (b \to c) \in J$ and $a \to b \in J$, we need only to prove $a \to c \in J$. In fact, if $(a \to c)' \in J$, then $a \in J$ by $(a \to c)' \leqslant a$. It follows that $b \to c \in J$, $b \in J$ and hence $c \in J$ and

$a \to c \in J$, *i.e.*, $J \in IF(L)$. On the other hand, if $a \vee b \in J$ and $a \notin J$, then $a' \in J$, it follows that $a \to b \in J$ and hence $b \in J$ by $(a \to b) \to b = a \vee b \in J$, *i.e.*, $J \in PF(L)$. Therefore, $J \in PF(L) \cap IF(L)$. □

For $S \subseteq L$ and $x \in L$, we denote

$$x^{-1}S = \{y \mid y \in L, x \vee y \in S\}.$$

Theorem 3.4.4 *[294] Let F be a filter of L and $x \in L$, then*

(1). $x^{-1}F = L$ if and only if $x \in F$;
(2). $x^{-1}F \subseteq y^{-1}F$ if $x \leqslant y$;
(3). if F_1 and F_2 are filters of L and $F_1 \subseteq F_2$, then for each $x \in L$, $x^{-1}F_1 \subseteq x^{-1}F_2$;
(4). $F \subseteq x^{-1}F$. □

Theorem 3.4.5 *[294] Let F be a filter of L. F is a prime filter if and only if for any $x \in L \setminus F$, $x^{-1}F = F$.*

Proof. Suppose that F is a prime filter and $x \in L \setminus F$. If $y \in x^{-1}F$, then $x \vee y \in F$ and hence $y \in F$, that is to say $x^{-1}F \subseteq F$.

Conversely, suppose that for any $x \in L \setminus F$, $x^{-1}F = F$. If $x \vee y \in F$, then $y \in x^{-1}F = F$, it follows that F is a prime filter. □

Corollary 3.4.2 *[294] Let L be a lattice implication algebra. The following statements are equivalent:*

(1). $\{I\}$ is a prime filter;
(2). all the filters of L are prime filters;
(3). $(L, \leqslant)$ is a chain. □

Theorem 3.4.6 *[294] Let L be a lattice implication algebra. Suppose that F is a proper filter of L, $L \supseteq S \neq \emptyset$ and S is $\vee$-closed. If $F \cap S = \emptyset$, then there exists a prime filter P of L such that $F \subseteq P$ and $S \cap P = \emptyset$.*

Proof. Let $\Gamma = \{A \in F(L) \mid F \subseteq A, A \cap S = \emptyset\}$. $\Gamma \neq \emptyset$ is clear. By Zorn's Lemma, there is a maximal element P w.r.t. $\subseteq$ in Γ. We need only to prove that P is a prime filter. Otherwise, there exist $x, y \in L$ such that $x \vee y \in P, x \notin P$, and $y \notin P$, it follows that $S \cap [P \cup \{x\}) \neq \emptyset$ and $S \cap [P \cup \{y\}) \neq \emptyset$. Suppose $a \in S \cap [P \cup \{x\})$ and $b \in S \cap [P \cup \{y\})$, it follows that $a \vee b \in S$, $a \vee b \in S \cap [P \cup \{x\})$ and $a \vee b \in S \cap [P \cup \{y\})$. By

$$[P \cup \{x\}) \cap [P \cup \{y\}) = [P \cup \{x \vee y\}) = [P) = P,$$

$a \vee b \in P$, this is a contradiction to $P \cap S = \emptyset$. □

Corollary 3.4.3 *(1). If $F \in F(L)$ and $x \in L \setminus F$, then there exists a prime filter P such that $x \notin P$ and $F \subseteq P$;*
(2). If $a \in L$ and $a \neq I$, then there exists a prime filter P such that $a \notin P$;

(3). Let F be a proper filter of L, then

$$F = \bigcap\{P \in PF(L) \mid F \subseteq P\}. \square$$

Definition 3.4.3 *Let L be a lattice implication algebra. A filter J of L is called an ultra-filter if for any $x \in L$, $x \in J$ if and only if $x' \notin J$.*

Corollary 3.4.4 *An ultra-filter is a prime filter, and also an implicative filter.* $\square$

Theorem 3.4.7 *Let L be a lattice implication algebra, $A \subseteq L$. The following statements are equivalent:*

(1). A is an ultra-filter;
(2). A is a maximal proper filter.

Proof. ((1)$\Longrightarrow$(2)). Suppose that A is an ultra-filter, then $O \notin A$ by $I \in A$, and hence A is a proper filter. If $A \subseteq B$ and B is also a proper filter, we need to prove $A = B$. In fact, if there exists $x \in B$ such that $x \notin A$, then $x \to O = x' \in A \subseteq B$, it follows that $O \in B$ and $B = L$, which is a contradiction.

((2)$\Longrightarrow$(1)). Suppose that A is a maximal proper filter. We need to prove that for any $x \in L$, $x \in A$ if and only if $x' \notin A$. If $x' \notin A$, let $B = A \cup \{x\}$ and $D = [B)$, we shall prove that B has the finite intersection property. In fact, for any $y_1, \cdots, y_n \in B$,

(a) If $y_1, \cdots, y_n \in A$, it follows that $y_1 \otimes \cdots \otimes y_n > O$ because A is a proper filter.
(b) If there exists $i \leqslant n$ such that $y_i = x$, without losing generality, suppose $y_1 = x$. If $y_1 \otimes \cdots \otimes y_n = x \otimes y_2 \otimes \cdots \otimes y_n = (x \to (y_2 \otimes \cdots \otimes y_n)')' = O$, then $x' \geqslant y_2 \otimes \cdots \otimes y_n$ and hence $x' \in A$, a contradiction.

By (a) and (b), we have proved that B has the finite intersection property, it follows that D is a proper filter and hence $A = [A \cup \{x\})$, that is, $x \in A$. If $x \in A$, then $x' \notin A$. Otherwise, $O = x \otimes x' \in A$, and A is not proper. $\square$

Theorem 3.4.8 *Let L be a lattice implication algebra, $A \subseteq L$. If A has the finite intersection property, then there exists an ultra-filter B such that $A \subseteq B$.*

Proof. Let

$$E = \{B \mid A \subseteq B, B \text{ is a proper filter of } L\}.$$

It follows that $E \neq \emptyset$ because $[A) \in E$. Suppose that $B_i \in E$ for any $i < \kappa$ such that:

$$B_1 \subseteq B_2 \subseteq \cdots \subseteq B_i \subseteq \cdots,$$

let $B = \bigcup_{i<\kappa} B_i$, it follows that:

(1). $A \subseteq B$;

(2). $O \notin B$ because $O \notin B_i$ for any $i < \kappa$;
(3). $I \in B$;
(4). if x, $x \to y \in B$, then there exists $i < \kappa$ such that x, $x \to y \in B_i$, it follows that $y \in B_i \subseteq B$.

So, B is a proper filter and $B \in E$. It follows by Zorn's Lemma that E has a maximal element B. Thus B is an ultra-filter such that $A \subseteq B$. □

Corollary 3.4.5 *Any proper filter of L can be extended to an ultra-filter.* □

Theorem 3.4.9 *Let L be a lattice implication algebra, A a proper filter of L. A is an ultra-filter if and only if $a \oplus b \in A$ implies $a \in A$ or $b \in A$ for any $a, b \in L$.*

Proof. We suppose that A is an ultra-filter and $a \oplus b \in A$. It follows that

$$a' \otimes b' = (a' \to b)' = (a \oplus b)' \notin A$$

and hence $a' \notin A$ or $b' \notin A$, *i.e.*, $a \in A$ or $b \in A$.

Conversely, we suppose that $a \oplus b \in A$ implies $a \in A$ or $b \in A$ for any $a, b \in L$. If $a' \notin A$, then $a \in A$ by $a' \oplus a = a \to a = I \in A$, it follows that A is an ultra-filter. □

Theorem 3.4.10 *Let L be a lattice implication algebra and P a filter of L. Then the following statements are equivalent:*

(1). P is a prime filter;
(2). for any filters A, B of L, $A \vee B \subseteq P$ implies $A \subseteq P$ or $B \subseteq P$.

Proof. Suppose that P is a prime filter such that $A \vee B \subseteq P$, where A and B are filters of L, and neither $A \subseteq P$ nor $B \subseteq P$. Then there exist $a \in A$ and $b \in B$ such that $a \notin P$ and $b \notin P$. Since $a \vee b \in A \vee B \subseteq P$, P being prime filter implies $a \in P$ or $b \in P$, which is a contradiction.

Let $a, b \in L$, $a \vee b \in P$, we consider $A(a)$ and $A(b)$, which are filters of L. We claim that $A(a) \vee A(b) \subseteq P$. Let $x, y \in L$, $x \in A(a)$ and $y \in A(b)$, then $a \vee b \leqslant x \vee y$ and hence $x \vee y \in P$, it follows that $A(a) \vee A(b) \subseteq P$. By the assumption, we know that $A(a) \subseteq P$ or $A(b) \subseteq P$, *i.e.*, $a \in P$ or $b \in P$. □

Theorem 3.4.11 *If P is a prime filter of lattice H implication algebra L, then P is an ultra-filter.*

Proof. Suppose that P is not an ultra-filter of L. Then there exists a proper filter F of L such that $P \subseteq F$ and $P \neq F$. It follows that there exists an element $a \in F$ such that $a \notin P$. Let b be an arbitrary element in L, we have

$$a \vee (a \to b) = ((a \to b) \to a) \to a = a \to a = I \in P,$$

it follows that $a \to b \in P \subseteq F$ and hence $b \in F$. This implies that $L \subseteq F$, which contradicts to that F is a proper filter. □

Corollary 3.4.6 *In a lattice H implication algebra, the concepts of prime filters and ultra-filters coincide.* □

3.5 *I*-Filters, Involution Filters and Obstinate Filters

Let L be a lattice implication algebra, $a \in L$, and

$$F_a = \{x \mid x \in L, x \vee a = I\}.$$

Lemma 3.5.1 *For any $a \in L$, F_a is a filter of L.*

Proof. $I \in F_a$ is trivial. Suppose that $x \in F_a$ and $x \to y \in F_a$, then $x \to a = a$, and $(x \to y) \to a = a$, it follows by

$$y \to a \geqslant a = (x \to y) \to a = (x \to y) \to (x \to a) = (x \wedge y) \to a \geqslant y \to a$$

that $y \to a = a$ and hence $y \in F_a$. □

Definition 3.5.1 *For any $a \in L, F_a$ is called the I-filter determined by a.*

Theorem 3.5.1 *Let L be a lattice implication algebra, $a \in L$.*

(1). $F_a \cap [a) = \{I\}$;
(2). if a is a dual atom, then F_a is a prime filter and $F_a = L \setminus [O, a]$.

Proof. (1). $I \in F_a \cap [a)$ is trivial. If $x \in F_a \cap [a)$, then $a \to x = x$ and there exists $n \in \mathbb{N}^+$ such that $[a, x]^n = I$, it follows that $x = I$.

(2). Suppose that a is a dual atom. First, we prove that $F_a = L \setminus [O, a]$. In fact, for any $x \in L \setminus [O, a]$, if $x \vee a = a$, then $x \leqslant a$, which is a contradiction. It follows that $x \vee a = I$ and $x \in F_a$. Conversely, if $x \in F_a$, then $x \vee a = I$ and $x \in L \setminus [O, a]$.

Second, we prove F_a is a prime filter. If $x \vee y \in F_a$, $x \notin F_a$, then

$$\begin{aligned} I &= (x \vee y) \vee a \\ &= (x \vee a) \vee y \\ &= a \vee y, \end{aligned}$$

i.e., $y \in F_a$. □

In what follows, for $A \subseteq L$, we denote A^* the set $\{x \mid x \vee a = I \text{ for any } a \in A\}$. It is clear that $A^* = \bigcap_{a \in A} F_a$ and A^* is a filter. We call A^* the I-filter determined by A.

Theorem 3.5.2 *Let L be a lattice implication algebra, $A, B \subseteq L$.*

(1). $A \subseteq B$ implies $B^* \subseteq A^*$;
(2). $A \subseteq (A^*)^*$;
(3). $A^* = ((A^*)^*)^*$;
(4). $(A \cup B)^* = A^* \cap B^*$. □

Definition 3.5.2 *A filter A of L is called an involution filter if $A = (A^*)^*$.*

Theorem 3.5.3 *Let L be a lattice implication algebra, A, B filters of L. Then $A \cap B = \{I\}$ if and only if $A \subseteq B^*$.* □

Proof. If $A \subseteq B^*$, then by $B \cap B^* = \{I\}$, it follows that $A \cap B \subseteq B^* \cap B = \{I\} \subseteq A \cap B$.

Conversely, assume $A \cap B = \{I\}$, then for any $y \in A$ and $b \in B$, it follows that $y \vee b \in A$ and $y \vee b \in B$ because A and B are filters, that is, $y \vee b = I$ and $y \in B^*$. □

Theorem 3.5.4 *Let L be a lattice implication algebra, $A \subseteq L$.*

(1). $A^* \cap [A) = \{I\}$;
(2). $A^* = [A)^*$.

Proof. (1). $A^* \cap [A) \supseteq \{I\}$ is trivial. Conversely, if $x \in A^* \cap [A)$, then there exist $a_1, \cdots, a_n \in A$ such that $[a_1, \cdots, a_n, x] = a_1 \to (\cdots \to (a_n \to x) \cdots) = I$, by $a_i \to x = x$ for any $i \leqslant n$, it follows that $x = I$ and hence $A^* \cap [A) \subseteq \{I\}$.

(2). By Theorem 3.5.2, $A^* \supseteq [A)^*$. Conversely, if $x \in A^*$, then for any $a \in A$, $a \to x = x$. For any $y \in [A)$, there exist $a_1, \cdots, a_n \in A$ such that $y \geqslant a_1 \otimes \cdots \otimes a_n$, it follows that

$$y \to x \leqslant a_1 \otimes \cdots \otimes a_n \to x = a_1 \to (\cdots \to (a_n \to x) \cdots) = x,$$

and hence $y \to x = x$, that is $x \in [A)^*$. □

Theorem 3.5.5 *Let L be a lattice implication algebra and each its filter is an involution filter. Then $(A \cap B)^* = [A^* \cup B^*)$ for any filters A and B of L.*

Proof. By $(A \cap B)^* \supseteq A^* \cup B^*$ and $(A \cap B)^*$ is a filter, it follows that

$$(A \cap B)^* \supseteq [A^* \cup B^*).$$

Conversely, $A^*, B^* \subseteq A^* \cup B^* \subseteq [A^* \cup B^*)$, *i.e.*, $[A^* \cup B^*)^* \subseteq A$, $[A^* \cup B^*)^* \subseteq B$, and hence $[A^* \cup B^*)^* \subseteq A \cap B$, it follows that $(A \cap B)^* \subseteq [A^* \cup B^*)$. □

Theorem 3.5.6 *Let L be a lattice implication algebra and each its filter an involution filter. Then $[A) = (A^*)^*$.*

Proof. By $A \subseteq (A^*)^*$, it follows that $[A) \subseteq [(A^*)^*) = (A^*)^*$. By $A \subseteq [A)$, it follows that $[A)^* \subseteq A^*$ and hence $(A^*)^* \subseteq ([A)^*)^* = [A)$. □

Definition 3.5.3 *Let L be a lattice implication algebra, J a filter of L. J is called an obstinate filter if $x \notin J, y \notin J$ imply $x \to y \in J, y \to x \in J$ for any $x, y \in L$.*

Corollary 3.5.1 *Let L be a lattice implication algebra, J_1, J_2 filters of L and $J_1 \subseteq J_2$. If J_1 is an obstinate filter, then J_2 is also an obstinate filter.* □

Theorem 3.5.7 *Let L be a lattice implication algebra. If J is an ultra-filter, then J is an obstinate filter.*

Proof. Suppose that $x, y \in L$ such that $x \notin J$ and $y \notin J$, then $x' \in J$ and $y' \in J$, it follows that $x' \to y' \in J$ and $y' \to x' \in J$, that is, $y \to x \in J$ and $x \to y \in J$. □

Theorem 3.5.8 *Let L be a lattice implication algebra, J a filter of L. If J is an obstinate filter, then J is a prime filter.*

Proof. Note that for any $x, y \in L$, $y \to x \in J$ or $x \to y \in J$. In fact,

(1). If $x \in J$ and $y \notin J$, then $y \to x \in J$ because J is a filter and $y \to x \geqslant x$.
(2). If $x \notin J$ and $y \in J$, then $x \to y \in J$ because J is a filter and $x \to y \geqslant y$.
(3). If $x \in J$ and $y \in J$, then $y \to x \in J$ and $x \to y \in J$ because J is a filter.
(4). If $x \notin J$ and $y \notin J$, then $y \to x \in J$ and $x \to y \in J$ because J is an obstinate filter.

For any $x, y \in L$, we suppose $x \vee y \in J$. If $y \to x \in J$, by $x \vee y = (y \to x) \to x$, it follows that $x \in J$; if $x \to y \in J$, by $x \vee y = (x \to y) \to y$, it follows that $y \in J$. □

Theorem 3.5.9 *Let L be a lattice implication algebra, J a filter of L. The following statements are equivalent.*

(1). J is a maximal positive implicative filter;
(2). J is a maximal implicative filter;
(3). J is an obstinate filter.

Proof. ((1)$\Longrightarrow$(2)). Trivial.

((2)$\Longrightarrow$(3)). Suppose $x \notin J$ and $y \notin J$. $[J \cup \{x\}) = \{z \mid x \to z \in J, z \in L\}$ is an implicative filter by Theorems 3.1.10 and 3.2.1. It follows that $[J \cup \{x\}) = L$, and hence $y \in [J \cup \{x\})$, *i.e.*, $x \to y \in J$. $y \to x \in J$ can be proved similarly. Hence, J is an obstinate filter.

((3)$\Longrightarrow$(1)). Firstly, we prove that J is a positive implicative filter. Otherwise, there exist $x, y \in L$ such that $(x \to y) \to x \in J$ and $x \notin J$. If $y \notin J$, then $x \to y \in J$, so $x \in J$ because J is a filter, which is a contradiction; if $y \in J$, then $x \to y \in J$, so $x \in J$, which is a contradiction.

Secondly, we prove that J is maximal. Suppose that $J \subseteq J_1$, J is an implicative filter and there exists $x \in J_1$ such that $x \notin J$, we need only to prove $J_1 = L$. In fact, for any $y \in L$, if $y \in J$, then $y \in J_1$; if $y \notin J$, then $x \to y \in J$ because J is an obstinate filter, it follows that $x \to y \in J_1$ and $y \in J_1$. □

Theorem 3.5.10 *Let J be a proper filter of lattice implication algebra L. For each lattice implication algebra L_1, there exists an implication homomorphism f from L to L_1 such that D-$\ker(f) = J$ if and only if J is an obstinate filter of L.*

Proof. Suppose that J is an obstinate filter and L_1 is a lattice implication algebra, we define $f: L \longrightarrow L_1$ by

$$f(x) = \begin{cases} I, & x \in J, \\ a, & x \notin J, \end{cases}$$

where $a \in L_1$ and $a \neq I$. We prove that f is an implication homomorphism, *i.e.*,

$$f(x \to y) = f(x) \to f(y)$$

for any $x, y \in L$. In fact,

(1). if $x, y \in J$, then $x \to y \in J$ and $f(x \to y) = I = I \to I = f(x) \to f(y)$;
(2). if $x, y \notin J$, then $x \to y \in J$ because J is an obstinate filter and hence

$$f(x \to y) = I = a \to a = f(x) \to f(y);$$

(3). if $x \notin J$ and $y \in J$, then $x \to y \in J$ because J is a filter and hence

$$f(x \to y) = I = a \to I = f(x) \to f(y);$$

(4). if $x \in J$ and $y \notin J$, then $x \to y \notin J$ because J is a filter and hence

$$f(x \to y) = a = I \to a = f(x) \to f(y).$$

D-$\ker(f) = J$ is trivial.

Conversely, let $L_1 = \{O, I\}$, it follows that there exists an implication homomorphism f from L to L_1 such that D-$\ker(f) = J$ and hence

$$f(x) = \begin{cases} I, & x \in J, \\ O, & x \notin J. \end{cases}$$

For any $x, y \in L$, if $x \notin J$ and $y \notin J$, then

$$f(x \to y) = f(x) \to f(y) = O \to O = I,$$

it follows that $x \to y \in J$. $y \to x \in J$ can be proved similarly. Hence, J is an obstinate filter. □

3.6 Fuzzy Filters

Since the introduction of Zadeh's theory of fuzzy sets in 1965, it has been developed rapidly. In this section, we apply the concept of fuzzy subsets to filter theory in lattice implication algebras. In what follows we denote by $\mathscr{F}(X)$ the set of all fuzzy subsets of X. For $A \in \mathscr{F}(X)$, let

$$A_\lambda \triangleq \{x \mid x \in X, A(x) \geqslant \lambda\}. \ (\lambda \in [0,1])$$

Definition 3.6.1 *Let L be a lattice implication algebra, $A \in \mathscr{F}(L)$ and $A \neq \emptyset$. A is called a fuzzy filter of L if it satisfies: for any $x, y \in L$,*

(1). $A(x) \leqslant A(I)$;
(2). $A(y) \geqslant \min\{A(x \to y), A(x)\}$.

Theorem 3.6.1 *Let L be a lattice implication algebra and A a fuzzy filter of L. Then for any $x, y \in L$, $x \leqslant y$ implies $A(x) \leqslant A(y)$.* □

Theorem 3.6.2 *Let L be a lattice implication algebra, $A \in \mathscr{F}(L)$ and $A \neq \emptyset$. A is a fuzzy filter of L if and only if for any $\lambda \in [0,1]$, A_λ is a filter of L when $A_\lambda \neq \emptyset$.*

Proof. Suppose that A is a fuzzy filter of L and $\lambda \in [0,1]$ such that $A_\lambda \neq \emptyset$. Then there exists $x \in A_\lambda$, and hence $A(I) \geqslant A(x) \geqslant \lambda$, *i.e.*, $I \in A_\lambda$. If $x \to y \in A_\lambda$ and $x \in A_\lambda$, then $A(x \to y) \geqslant \lambda$, $A(x) \geqslant \lambda$, it follows that

$$A(y) \geqslant \min\{A(x \to y), A(x)\} \geqslant \lambda,$$

i.e., $y \in A_\lambda$. So, A_λ is a filter of L.

Conversely, suppose that for any $\lambda \in [0,1]$, A_λ is a filter of L when $A_\lambda \neq \emptyset$. For any $x \in L$, $x \in A_{A(x)}$ and hence $A_{A(x)}$ is a filter. By $I \in A_{A(x)}$ it follows that $A(x) \leqslant A(I)$. For any $x, y \in L$, let

$$\lambda = \min\{A(x \to y), A(x)\},$$

it follows that $A_\lambda \neq \emptyset$ and hence A_λ is a filter. Since $x \to y \in A_\lambda$, $x \in A_\lambda$, this implies $y \in A_\lambda$, *i.e.*,

$$A(y) \geqslant \lambda = \min\{A(x \to y), A(x)\}. \ \square$$

Corollary 3.6.1 *Let L be a lattice implication algebra.*

(1). If A is a fuzzy filter of L, then $\{x \mid x \in L, A(x) = A(I)\}$ is a filter of L.
(2). If A is a filter of L, then χ_A is a fuzzy filter of L, where χ_A is the characteristic function of A. □

Theorem 3.6.3 *Let L_1 and L_2 be lattice implication algebras, $f : L_1 \longrightarrow L_2$ a lattice implication homomorphism from L_1 to L_2, and $A \in \mathscr{F}(L_1)$, $B \in \mathscr{F}(L_2)$.*

(1). If B is a fuzzy filter of L_2, then $f^{-1}(B)$ is a fuzzy filter of L_1.
(2). If A is a fuzzy filter of L_1 and f is a bijection, then $f(A)$ is a fuzzy filter of L_2.

Proof. (1). For any $x, y \in L_1$, by $f(I) = I$, it follows that

$$\begin{aligned} f^{-1}(B)(x) &= B(f(x)) \\ &\leqslant B(I) \\ &= B(f(I)) \\ &= f^{-1}(B)(I), \end{aligned}$$

$$
\begin{aligned}
f^{-1}(B)(y) &= B(f(y)) \\
&\geqslant \min\{B(f(x) \to f(y)), B(f(x))\} \\
&= \min\{B(f(x \to y)), B(f(x))\} \\
&= \min\{f^{-1}(B)(x \to y), f^{-1}(B)(x)\},
\end{aligned}
$$

i.e., $f^{-1}(B)$ is a fuzzy filter of L_1.

(2). For any $x, y \in L_2$, there exist $u, v \in L_1$ such that $f(u) = x$ and $f(v) = y$, it follows that

$$
\begin{aligned}
f(A)(x) &= \bigvee_{f(t)=x} A(t) \\
&\leqslant A(I) \\
&= \bigvee_{f(t)=I} A(t) \\
&= f(A)(I), \\
f(A)(y) &= \bigvee_{f(t)=y} A(t) \\
&= A(v) \\
&\geqslant \min\{A(u \to v), A(u)\} \\
&= \min\{f(A)(x \to y), f(A)(x)\}. \ \square
\end{aligned}
$$

Definition 3.6.2 *Let L be a lattice implication algebra, $A \in \mathscr{F}(L)$ and $A \neq \emptyset$. A is called a fuzzy implicative filter of L if it satisfies for any $x, y, z \in L$,*

(1). $A(x) \leqslant A(I)$;
(2). $A(x \to z) \geqslant \min\{A(x \to y), A(x \to (y \to z))\}$.

Corollary 3.6.2 *Let L be a lattice implication algebra and A a fuzzy implicative filter of L. Then A is a fuzzy filter.* □

Theorem 3.6.4 *Let L be a lattice implication algebra, $A \in \mathscr{F}(L)$ and $A \neq \emptyset$. A is a fuzzy implicative filter of L if and only if for any $\lambda \in [0, 1]$, A_λ is an implicative filter of L when $A_\lambda \neq \emptyset$.* □

Theorem 3.6.5 *Let L be a lattice implication algebra. Then L is a lattice H implication algebra if and only if each fuzzy filter of L is a fuzzy implicative filter.*

Proof. Suppose that L is a lattice H implication algebra and A is a fuzzy filter of L, then

$$x \to (y \to z) = (x \to y) \to (x \to z),$$

it follows that

$$\begin{aligned}&\min\{A(x \to (y \to z)), A(x \to y)\}\\ =&\min\{A((x \to y) \to (x \to z)), A(x \to y)\}\\ \leqslant&\, A(x \to z),\end{aligned}$$

and hence A is a fuzzy implicative filter of L.

Conversely, suppose that each fuzzy filter of L is a fuzzy implicative filter. Define $A: \ L \longrightarrow [0,1]$ satisfying:

$$A(x) = \begin{cases} 0.5, & x = I, \\ 0, & x \neq I. \end{cases}$$

Then A is a fuzzy filter and hence a fuzzy implicative filter. It follows that $\{I\} = A_{0.5}$ is an implicative filter of L. This implies that L is a lattice H implication algebra. □

Similarly as for fuzzy filters, we can prove the following theorems about fuzzy implicative filters.

Theorem 3.6.6 *Let L be a lattice implication algebra.*

(1). If A is a fuzzy implicative filter of L, then $\{x \mid x \in L, A(x) = A(I)\}$ is an implicative filter of L.
(2). If A is an implicative filter of L, then χ_A is a fuzzy implicative filter of L. □

Theorem 3.6.7 *Let L_1 and L_2 be a lattice implication algebras, $f: \ L_1 \to L_2$ a lattice implication homomorphism from L_1 to L_2, and $A \in \mathscr{F}(L_1)$, $B \in \mathscr{F}(L_2)$.*

(1). If B is a fuzzy implicative filter of L_2, then $f^{-1}(B)$ is a fuzzy implicative filter of L_1.
(2). If A is a fuzzy implicative filter of L_1 and f is a bijection, then $f(A)$ is a fuzzy implicative filter of L_2. □

In this section, we fuzzified the concepts of filters and implicative filters and discussed their basic properties. Similarly, we can fuzzify the concepts of positive implicative filters, associative filters, prime filters, ultra-filters, I-filters and involution filters and discuss their basic properties. Readers can refer to [215] and [378].

4 *LI*-Ideals

The concept of LI-ideal is dual to filter of lattice implication algebra. In this chapter, we establish the concepts of LI-ideal, fuzzy LI-ideal, normal fuzzy LI-ideal, intuitionistic fuzzy LI-ideal and discuss their basic properties. Similarly, we can establish other type of LI-ideals.

4.1 *LI*-Ideals

In this section, we define LI-ideal of lattice implication algebra and show that every LI-ideal is a lattice ideal. We can give an example that a lattice ideal is not an LI-ideal, but every lattice ideal is an LI-ideal in a lattice H implication algebra. We also discuss the relation between filters and LI-ideals and study how to generate an LI-ideal by a set.

Definition 4.1.1 *Let L be a lattice implication algebra. An LI-ideal A is a non-empty subset of L such that for any $x, y \in L$,*

(1). $O \in A$;
(2). $(x \to y)' \in A$ and $y \in A$ imply $x \in A$.

From Definition 4.1.1, $\{O\}$ and L are trivial LI-ideals of lattice implication algebra. The following example gives a non-trivial LI-ideal of a lattice implication algebra.

Example 4.1.1 *Let L be the lattice implication algebra defined in Example 2.1.4. $A = \{O, c\}$ is an LI-ideal of L.*

Theorem 4.1.1 *Let L be a lattice implication algebra. Every LI-ideal A of L is a lattice ideal.*

Proof. Suppose $x, y \in L$. If $x \in A$, $y \leqslant x$, then $y \in A$ by $(y \to x)' = I' = O \in A$.

Let $x, y \in A$. Since

$$((x \vee y) \to y)' = (x \to y)' \leqslant x,$$

then $x \vee y \in A$, *i.e.*, A is a lattice ideal. □

Remark 8. The converse of Theorem 4.1.1 is not true. In fact, in Example 2.1.4, $A = \{O, d\}$ is a lattice ideal, but not an LI-ideal, since $(a \to d)' = d \in A$ and $a \notin A$.

Theorem 4.1.2 *In a lattice H implication algebra L, every lattice ideal is an LI-ideal.*

Proof. Let A be a lattice ideal of L, $x, y \in L$. Suppose $(x \to y)' \in A$ and $y \in A$. Note that

$$y \vee (x \to y)' = x \vee y.$$

It follows that $x \vee y = y \vee (x \to y)' \in A$. Since $x \leqslant x \vee y$, we have $x \in A$. Clearly $O \in A$. Hence A is an LI-ideal of L. □

For any non-empty subset A of a lattice implication algebra L, let

$$A' = \{a' \mid a \in A\}.$$

We show the relation between LI-ideal and filter of lattice implication algebra.

Theorem 4.1.3 *Let A be a non-empty subset of a lattice implication algebra L. Then A is a filter of L if and only if A' is an LI-ideal of L.*

Proof. Suppose that A is a filter of L. Then $I \in A$, and so $O = I' \in A'$. Suppose $(x \to y)' \in A'$ and $y \in A'$. Then $(x \to y)' = u'$ and $y = v'$ for some $u, v \in A$.

$$\begin{aligned} v \to x' &= x \to v' \\ &= x \to y \\ &= ((x \to y)')' \\ &= (u')' \\ &= u \in A. \end{aligned}$$

Since A is a filter, we have $x' \in A$, and so $x = (x')' \in A'$, *i.e.*, A' is an LI-ideal of L.

Conversely, $I \in A$ is clear. Let $x, y \in L$, $x \in A$, $x \to y \in A$, then $x' \in A'$ and

$$(y' \to x')' = (x \to y)' \in A'.$$

As A' is an LI-ideal, it follows that $y' \in A'$, *i.e.*, $y \in A$. Hence A is a filter of L. □

Theorem 4.1.4 *Suppose $\mathcal{A}$ is a non-empty family of LI-ideals of a lattice implication algebra L. Then $\bigcap \mathcal{A}$ is also an LI-ideal of L.* □

Let A be a subset of a lattice implication algebra L. The least LI-ideal containing A is called the LI-ideal generated by A, denoted by $\langle A \rangle$. Specially, if $A = \{a\}$, we write $\langle A \rangle$ as $\langle a \rangle$.

Theorem 4.1.5 *If A is a non-empty subset of a lattice implication algebra L, then $\langle A\rangle = [A')'$.*

Proof. By Theorem 4.1.3, $[A')'$ is an LI-ideal. For any $a \in A$, $a' \in A' \subseteq [A')$, it follows that $a \in [A')'$, *i.e.*, $[A')'$ is an LI-ideal containing A.

On the other hand, if B is an LI-ideal and $B \supseteq A$, then $B' \supseteq A'$ and B' is a filter, it follows that $[A') \subseteq B'$, *i.e.*, $[A')' \subseteq (B')' = B$. So, $\langle A\rangle = [A')'$. □

The following corollaries can be immediately obtained from Theorems 4.1.5 and 3.2.2.

Corollary 4.1.1 *If A is a non-empty subset of a lattice implication algebra L, then*

$$\langle A\rangle = \{x \in L \mid [a_1', \cdots, a_n', x'] = I, \text{ for some } a_1, \cdots, a_n \in A\}. \square$$

Corollary 4.1.2 *For any element b of a lattice implication algebra L, we have*

$$\langle b\rangle = \{x \in L \mid [b', x']^n = I, n \in \mathbb{N}^+\}. \square$$

Theorem 4.1.6 *Let $f: L_1 \longrightarrow L_2$ be an implication homomorphism of lattice implication algebras. If $\ker(f) \neq \emptyset$, then $\ker(f)$ is an LI-ideal of L_1.*

Proof. Since $\ker(f) \neq \emptyset$, then $O \in \ker(f)$. Let $(x \to y)', y \in \ker(f)$. Then $f((x \to y)') = O$ and $f(y) = O$. Hence

$$\begin{aligned} O &= f((x \to y)') \\ &= (f(x) \to f(y))' \\ &= (f(x) \to O)' \\ &= f(x), \end{aligned}$$

which implies $x \in ker(f)$. Therefore, $\ker(f)$ is an LI-ideal of L_1. □

Theorem 4.1.7 *Let $f: L \longrightarrow \{O, I\}$ be an implication epimorphism of lattice implication algebras. Then $\ker(f)$ is a maximal proper LI-ideal of L.*

Proof. Since f is onto, we know that $\ker(f) \neq \emptyset$. Hence, by Theorem 4.1.6, $\ker(f)$ is an LI-ideal of L. Suppose $\ker(f)$ is not maximal proper LI-ideal. Then there is a proper LI-ideal A containing $\ker(f)$. Therefore there exist $x, y \in L$ such that $x \in L \setminus A$ and $y \in A \setminus \ker(f)$. Thus $f(x) = f(y) = I$, and so

$$f(x \to y) = f(x) \to f(y) = I.$$

It follows that

$$\begin{aligned} f((x \to y)') &= (f(x \to y))' \\ &= I' \\ &= O. \end{aligned}$$

So $(x \to y)' \in \ker(f) \subseteq A$. Since $y \in A$, we have $x \in A$, which is a contradiction. □

4.2 Fuzzy LI-Ideals

In this section we consider the fuzzy LI-ideal which is a fuzzification of LI-ideal of lattice implication algebras, and discuss the relation between fuzzy LI-ideals and fuzzy lattice ideals.

Definition 4.2.1 *Let L be a lattice implication algebra. A fuzzy subset μ of L is called a fuzzy LI-ideal of L if for any $x, y \in L$,*

(1). $\mu(O) \geqslant \mu(x)$;
(2). $\mu(x) \geqslant \min\{\mu((x \to y)'), \mu(y)\}$.

Example 4.2.1 *Let L be the lattice implication algebra defined in Example 2.1.4 and a fuzzy subset μ of L defined by $\mu(O) = \mu(c)$ and $\mu(O) \geqslant \mu(x)$ for any $x \in L \setminus \{O, c\}$, then μ is a fuzzy LI-ideal of L.*

Theorem 4.2.1 *Every fuzzy LI-ideal μ of a lattice implication algebra L is order reversing.*

Proof. If $x, y \in L$ and $y \leqslant x$, then $(y \to x)' = I' = O$, and so

$$\begin{aligned} \mu(y) &\geqslant \min\{\mu((y \to x)'), \mu(x)\} \\ &= \min\{\mu(O), \mu(x)\} \\ &= \mu(x). \end{aligned}$$

This shows that μ is order reversing. □

Definition 4.2.2 *A fuzzy subset μ of a lattice L is called a fuzzy lattice ideal if for any $x, y \in L$,*

(1). if $y \leqslant x$, then $\mu(y) \geqslant \mu(x)$;
(2). $\mu(x \vee y) \geqslant \min\{\mu(x), \mu(y)\}$.

Example 4.2.2 *Let L be a lattice implication algebra defined in Example 2.1.4. $\{O, d\}$ is a non-trivial lattice ideal, and a fuzzy subset μ of L defined by*

$$\mu(O) = \mu(d) \geqslant \mu(x)$$

for any $x \in L \setminus \{O, d\}$ is a non-trivial fuzzy lattice ideal of L.

Theorem 4.2.2 *Let L be a lattice implication algebra. Every fuzzy LI-ideal of L is a fuzzy lattice ideal.*

Proof. Suppose μ is a fuzzy LI-ideal of L. Theorem 4.2.1 shows that $\mu(y) \geqslant \mu(x)$ if $y \leqslant x$. By

$$\begin{aligned} ((x \vee y) \to y)' &= ((x \to y) \wedge (y \to y))' \\ &= (x \to y)' \\ &\leqslant x, \end{aligned}$$

we get

$$\begin{aligned}\mu(x \vee y) &\geqslant \min\{\mu(((x \vee y) \to y)'), \mu(y)\} \\ &\geqslant \min\{\mu(x), \mu(y)\}. \square\end{aligned}$$

The following example shows that the converse of Theorem 4.2.2 is not true.

Example 4.2.3 *Let L be the lattice implication algebra defined in Example 2.1.4 and μ a fuzzy subset of L defined by*

$$\mu(x) = \begin{cases} 0.8, & x \in \{O, d\}, \\ 0.3, & x \in L \setminus \{O, d\}. \end{cases}$$

Then μ is a fuzzy lattice ideal of L, but not a fuzzy LI-ideal for

$$\mu(a) \ngeqslant \min\{\mu((a \to d)'), \mu(d)\}.$$

Theorem 4.2.3 *In a lattice H implication algebra L, every fuzzy lattice ideal is a fuzzy LI-ideal.*

Proof. Since $O \leqslant x$, it follows that $\mu(O) \geqslant \mu(x)$ for any $x \in L$. Let $x, y \in L$, we have

$$\begin{aligned}\mu(x) &\geqslant \mu(x \vee y) \\ &= \mu(y \vee (x' \vee y)') \\ &= \mu(y \vee (x \to y)') \\ &\geqslant \min\{\mu(y), \mu((x \to y)')\}.\end{aligned}$$

Hence μ is a fuzzy LI-ideal. $\square$

Theorem 4.2.4 *Let μ be a fuzzy subset of a lattice implication algebra L. Then μ is a fuzzy LI-ideal if and only if μ_t is an LI-ideal when $\mu_t \neq \emptyset$, $t \in [0, 1]$.*

Proof. Assume that μ is a fuzzy LI-ideal of L and $t \in [0, 1]$ such that $\mu_t \neq \emptyset$. Clearly $O \in \mu_t$. Suppose $x, y \in L$, $(x \to y)' \in \mu_t$ and $y \in \mu_t$. Then $\mu((x \to y)') \geqslant t$ and $\mu(y) \geqslant t$. It follows that

$$\mu(x) \geqslant \min\{\mu((x \to y)'), \mu(y)\} \geqslant t,$$

so that $x \in \mu_t$. Hence μ_t is an LI-ideal of L.

Conversely, suppose μ_t ($t \in [0, 1]$) is an LI-ideal of L when $\mu_t \neq \emptyset$. For any $x \in L$, $x \in \mu_{\mu(x)}$, it follows that $\mu_{\mu(x)}$ is an LI-ideal of L and hence $O \in \mu_{\mu(x)}$, that is $\mu(O) \geqslant \mu(x)$. For any $x, y \in L$, let $t = \min\{\mu((x \to y)'), \mu(y)\}$, it follows that μ_t is an LI-ideal and $(x \to y)' \in \mu_t$, $y \in \mu_t$, this implies that $x \in \mu_t$ and $\mu(x) \geqslant t = \min\{\mu((x \to y)'), \mu(y)\}$. $\square$

4.3 Normal Fuzzy *LI*-Ideals

In this section we introduce the concept of maximal fuzzy LI-ideals and completely normal fuzzy LI-ideals in lattice implication algebras. We show that every maximal fuzzy LI-ideal of a lattice implication algebra is completely normal.

Definition 4.3.1 *A fuzzy LI-ideal μ of a lattice implication algebra L is said to be normal if there exists $x \in L$ such that $\mu(x) = 1$.*

Example 4.3.1 *Let L be a lattice implication algebra in Example 2.1.4 and define a fuzzy subset μ of L by*

$$\mu(x) = \begin{cases} 1, & x = O, c; \\ 0.3, & x \in L \setminus \{O, c\}. \end{cases}$$

Then μ is a normal fuzzy LI-ideal of L.

We notice that if μ is a normal fuzzy LI-ideal of L, then $\mu(O) = 1$, hence μ is normal if and only if $\mu(O) = 1$.

Let L be a lattice implication algebra, $\mu \in \mathscr{F}(L)$,

$$L_\mu \triangleq \{x \in L \mid \mu(x) = \mu(O)\}.$$

Theorem 4.3.1 *Given a fuzzy LI-ideal μ of L, let μ^* be a fuzzy subset of L defined by*

$$\mu^*(x) = \mu(x) + 1 - \mu(O)$$

for any $x \in L$. Then μ^* is a normal fuzzy LI-ideal of L containing μ.

Proof. Notice that for any LI-ideal A of L the characteristic function χ_A of A is a normal fuzzy LI-ideal of L. It is clear that μ is normal if and only if $\mu = \mu^*$. If μ is a fuzzy LI-ideal of L, then $\mu^* = (\mu^*)^*$. Moreover if μ is normal, then $\mu = (\mu^*)^*$.

According to Definitions 4.2.1 and 4.3.1, we know that if μ and ν are fuzzy LI-ideals of L such that $\mu \subseteq \nu$ and $\mu(O) = \nu(O)$, then $L_\mu \subseteq L_\nu$. If μ is a fuzzy LI-ideal of L and there exists a fuzzy LI-ideal ν of L such that $\nu^* \subseteq \mu$, then μ is normal. □

Theorem 4.3.2 *Let μ be a fuzzy LI-ideal of L and $f : [0, \mu(O)] \longrightarrow [0, 1]$ an increasing function. Then the fuzzy subset $\mu_f(x) = f(\mu(x))$ for any $x \in L$ is a fuzzy LI-ideal of L. In particular, if $f(\mu(O)) = 1$ then μ_f is normal; and if $f(t) \geqslant t$ for any $t \in [0, \mu(O)]$, then μ is contained in μ_f.*

Proof. Notice that

$$\mu(x) \leqslant \mu(O)$$

for any $x \in L$. Since f is an increasing function it follows that

$$\begin{aligned}\mu_f(O) &= f(\mu(O))\\ &\geqslant f(\mu(x))\\ &= \mu_f(x)\end{aligned}$$

for any $x \in L$. Let $x, y \in L$. Then

$$\begin{aligned}\min\{\mu_f((x \to y)'), \mu_f(y)\} &= \min\{f(\mu((x \to y)')), f(\mu(y))\}\\ &= f(\min\{\mu((x \to y)'), \mu(y)\})\\ &\leqslant f(\mu(x))\\ &= \mu_f(x).\end{aligned}$$

Hence μ_f is a fuzzy LI-ideal of L. If $f(\mu(O)) = 1$, then μ_f is normal. If $f(t) \geqslant t$ for any $t \in [0, \mu(O)]$, then $\mu_f(x) = f(\mu(x)) \geqslant \mu(x)$ for any $x \in L$, so, $\mu \subseteq \mu_f$. □

Denote $N(L)$ as the set of all normal fuzzy LI-ideal of L. Note that $N(L)$ is a poset under the set inclusion.

Theorem 4.3.3 *If $\mu(\neq \emptyset)$ and $\nu(\neq \emptyset)$ are fuzzy LI-ideals of L such that for any $x, y \in L$,*

$$\mu(x) \leqslant \mu(y) \qquad \textit{iff} \qquad \nu(x) \leqslant \nu(y),$$

then $\mu \circ \nu$ is also a fuzzy LI-ideal of L and

$$L_{\mu\circ\nu} = L_\mu \cap L_\nu,$$

where $(\mu \circ \nu)(x) = \mu(x)\nu(x)$ for any $x \in L$.

Proof. It is easy to prove that $\mu\circ\nu$ is a fuzzy LI-ideal of L and $L_{\mu\circ\nu} \supseteq L_\mu \cap L_\nu$. Let $x \in L_{\mu\circ\nu}$. Then

$$(\mu \circ \nu)(x) = (\mu \circ \nu)(O),$$

i.e.,

$$\mu(x)\nu(x) = \mu(O)\nu(O),$$

hence $\mu(x) \neq 0$ and $\nu(x) \neq 0$. If $\mu(x) < \mu(O)$, then

$$\mu(x)\nu(x) < \mu(O)\nu(O).$$

This is a contradiction. Similarly, it is also a contradiction when $\nu(x) < \nu(O)$. Hence $\mu(x) = \mu(O)$ and $\nu(x) = \nu(O)$, *i.e.*, $x \in L_\mu \cap L_\nu$. □

A fuzzy LI-ideal is called a maximal LI-ideal if it is not L, and it is a maximal element of the set of all fuzzy LI-ideals w.r.t. fuzzy set inclusion.

Theorem 4.3.4 *If μ is a maximal fuzzy LI-ideal of L, then*

(1). μ is normal;
(2). μ takes only the values 0 and 1;

(3). L_μ is a maximal LI-ideal of L.

Proof. (1). If μ is not normal, then for any $x \in L$,

$$\mu(x) \leqslant \mu(x) + 1 - \mu(O) = \mu^*(x),$$

$$\mu(O) < 1 = \mu(O) + 1 - \mu(O) = \mu^*(O),$$

it follows that μ is not a maximal fuzzy LI-ideal of L, which is a contradiction.

(2). Note that $\mu(O) = 1$ since μ is normal. Let $a \in L$ and $\mu(a) \neq 1$. We claim that $\mu(a) = 0$. If not, then

$$0 < \mu(a) < 1.$$

Let ν be a fuzzy subset of L defined by

$$\nu(x) = \frac{1}{2}(\mu(x) + \mu(a))$$

for any $x \in L$. Then

$$\begin{aligned}\nu(0) &= \frac{1}{2}(\mu(0) + \mu(a)) \\ &= \frac{1}{2}(1 + \mu(a)) \\ &\geqslant \frac{1}{2}(\mu(x) + \mu(a)) \\ &= \nu(x).\end{aligned}$$

For any $x, y \in L$, we obtain

$$\begin{aligned}\nu(x) &= \frac{1}{2}(\mu(x) + \mu(a)) \\ &\geqslant \frac{1}{2}(\min\{\mu((x \to y)'), \mu(y)\} + \mu(a)) \\ &= \min\{\frac{1}{2}(\mu((x \to y)') + \mu(a)), \frac{1}{2}(\mu(y) + \mu(a))\} \\ &= \min\{\nu((x \to y)'), \nu(y)\}.\end{aligned}$$

Hence ν is a fuzzy LI-ideal of L. It follows from Theorem 4.3.1 that for any $x \in L$,

$$\begin{aligned}\nu^*(x) &= \frac{1}{2}(\mu(x) + \mu(a)) + 1 - \frac{1}{2}(\mu(O) + \mu(a)) \\ &= \frac{1}{2}(\mu(x) + 1) \\ &\geqslant \mu(x).\end{aligned}$$

Note that

$$\begin{aligned}\nu^*(a) &= \nu(a) + 1 - \nu(O) \\ &= \frac{1}{2}(\mu(a) + 1) \\ &< 1 \\ &= \nu^*(O).\end{aligned}$$

Hence ν^* is non-constant and μ is not a maximal element of $N(L)$. This is a contradiction.

(3). L_μ is an LI-ideal of L. If $L_\mu \subseteq A \neq L$ and A is an LI-ideal, then the characteristic functions χ_A and χ_{L_μ} are fuzzy LI-ideals and

$$\mu = \chi_{L_\mu} \subseteq \chi_A,$$

it follows that $\mu = \chi_A$ because μ is a maximal fuzzy LI-ideal and hence $L_\mu = A$. This proved that L_μ is a maximal LI-ideal of L. □

Definition 4.3.2 *A normal fuzzy LI-ideal μ of L is said to be completely normal if there exists $x \in L$ such that $\mu(x) = 0$. Denote by $C(L)$ the set of all completely normal fuzzy LI-ideals of L.*

Note that if μ is a completely normal fuzzy LI-ideal of L, then clearly $\mu(I) = 0$; if μ is a normal fuzzy LI-ideal of L, then μ is completely normal if and only if $\mu(I) = 0$.

Theorem 4.3.5 *Let μ be a non-constant fuzzy LI-ideal of L and define a fuzzy subset $\widehat{\mu}$ of L by*

$$\hat{\mu}(x) = \frac{\mu(x) - \mu(I)}{\mu(O) - \mu(I)}$$

for any $x \in L$. Then $\widehat{\mu} \in C(L)$.

Proof. By Theorem 4.2.1, μ is order-reversing, then $\widehat{\mu}$ is well defined. Clearly $\widehat{\mu}(I) = 0$ and $\widehat{\mu}(O) = 1 \geqslant \widehat{\mu}(x)$ for any $x \in L$. Let $x, y \in L$. Then

$$\begin{aligned}&\min\{\hat{\mu}((x \to y)'), \hat{\mu}(y)\} \\ &= \min\{\frac{\mu((x \to y)') - \mu(I)}{\mu(O) - \mu(I)}, \frac{\mu(y) - \mu(I)}{\mu(O) - \mu(I)}\} \\ &= \frac{1}{\mu(O) - \mu(I)} \min\{\mu((x \to y)') - \mu(I), \mu(y) - \mu(I)\} \\ &= \frac{1}{\mu(O) - \mu(I)} (\min\{\mu((x \to y)'), \mu(y)\} - \mu(I)) \\ &\leqslant \frac{1}{\mu(O) - \mu(I)} (\mu(x) - \mu(I)) \\ &= \hat{\mu}(x),\end{aligned}$$

i.e., $\widehat{\mu} \in C(L)$. □

Theorem 4.3.6 *If $\mu \in C(L)$, then $\widehat{\mu} = \mu$.* □

Note that $C(L) \subseteq N(L)$ and the restriction of the partial ordering $\subseteq$ of $N(L)$ gives a partial ordering of $C(L)$.

Theorem 4.3.7 *Every non-constant maximal element of $(N(L), \subseteq)$ is also a maximal element of $(C(L), \subseteq)$.*

Proof. Let μ be a non-constant maximal element of $(N(L), \subseteq)$. By Theorem 4.3.3, μ takes only the values 0 and 1, and $\mu(O) = 1$, $\mu(I) = 0$. Hence

$$\mu \in C(L).$$

Assume that there exists $\nu \in C(L)$ such that $\mu \subseteq \nu$. It follows that

$$\mu \subseteq \nu \in N(L).$$

If ν is constant, then $\nu = 1$. So μ is a maximal element of $(C(L), \subseteq)$; if ν is not constant, since μ is non-constant maximal in $(N(L), \subseteq)$, therefore $\mu = \nu$. Thus μ is a maximal element of $(C(L), \subseteq)$. □

Theorem 4.3.8 *Every maximal fuzzy LI-ideal of L is completely normal.*

Proof. Let μ be a maximal fuzzy LI-ideal of L. By Theorem 4.3.3, μ is normal and $\mu = \mu^*$ takes only the values 0 and 1. Now we claim that $\mu(I) = 0$. If not, then $\mu(I) = 1$. Hence $\mu(x) \geqslant \mu(I) = 1$ and so $\mu(x) = 1$ for any $x \in L$. This means that $\mu = 1$, which is a contradiction. □

4.4 Intuitionistic Fuzzy *LI*-Ideals

Based on Atanassov's idea, we consider the intuitionistic fuzzification of LI-ideals of a lattice implication algebra, called intuitionistic fuzzy LI-ideal. We also discuss the intuitionistic fuzzification of lattice ideals of a lattice, called intuitionistic fuzzy lattice ideal, and study the relations between intuitionistic fuzzy LI-ideals and intuitionistic fuzzy lattice ideals. We establish some characterizations of intuitionistic fuzzy LI-ideals.

An intuitionistic fuzzy subset (briefly, IFS) A of a non-empty set X is an object having the form

$$A = \{(x, \alpha_A(x), \beta_A(x)) \mid x \in X\} \triangleq (\alpha_A, \beta_A)$$

where the function $\alpha_A, \beta_A \in \mathscr{F}(X)$ denote the degree of membership and nonmembership respectively, and $0 \leqslant \alpha_A(x) + \beta_A(x) \leqslant 1$ for any $x \in X$.

Definition 4.4.1 *Let L be a lattice implication algebra. An IFS $A = (\alpha_A, \beta_A)$ of L is called an intuitionistic fuzzy LI-ideals of L if it satisfies the following inequalities for any $x, y \in L$,*

(1). $\alpha_A(O) \geqslant \alpha_A(x)$ *and* $\beta_A(O) \leqslant \beta_A(x)$*;*
(2). $\alpha_A(x) \geqslant \min\{\alpha_A((x \to y)'), \alpha_A(y)\}$*;*
(3). $\beta_A(x) \leqslant \max\{\beta_A((x \to y)'), \beta_A(y)\}$*.*

Example 4.4.1 *Let L be the lattice implication algebra as defined in Example 2.1.4. Define an IFS $A = (\alpha_A, \beta_A)$ of L as follows*

(1). $\alpha_A(O) = \alpha_A(c) = 1$,
(2). $\alpha_A(x) = s$ *for any* $x \in L \setminus \{O, c\}$*;*
(3). $\beta_A(O) = \beta_A(c) = 0$,
(4). $\beta_A(x) = t$ *for any* $x \in L \setminus \{O, c\}$,

where $t, s \in [0,1]$ and $s + t \leqslant 1$. We know that $A = (\alpha_A, \beta_A)$ is an intuitionistic fuzzy LI-ideal of L.

Theorem 4.4.1 *Every intuitionistic fuzzy LI-ideal $A = (\alpha_A, \beta_A)$ of a lattice implication algebra L is intuitionistic monotonic, i.e., if $x \leqslant y$, then $\alpha_A(x) \geqslant \alpha_A(y)$ and $\beta_A(x) \leqslant \beta_A(y)$.*

Proof. Let $x, y \in L$, $x \leqslant y$. Then $(x \to y)' = I' = O$,

$$\begin{aligned}\alpha_A(x) &\geqslant \min\{\alpha_A((x \to y)'), \alpha_A(y)\}\\ &= \min\{\alpha_A(O), \alpha_A(y)\}\\ &= \alpha_A(y),\end{aligned}$$

and

$$\begin{aligned}\beta_A(x) &\leqslant \max\{\beta_A((x \to y)'), \beta_A(y)\}\\ &= \max\{\beta_A(O), \beta_A(y)\}\\ &= \beta_A(y). \ \square\end{aligned}$$

Theorem 4.4.2 *Let IFS $A = (\alpha_A, \beta_A)$ be an intuitionistic fuzzy LI-ideal of lattice implication algebra L. For any $x, y, z \in L$ which satisfies $x \leqslant y' \to z$, then $\alpha_A(x) \geqslant \min\{\alpha_A(y), \alpha_A(z)\}$ and $\beta_A(x) \leqslant \max\{\beta_A(y), \beta_A(z)\}$.*

Proof. Let $x, y, z \in L$, $x \leqslant y' \to z$. Then

$$\begin{aligned}I &= x \to (y' \to z)\\ &= z' \to (x \to y)\\ &= (x \to y)' \to z,\end{aligned}$$

and so $((x \to y)' \to z)' = O$. It follows that

$$\begin{aligned}\alpha_A(x) &\geqslant \min\{\alpha_A((x \to y)'), \alpha_A(y)\}\\ &\geqslant \min\{\min\{\alpha_A(((x \to y)' \to z)'), \alpha_A(z)\}, \alpha_A(y)\}\\ &= \min\{\min\{\alpha_A(O), \alpha_A(z)\}, \alpha_A(y)\}\\ &= \min\{\alpha_A(y), \alpha_A(z)\},\end{aligned}$$

and

$$\begin{aligned}\beta_A(x) &\leqslant \max\{\beta_A((x\to y)'),\beta_A(y)\}\\ &\leqslant \max\{\max\{\beta_A(((x\to y)'\to z)'),\beta_A(z)\},\beta_A(y)\}\\ &= \max\{\max\{\beta_A(O),\beta_A(z)\},\beta_A(y)\}\\ &= \max\{\beta_A(y),\beta_A(z)\}.\ \square\end{aligned}$$

Definition 4.4.2 *An IFS $A=(\alpha_A,\beta_A)$ of a lattice L is called an intuitionistic fuzzy lattice ideal of L if it satisfies the following conditions*

(1). $A=(\alpha_A,\beta_A)$ is intuitionistic monotonic;
(2). $\alpha_A(x\vee y)\geqslant \min\{\alpha_A(x),\alpha_A(y)\}$;
(3). $\beta_A(x\vee y)\leqslant \max\{\beta_A(x),\beta_A(y)\}$.

In Definition 4.4.2, (2) and (3) can be equivalently replaced by

$$\alpha_A(x\vee y)=\min\{\alpha_A(x),\alpha_A(y)\}$$

and

$$\beta_A(x\vee y)=\max\{\beta_A(x),\beta_A(y)\}$$

respectively by (1).

Example 4.4.2 *Let L be the lattice implication algebra defined in Example 2.1.4 and let $A=(\alpha_A,\beta_A)$ be an IFS of L defined by*

$$\alpha_A(x)=\begin{cases}1, & x\in\{O,d\},\\ s, & x\in L\setminus\{O,d\},\end{cases}$$

$$\beta_A(x)=\begin{cases}0, & x\in\{O,d\},\\ t, & x\in L\setminus\{O,d\},\end{cases}$$

where $t,s\in[0,1]$ and $s+t\leqslant 1$. By routine calculation we know that $A=(\alpha_A,\beta_A)$ is an intuitionistic fuzzy lattice ideal of L.

Theorem 4.4.3 *Let L be a lattice implication algebra. Every intuitionistic fuzzy LI-ideal of L is an intuitionistic fuzzy lattice ideal of L.*

Proof. Let $A=(\alpha_A,\beta_A)$ be an intuitionistic fuzzy LI-ideal of L. Theorem 4.4.1 shows that $A=(\alpha_A,\beta_A)$ is intuitionistic monotonic. Note that

$$\begin{aligned}((x\vee y)\to y)' &= (x\to y)'\\ &\leqslant (x')'\\ &= x\end{aligned}$$

for any $x,y\in L$, it follows that

$$\begin{aligned}\alpha_A(x\vee y) &\geqslant \min\{\alpha_A(((x\vee y)\to y)'),\alpha_A(y)\}\\ &\geqslant \min\{\alpha_A(x),\alpha_A(y)\},\end{aligned}$$

and

$$\begin{aligned}\beta_A(x \vee y) &\leqslant \max\{\beta_A(((x \vee y) \to y)'), \beta_A(y)\} \\ &\leqslant \max\{\beta_A(x), \beta_A(y)\}.\end{aligned}$$

Hence $A = (\alpha_A, \beta_A)$ is an intuitionistic fuzzy lattice ideal of L. □

The following example shows that the converse of Theorem 4.4.3 is not true.

Example 4.4.3 *Let L be the lattice implication algebra defined in Example 2.1.4, $A = (\alpha_A, \beta_A)$ an IFS of L defined by*

$$\alpha_A(x) = \begin{cases} 0.8, \ x = O, d, \\ 0.3, \ x \in L \setminus \{O, d\}, \end{cases}$$

$$\beta_A(x) = \begin{cases} 0.03, \ x = O, d, \\ 0.08, \ x \in L \setminus \{O, d\}. \end{cases}$$

Then $A = (\alpha_A, \beta_A)$ is an intuitionistic fuzzy lattice ideal of L, but not an intuitionistic fuzzy LI-ideal of L for

$$\alpha_A(a) < \min\{\alpha_A((a \to d)'), \alpha_A(d)\}$$

and

$$\beta_A(a) > \max\{\beta_A((a \to d)'), \beta_A(d)\}.$$

Theorem 4.4.4 *In a lattice H implication algebra L, every intuitionistic fuzzy lattice ideal of L is an intuitionistic fuzzy LI-ideal of L.*

Proof. Let $A = (\alpha_A, \beta_A)$ is an intuitionistic fuzzy lattice ideal of L. Clearly, $\alpha_A(O) \geqslant \alpha_A(x)$ and $\beta_A(O) \leqslant \beta_A(x)$ for any $x \in L$. Note that $x \leqslant x \vee y$, it follows that

$$\begin{aligned}\alpha_A(x) &\geqslant \alpha_A(x \vee y) \\ &= \alpha_A(y \vee (x' \vee y)') \\ &= \alpha_A(y \vee (x \to y)') \\ &\geqslant \min\{\alpha_A(y), \alpha_A((x \to y)')\},\end{aligned}$$

and

$$\begin{aligned}\beta_A(x) &\leqslant \beta_A(x \vee y) \\ &= \beta_A(y \vee (x' \vee y)') \\ &= \beta_A(y \vee (x \to y)') \\ &\leqslant \max\{\alpha_A(y), \alpha_A((x \to y)')\}.\end{aligned}$$ □

Theorem 4.4.5 *Let L be a lattice implication algebra. An IFS $A = (\alpha_A, \beta_A)$ is an intuitionistic fuzzy LI-ideal of L if and only if the fuzzy subsets α_A and $\bar{\beta}_A$ are fuzzy LI-ideal of L, where $\bar{\beta}_A(x) = 1 - \beta_A(x)$ for any $x \in L$.*

Proof. Let $A = (\alpha_A, \beta_A)$ be an intuitionistic fuzzy LI-ideal of L. Clearly α_A is a fuzzy LI-ideal of L. For any $x, y \in L$, we have

$$\begin{aligned}\bar{\beta}_A(O) &= 1 - \beta_A(O)\\ &\geqslant 1 - \beta_A(x)\\ &= \bar{\beta}_A(x)\end{aligned}$$

and

$$\begin{aligned}\bar{\beta}_A(x) &= 1 - \beta_A(x)\\ &\geqslant 1 - \max\{\beta_A((x \to y)'), \beta_A(y)\}\\ &= \min\{1 - \beta_A((x \to y)'), 1 - \beta_A(y)\}\\ &= \min\{\bar{\beta}_A((x \to y)'), \bar{\beta}_A(y)\}.\end{aligned}$$

Hence $\bar{\beta}_A$ is a fuzzy LI-ideal of L.

Conversely assume that α_A and $\bar{\beta}_A$ are fuzzy LI-ideals of L. For any $x, y \in L$, we get $\alpha_A(O) \geqslant \alpha_A(x)$ and

$$\begin{aligned}1 - \beta_A(O) &= \bar{\beta}_A(O)\\ &\geqslant \bar{\beta}_A(x)\\ &= 1 - \beta_A(x),\end{aligned}$$

i.e.,

$$\beta_A(O) \leqslant \beta_A(x),$$
$$\alpha_A(x) \geqslant \min\{\alpha_A((x \to y)'), \alpha_A(y)\},$$

and

$$\begin{aligned}1 - \beta_A(x) &= \bar{\beta}_A(x)\\ &\geqslant \min\{\bar{\beta}_A((x \to y)'), \bar{\beta}_A(y)\}\\ &= \min\{1 - \beta_A((x \to y)'), 1 - \beta_A(y)\}\\ &= 1 - \max\{\beta_A((x \to y)'), \beta_A(y)\},\end{aligned}$$

i.e.,

$$\beta_A(x) \leqslant \max\{\beta_A((x \to y)'), \beta_A(y)\}.$$

Hence $A = (\alpha_A, \beta_A)$ is an intuitionistic fuzzy LI-ideal of L. □

Theorem 4.4.6 *Let L be a lattice implication algebra and let $A = (\alpha_A, \beta_A)$ be an intuitionistic fuzzy LI-ideal of L. Then $A = (\alpha_A, \beta_A)$ is an intuitionistic fuzzy LI-ideal of L if and only if $(\alpha_A, \bar{\alpha}_A)$ and $(\bar{\beta}_A, \beta_A)$ are intuitionistic fuzzy LI-ideals of L.*

Proof. If $A = (\alpha_A, \beta_A)$ is an intuitionistic fuzzy LI-ideal of L, then α_A and $\bar{\beta}_A$ are fuzzy LI-ideals of L from Theorem 4.4.5. Hence $(\alpha_A, \bar{\alpha}_A)$ and $(\bar{\beta}_A, \beta_A)$ are intuitionistic fuzzy LI-ideals of L.

Conversely, if $(\alpha_A, \bar{\alpha}_A)$ and $(\bar{\beta}_A, \beta_A)$ are intuitionistic fuzzy LI-ideals of L then the fuzzy subsets α_A and $\bar{\beta}_A$ are fuzzy LI-ideals of L, hence $A = (\alpha_A, \beta_A)$ is an intuitionistic fuzzy LI-ideal of L. □

Theorem 4.4.7 *Let L be a lattice implication algebra and $A = (\alpha_A, \beta_A)$ an IFS of L. $A = (\alpha_A, \beta_A)$ is an intuitionistic fuzzy LI-ideal of L if and only if for any $s, t \in [0,1]$, the set $(\alpha_A)_t$ is either empty set or LI-ideal of L and so is $\{x \mid \beta_A(x) \leqslant s\}$.* □

Theorem 4.4.8 *Let L be a lattice implication algebra, Λ a non-empty subset of $[0,1]$, and $\{I_t \mid t \in \Lambda\}$ a collection of LI-ideals of L such that*

(1). $L = \bigcup_{t\in\Lambda} I_t$;
(2). $s > t$ if and only if $I_s \subseteq I_t$ for any $s, t \in \Lambda$.

Then the IFS $A = (\alpha_A, \beta_A)$ of L defined by

$$\alpha_A(x) = \sup\{t \in \Lambda \mid x \in I_t\}$$

and

$$\beta_A(x) = \inf\{t \in \Lambda \mid x \in I_t\}$$

for any $x \in L$ is an intuitionistic fuzzy LI-ideals of L.

Proof. According to Theorem 4.4.5, it is sufficient to show that α_A and $\bar{\beta}_A$ are fuzzy LI-ideals of L. In fact, for any $x \in L$,

$$\begin{aligned}\alpha_A(O) &= \sup\{t \in \Lambda \mid O \in I_t\}\\ &= \sup \Lambda\\ &\geqslant \alpha_A(x).\end{aligned}$$

If there exist $x, y \in L$ such that

$$\alpha_A(x) < \min\{\alpha_A((x \to y)'), \alpha_A(y)\},$$

then there exists t_1 such that

$$\alpha_A(x) < t_1 < \min\{\alpha_A((x \to y)'), \alpha_A(y)\},$$

it follows that $t_1 < \alpha_A((x \to y)')$, $t_1 < \alpha_A(y)$ and hence there exist $t_2, t_3 \in \Lambda$, $t_2 > t_1$, $t_3 > t_1$, $(x \to y)' \in I_{t_2}$ and $y \in I_{t_3}$. It follows that $(x \to y)' \in I_{t_2 \wedge t_3}$ and $y \in I_{t_2\wedge t_3}$, hence $x \in I_{t_2\wedge t_3}$, *i.e.*,

$$\begin{aligned}\alpha_A(x) &= \sup\{t \in \Lambda \mid x \in I_t\}\\ &\geqslant t_2 \wedge t_3\\ &> t_1,\end{aligned}$$

this is a contradiction. So we proved that α_A is a fuzzy LI-ideal of L. $\bar{\beta}_A$ is a fuzzy LI-ideal, which can be proved similarly. □

5 Homomorphisms and Representations

Homomorphisms are mappings that keep the operations of algebraic structures. They play an important role in studying algebraic structures, such as groups, rings, fields, and modules, *etc.* By discussing homomorphisms, we obtain relations among algebraic structures and their important properties. Studying representations of algebraic structures is another important issue.

In this chapter, we discuss homomorphisms and representations of lattice implication algebra. In Section 5.1, we give the definitions and discuss some basic properties of congruence relations on lattice implication algebras, then define several special congruence relations, induced by filters, fuzzy filters, LI-ideals, and fuzzy LI-ideals, respectively, on lattice implication algebras. In Section 5.2, we focus on the characteristic of proper lattice implication algebras. In Section 5.3, we study the representations of a class of complete and quasi-atomic lattice implication algebras.

5.1 Congruence Relations

In this section, we discuss congruence relations on lattice implication algebras, then we further describe the structure of lattice implication algebras.

5.1.1 Congruence Relations Induced by Filters

Definition 5.1.1 *Let L be a lattice implication algebra. $\equiv_\theta \subseteq L \times L$ is said to be a congruence relation on L, if $\equiv_\theta$ satisfies: for any $x, y, z \in L$,*

(1). $x \equiv_\theta x$;
(2). $x \equiv_\theta y$ implies $y \equiv_\theta x$;
(3). $x \equiv_\theta y$, $y \equiv_\theta z$ imply $x \equiv_\theta z$;
(4). $x \equiv_\theta y$ implies $x \to z \equiv_\theta y \to z$.

Definition 5.1.1 shows that $\equiv_\theta$ is an equivalence relation on L, and $\equiv_\theta$ is compatible w.r.t. $\to$. The following theorem shows that $\equiv_\theta$ is compatible w.r.t. the other three operators $\vee$, $\wedge$, $'$.

Theorem 5.1.1 *If $\equiv_\theta$ is a congruence relation on a lattice implication algebra L, then for any $x, y \in L$,*

(1). $x \equiv_\theta y$ if and only if $x' \equiv_\theta y'$;
(2). $x \equiv_\theta y$ if and only if $z \to x \equiv_\theta z \to y$ for any $z \in L$;
(3). $x \equiv_\theta y$ if and only if $x \vee z \equiv_\theta y \vee z$ for any $z \in L$;
(4). $x \equiv_\theta y$ if and only if $x \wedge z \equiv_\theta y \wedge z$ for any $z \in L$.

Proof. (1). If $x \equiv_\theta y$, we have $x' = x \to O \equiv_\theta y \to O = y'$. Conversely, we can get $x \equiv_\theta y$ from $x' \equiv_\theta y'$ similarly.

The sufficiency of (2), (3) and (4) can be proved by putting $z = I, O, I$ respectively. Now we prove the necessity.

(2). If $x \equiv_\theta y$, then by (1), $x' \equiv_\theta y'$ and for any $z \in L$,

$$z \to x = x' \to z' \equiv_\theta y' \to z' = z \to y.$$

(3). If $x \equiv_\theta y$, then for any $z \in L$, we have

$$x \to z \equiv_\theta y \to z.$$

Furthermore,

$$(x \to z) \to z \equiv_\theta (y \to z) \to z,$$

i.e.,

$$x \vee z \equiv_\theta y \vee z.$$

(4). If $x \equiv_\theta y$, then by (1), we have $x' \equiv_\theta y'$, it follows from (3) that

$$x' \vee z' \equiv_\theta y' \vee z'$$

holds for any $z \in L$, *i.e.*,

$$(x \wedge z)' \equiv_\theta (y \wedge z)'.$$

Furthermore, by (1),

$$x \wedge z \equiv_\theta y \wedge z. \quad \square$$

Given a congruence relation $\equiv_\theta$ on a lattice implication algebra L, we can construct another lattice implication algebra by $\equiv_\theta$. For any $x \in L$, denote

$$[x]_\theta = \{a \mid a \in L, x \equiv_\theta a\},$$

which is the congruence class of x w.r.t. the congruence relation $\equiv_\theta$. In the following, we sometimes write $[x]_\theta$ as $[x]$ for short. Denote

$$L/_{\equiv_\theta} = \{[x] \mid x \in L\},$$

which is the set of congruence classes of L w.r.t. $\equiv_\theta$.

Theorem 5.1.2 *Let L be a lattice implication algebra and $\equiv_\theta$ a congruence relation on it. In $L/_{\equiv_\theta}$, for any $x, y \in L$,*

$$\begin{aligned}[x] \to [y] &\triangleq [x \to y],\\ [x] \cup [y] &\triangleq [x \vee y],\\ [x] \cap [y] &\triangleq [x \wedge y],\\ [x]' &\triangleq [x'],\end{aligned}$$

then $(L/_{\equiv_\theta}, \cap, \cup, ', \to)$ is a lattice implication algebra. $\square$

The lattice implication algebra $L/_{\equiv_\theta}$ obtained from Theorem 5.1.2 is said to be a lattice implication quotient algebra of L w.r.t. $\equiv_\theta$. Define a mapping

$$\pi\colon\ L \longrightarrow L/_{\equiv_\theta}$$

satisfying: for any $x \in L$,

$$\pi(x) = [x]_\theta,$$

then π is a natural lattice implication homomorphism.

Definition 5.1.2 *Let $(L_\alpha, \vee_\alpha, \wedge_\alpha, '_{(\alpha)}, \to_\alpha, O_\alpha, I_\alpha)$ $(\alpha \in J)$ be a family of lattice implication algebras, where J is an index set. Define*

$$\begin{aligned}A &= \prod\{L_\alpha \mid \alpha \in J\}\\ &= \{f \mid f\colon\ J \longrightarrow \bigcup_{\alpha\in J} L_\alpha,\ \textit{s.t., for any } \alpha \in J, f(\alpha) \in L_\alpha\}.\end{aligned}$$

For any $f, g \in A$, $\alpha \in J$, binary operations $\vee$, $\wedge$, $\to$ and an unary operation $'$ on A are defined as follows:

$$\begin{aligned}(f \vee g)(\alpha) &= f(\alpha) \vee_\alpha g(\alpha),\\ (f \wedge g)(\alpha) &= f(\alpha) \wedge_\alpha g(\alpha),\\ f'(\alpha) &= (f(\alpha))'^{(\alpha)},\\ (f \to g)(\alpha) &= f(\alpha) \to_\alpha g(\alpha),\end{aligned}$$

and $O(\alpha) = O_\alpha, I(\alpha) = I_\alpha$. It can be proved that these operations are well defined and A is a lattice implication algebra. It is called the direct product or a lattice implication product algebra of L_α $(\alpha \in J)$. Specially, if $|J| = 2$, then $A = L_1 \times L_2$ is a lattice implication product algebra of L_1 and L_2.

Lemma 5.1.1 *Suppose $p_1 :\ L_1 \times L_2 \longrightarrow L_1$ and $p_2 :\ L_1 \times L_2 \longrightarrow L_2$ are two mappings defined as follows*

$$p_1(x, y) = x,$$

$$p_2(x, y) = y.$$

Then p_1 and p_2 are lattice implication homomorphisms.

Proof. For any $(x_1, y_1), (x_2, y_2) \in L_1 \times L_2$, since

$$(x_1, y_1) \to (x_2, y_2) = (x_1 \to_1 x_2, y_1 \to_2 y_2),$$

we have

$$\begin{aligned} p_1((x_1, y_1) \to (x_2, y_2)) &= x_1 \to_1 x_2 \\ &= p_1(x_1, y_1) \to_1 p_1(x_2, y_2). \end{aligned}$$

It is trivial that $p_1(O_1, O_2) = O_1$. By Theorem 2.4.1, we know that

$$p_1 \in \hom(L_1 \times L_2, L_1).$$

Similarly, we can prove $p_2 \in \hom(L_1 \times L_2, L_2)$. □

Theorem 5.1.3 *The set of filters of the lattice implication product algebra $L_1 \times L_2$ can be obtained as follows:*

$$F(L_1 \times L_2) = \{F_1 \times F_2 \mid F_i \in F(L_i), i = 1, 2\}.$$

Proof. For any $F_1 \in F(L_1)$ and $F_2 \in F(L_2)$, $(I_1, I_2) \in F_1 \times F_2$ holds obviously. If $(x_1, y_1), (x_1, y_1) \to (x_2, y_2) \in F_1 \times F_2$, then $x_1, x_1 \to_1 x_2 \in F_1$ and $y_1, y_1 \to_2 y_2 \in F_2$. Hence, $x_2 \in F_1, y_2 \in F_2$, which shows that $(x_2, y_2) \in F_1 \times F_2$, thus, $F_1 \times F_2 \in F(L_1 \times L_2)$.

On the other hand, for any $F \in F(L_1 \times L_2)$, let F_1 and F_2 be the projections of F on L_1 and L_2, respectively, *i.e.*, $F_i = p_i(F), i = 1, 2$. Since $(I_1, I_2) \in F$, it follows that $I_1 = p_1(I_1, I_2) \in F_1$. For any $x, x \to_1 y \in F_1$, then

$$(x, I_2), (x \to_1 y, I_2) \in F,$$

and since

$$(x \to_1 y, I_2) = (x, I_2) \to (y, I_2),$$

we have $(y, I_2) \in F$. Further, $y \in F_1$, so $F_1 \in F(L_1)$. Similarly, we can get $F_2 \in F(L_2)$.

Now we prove $F = F_1 \times F_2$. In fact, $F \subseteq F_1 \times F_2$ holds obviously. On the other hand, if $(x, y) \in F_1 \times F_2$, then $x \in F_1, y \in F_2$ and hence $(x, I_2), (I_1, y) \in F$, it follows that

$$(x, I_2) \to (x, y) = (I_1, y) \in F.$$

Since F is a filter of $L_1 \times L_2$, thus $(x, y) \in F$, and therefore, $F_1 \times F_2 \subseteq F$. Consequently, $F = F_1 \times F_2$ holds. □

Now we discuss congruence relations on lattice implication product algebras.

Theorem 5.1.4 *Let L_1 and L_2 be lattice implication algebras and $\equiv_\theta, \equiv_\varphi$ congruence relations on them, respectively. A binary relation $\equiv_\rho$ is defined as follows: for any $(x_1, y_1), (x_2, y_2) \in L_1 \times L_2$,*

$$(x_1, y_1) \equiv_\rho (x_2, y_2) \quad \textit{iff} \quad x_1 \equiv_\theta x_2 \textit{ and } y_1 \equiv_\varphi y_2,$$

then $\equiv_\rho$ *is a congruence relation on* $L_1 \times L_2$*, denoted by* $\equiv_\theta \times \equiv_\varphi$*, and any congruence relation on* $L_1 \times L_2$ *has this form.*

Proof. We first prove $\equiv_\rho$ is a congruence relation on $L_1 \times L_2$. In fact, for any $(x, y) \in L_1 \times L_2$, by the reflexivity of θ and φ, we have $x \equiv_\theta x, y \equiv_\varphi y$. And then $(x, y) \equiv_\rho (x, y)$, *i.e.*, $\equiv_\rho$ is reflexive.

For any $(x_1, y_1), (x_2, y_2) \in L_1 \times L_2$, if $(x_1, y_1) \equiv_\rho (x_2, y_2)$, then $x_1 \equiv_\theta x_2, y_1 \equiv_\varphi y_2$. By the symmetry of θ and φ, we have $x_2 \equiv_\theta x_1, y_2 \equiv_\varphi y_1$. Hence, $(x_2, y_2) \equiv_\rho (x_1, y_1)$, *i.e.*, $\equiv_\rho$ is symmetric.

For any $(x_1, y_1), (x_2, y_2), (x_3, y_3) \in L_1 \times L_2$, if

$$(x_1, y_1) \equiv_\rho (x_2, y_2) \text{ and } (x_2, y_2) \equiv_\rho (x_3, y_3),$$

then $x_1 \equiv_\theta x_2$, $y_1 \equiv_\varphi y_2$, $x_2 \equiv_\theta x_3$, $y_2 \equiv_\varphi y_3$. By the transitivity of θ and φ, we can get $x_1 \equiv_\theta x_3$ and $y_1 \equiv_\varphi y_3$, hence $(x_1, y_1) \equiv_\rho (x_3, y_3)$, so $\equiv_\rho$ is transitive.

For any $(x_1, y_1), (x_2, y_2), (x, y) \in L_1 \times L_2$, if $(x_1, y_1) \equiv_\rho (x_2, y_2)$, then $x_1 \equiv_\theta x_2, y_1 \equiv_\varphi y_2$. Furthermore, $x_1 \to x \equiv_\theta x_2 \to x, y_1 \to y \equiv_\varphi y_2 \to y$, hence,

$$(x_1 \to x, y_1 \to y) \equiv_\rho (x_2 \to x, y_2 \to y),$$

i.e.,

$$(x_1, y_1) \to (x, y) \equiv_\rho (x_2, y_2) \to (x, y).$$

Sum up, we know that $\equiv_\rho$ is a congruence relation on $L_1 \times L_2$.

Let $\equiv_\rho$ be a congruence relation on $L_1 \times L_2$, $\equiv_\theta$, $\equiv_\varphi$ be defined respectively as follows: for any $x_1, x_2 \in L_1$ and $y_1, y_2 \in L_2$,

$$x_1 \equiv_\theta x_2 \quad \text{iff} \quad (x_1, I_2) \equiv_\rho (x_2, I_2),$$

$$y_1 \equiv_\varphi y_2 \quad \text{iff} \quad (I_1, y_1) \equiv_\rho (I_1, y_2),$$

where I_1 and I_2 are the greatest elements of L_1 and L_2, respectively.

It is easy to prove that $\equiv_\theta$ and $\equiv_\varphi$ are congruence relations of L_1 and L_2, respectively.

The remainder of the proof shows $\equiv_\rho = \equiv_\theta \times \equiv_\varphi$. Suppose

$$((x_1, y_1), (x_2, y_2)) \in \equiv_\theta \times \equiv_\varphi,$$

then $x_1 \equiv_\theta x_2$ and $y_1 \equiv_\varphi y_2$. It follows that $(x_1, I_2) \equiv_\rho (x_2, I_2)$. Combining $(x_2, I_2) \equiv_\rho (x_2, I_2)$, we have

$$(x_2 \to x_1, I_2 \to I_2) \equiv_\rho (x_2 \to x_2, I_2 \to I_2),$$

i.e.,

$$(x_2 \to x_1, I_2) \equiv_\rho (I_2, I_2).$$

Using $(x_1, y_1) \equiv_\rho (x_1, y_1)$, we can get

$$((x_2 \to x_1) \to x_1, y_1) \equiv_\rho (x_1, y_1).$$

In the same way we have

$$((x_1 \to x_2) \to x_2, y_1) \equiv_\rho (x_2, y_1).$$

Since $(x_1 \to x_2) \to x_2 = (x_2 \to x_1) \to x_1$, it follows that $(x_1, y_1) \equiv_\rho (x_2, y_1)$. Likewise for $(x_2, y_1) \equiv_\rho (x_2, y_2)$. Thus $(x_1, y_1) \equiv_\rho (x_2, y_2)$, so $\equiv_\theta \times \equiv_\varphi \subseteq \equiv_\rho$.

Conversely, if $((x_1, y_1), (x_2, y_2)) \in \equiv_\rho$, *i.e.*, $(x_1, y_1) \equiv_\rho (x_2, y_2)$. Using $(I_1, y_1) \equiv_\rho (I_1, y_1)$, we have

$$(I_1 \to x_1, y_1 \to y_1) \equiv_\rho (I_1 \to x_2, y_1 \to y_2),$$

i.e., $(x_1, I_2) \equiv_\rho (x_2, y_1 \to y_2)$. Since

$$(I_1, I_2) = (x_1, y_1) \to (x_1, y_1),$$
$$(x_1, y_1) \to (x_2, y_2) = (x_1 \to x_2, y_1 \to y_2)$$

and

$$(x_1, y_1) \to (x_1, y_1) \equiv_\rho (x_1, y_1) \to (x_2, y_2),$$

we have

$$(I_1, I_2) \equiv_\rho (x_1 \to x_2, y_1 \to y_2).$$

Furthermore,

$$(x_1, I_2) \equiv_\rho ((x_1 \to x_2) \to x_2, I_2).$$

It follows that $x_1 \equiv_\theta (x_1 \to x_2) \to x_2$. In the same way we have $x_2 \equiv_\theta (x_2 \to x_1) \to x_1$. Using $(x_1 \to x_2) \to x_2 = (x_2 \to x_1) \to x_1$, $x_1 \equiv_\theta x_2$ followed. Similarly, we can prove $y_1 \equiv_\varphi y_2$. Therefore,

$$((x_1, y_1), (x_2, y_2)) \in \equiv_\theta \times \equiv_\varphi,$$

which means that $\equiv_\rho \subseteq \equiv_\theta \times \equiv_\varphi$.

Sum up, we obtain $\equiv_\rho = \equiv_\theta \times \equiv_\varphi$. □

Denote $CR(L)$ as the set of all congruence relations on the lattice implication algebra L.

Lemma 5.1.2 *(1). For any $\equiv_\theta \in CR(L)$, $[I]_\theta = \{x \in L \mid x \equiv_\theta I\} \in F(L)$.*

(2). For any $F \in F(L)$, $\equiv_F$ is defined as follows: for any $x, y \in L$,

$$x \equiv_F y \qquad \textit{iff} \qquad x \to y, y \to x \in F.$$

Then $\equiv_F \in CR(L)$, $[I]_F = F$.

In what follows $(L/\equiv_F, \cup, \cap, ', \to)$ is called the lattice implication algebra induced by F and denoted by $(L/F, \cup, \cap, ', \to)$.

Proof. (1). $I \in [I]_\theta$ holds obviously. If $x, x \to y \in [I]_\theta$, then $x \equiv_\theta I, x \to y \equiv_\theta I$. So we have $x \to y \equiv_\theta I \to y$, *i.e.*, $x \to y \equiv_\theta y$, hence, $y \equiv_\theta I$, *i.e.*, $y \in [I]_\theta$. This proved that $[I]_\theta$ is a filter of L.

(2). The proof of $\equiv_F \in CR(L)$ is trivial.

Now we prove $[I]_F = F$. Given any $x \in [I]_F$, then $x \equiv_F I, x = I \to x \in F$, so $[I]_F \subseteq F$. Conversely, for any $x \in F$, since $x \to I = I \in F$ and $I \to x = x \in F$, so $x \equiv_F I$, *i.e.*, $x \in [I]_F$, which means $F \subseteq [I]_F$. Therefore, $[I]_F = F$. □

Theorem 5.1.5 *Let L_1 and L_2 be two lattice implication algebras, $L_1 \times L_2$ their lattice implication product algebra. For any $F_1 \in F(L_1)$, $F_2 \in F(L_2)$, put*

$$F = F_1 \times F_2.$$

Then for any $(x_1, y_1), (x_2, y_2) \in L_1 \times L_2$,

$$(x_1, y_1) \equiv_F (x_2, y_2) \quad \textit{iff} \quad x_1 \equiv_{F_1} x_2 \textit{ and } y_1 \equiv_{F_2} y_2,$$

i.e.,

$$\equiv_F = \equiv_{F_1} \times \equiv_{F_2}. \ \square$$

We give a lattice implication quotient algebra induced by an implicative filter.

Theorem 5.1.6 *Let L be a lattice implication algebra and F a filter of L. Then F is an implicative filter if and only if the lattice implication quotient algebra L/F is a lattice H implication algebra.*

Proof. We suppose F is an implicative filter. For any $x, y, z \in L$, let

$$u = x \to (y \to z),$$

it follows that

$$x \to (y \to (u \to z)) = u \to (x \to (y \to z)) = I \in F,$$

and hence $(x \to y) \to (x \to (u \to z)) \in F$ by Theorem 3.1.8. By

$$\begin{aligned} &(x \to y) \to (x \to (u \to z)) \\ &= (x \to y) \to (u \to (x \to z)) \\ &= u \to ((x \to y) \to (x \to z)) \\ &= (x \to (y \to z)) \to ((x \to y) \to (x \to z)), \end{aligned}$$

it follows that

$$(x \to (y \to z)) \to ((x \to y) \to (x \to z)) \in F$$

and hence

$$\begin{aligned}&([x] \to ([y] \to [z])) \to (([x] \to [y]) \to ([x] \to [z]))\\ &= [(x \to (y \to z)) \to ((x \to y) \to (x \to z))]\\ &= [I].\end{aligned}$$

On the other hand, by

$$\begin{aligned}(x \to y) \to (x \to z) &= x \to ((x \to y) \to z)\\ &\leqslant x \to (y \to z),\end{aligned}$$

it follows that

$$((x \to y) \to (x \to z)) \to (x \to (y \to z)) = I \in F,$$

and hence

$$\begin{aligned}&(([x] \to [y]) \to ([x] \to [z])) \to ([x] \to ([y] \to [z]))\\ &= [((x \to y) \to (x \to z)) \to (x \to (y \to z))]\\ &= [I].\end{aligned}$$

So, we have proved that

$$[x] \to ([y] \to [z]) = ([x] \to [y]) \to ([x] \to [z])$$

and L/F is a lattice H implication algebra.

Conversely, we suppose that L/F is a lattice H implication algebra. It follows that $\{[I]\}$ is an implicative filter of L/F because $\{[I]\}$ is a filter. For any $x, y, z \in L$, if $x \to (y \to z) \in F$, then $[x] \to ([y] \to [z]) = [x \to (y \to z)] = [I]$, $([x] \to [y]) \to ([x] \to [z]) = [I]$, *i.e.*, $(x \to y) \to (x \to z) \in F$, so F is an implicative filter by Theorem 3.1.8. □

Corollary 5.1.1 *If F is a filter of lattice H implication algebra L, then the lattice implication quotient algebra L/F is also a lattice H implication algebra.* □

Theorem 5.1.7 *Let L be a lattice implication algebra and P a proper filter of L. Then L/P is a totally ordered lattice implication algebra if and only if P is a prime filter of L.*

Proof. Assume L/P is a totally ordered lattice implication algebra and $x \vee y \in P$, where $x, y \in L$. Since $[x] \cup [y] = [x \vee y] = [I]$, then $[x] = [I]$ or $[y] = [I]$, *i.e.*, $x \in P$ or $y \in P$, so P is a prime filter of L.

Conversely, we assume P is a prime filter of L. Since

$$(x \to y) \vee (y \to x) = I$$

holds for any $x, y \in L$, then

$$x \to y \in P \quad \text{or} \quad y \to x \in P,$$

i.e.,

$$x \to y \equiv_P I \quad \text{or} \quad y \to x \equiv_P I.$$

If $x \to y \equiv_P I$ holds, then $(x \to y) \to y \equiv_P y$, *i.e.*, $x \vee y \equiv_P y$ which implies $[x] \cup [y] = [y]$, so $[x] \leqslant [y]$.

Similarly, we can prove that $[y] \leqslant [x]$ whenever $y \to x \equiv_P I$ holds. Therefore, L/P is a totally ordered lattice implication algebra. □

Lemma 5.1.3 *Let L be a lattice implication algebra, F a filter of L and L/F the lattice implication quotient algebra induced by F.*

(1). If J is a lattice ideal (or a dual lattice ideal) of L, then

$$J/F = \{[x]_F \mid x \in J\}$$

is a lattice ideal (or a dual lattice ideal) of L/F.

(2). If T is a lattice ideal (or a dual lattice ideal) of L/F, then

$$J = \{x \mid x \in L, [x]_F \in T\}$$

is a lattice ideal (or a dual lattice ideal) of L.

(3). Let

$$F(L, F) = \{J \mid F \subseteq J \subseteq L, J \text{ is a filter of } L\}.$$

Then for any $J \in F(L, F)$, J/F is a filter of L/F. If $J \in F(L, F)$ is an implicative filter of L, then J/F is an implicative filter of L/F.

(4). $J_1 \neq J_2$ if and only if $J_1/F \neq J_2/F$ for any $J_1, J_2 \in F(L, F)$.

(5). If T is a filter (or an implicative filter) of L/F, and $J = \{x \mid x \in L, [x]_F \in T\}$, then $F \subseteq J \subseteq L$ and J is a filter (or an implicative filter) of L and $J/F = T$.

Proof. (1). If J is a lattice ideal of L, then for any $[x]_F, [y]_F \in J/F$, there exist $x_1, x_2 \in J$, $[x_1]_F = [x]_F, [x_2]_F = [y]_F$, *i.e.*, $x_1 \equiv_F x, x_2 \equiv_F y$. Hence, by $x_1 \vee x_2 \in J$, $x_1 \vee x_2 \equiv_F x \vee y$, we have

$$[x]_F \cup [y]_F = [x_1 \vee x_2]_F \in J/F.$$

If $[x]_F, [y]_F \in L/F$ and $[x]_F \leqslant [y]_F, [y]_F \in J/F$, then there exists $y_1 \in J$, $[y]_F = [y_1]_F$. So $[x]_F \leqslant [y_1]_F$ and by $y_1 \wedge x \in J$, we have

$$[x]_F = [x]_F \cap [y_1]_F = [x \wedge y_1]_F \in J/F.$$

This means that J/F is a lattice ideal of L/F. The proof of the case for the dual lattice ideals is similar.

(2). If T is a lattice ideal of L/F, $J = \{x \mid x \in L, [x]_F \in T\}$, then for any $x_1, x_2 \in J$, $[x_1 \vee x_2]_F = [x_1]_F \cup [x_2]_F \in T$, hence $x_1 \vee x_2 \in J$.

If $x, y \in L$ such that $x \leqslant y \in J$, then $[x]_F \leqslant [y]_F \in T$ and hence $[x]_F \in T, x \in J$. Therefore, J is a lattice ideal of L. The proof of the case for the dual lattice ideals is similar.

(3). $F = [I]_F$ is the greatest element of L/F, by $I \in J$, we have $[I]_F \in J/F$.

If $[x]_F, [x]_F \to [y]_F \in J/F$, *i.e.*, $[x]_F, [x \to y]_F \in J/F$, then there exist $x_1, x_2 \in J$ such that $[x_1]_F = [x \to y]_F, [x_2]_F = [x]_F$. So $x_2 \equiv_F x$ and $x_2 \to x \in F$. It follows that $x_2 \to x \in J$. Since $x_2 \in J$ and J is a filter of L, we have $x \in J$. Similarly, we can prove $x \to y \in J$. Hence, $y \in J$, $[y]_F \in J/F$. Therefore, J/F is a filter of L/F.

The statement about the implicative filters can be proved similarly.

(4). Suppose $J_1, J_2 \in F(L, F)$. If $J_1/F \neq J_2/F$, then $J_1 \neq J_2$ is clear.

If $J_1/F = J_2/F$, but $J_1 \neq J_2$, let $x_1 \in J_1, x_1 \notin J_2$, then, by $[x_1]_F \in J_1/F = J_2/F$, there exists $x_2 \in J_2$ such that $[x_1]_F = [x_2]_F$, hence $x_2 \to x_1 \in F \subseteq J_2$ and $x_1 \in J_2$. This contradiction shows that the conclusion holds.

(5). If T is a filter of L/F, then $I \in J$ and $F \subseteq J$ by $[x]_F = [I]_F$ for any $x \in F$.

If $x, x \to y \in J$, then $[x]_F, [x \to y]_F \in T$, *i.e.*, $[x]_F, [x]_F \to [y]_F \in T$. Hence $[y]_F \in T$ and $y \in J$. So J is a filter of L. $J/F = T$ is clear. Similarly, we can prove the statement about implicative filters. □

Corollary 5.1.2 *Let $F \in F(L)$, and $\equiv_F$ be the congruence relation of L defined in Lemma 5.1.2(2). Define a mapping*

$$P\colon\ F(L, F) \longrightarrow F(L/F)$$

satisfying

$$P(J) = J/F,$$

then P is a bijection. □

Corollary 5.1.3 *Let L be a lattice implication algebra, $A \subseteq L$ and $A \neq \emptyset$. Then*

$$\{J \mid A \subseteq J \subseteq L, J \textit{ is a filter of } L\} = F(L) \qquad \textit{iff} \qquad A = \{I\}.$$

Proof. If $\{J \mid A \subseteq J \subseteq L, J$ is a filter of $L\} = F(L)$, then for any $F \in F(L)$, $A \subseteq F$. Specially, for $\{I\}$, $A \subseteq \{I\}$, so $A = \{I\}$. Conversely, it is trivial. □

Lemma 5.1.4 *Let $f\colon\ CR(L) \longrightarrow F(L)$ be a mapping and satisfy*

$$f(\equiv_\theta) = [I]_\theta.$$

Then f is a bijection.

Proof. According to Lemma 5.1.2(2), f is surjective obviously.

Now we need to prove that f is injective, *i.e.*, for any $\equiv_\theta, \equiv_\varphi \in CR(L)$, if $[I]_\theta = [I]_\varphi$, then $\equiv_\theta = \equiv_\varphi$. In fact, for any $x, y \in L$, if $x \equiv_\theta y$, then

$x \to y \equiv_\theta I$, *i.e.*, $x \to y \in [I]_\theta = [I]_\varphi$, which implies $x \to y \equiv_\varphi I$. By (4) of Definition 5.1.1, we can get $(x \to y) \to y \equiv_\varphi I \to y$, *i.e.*, $x \vee y \equiv_\varphi y$. In the same way, we can prove $x \vee y \equiv_\varphi x$. Therefore, $x \equiv_\varphi y$.

Conversely, if $x \equiv_\varphi y$, similar to the above proof, we can get $x \equiv_\theta y$. Hence, $\equiv_\theta = \equiv_\varphi$. □

Theorem 5.1.8 *Assume* $\equiv_\theta \in CR(L)$. *Then* $L/_{\equiv_\theta}$ *is a totally ordered set if and only if for any* $x, y \in L$, *if* $x \vee y \equiv_\theta I$, *then* $x \equiv_\theta I$ *or* $y \equiv_\theta I$.

Proof. Assume $L/_{\equiv_\theta}$ is a totally ordered set and $x \vee y \equiv_\theta I$. If $[x]_\theta \leqslant [y]_\theta$, then

$$[I]_\theta = [x \vee y]_\theta = [x]_\theta \cup [y]_\theta = [y]_\theta,$$

i.e., $y \equiv_\theta I$. If $[y]_\theta \leqslant [x]_\theta$, we can prove $x \equiv_\theta I$ similarly.

Conversely, since

$$(x \to y) \vee (y \to x) = I,$$

it follows that $x \to y \equiv_\theta I$ or $y \to x \equiv_\theta I$.

If $x \to y \equiv_\theta I$, then $(x \to y) \to y \equiv_\theta y$, *i.e.*, $x \vee y \equiv_\theta y$, which implies $[x]_\theta \cup [y]_\theta = [y]_\theta$, so $[x]_\theta \leqslant [y]_\theta$.

If $y \to x \equiv_\theta I$, we can prove $[y]_\theta \leqslant [x]_\theta$ similarly. Therefore, $L/_{\equiv_\theta}$ is a totally ordered set. □

5.1.2 Congruences Relations Induced by *LI*-ideals

In Section 4.1, we discussed *LI*-ideals and their basic properties. Now we begin to discuss the congruence relations and lattice implication quotient algebras induced by *LI*-ideals.

Let A be an *LI*-ideal of a lattice implication algebra L. Define a binary relation $\equiv_A$ on L as follows: for any $x, y \in L$,

$$x \equiv_A y \qquad \text{iff} \qquad (x \to y)' \in A, (y \to x)' \in A.$$

Lemma 5.1.5 $\equiv_A$ *is an equivalence relation on* L.

Proof. Clearly $\equiv_A$ is both reflexive and symmetric.

To prove the transitivity, assume $x \equiv_A y$ and $y \equiv_A z$. Then

$$(x \to y)' \in A, (y \to x)' \in A$$

and

$$(y \to z)' \in A, (z \to y)' \in A.$$

Since

$$((x \to z)' \to (x \to y)')' = ((x \to y) \to (x \to z))' \leqslant (y \to z)',$$

it follows that

$$((x \to z)' \to (x \to y)')' \in A.$$

Furthermore, $(x \to z)' \in A$ because $(x \to y)' \in A$ and A is an LI-ideal of L. Similarly, we have $(z \to x)' \in A$ and hence $x \equiv_A z$.

Assume that $x \equiv_A u$ and $y \equiv_A v$. Then

$$(x \to u)' \in A, (u \to x)' \in A$$

and

$$(y \to v)' \in A, (v \to y)' \in A.$$

Since

$$((x \to y) \to (x \to v))' \leqslant (y \to v)'$$

and

$$((x \to v) \to (x \to y))' \leqslant (v \to y)',$$

it follows that

$$((x \to y) \to (x \to v))' \in A$$

and

$$((x \to v) \to (x \to y))' \in A,$$

which means that

$$x \to y \equiv_A x \to v.$$

Similarly, we get $x \to v \equiv_A u \to v$. By the transitivity of $\equiv_A$, we conclude that $x \to y \equiv_A u \to v$. □

Let

$$[x]_A = \{y \in L \mid x \equiv_A y\}, \quad L/A = \{[x]_A \mid x \in L\}.$$

It is clear that $[O]_A = A$ and $[I]_A = \{y \in L \mid y' \in A\}$. Define binary operations $\sqcup$, $\sqcap$, $\rightrightarrows$ and an unary operation N on L/A as follows:

$$\begin{aligned} [x]_A \sqcup [y]_A &= [x \vee y]_A, \\ [x]_A \sqcap [y]_A &= [x \wedge y]_A, \\ [x]_A \rightrightarrows [y]_A &= [x \to y]_A, \\ [x]_A^N &= [x']_A, \end{aligned}$$

for any $[x]_A, [y]_A \in L/A$. It can be easily verified that $(L/A, \sqcup, \sqcap, O_A, I_A)$ is a bounded lattice. Moreover $(L/A, \sqcup, \sqcap, ^N, \rightrightarrows)$ is a lattice implication algebra, which is called the lattice implication quotient algebra of L induced by the LI-ideal A.

By Homomorphism Theorem of universal algebra [42], we have the following theorem.

Theorem 5.1.9 *Let L_1 and L_2 be lattice implication algebras and $f\colon L_1 \longrightarrow L_2$ be an implication epimorphism. Then $L_1/_{\ker(f)}$ is isomorphic to L_2.* □

5.1.3 Congruence Relations Induced by Fuzzy Filters

In Chapter 3, some properties of fuzzy filters of lattice implication algebras have been discussed in detail. In this section, we study the congruence relations induced by fuzzy filters.

Let A be a fuzzy filter of a lattice implication algebra L. For any $x, y \in L$, define a binary relation $\equiv_A$ on L by

$$x \equiv_A y \qquad \text{iff} \qquad A(x \to y) = A(I), A(y \to x) = A(I). \tag{5.1}$$

Lemma 5.1.6 *$\equiv_A$ is an equivalent relation on L.*

Proof. It is clear that $\equiv_A$ is reflexive and symmetric.

For any $x, y, z \in L$, if $x \equiv_A y, y \equiv_A z$, then

$$\begin{aligned} A(x \to y) &= A(y \to x) \\ &= A(y \to z) \\ &= A(z \to y) \\ &= A(I). \end{aligned}$$

Since $y \to z \leqslant (x \to y) \to (x \to z)$ and $y \to x \leqslant (z \to y) \to (z \to x)$, then by Definition 3.6.1, we have

$$\begin{aligned} A(x \to z) &\geqslant \min\{A(x \to y), A((x \to y) \to (x \to z))\} \\ &\geqslant \min\{A(x \to y), A(y \to z)\} \\ &= A(I) \end{aligned}$$

and

$$\begin{aligned} A(z \to x) &\geqslant \min\{A(z \to y), A((z \to y) \to (z \to x))\} \\ &\geqslant \min\{A(z \to y), A(y \to x)\} \\ &= A(I), \end{aligned}$$

so $A(x \to z) = A(I)$, $A(z \to x) = A(I)$, *i.e.*, $x \equiv_A z$. Hence, $\equiv_A$ is an equivalent relation on L. □

Lemma 5.1.7 *$x \equiv_A y$ implies $x \to z \equiv_A y \to z$ for any $x, y, z \in L$.*

Proof. If $x \equiv_A y$, then we have $A(x \to y) = A(y \to x) = A(I)$. Since $x \to y \leqslant (y \to z) \to (x \to z)$ and $y \to x \leqslant (x \to z) \to (y \to z)$, we get

$$A(I) = A(y \to x) \leqslant A((x \to z) \to (y \to z))$$

and

$$A(I) = A(x \to y) \leqslant A((y \to z) \to (x \to z)),$$

so $A((x \to z) \to (y \to z)) = A(I)$ and $A((y \to z) \to (x \to z)) = A(I)$. Hence we obtain $x \to z \equiv_A y \to z$. □

Sum up, we have the following theorems.

Theorem 5.1.10 *$\equiv_A$ is a congruence relation on L.* □

Theorem 5.1.11 *Let A be a fuzzy filter of a lattice implication algebra L. Let*

$$[x]_A = \{y \in L \mid x \equiv_A y\}$$

and

$$L/A = \{[x]_A \mid x \in L\}.$$

The operations $\sqcup$, $\sqcap$, N, $\rightarrow$ on L/A are defined by

$$\begin{aligned} [x]_A \sqcup [y]_A &= [x \vee y]_A, \\ [x]_A \sqcap [y]_A &= [x \wedge y]_A, \\ [x]_A \rightarrow [y]_A &= [x \rightarrow y]_A, \\ [x]_A^N &= [x']_A. \end{aligned}$$

Then $(L/A, \sqcup, \sqcap, N, \rightarrow)$ is a lattice implication algebra, moreover $[O]_A$ and $[I]_A$ are the smallest and the greatest elements of L/A, respectively. □

$(L/A, \sqcup, \sqcap, N, \rightarrow)$ is called the lattice implication quotient algebra induced by the fuzzy filter A.

In Lemma 5.1.2, a congruence relation $\equiv_F$ induced by a filter F of a lattice implication algebra L was given, *i.e.*, for any $x, y \in L$,

$$x \equiv_F y \quad \text{iff} \quad x \rightarrow y, y \rightarrow x \in F.$$

Theorem 5.1.12 *Let F be a filter of a lattice implication algebra L and $\mathcal{X}_F$ be the characteristic function of F. Then for any $x, y \in L$,*

$$x \equiv_F y \quad \textit{iff} \quad x \equiv_{\mathcal{X}_F} y.$$

Proof. By definitions of $\equiv_F$ and $\equiv_{\mathcal{X}_F}$, we obtain

$$\begin{aligned} x \equiv_F y \ &\text{iff} \ x \rightarrow y, y \rightarrow x \in F \\ &\text{iff} \ \mathcal{X}_F(x \rightarrow y) = 1, \mathcal{X}_F(x \rightarrow y) = 1 \\ &\text{iff} \ \mathcal{X}_F(x \rightarrow y) = \mathcal{X}_F(I), \mathcal{X}_F(x \rightarrow y) = \mathcal{X}_F(I) \\ &\text{iff} \ x \equiv_{\mathcal{X}_F} y. \ \square \end{aligned}$$

In the following, let $L/_{\equiv_{\nu \circ f}} \triangleq L/_{(\nu \circ f)}$ and $L^*/_{\equiv_\nu} \triangleq L^*/_\nu$.

Theorem 5.1.13 (Fuzzy Homomorphism Fundamental) *Let L and L^* be lattice implication algebras, $f\colon L \longrightarrow L^*$ a lattice implication epimorphism and ν a fuzzy filter in L^*. Then*

$$L/_{(\nu \circ f)} \cong L^*/_\nu.$$

Proof. By Theorem 3.6.3, we know that $\nu\circ f$ is a fuzzy filter of L, and then by Theorem 5.1.11, $L/_{(\nu\circ f)} = (L/_{(\nu\circ f)}, \vee, \wedge, N, \rightarrow)$ and $L^*/_\nu = (L/_\nu, \vee, \wedge, N, \rightarrow)$ are lattice implication algebras. Define η: $L/_{(\nu\circ f)} \longrightarrow L^*/_\nu$ by

$$\eta([x]_{(\nu\circ f)}) = [f(x)]_\nu.$$

(1). η is well defined. In fact, for any x, $y \in L$,

$$\begin{aligned}[x]_{(\nu\circ f)} = (\nu\circ f)_y \ &\text{iff}\ (\nu\circ f)(x\rightarrow y) = (\nu\circ f)(y\rightarrow x) = (\nu\circ f)(I)\\ &\text{iff}\ \nu(f(x)\rightarrow f(y)) = \nu(f(y)\rightarrow f(x)) = \nu(I)\\ &\text{iff}\ [f(x)]_\nu = [f(y)]_\nu\\ &\text{iff}\ \eta([x]_{(\nu\circ f)}) = \eta([y]_{(\nu\circ f)}).\end{aligned}$$

(2). η is an implication homomorphism. In fact, for any x, $y \in L$,

$$\begin{aligned}\eta([x]_{(\nu\circ f)} \rightarrow [y]_{(\nu\circ f)}) &= \eta([x\rightarrow y]_{(\nu\circ f)})\\ &= [f(x\rightarrow y)]_\nu\\ &= [f(x)\rightarrow f(y)]_\nu\\ &= [f(x)]_\nu \rightarrow [f(y)]_\nu\\ &= \eta([x]_{(\nu\circ f)}) \rightarrow \eta([y]_{(\nu\circ f)}).\end{aligned}$$

(3). For any $\nu_z \in L^*/_\nu$, since f is an epimorphism, there exists $x \in L$ such that $f(x) = z$. So $\eta([x]_{(\nu\circ f)}) = [f(x)]_\nu = [z]_\nu$. Hence η is an epimorphism. By (1), η is one-to-one and onto and hence an isomorphism. □

As an application of Theorem 5.1.13, we can prove the Homomorphism Fundamental of lattice implication algebras.

Corollary 5.1.4 (Homomorphism Fundamental) *Let L and L^* be lattice implication algebras, f: $L \longrightarrow L^*$ a lattice implication epimorphism and F^* a filter of L^*. Then*

$$L/_{f^{-1}(F^*)} \cong L^*/_{F^*}.$$

Proof. By Theorem 3.1.12, we know that $f^{-1}(F^*)$ is a filter of L. Consider

$$\mathcal{X}_{f^{-1}(F^*)}(x) = \begin{cases}1,\ x\in f^{-1}(F^*),\\ 0,\ \text{otherwise},\end{cases}$$

it follows that

$$\mathcal{X}_{f^{-1}(F^*)}(x) = (\mathcal{X}_{F^*}\circ f)(x).$$

Hence,

$$L/_{(\mathcal{X}_{F^*}\circ f)} = L/_{\mathcal{X}_{f^{-1}(F^*)}}.$$

Furthermore, we have

$$L/_{f^{-1}(F^*)} = L/_{(\mathcal{X}_{F^*}\circ f)}$$

and

$$L^*/_{F^*} = L^*/_{\chi_{F^*}}.$$

By Theorem 5.1.13, we obtain

$$L/_{(\chi_{F^*} \circ f)} \cong L^*/_{\chi_{F^*}}.$$

Therefore,

$$L/_{f^{-1}(F^*)} \cong L^*/_{F^*}. \quad \square$$

Corollary 5.1.5 *Let L and L^* be lattice implication algebras and f be a lattice implication epimorphism from L to L^*. Then*

$$L/_{D\text{-}\ker(f)} \cong L^*. \ \square$$

Let L be a lattice implication algebra. Now we generalize the congruence relation induced by a fuzzy filter of L. This generalization shows that more congruence relations on L can be induced by a fuzzy filter of L. Moreover, the above congruence relation is a special case of the following congruence relations to be defined.

Let L be a lattice implication algebra, $A \in \mathscr{F}(L)$, $\alpha \in [0,1]$. A binary relation $\tilde{A}^\alpha$ on L is defined as follows:

$$\tilde{A}^\alpha = \{(x,y) \mid (x,y) \in L \times L, A(x \to y) > \alpha, A(y \to x) > \alpha\}. \tag{5.2}$$

Lemma 5.1.8 *Let A be a fuzzy filter of L, $\alpha \in [0,1)$. If $\tilde{A}^\alpha \neq \emptyset$, then $A(I) > \alpha$.*

Proof. If $\tilde{A}^\alpha \neq \emptyset$, then there exists $(x,y) \in L \times L$, such that $A(x \to y) > \alpha$. By Definition 3.6.1, we have $A(I) \geqslant A(x \to y) > \alpha$. $\square$

Corollary 5.1.6 *Let A be a fuzzy filter of L, $\alpha \in [0,1)$. If $\tilde{A}^\alpha \neq \emptyset$, then $\tilde{A}^\alpha$ is a reflexive relation on L.* $\square$

Lemma 5.1.9 *Let A be a fuzzy filter of L, $\alpha \in [0,1)$. If $\tilde{A}^\alpha \neq \emptyset$, then for any $(x,y) \in \tilde{A}^\alpha, z \in L$,*

$$A((x \to z) \to (y \to z)) > \alpha, \tag{5.3}$$
$$A((y \to z) \to (x \to z)) > \alpha, \tag{5.4}$$
$$A((z \to x) \to (z \to y)) > \alpha, \tag{5.5}$$
$$A((z \to y) \to (z \to x)) > \alpha. \tag{5.6}$$

Proof. Since

$$A((y \to x) \to ((x \to z) \to (y \to z))) = A(I),$$

by Lemma 5.1.8, we have

$$A((y \to x) \to ((x \to z) \to (y \to z))) > \alpha. \tag{5.7}$$

And since $(x, y) \in \tilde{A}^\alpha$, then $A(y \to x) > \alpha$. We have

$$\begin{aligned} &A((x \to z) \to (y \to z)) \\ \geqslant\ &\min\{A((y \to x) \to ((x \to z) \to (y \to z))), A(y \to x)\} \\ >\ &\alpha. \end{aligned}$$

which is Eq. (5.3).

Similarly, we can prove Eqs. (5.4), (5.5), and (5.6). □

Corollary 5.1.7 *Let A be a fuzzy filter of L, $\alpha \in [0, 1)$. If $\tilde{A}^\alpha \neq \emptyset$, then for any $(x, y) \in \tilde{A}^\alpha$, $z \in L$,*

$$\begin{aligned} &(x \to z, y \to z) \in \tilde{A}^\alpha, \\ &(z \to x, z \to y) \in \tilde{A}^\alpha. \quad \square \end{aligned}$$

Theorem 5.1.14 *Let A be a fuzzy filter of L, $\alpha \in [0, 1)$. If $\tilde{A}^\alpha \neq \emptyset$, then $\tilde{A}^\alpha$ is an equivalent relation on L.*

Proof. Corollary 5.1.6 shows the reflexivity of $\tilde{A}^\alpha$. It is trivial that $\tilde{A}^\alpha$ is symmetric. Hence, we need to prove the transitivity of $\tilde{A}^\alpha$. Let $(x, y), (y, z) \in \tilde{A}^\alpha$. By Lemma 5.1.9, we have

$$A((y \to z) \to (x \to z)) > \alpha.$$

Since $(y, z) \in \tilde{A}^\alpha$, then $A(y \to z) > \alpha$. And since A is a fuzzy filter of L, we have

$$\begin{aligned} &A(x \to z) \\ \geqslant\ &\min\{A((y \to z) \to (x \to z)), A(y \to z)\} \\ >\ &\alpha. \end{aligned}$$

Similarly, $A(z \to x) > \alpha$ can be proved. Therefore, $(x, z) \in \tilde{A}^\alpha$, which means $\tilde{A}^\alpha$ is transitive. □

Theorem 5.1.15 *Let A be a fuzzy filter of L, $\alpha \in [0, 1)$. If $\tilde{A}^\alpha \neq \emptyset$, then $\tilde{A}^\alpha$ is a congruence relation on L.*

Proof. For any $(x_1, y_1), (x_2, y_2) \in \tilde{A}^\alpha$, by Corollary 5.1.7, we can get

$$(x_1 \to x_2, y_1 \to x_2) \in \tilde{A}^\alpha, (y_1 \to x_2, y_1 \to y_2) \in \tilde{A}^\alpha.$$

And by the transitivity of $\tilde{A}^\alpha$, we have

$$(x_1 \to x_2, y_1 \to y_2) \in \tilde{A}^\alpha.$$

Hence, $\tilde{A}^\alpha$ is a congruence relation on L. □

Theorem 5.1.16 *Let A be a fuzzy filter of L, $\alpha \in [0,1)$. If $\alpha < A(I)$, then $\tilde{A}^{\alpha}$ is a congruence relation on L.*

Proof. If $\alpha < A(I)$, then for any $x \in L$, $A(x \to x) = A(I) > \alpha$, thus $(x,x) \in \tilde{A}^{\alpha}$, *i.e.*, $\tilde{A}^{\alpha} \neq \emptyset$. It follows from Theorem 5.1.15 that $\tilde{A}^{\alpha}$ is a congruence relation on L . □

Remark 9. Assume A is a fuzzy filter of L. If $\alpha = A(I)$, then $\tilde{A}^{\alpha} = \emptyset$. In this case, we can modify the definition of $\tilde{A}^{\alpha}$ as follows:

$$\tilde{A}^{\alpha} = \{(x,y) \mid (x,y) \in L \times L, A(x \to y) = A(I), A(y \to x) = A(I)\}. \quad (5.8)$$

Clearly, Eq. (5.8) is namely the congruence relation $\equiv_A$ defined in Eq. (5.1).

If $\alpha = A(I)$, then A_{α} at least contains the greatest element I. Thus A_{α} is a filter of L. In addition,

$$\begin{aligned}\tilde{A}^{\alpha} &= \{(x,y) \mid (x,y) \in L \times L, A(x \to y) = \alpha, A(y \to x) = \alpha\}\\ &= \{(x,y) \mid (x,y) \in L \times L, A(x \to y) \geqslant \alpha, A(y \to x) \geqslant \alpha\}\\ &= \{(x,y) \mid (x,y) \in L \times L\, x \to y \in A_{\alpha}, y \to x \in A_{\alpha}\},\end{aligned}$$

which shows that $\tilde{A}^{\alpha}$ is namely the congruence relation induced by the filter A_{α}.

Suppose A is a fuzzy filter of a lattice implication algebra L, according to the above discussion, when α ranges over the interval $[0, A(I)]$, $\tilde{A}^{\alpha}$ compose a family of congruence relations on L. Therefore, by Theorem 5.1.2, this family of congruence relations can induce a family of lattice implication quotient algebras of L. The following theorems discuss their properties and relations among them.

Theorem 5.1.17 *Let A be a fuzzy filter of a lattice implication algebra L, then for any $\alpha \in [0, A(I)]$, a mapping*

$$f_{\alpha}\colon\ L \longrightarrow L/_{\tilde{A}^{\alpha}}$$

satisfying

$$f_{\alpha}(x) = [x]_{\tilde{A}^{\alpha}}$$

is a lattice implication epimorphism. □

Theorem 5.1.18 *Let A be a fuzzy filter of a lattice implication algebra L, $\alpha, \beta \in [0, A(I)]$. If $\beta \leqslant \alpha$, then a mapping*

$$f_{\alpha,\beta}\colon\ L/_{\tilde{A}^{\alpha}} \longrightarrow L/_{\tilde{A}^{\beta}}$$

satisfying

$$f_{\alpha,\beta}([x]_{\tilde{A}^{\alpha}}) = [x]_{\tilde{A}^{\beta}}$$

is a lattice implication epimorphism. □

Theorem 5.1.19 *Let A and B be two fuzzy filters of a lattice implication algebra L, if $A \subseteq B$, then for any $\alpha \in [0, A(I)]$, a mapping*

$$f_\alpha \colon\ L/_{\tilde{A}^\alpha} \longrightarrow L/_{\tilde{B}^\alpha}$$

satisfying

$$f_\alpha([x]_{\tilde{A}^\alpha}) = [x]_{\tilde{B}^\alpha}$$

is a lattice implication epimorphism. □

5.1.4 Congruence Relations Induced by Fuzzy *LI*-ideals

The aim of this section is to construct the lattice implication quotient algebras induced by fuzzy LI-ideals and to study the relation between lattice implication quotient algebras and fuzzy LI-ideals.

Assume L is a lattice implication algebra. In [210], a congruence relation $\rhd_A$ induced by a fuzzy LI-ideal A of L is defined as follows:

$$x \rhd_A y \qquad \text{iff} \qquad A((x \to y)') > 0, A((y \to x)') > 0$$

for any $x, y \in L$. Now we generalize this relation and obtain a family of congruence relations induced by a fuzzy LI-ideal of L.

If A is a fuzzy LI-ideal of a lattice implication algebra L, we know that $A(O)$ is the maximum of A, *i.e.*,

$$A(O) = \max\{A(x) \mid x \in L\},$$

i.e., for any $x \in L$, $A(x) \in [0, A(O)]$.

Let L be a lattice implication algebra and A a fuzzy LI- ideal of L. For any $\alpha \in [0, A(O)]$, a binary relation $\tilde{A}^\alpha$ on L is defined as follows: when $\alpha < A(O)$,

$$\tilde{A}^\alpha = \{(x, y) \mid (x, y) \in L \times L, A((x \to y)') > \alpha, A((y \to x)') > \alpha\}, \quad (5.9)$$

when $\alpha = A(O)$,

$$\tilde{A}^\alpha = \{(x, y) \mid (x, y) \in L \times L, A((x \to y)') = \alpha, A((y \to x)') = \alpha\}. \quad (5.10)$$

Lemma 5.1.10 *Let A be a fuzzy LI-ideal of a lattice implication algebra L. For any $\alpha \in [0, A(O)]$, $\tilde{A}^\alpha \neq \emptyset$.*

Proof. For any $\alpha \in [0, A(O)]$, $(O, O) \in \tilde{A}^\alpha$ and hence $\tilde{A}^\alpha \neq \emptyset$.□

Lemma 5.1.11 *Let A be a fuzzy LI-ideal of a lattice implication algebra L. If $\alpha = A(O)$, then $\tilde{A}^\alpha$ is a congruence relation on L.*

Proof. By $\alpha = A(O)$, A_α is an LI-ideal of L. By Eq. (5.10), we have

$$\begin{aligned}\tilde{A}^\alpha &= \{(x,y) \mid (x,y) \in L \times L, A((x \to y)') = \alpha, A((y \to x)') = \alpha\} \\ &= \{(x,y) \mid (x,y) \in L \times L, A((x \to y)') \geqslant \alpha, A((y \to x)') \geqslant \alpha\} \\ &= \{(x,y) \mid (x,y) \in L \times L, (x \to y)' \in A_\alpha, (y \to x)' \in A_\alpha\}.\end{aligned}$$

Hence, when $\alpha = A(O)$, $\tilde{A}^\alpha$ is namely the congruence relation on L induced by the LI-ideal A_α. □

Lemma 5.1.12 *If A is a fuzzy LI-ideal of a lattice implication algebra L, then for any $\alpha \in [0, A(O))$, $\tilde{A}^\alpha$ is reflexive and symmetric.*

Proof. Since for any $x \in L$,

$$A((x \to x)') = A(I') = A(O) > \alpha,$$

then $(x,x) \in \tilde{A}^\alpha$, which shows that $\tilde{A}^\alpha$ is reflexive. The symmetry of $\tilde{A}^\alpha$ is obvious. □

Lemma 5.1.13 *Let A be a fuzzy LI-ideal of L, $\alpha \in [0, A(O))$. Then for any $(x,y) \in \tilde{A}^\alpha, z \in L$,*

$$A(((x \to z) \to (y \to z))') > \alpha, \tag{5.11}$$

$$A(((y \to z) \to (x \to z))') > \alpha, \tag{5.12}$$

$$A(((z \to x) \to (z \to y))') > \alpha, \tag{5.13}$$

$$A(((z \to y) \to (z \to x))') > \alpha. \tag{5.14}$$

Proof. For any $(x,y) \in \tilde{A}^\alpha$, $z \in L$,

$$\begin{aligned}&A((((x \to z) \to (y \to z))' \to (y \to x)')') \\ &= A(((y \to x) \to ((x \to z) \to (y \to z)))') \\ &= A(I') = A(O) \\ &> \alpha.\end{aligned}$$

Since $(x,y) \in \tilde{A}^\alpha$, then $A((y \to x)') > \alpha$. By Definition 4.2.1, we have

$$\begin{aligned}&A(((x \to z) \to (y \to z))') \\ &\geqslant \min\{A((((x \to z) \to (y \to z))' \to (y \to x)')'), A((y \to x)')\} \\ &> \alpha,\end{aligned}$$

which is Eq. (5.11).

Similarly, we can prove Eqs. (5.12), (5.13), and (5.14). □

Corollary 5.1.8 *Let A be a fuzzy LI-ideal of a lattice implication algebra L, $\alpha \in [0, A(O))$. Then for any $(x,y) \in \tilde{A}^\alpha, z \in L$,*

$$\begin{aligned}&(x \to z, y \to z) \in \tilde{A}^\alpha, \\ &(z \to x, z \to y) \in \tilde{A}^\alpha. \quad \square\end{aligned}$$

Theorem 5.1.20 *Let A be a fuzzy LI-ideal of a lattice implication algebra L, $\alpha \in [0, A(O))$. Then $\tilde{A}^\alpha$ is an equivalent relation on L.*

Proof. Lemma 5.1.12 shows the reflexivity and the symmetry of $\tilde{A}^\alpha$. Hence, we need to prove the transitivity of $\tilde{A}^\alpha$. Let $(x, y), (y, z) \in \tilde{A}^\alpha$. By Lemma 5.1.13, we have

$$\begin{aligned} & A(((x \to z)' \to (y \to z)')') \\ = & A(((y \to z) \to (x \to z))') \\ > & \alpha. \end{aligned}$$

Since $(y, z) \in \tilde{A}^\alpha$, then $A((y \to z)') > \alpha$. And since A is a fuzzy LI-ideal of L, then

$$\begin{aligned} & A((x \to z)') \\ \geqslant & \min\{A(((x \to z)' \to (y \to z)')'), A((y \to z)')\} \\ > & \alpha. \end{aligned}$$

Similarly, $A((z \to x)') > \alpha$ can be proved. Therefore, $(x, z) \in \tilde{A}^\alpha$, which means $\tilde{A}^\alpha$ is transitive. □

Theorem 5.1.21 *Let A be a fuzzy LI-ideal of a lattice implication algebra L. Then for any $\alpha \in [0, A(O)]$, $\tilde{A}^\alpha$ is a congruence relation on L.*

Proof. If $\alpha = A(O)$, by Lemma 5.1.11, we know $\tilde{A}^\alpha$ is a congruence relation on L.

For any $\alpha \in [0, A(O))$, $(x_1, y_1), (x_2, y_2) \in \tilde{A}^\alpha$, according to Corollary 5.1.8, we have

$$(x_1 \to x_2, y_1 \to x_2) \in \tilde{A}^\alpha,$$
$$(y_1 \to x_2, y_1 \to y_2) \in \tilde{A}^\alpha.$$

By the transitivity of $\tilde{A}^\alpha$, we have

$$(x_1 \to x_2, y_1 \to y_2) \in \tilde{A}^\alpha.$$

Hence, $\tilde{A}^\alpha$ is a congruence relation on L. □

Suppose A is a fuzzy LI-ideal of a lattice implication algebra L, according to the above discussion, when α ranges over the interval $[0, A(O)]$, $\tilde{A}^\alpha$ compose a family of congruence relations on L. Therefore, by Theorem 5.1.2, this family of congruence relations can induce a family of lattice implication quotient algebras of L. The following theorems discuss their properties and the relations among them.

Theorem 5.1.22 *Let A be a fuzzy LI-ideal of a lattice implication algebra L, then for any $\alpha \in [0, A(O)]$, a mapping*

$$f_\alpha\colon\ L \longrightarrow L/_{\tilde{A}^\alpha}$$

satisfying

$$f_\alpha(x) = [x]_{\tilde{A}^\alpha}$$

is a lattice implication epimorphism. □

Theorem 5.1.23 *Let A be a fuzzy LI-ideal of a lattice implication algebra L, $\alpha, \beta \in [0, A(O)]$. If $\beta \leqslant \alpha$, then a mapping*

$$f_{\alpha,\beta}\colon L/_{\tilde{A}^\alpha} \longrightarrow L/_{\tilde{A}^\beta}$$

satisfying

$$f_{\alpha,\beta}([x]_{\tilde{A}^\alpha}) = [x]_{\tilde{A}^\beta}$$

is a lattice implication epimorphism. □

Theorem 5.1.24 *Let A and B be two fuzzy LI-ideals of a lattice implication algebra L, if $A \subseteq B$, then for any $\alpha \in [0, A(O)]$, a mapping*

$$f_\alpha\colon\ L/_{\tilde{A}^\alpha} \longrightarrow L/_{\tilde{B}^\alpha}$$

satisfying

$$f_\alpha([x]_{\tilde{A}^\alpha}) = [x]_{\tilde{B}^\alpha}$$

is a lattice implication epimorphism. □

5.2 Proper Lattice Implication Algebras

A proper lattice implication algebra is a lattice implication algebra which is not a lattice H implication algebra. In this section, we discuss the characteristics of proper lattice implication algebras.

Definition 5.2.1 *Let L_1 be a lattice implication algebra and L_2 a lattice H implication algebra. A lattice implication epimorphism $f\colon\ L_1 \longrightarrow L_2$ is called a representation of L_1 on L_2; if f is a lattice implication isomorphism, then f is called an isomorphism representation of L_1 on L_2.*

Corollary 5.2.1 *Let L_1 be a lattice implication algebra and L_2 a lattice H implication algebra. If there exists an isomorphism representation of L_1 on L_2, then L_1 is a lattice H implication algebra.* □

Let L and L_1 be lattice implication algebras, $f: L \longrightarrow L_1$ is a lattice implication homomorphism, by Theorem 3.1.13, we know that D-ker(f) is a filter of L and we can define a congruence relation $\equiv_f$ on L by:

$$x \equiv_f y \quad \text{iff} \quad x \to y, y \to x \in \text{D-ker}(f).$$

This relation is called the congruence relation induced by f.

Let

$$f_1 : L \longrightarrow L_1,$$
$$f_2 : L \longrightarrow L_2$$

be representations of L on L_1 and L_2, respectively. If there exists a bijection

$$\varphi : L_1 \longrightarrow L_2$$

such that for any $a \in L$,

$$f_2(a) = \varphi(f_1(a)),$$

then f_1 and f_2 are called equivalent representations; if the congruence relations induced by f_1 and f_2 are equal, then f_1 and f_2 are called equivalent representations.

Theorem 5.2.1 *Let $f_1 : L \longrightarrow L_1$, $f_2 : L \longrightarrow L_2$ be two representations of L,*

(1). if f_1 and f_2 are equivalent representations, then f_1 and f_2 are equivalent representations;
(2). if f_1 and f_2 are equivalent representations, then $L_1 \cong L_2$.

Proof. (1). Let I_1 and I_2 be the greatest elements of L_1 and L_2, respectively. If f_1 and f_2 are equivalent representations, then there exists a bijection $\varphi : L_1 \longrightarrow L_2$, such that for any $a \in L$,

$$f_2(a) = \varphi(f_1(a)).$$

Suppose $\equiv_{f_1}$ and $\equiv_{f_2}$ are congruence relations induced by f_1 and f_2, respectively. For any $x, y \in L$, if $x \equiv_{f_1} y$, then $x \to y$, $y \to x \in$ D-ker(f_1), *i.e.*,

$$f_1(x) \to f_1(y) = f_1(x \to y) = I_1,$$
$$f_1(y) \to f_1(x) = f_1(y \to x) = I_1,$$

so $f_1(x) = f_1(y)$, and hence

$$\begin{aligned} f_2(x) &= \varphi(f_1(x)) \\ &= \varphi(f_1(y)) \\ &= f_2(y), \end{aligned}$$

and $f_2(x \to y) = f_2(y \to x) = I_2$, so $x \to y, y \to x \in \text{D-}\ker(f_2)$, *i.e.*, $x \equiv_{f_2} y$. Similarly, it can be proved that $x \equiv_{f_2} y$ implies $x \equiv_{f_1} y$. Therefore, f_1 and f_2 are equivalent representations.

(2). Because $f_1 : L \longrightarrow L_1$ is a lattice implication epimorphism, by Corollary 5.1.5,

$$L_1 \cong L/_{\text{D-}\ker(f_1)}.$$

Similarly, we have $L_2 \cong L/_{\text{D-}\ker(f_2)}$. If f_1 and f_2 are equivalent representations, then f_1 and f_2 induce the same congruence relations, and hence

$$\text{D-}\ker(f_1) = \text{D-}\ker(f_2),$$

so we have

$$L_1 \cong L/_{\text{D-}\ker(f_1)} = L/_{\text{D-}\ker(f_2)} \cong L_2. \quad \square$$

Definition 5.2.2 *Let f be a representation of a lattice implication algebra L on a subalgebra of a lattice H implication algebra $(P(X), \vee, \wedge, ', \to)$, where X is a nonempty set, $P(X)$ is the power set of X, and for any $A, B \in P(X)$,*

$$\begin{aligned} A \vee B &\triangleq A \cup B, \\ A \wedge B &\triangleq A \cap B, \\ A' &\triangleq A^c = X \backslash A, \\ A \to B &\triangleq A' \vee B. \end{aligned}$$

If f satisfies

(1). $\bigcap_{a \in L} f(a) = \emptyset$, $\bigcup_{a \in L} f(a) = X$;
(2). for any $x, y \in X$, if $x \neq y$, then there exists $a \in L$, such that $x \in f(a)$ and $y \notin f(a)$, or $x \notin f(a)$ and $y \in f(a)$,

then f is called a meaning representation of L.

Theorem 5.2.2 *Let L be a lattice implication algebra and*

$$\Omega_1 = \{J \mid J \text{ is a maximal implicative filter of } L\},$$

then L is a proper lattice implication algebra if and only if there exist x, $y \in L$, $x \neq y$ such that for any $J \in \Omega_1$, $J \cap \{x, y\} = \emptyset$ or $J \cap \{x, y\} = \{x, y\}$.

Proof. Suppose there exist $x, y \in L$, $x \neq y$ and for any $J \in \Omega_1$, $J \cap \{x, y\} = \emptyset$ or $J \cap \{x, y\} = \{x, y\}$. If L is not a proper lattice implication algebra, then L is a lattice H implication algebra, and hence L is a Boolean algebra. Since $x \neq y$, $x \wedge y < x \vee y$, by Theorem 3 of Section 7.2 in [189] there exists a prime dual ideal J of L such that $x \wedge y \notin J$ and $x \vee y \in J$, so there exists exactly one element of $\{x, y\}$ belonging to J, *i.e.*, $J \cap \{x, y\} = \{x\}$ or $J \cap \{x, y\} = \{y\}$. Since L is a lattice H implication algebra and J is an implicative filter of L, thus J is a maximal implicative filter, which is a contradiction. Hence, L is a proper lattice implication algebra.

Conversely, if L be a proper lattice implication algebra, suppose

$$\Omega_0 = \{P \mid P \text{ is a prime dual lattice ideal of } L\}.$$

We first prove that $\Omega_1 \subseteq \Omega_0$. For any $P \in \Omega_1$, P is a maximal implicative filter, so P is a dual lattice ideal of L. If $x \vee y \in P$ and $x \notin P$, since P is maximal implicative filter and by Theorem 3.2.1, we have $[P \cup \{x\}) = L$, *i.e.*, $\{z \mid z \in L, x \to z \in P\} = L$, so $x \to y \in P$ and hence $y \in P$ by $(x \to y) \to y = x \vee y \in P$. Therefore, P is a prime dual lattice ideal, *i.e.*,

$$\Omega_1 \subseteq \Omega_0.$$

Let

$$\varphi\colon\ L \longrightarrow P(\Omega_0)$$

satisfy: for any $a \in L$,

$$\varphi(a) = \{P \mid P \in \Omega_0, a \in P\}.$$

By Theorem 3.6.2 in [189], φ is a meaning representation of the lattice $(L, \vee, \wedge)$.

Let

$$f\colon\ L \longrightarrow P(\Omega_1)$$

satisfy: for any $a \in L$,

$$\begin{aligned} f(a) &= \varphi(a) \cap \Omega_1 \\ &= \{J \mid J \in \Omega_1, a \in J\}. \end{aligned}$$

Then for any $a, b \in L$,

(1).

$$\begin{aligned} f(a \vee b) &= \varphi(a \vee b) \cap \Omega_1 \\ &= (\varphi(a) \cup \varphi(b)) \cap \Omega_1 \\ &= (\varphi(a) \cap \Omega_1) \cup (\varphi(b) \cap \Omega_1) \\ &= f(a) \vee f(b). \end{aligned}$$

(2).

$$\begin{aligned} f(a \wedge b) &= \varphi(a \wedge b) \cap \Omega_1 \\ &= (\varphi(a) \cap \varphi(b)) \cap \Omega_1 \\ &= (\varphi(a) \cap \Omega_1) \cap (\varphi(b) \cap \Omega_1) \\ &= f(a) \wedge f(b). \end{aligned}$$

(3). For any $J \in f(a')$, we have $a' \in J$. If $a \in J$, it follows from $a' = a \to O \in J$ that $O \in J$, thus $J = L$, which is a contradiction with the fact that J is a proper filter of L. Therefore $a \notin J$, *i.e.*, $J \notin f(a)$, and hence

$$J \in (f(a))' = \Omega_1 - f(a),$$

$$f(a') \subseteq (f(a))'.$$

If $J \in (f(a))'$, then $J \notin f(a), a \notin J$. By Theorem 3.2.1, we have

$$[J \cup \{a\}) = L,$$

i.e.,

$$\{z \mid z \in L, a \to z \in J\} = L,$$

so $a \to O \in J$, *i.e.*, $a' \in J, J \in f(a')$, and hence $(f(a))' \subseteq f(a')$. So we have $f(a') = (f(a))'$.

(4). For any $J \in \Omega_1$, if $J \in f(b)$, then $b \in J$ and hence $a \to b \in J, J \in f(a \to b)$, *i.e.*, $f(b) \subseteq f(a \to b)$. Similarly, we have $(f(a))' \subseteq f(a \to b)$, so $(f(a))' \cup f(b) \subseteq f(a \to b)$.
Suppose $J \in f(a \to b)$, then $a \to b \in J$. If $J \notin f(b)$, then $b \notin J$ and hence $a \notin J, J \notin f(a), J \in (f(a))'$. So $f(a \to b) \subseteq (f(a))' \cup f(b)$ and we have proved

$$f(a \to b) = (f(a))' \cup f(b) = f(a) \to f(b).$$

(5). $f(O) = \{J \mid J \in \Omega_1, O \in J\} = \emptyset$.

By (1), (2), (3), (4), (5) and Theorem 2.3.5, $f(L)$ is a lattice implication subalgebra of $P(\Omega_1)$ and hence a lattice H implication algebra, so f is a representation of L on $f(L)$.

Since L is a proper lattice implication algebra, then $f\colon L \longrightarrow f(L)$ is not a lattice isomorphism, *i.e.*, there exist $x, y \in L$, such that $x \neq y, f(x) = f(y)$, so

$$\{J \mid J \in \Omega_1, x \in J\} = \{J \mid J \in \Omega_1, y \in J\}$$

and hence for any $J \in \Omega_1$, $x \in J$ if and only if $y \in J$, *i.e.*, $J \cap \{x, y\} = \emptyset$ or $J \cap \{x, y\} = \{x, y\}$. □

Corollary 5.2.2 *A lattice implication algebra L is a lattice H implication algebra if and only if for any $x, y \in L, x \neq y$, there exists $J \in \Omega_1$ such that $x \in J$ and $y \notin J$, or $x \notin J$ and $y \in J$, where Ω_1 is defined in Theorem 5.2.2.*
□

Corollary 5.2.3 *Let L be a lattice implication algebra, Ω_1 defined in Theorem 5.2.2 and $f\colon L \longrightarrow P(\Omega_1)$ satisfy for any $a \in L$,*

$$f(a) = \{J \mid J \in \Omega_1, a \in J\}.$$

If we define an equivalent relation $\equiv_f$ on L by:

$$x \equiv_f y \quad \textit{iff} \quad f(x) = f(y),$$

then

$$L/_{\equiv_f} \cong f(L). \quad \square$$

Theorem 5.2.3 *Let L be a lattice implication algebra and $f\colon\ L \longrightarrow P(\Omega_1)$ satisfy: for any $a \in L$,*

$$f(a) = \{J \mid J \in \Omega_1, a \in J\},$$

where Ω_1 is defined in Theorem 5.2.2, then f is a meaning representation of L.

Proof. By the proof of Theorem 5.2.2, f is clearly a representation of L. If $\bigcap_{a \in L} f(a) \neq \emptyset$, then there exists $J \in \Omega_1$, such that $J \in \bigcap_{a \in L} f(a)$, *i.e.*, for any $a \in L$, $J \in f(a), a \in J$ and hence $J = L$, this is a contradiction.

For any $J \in \Omega_1$, $J \neq \emptyset$, suppose $a \in J$, then $J \in f(a)$, and hence

$$J \in \bigcup_{a \in L} f(a), \bigcup_{a \in L} f(a) = \Omega_1.$$

If $J_1, J_2 \in \Omega_1$, $J_1 \neq J_2$, suppose $J_1 \backslash J_2 \neq \emptyset$, then there exists $a \in L$ such that $a \in J_1$ and $a \notin J_2$, *i.e.*, $J_1 \in f(a)$ and $J_2 \notin f(a)$.

Therefore, by Definition 5.2.2, f is a meaning representation of L. $\square$

5.3 Representations

In Section 5.2, we discussed the characteristics of proper lattice implication algebras using the concept of representation of lattice implication algebras. In this section, we further study the representation of lattice implication algebras. We discuss the direct representation of a class of complete and quasi-atomic lattice implication algebras, and prove that a complete and quasi-atomic lattice implication algebra is a direct product of finite Łukasiewicz chains.

Definition 5.3.1 *Let L be a lattice implication algebra. L is said to be complete if it is a complete lattice; L is said to be quasi-atomic if L is a quasi-atomic lattice, i.e., for any $y \in L$, $y \neq O$, there exists an atom $x \in L$ such that $x \leqslant y$.*

Lemma 5.3.1 *Let L be a lattice implication algebra, $x \in L$, then x is an atom of L if and only if x' is a dual atom of L.* $\square$

For the Łukasiewicz implication algebra L defined in Example 2.1.2, given any $n \in \mathbb{N}^+$, there exists a lattice implication subalgebra

$$\{0, 1/n, 2/n, \cdots, (n-1)/n, 1\},$$

denoted by $L(n)$, where

$$(i/n) \to (j/n) = \begin{cases} (n+j-i)/n, & i > j, \\ 1, & i \leqslant j. \end{cases}$$

In fact, $L(n)$ is a finite Łukasiewicz chain.

Definition 5.3.2 *Let L be a lattice implication algebra, $a \in L$. The smallest p, $p \in \mathbb{N}^+$, satisfying*

$$a^{p+1} = a^p$$

is said to be the order of a, denoted by $ord(a)$. If such p does not exist, then the order of a is said to be infinite and denoted by $ord(a) = \infty$.

Theorem 5.3.1 *Suppose L is a lattice implication algebra which is neither a chain nor a Boolean algebra. If L has a dual atom, then it contains at least two dual atoms, i.e., if it has the unique dual atom, then it is a chain.*

Proof. If this statement is not true, we suppose a is the unique dual atom of L and L is not a chain, then there exist two incomparable elements $b, c \in L$ and $d \in L$, such that

$$b \vee c = d \leqslant a$$

and $b < d, c < d$. Obviously, we know that $\{I\}$ is a prime filter of L. Note that

$$(b \to c) \vee (c \to b) = I \in \{I\}.$$

Thus, $b \to c = I$ or $c \to b = I$. We can assume $b \to c = I$, then

$$(b \to c) \to c = b \vee c = d. \tag{5.15}$$

But

$$(b \to c) \to c = I \to c = c.$$

This contradicts to Eq. (5.15). Hence, there does not exist incomparable elements in L, *i.e.*, L is a chain. □

Theorem 5.3.2 *If L is a finite chain and a lattice implication algebra, then it is a finite Łukasiewicz chain.*

Proof. Suppose $L = \{a_i \mid 1 \leqslant i \leqslant n\}$ and $O = a_1 < \cdots < a_n = I$.

Since $\prime\colon\ L \longrightarrow L$ is an order-reversing involution, we have

$$(a_i)' = a_{n+1-i}, \qquad 1 \leqslant i \leqslant n.$$

For any i, j, $1 \leqslant i, j \leqslant n$, we have

(1). if $i \leqslant j$, then $a_i \to a_j = I = a_n$;

(2). if $i > j$, then $a_i' \leqslant a_i \rightarrow a_j < a_n$, *i.e.*,

$$a_{n+1-i} \leqslant a_i \rightarrow a_j \leqslant a_{n-1}.$$

Moreover, for any $m_1, m_2 < i$, there holds

$$a_{m_1}' \rightarrow a_i' = a_{m_2}' \rightarrow a_i',$$

when $a_i \rightarrow a_{m_1} = a_i \rightarrow a_{m_2}$. Therefore,

$$\begin{aligned}(a_{m_1}' \rightarrow a_i') \rightarrow a_i' &= (a_{m_2}' \rightarrow a_i') \rightarrow a_i',\\ a_{m_1}' = a_{m_1}' \vee a_i' &= a_{m_2}' \vee a_i' = a_{m_2}'.\end{aligned}$$

Hence $a_{m_1} = a_{m_2}$. This means operation $a_i \rightarrow a_j$ is monotone increasing for a_j. Notice that there exactly exist $i-1$ elements from a_{n+1-i} to a_{n-1}, we have

$$\begin{aligned} a_i \rightarrow a_1 &= a_{n+1-i},\\ a_i \rightarrow a_2 &= a_{n+2-i},\\ \cdots \quad & \cdots,\\ a_i \rightarrow a_{i-1} &= a_{n+(i-1)-i} = a_{n-1}.\end{aligned}$$

Sum up, there exists a unique operation $\rightarrow$ and a unique $'$ on L, such that L is a lattice implication algebra. Since the Łkasiewicz implication algebra defined in Example. 2.1.2 is a lattice implication algebra, L is a finite Łukasiewicz chain. □

Corollary 5.3.1 *Let L be a lattice implication algebra, a the unique dual atom of L, and $ord(a) = p(finite)$.*

(1). For any $m, n \leqslant p$,

$$a^m \rightarrow a^n = \begin{cases} a^{n-m}, & m < n \leqslant p,\\ I, & n \leqslant m \leqslant p.\end{cases}$$

(2). $L = [a) = \{a^n \mid n = 0, 1, \cdots, p\}$.

Corollary 5.3.2 *Let L be a lattice implication algebra and a the unique dual atom of L. If there exists $p \in \mathbb{N}$, such that $a^p = O$, then $L \cong L(p)$.*

Proof. It follows from Corollary 5.3.1 that $L = [a) = \{a^n \mid n = 0, 1, \cdots, p\}$ and

$$a^k \rightarrow a^h = \begin{cases} a^{h-k}, & h > k,\\ 1, & h \leqslant k.\end{cases}$$

Then we can define a function $f\colon\ L \longrightarrow L(p)$ satisfying:

$$f(a^i) = (p-i)/p \ (O \leqslant i \leqslant p).$$

It is easily verified that f is an isomorphism from L to $L(p)$. □

As what follows, by using the concepts of universal algebra [42] and lattice theory [26], we introduce the concept of subdirect product and direct product decompositions of lattice implication algebras.

Definition 5.3.3 *Let* $A = (A, \vee, \wedge, ', \rightarrow, O, I)$ *be the product algebra of a family of lattice implication algebras* $(L_\alpha, \vee_\alpha, \wedge_\alpha, '_\alpha, \rightarrow_\alpha, O_\alpha, I_\alpha)$ $(\alpha \in J)$*. Then a lattice implication subalgebra* L *of* A *is called a subdirect product of* $L_\alpha(\alpha \in J)$*, if for any* $\alpha \in J$*,* $a \in L_\alpha$*, there exists* $f \in L$*, such that* $f(\alpha) = a$*.*

If a lattice implication algebra L *is a subdirect product of a family of lattice implication algebras, then* L *is called subdirectly representable, or has a subdirect product decomposition. If a lattice implication algebra* L *is a direct product of a family of lattice implication algebras, then* L *is called directly representable or has a direct product decomposition.*

Lemma 5.3.2 *Let* L *be a countably complete lattice implication algebra chain with atoms,* a *the dual atom of* L*. Then there exists* $p \in \mathbb{N}$*, such that* $a^p = O$*.*

Proof. Assume that the conclusion does not hold, *i.e.*, for any $n \in \mathbb{N}$, $a^n > O$. Furthermore, we can obtain that $a^n > a'$ for any $n \in \mathbb{N}$. Otherwise, since L is a chain, there exists $m \in \mathbb{N}$, $a^m \leqslant a'$, *i.e.*, $a^m \leqslant a \rightarrow O$, then $a^{m+1} = O$. this contradicts to the assumption.

Since L is countably complete, we set

$$s = \bigwedge\{a^n \mid n \in \mathbb{N}\}.$$

Then for any $n \in \mathbb{N}$, $a^{n+1} \geqslant s$, it follows from the monotonicity of $\rightarrow$ that

$$a \rightarrow a^{n+1} \geqslant a \rightarrow s.$$

But

$$\begin{aligned} a \rightarrow a^{n+1} &= a \rightarrow (a^n \otimes a) \\ &= a' \vee a^n \\ &= a^n, \end{aligned}$$

hence for any $n \in \mathbb{N}$, $a^n \geqslant a \rightarrow s$. Furthermore, from the definition of s, we have $s \geqslant a \rightarrow s$, so $a \vee s = I$, and since $a \geqslant s$, then $a = I$, which is a contradiction to the fact that a is a dual atom. □

Theorem 5.3.3 *A complete and quasi-atomic lattice implication algebra* L *is a direct product of finite Łukasiewicz chains.*

Proof. Assume that $Dat(L)$ is the set of dual atoms of L. By (2) of Corollary 3.4.3, we know that there exists a prime filter P_a for any $a \in Dat(L)$, such that $a \notin P_a$. Hence, if $b \in Dat(L) \setminus \{a\}$, then $a \vee b = I \in P_a$. Since P_a is prime, $b \in P_a$, thus $Dat(L) \setminus \{a\} \subseteq P_a$. We can construct a lattice implication quotient algebra $L/_{P_a}$. It follows from Theorem 5.1.7 that $L/_{P_a}$ is a totally ordered lattice implication quotient algebra, and $[a]_{P_a}$ is a dual atom of $L/_{P_a}$. In fact, let $x \in L$, if we assume that

$$[a]_{P_a} \leqslant [x]_{P_a} < [I]_{P_a},$$

then by the definition of lattice implication quotient algebra, we have

$$[x]_{P_a} \cup [a]_{P_a} = [a \vee x]_{P_a}.$$

Since a is a dual atom, then $a \vee x = I$ or $a \vee x = a$. Obviously, $a \vee x \neq I$, so we have $a \vee x = a$, *i.e.*, $a \geqslant x$, then

$$[a]_{P_a} \geqslant [x]_{P_a},$$

thus,

$$[a]_{P_a} = [x]_{P_a},$$

hence $[a]_{P_a}$ is a dual atom of $L/_{P_a}$. Furthermore, $[a]_{P_a}$ is the unique dual atom of $L/_{P_a}$.

Since L/P_a is countably complete, then by Lemma 5.3.2, there exists $n(a) \in \mathbb{N}$, such that $([a]_{P_a})^{n(a)} = O$. Suppose that n_a is the smallest $n(a)$ such that $([a]_{P_a})^{n(a)} = O$.

By Corollary 5.3.2, we have

$$L/_{P_a} \cong L(n_a),$$

and by Corollary 5.3.1,

$$L/P_a = [[a]_{P_a}) = \{([a]_{P_a})^m | m = 0, 1, \cdots, n_a\}.$$

By the proof of Corollary 5.3.2, we can give an isomorphism $\varphi_a: L/_{P_a} \longrightarrow L(n_a)$ and a mapping $S_a: L \longrightarrow \mathbb{N}$ as follows: for any $x \in L$, $[x]_{P_a} \in L/P_a$, if $[x]_{P_a} = ([a]_{P_a})^{m_x}$, $m_x \in \{0, 1, \cdots, n_a\}$, then we define

$$\varphi_a([x]_{P_a}) = S_a(x)/n_a = (n_a - m_x)/n_a,$$

$$S_a(x) = n_a - m_x.$$

Obviously, $\varphi_a([a]_{P_a}) = (n_a - 1)/n_a$, $S_a(a) = n_a - 1$ and $S_a(I) = n_a$.

Let a mapping

$$\varphi\colon L \longrightarrow \prod\{L(n_a) \mid a \in Dat(L)\}$$

be given by the following equality:

$$\varphi(x)(a) = \varphi_a([x]_{P_a}) = S_a(x)/_{n_a} = (n_a - m_x)/_{n_a}, \tag{5.16}$$

where $x \in L$, for any $a \in Dat(L)$.

It is easily proved that

$$\begin{aligned} m_{x\vee y} &= \min\{m_x, m_y\}, \\ m_{x\wedge y} &= \max\{m_x, m_y\}, \\ m_{x'} &= n_a - m_x, \\ m_{x\to y} &= \begin{cases} n_a + m_x - m_y, & m_x \geqslant m_y, \\ 1, & m_x < m_y. \end{cases} \end{aligned}$$

First, we prove that φ is a lattice implication homomorphism. In fact, for any $x, y \in L$, $a \in Dat(L)$,

$$\begin{aligned} \varphi(x \vee y)(a) &= \varphi_a([x \vee y]_{P_a}) \\ &= S_a(x \vee y)/n_a \\ &= \max\{S_a(x)/n_a, S_a(y)/n_a\} \\ &= \max\{\varphi(x)(a), \varphi(y)(a)\} \\ &= (\varphi(x) \vee \varphi(y))(a), \end{aligned}$$

hence, $\varphi(x\vee y) = \varphi(x)\vee\varphi(y)$. Similarly, we can prove $\varphi(x\wedge y) = \varphi(x)\wedge\varphi(y)$, $\varphi(x') = (\varphi(x))'$, $\varphi(x \to y) = \varphi(x) \to \varphi(y)$. So, φ is a lattice implication homomorphism.

Second, we prove that φ is injective. In fact, let $x_1, x_2 \in L$ and $x_1 \neq x_2$. Note that

$$\bigcap\{P_a \mid a \in Dat(L)\} = \{I\}.$$

Otherwise, there exists $x \neq I$, $x \in \cap\{P_a \mid a \in Dat(L)\}$. Since L is quasi-atomic, there exists $b \in Dat(L)$, $x \leqslant b$. Then $b \in P_b$, this is a contradiction. Hence, there exists $c \in Dat(L)$, $[x_1]_{P_c} \neq [x_2]_{P_c}$. Then we have

$$\begin{aligned} \varphi(x_1)(c) &= \varphi_c([x_1]_{P_c}) \\ &\neq \varphi_c([x_2]_{P_c}) \\ &= \varphi(x_2)(c). \end{aligned}$$

So $\varphi(x_1) \neq \varphi(x_2)$. Hence φ is a monomorphism.

Finally, we prove that φ is surjective. In fact, let $f \in \prod\{L(n_a) \mid a \in Dat(L)\}$, then for any $a \in Dat(L)$, there exists s_a such that $0 \leqslant s_a \leqslant n_a$ and $f(a) = s_a/_{n_a}$. Take $x_{f(a)} = a^{n_a - S_a}$. Note that for any $a \in Dat(L)$,

$$\varphi(x_{f(a)})(b) = \begin{cases} f(a), & b = a, \\ 1, & b \neq a, \end{cases}$$

where $b \in Dat(L)$.

In fact, for all $a \in Dat(L)$, for $b(\neq a)$, there exists a prime filter P_b such that $b \notin P_a$. Since $a \vee b = I \in P_b$, we have $a \in P_b$, hence $[a]_{P_b} = [I]_{P_b}$. And for any $n \in \mathbb{N}$, $a^n \in P_b$, so $[a^n]_{P_b} = [I]_{P_b}$, furthermore, we have $[a^{n_a - S_a}]_{P_b} = [I]_{P_b}$, *i.e.*, $\varphi(x_{f(a)})(b)$. If $b = a$, then

$$\begin{aligned}\varphi(x_{f(a)})(b) &= \varphi(x_{f(a)})(a)\\ &= \varphi_a([a^{n_a-s_a}]_{P_a})\\ &= (n_a-(n_a-s_a))/n_a\\ &= s_a/n_a\\ &= f(a).\end{aligned}$$

Now let

$$x = \bigwedge\{x_{f(a)} \mid a \in Dat(L)\}.$$

If $a \neq b$, then $a \vee x_{f(b)} = I$. In fact, $a \neq b$ implies $a \vee b = I$, *i.e.*, $b \in P_a$, and then $b^{n_b-s_b} \in P_a$, *i.e.*, $x_{f(b)} \in P_a$, hence $a \geqslant x_{f(b)}$ does not hold (otherwise, $a \in P_a$, this is a contradiction).

Let

$$m_a = \bigwedge\{x_{f(b)} \mid b \neq a, b \in Dat(L)\}.$$

It follows from the above remarks that $a \vee m_a = I$, which means $m_a \in P_a$. Consequently, for any $a \in Dat(L)$, we have

$$\begin{aligned}\varphi(x)(a) &= \varphi(x_{f(a)} \wedge m_a)(a)\\ &= \varphi(x_{f(a)})(a)\\ &= f(a),\end{aligned}$$

which shows that $\varphi(x) = f$ and means that φ is surjective.

Sum up, φ is a lattice implication isomorphism. Consequently, L is a direct product of a family of algebras $\{L(n_a) \mid a \in Dat(L)\}$, *i.e.*, L is a direct product of finite Łukasiewicz chains. □

Corollary 5.3.3 *A finite lattice implication algebra is a direct product of finite Łukasiewicz chains.* □

Theorem 5.3.4 *If L is a complete and quasi-atomic lattice implication algebra, $x \in L$, then*

$$x = \wedge\{a^{n_a-S_a(x)} \mid a \in Dat(L)\},$$

where $n_a = ord(a)$, $S_a(x) = n_a - m_x$, $m_x \in \{0, 1, \cdots, n_a\}$ such that $[x]_{P_a} = ([a]_{P_a})^{m_x}$, P_a is a prime filter of L not containing a.

Proof. For any $a \in Dat(L)$, let φ_a be an isomorphism in the proof of Theorem 5.3.3, *i.e.*, for any $x \in L$,

$$\varphi(x)(a) = \varphi_a([x]_{P_a}) = S_a(x)/_{n_a}.$$

Analogously to Theorem 5.3.3, we can prove that

$$\varphi(a^{n_a-S_a(x)})(b) = \begin{cases} S_a(x)/_{n_a}, & b = a,\\ 1, & b \neq a,\end{cases}$$

where $b \in Dat(L)$.

Furthermore, we know that $([a]_{P_a})^n = [a^n]_{P_a}$. Then

$$\begin{aligned}\varphi(a^{n_a - S_a(x)})(a) &= \varphi_a([a^{n_a - S_a(x)}]_{P_a}) \\ &= \varphi_a(([a]_{P_a})^{n_a - S_a(x)}) \\ &= [\varphi_a([a]_{P_a})]^{n_a - S_a(x)} \\ &= S_a(x)/_{n_a}.\end{aligned}$$

Therefore, for any $a \in Dat(L)$,

$$\varphi(a^{n_a - S_a(x)}) \geqslant \varphi(x).$$

Since φ is a lattice implication isomorphism, we have

$$a^{n_a - S_a(x)} \geqslant x$$

for any $a \in Dat(L)$. Suppose now that $y \in L$ is a lower bound of the set

$$\{a^{n_a - S_a(x)} \mid a \in Dat(L)\}.$$

It follows that

$$\varphi(x)(a) = \varphi(a^{n_a - S_a(x)})(a) \geqslant \varphi(y)(a)$$

for any $a \in Dat(L)$. Hence $\varphi(x) \geqslant \varphi(y)$ and then $x \geqslant y$. This shows that

$$x = \wedge\{a^{n_a - S_a(x)} \mid a \in Dat(L)\}. \quad \square$$

6 Topological Structure of Filter Spaces

In chapter 3, filters of lattice implication algebras have been studied from algebraic viewpoint. In this chapter, a topological space based on filters of a lattice implication algebra is constructed, and its topological properties, such as separability, compactness and connectedness are discussed. In the sequel, the product topology and quotient topology for filter spaces are studied. Finally, the prime spaces for lattice implication algebras and topological lattices generated by filters are presented.

6.1 Filter Spaces

If no speciality, L denotes a lattice implication algebra and $F(L)$ denotes the set of all the filters of L.

6.1.1 Basic Concepts

Since for any $F_1, F_2 \in F(L)$, $F_1 \cap F_2 \in F(L)$ and $\bigcup\{F \mid F \in F(L)\} = L$, $F(L)$ can be regarded as a topological base for L (taking $\emptyset$ as a filter).

Definition 6.1.1 *The topology generated by the base $F(L)$ is called the filter topology for L and denoted by $T_{F(L)}$, the topological space $(L, T_{F(L)})$ is called the filter space for L.*

Lemma 6.1.1 *For any $x \in L$, define*

$$\mathcal{U}_x = \{F \mid x \in F, F \in F(L)\},$$

then $\mathcal{U}_x$ is a neighborhood base of x w.r.t. the filter topology $T_{F(L)}$. □

Lemma 6.1.2 *For any $x \in L$, if $U, V \in \mathcal{U}_x$, then there exists $W \in \mathcal{U}_x$, such that $W \subseteq U \cap V$.* □

Note that for any topological space $(X, \mathcal{T})$, if every element x of X has a neighborhood base, then the union of all these neighborhood bases becomes a topological base for X. By Lemma 6.1.1, for any $x \in L$, it has a neighborhood base, so the set

$$\mathcal{B} = \bigcup_{x \in L} \mathcal{U}_x$$

is a topological base for L, the topology generated by it is said to be the one generated by the neighborhood base whose elements are $\mathcal{U}_x$, $x \in L$.

Since the topology generated by neighborhood bases is unique, by Lemmas 6.1.1 and 6.1.2, we know that the topology generated by $F(L)$ is identical with the one generated by $\mathcal{B}$. In fact, $\mathcal{B} = F(L)$.

Example 6.1.1 *Let $L = ([0,1], \vee, \wedge, ', \rightarrow)$ be the Łukasiewicz implication algebra defined in Example 2.1.2. It has been shown in Example 3.1.2 that*

$$F(L) = \{\{1\}, [0,1]\}.$$

So, we have

$$T_{F(L)} = \{\emptyset, \{1\}, [0,1]\}.$$

Example 6.1.2 *Let $L = \{O, a, b, I\}$ be a lattice implication algebra defined in Example 2.1.6. Note that*

$$F(L) = \{\{I\}, \{a, I\}, \{b, I\}, L\}.$$

Hence,

$$T_{F(L)} = \{\emptyset, \{a, b, I\}\} \cup F(L).$$

Now we discuss derived sets, closed sets, closures and interiors in the topological space $(L, T_{F(L)})$.

Theorem 6.1.1 *In the topological space $(L, T_{F(L)})$, for any $x \in L$, $[x)$ is the smallest element of $(N(x), \subseteq)$, where $N(x)$ is the neighborhood system of x.* □

In the topological space $(L, T_{F(L)})$, for any $A \subseteq L$, denote its derived set, closure and interior as $d(A)$, $c(A)$ and $i(A)$, respectively.

Theorem 6.1.2 *In the topological space $(L, T_{F(L)})$, for any $A \subseteq L$,*

$$d(A) = \{x \mid x \in L, [x) \cap (A \setminus \{x\}) \neq \emptyset\}.$$

Proof. Note that $d(A)$ consists of all the accumulation points of A, and x is an accumulation point of A if and only if for any $U \in N(x)$,

$$U \cap (A \setminus \{x\}) \neq \emptyset.$$

Then by Theorem 6.1.1, we can easily get the conclusion. □

We can get the following corollary from Theorem 6.1.2.

Corollary 6.1.1 *In the topological space $(L, T_{F(L)})$,*

(1). for any $A \subseteq L$, $I \notin d(A)$;

(2). for any $A \in F(L)$, if $x \in A$ and $x \neq I$, then $x \in d(A)$;
(3). for any $A \subseteq L$, if $I \in A$, then I is an isolated point of A. □

Theorem 6.1.3 *In the topological space $(L, T_{F(L)})$, for any $A \subseteq L$, A is a closed set if and only if for any $x \notin A$, $[x) \cap A = \emptyset$.*

Proof. Note that for any $A \subseteq L$, A is closed if and only if $d(A) \subseteq A$. Now we assume A is closed and there exists $x_0 \notin A$, such that

$$[x_0) \cap A \neq \emptyset,$$

then,

$$[x_0) \cap (A \setminus \{x_0\}) \neq \emptyset,$$

i.e., $x_0 \in d(A)$, which contradicts to $d(A) \subseteq A$.

Conversely, for any $x \in d(A)$, by Theorem 6.1.2,

$$[x) \cap (A \setminus \{x\}) \neq \emptyset.$$

If $x \notin A$, then

$$[x) \cap A = [x) \cap (A \setminus \{x\}) \neq \emptyset,$$

which contradicts to the fact that $[x) \cap A = \emptyset$ holds for any $x \notin A$. Hence, $x \in A$, $d(A) \subseteq A$, and thus, A is closed. □

Corollary 6.1.2 *If $I \in A$, $A \subset L$, then A is not a closed set.*

Proof. By $A \subset L$, it follows that there exists $x \in L$ such that $x \notin A$ and $[x) \cap A \neq \emptyset$, then A is not a closed set. □

Lemma 6.1.3 *For any $F \in F(L)$, $F = \bigcup_{x \in F} [x)$.* □

Theorem 6.1.4 *The set $\{[x) \mid x \in L\}$ is a base for the filter topological space $(L, T_{F(L)})$.*

Proof. For any $F \in F(L)$, by Lemma 6.1.3,

$$F = \bigcup_{x \in F} [x),$$

i.e., F is the union of some elements of the set $\{[x) \mid x \in L\}$. And by Theorem 3.2.6, $\{[x) \mid x \in L\}$ is closed w.r.t. the intersection. Hence, $\{[x) \mid x \in L\}$ is a base for $(L, T_{F(L)})$. □

Theorem 6.1.5 *In the topological space $(L, T_{F(L)})$, for any $A \subseteq L$,*

$$c(A) = \cap\{L \setminus [x) \mid x \in L, [x) \cap A = \emptyset\},$$
$$i(A) = \cup\{[x) \mid x \in L, [x) \subseteq A\}.$$

Proof. By Theorem 6.1.4, we can get

$$\begin{aligned} c(A) &= \cap\{B \mid A \subseteq B \subseteq L, B \text{ is closed}\} \\ &= \cap\{L \setminus F \mid F \in T_{F(L)}, A \cap F = \emptyset\} \\ &= \cap\{L \setminus F \mid F \in F(L), A \cap F = \emptyset\} \\ &= \cap\{L \setminus [x) \mid x \in L, [x) \cap A = \emptyset\}. \end{aligned}$$

And

$$\begin{aligned} i(A) &= \cup\{F \mid F \subseteq A, F \in T_{F(L)})\} \\ &= \cup\{F \mid F \subseteq A, F \in F(L)\} \\ &= \cup\{[x) \mid x \in L, [x) \subseteq A\}. \ \square \end{aligned}$$

Corollary 6.1.3 *In the topological space* $(L, T_{F(L)})$, *for any* $A \subseteq L$,

(1). if $I \in A$, *then* $c(A) = L$;
(2). if $I \notin A$, *then* $i(A) = \emptyset$.

6.1.2 Topological Properties

The first countability, the second countability, T_0-separability, T_1-separability, T_2-separability, and the connectedness of topological spaces are all topological properties, *i.e.*, topological invariance. Now we discuss these properties of filter spaces for lattice implication algebras.

Theorem 6.1.6 $(L, T_{F(L)})$ *is a topological space satisfying the first axiom of countability.*

Proof. Since for any $x \in L$, $\{[x)\}$ is a neighborhood base of x, so $(L, T_{F(L)})$ satisfies the first axiom of countability. $\square$

For the second countability, we have the following theorem

Theorem 6.1.7 $(L, T_{F(L)})$ *satisfies the second axiom of countability if and only if the set* $\{[x) \mid x \in L\}$ *is countable.*

Proof. By Theorem 6.1.4, the sufficiency is trivial. In what follows, we prove the necessity of this theorem. If $(L, T_{F(L)})$ satisfies the second axiom of countability, then there exists a countable base in $(L, T_{F(L)})$ denoted as

$$\mathcal{B} = \{B_1, \cdots, B_n, \cdots\}.$$

For any $x \in L$, since

$$[x) \in F(L) \subseteq T_{F(L)},$$

then there exist $B_k \in \mathcal{B}(k \in \Gamma)$, such that

$$[x) = \bigcup_{k \in \Gamma} B_k,$$

which implies that there exists $k_0 \in \Gamma$ such that $x \in B_{k_0} \subseteq [x)$, and hence $[x) = B_{k_0}$. Therefore,

$$|\{[x) \mid x \in L\}| \leqslant |\mathcal{B}|,$$

it follows that $\{[x) \mid x \in L\}$ is countable. □

A filter space for a lattice implication algebra which does not satisfy the second axiom of countability is given in the following example.

Example 6.1.3 *Suppose X is a uncountable set. Note that its power set lattice $P(X)$ is a Boolean lattice. By Example 2.1.1, for any $A, B \subseteq X$, if we define*

$$A \to B = A' \cup B,$$

then $(P(X), \cap, \cup, ', \to)$ is a lattice implication algebra, where $A' = X \setminus A$. Furthermore, by Corollary 2.2.2, we know that it is a lattice H implication algebra. Now we show that it does not satisfy the second axiom of countability.

Firstly, we prove that: for any $x, y \in X$, $x \neq y$ implies $[\{x\}) \neq [\{y\})$. In fact, if $[\{x\}) = [\{y\})$, then $\{y\} \in [\{x\})$, and since $[\{x\})$ is a filter and $\{x\} \in [\{x\})$, we have

$$\emptyset = \{x\} \cap \{y\} \in [\{x\}).$$

Hence, by Corollary 3.2.2, there exists $n \in \mathbb{N}$, such that

$$\{x\}^n \subseteq \{x\} \cap \{y\}.$$

By Corollary 2.2.3, $\{x\}^n = \{x\}$. Thus, $\{x\} \subseteq \emptyset$, which is a contradiction. Hence, $[\{x\}) \neq [\{y\})$. Since X is uncountable, the set $\{[\{x\}) \mid x \in X\}$ is uncountable. It follows from Theorem 6.1.7 that $(P(X), \cap, \cup, ', \to)$ does not satisfy the second axiom of countability.

The T_0-separability of the filter space is closely related to a lattice H implication algebra, which can be seen from the following theorem.

Theorem 6.1.8 *L is a lattice H implication algebra if and only if $(L, T_{F(L)})$ is a T_0-space.*

Proof. Suppose that L is a lattice H implication algebra. By Theorem 6.1.1, for any $x \in L$, $[x)$ is the smallest element in $(N(x), \subseteq)$, hence, in order to prove that $(L, T_{F(L)})$ is a T_0-space, we need to prove: for any $x, y \in L$, $x \in [y)$, $y \in [x)$ implies $x = y$. In fact, assume that $x \in [y)$, then by Corollary 3.2.2, there exists $n_1 \in \mathbb{N}$, such that $x \geqslant y^{n_1}$. Similarly, if $y \in [x)$, then by Corollary 3.2.2, there exists $n_2 \in \mathbb{N}$, such that $y \geqslant x^{n_2}$. Since L is lattice H implication algebra, then by Corollary 2.2.3, $y^{n_1} = y$, $x^{n_2} = x$, therefore, $x \geqslant y$, $y \geqslant x$, which implies $x = y$. Hence, $(L, T_{F(L)})$ is a T_0-space.

Conversely, suppose that $(L, T_{F(L)})$ is a T_0-space and L is not a lattice H implication algebra, then by Lemma 2.2.1, there exists $x \in L$, such that $x \neq x \otimes x$. However, $x \in [x \otimes x)$, $x \otimes x \in [x)$, which contradicts to the fact that $(L, T_{F(L)})$ is a T_0-space. Hence L is a lattice H implication algebra. □

Theorem 6.1.9 *$(L, T_{F(L)})$ is neither a T_1-space nor a T_2-space whenever $|L| \geqslant 2$.*

Proof. Suppose $|L| \geqslant 2$. In the filter space $(L, T_{F(L)})$, each non-empty open set contains the greatest element I, so $L \setminus \{I\}$ is not open, which implies that $\{I\}$ is not a closed set. Note that a topological space is a T_1-space if and only if its every singleton is closed. Hence, $(L, T_{F(L)})$ is not a T_1-space, furthermore, it is not a T_2-space. □

Theorem 6.1.10 *If $|L| \geqslant 2$, then $(L, T_{F(L)})$ is neither a regular space nor a normal space.*

Proof. By $|L| \geqslant 2$, we know that each non-empty open set of $(L, T_{F(L)})$ contains the greatest element I, so $(L, T_{F(L)})$ is neither a regular space nor a normal space. □

Theorem 6.1.11 *The following statements are equivalent:*

(1). $|L| = 1$;
(2). $(L, T_{F(L)})$ *is a trivial space;*
(3). $(L, T_{F(L)})$ *is a discrete space;*
(4). $(L, T_{F(L)})$ *is a T_2-space;*
(5). $(L, T_{F(L)})$ *is a T_1-space.*

Proof. By Theorem 6.1.9, it is easily proved (1) is equivalent to (4) and (5), respectively, so (1), (4) and (5) are equivalent each other. In addition, it is trivial that (1) implies (2) and (3). Now we need to prove (2) implies (1) and (3) implies (1).

Assume (2), $(L, T_{F(L)})$ is a trivial space, then

$$T_{F(L)} = \{\emptyset, L\}.$$

But $\{I\} \in F(L) \subseteq T_{F(L)}$, so $L = \{I\}$, *i.e.*, $|L| = 1$.

Assume (3), $(L, T_{F(L)})$ is a discrete space, then $T_{F(L)} = P(L)$, which is the power set of L, so $\{O\} \in T_{F(L)}$. But for any $F \in F(L)$, $O \in F$ if and only if $F = L$, hence $L = \{O\}$, *i.e.*, $|L| = 1$. □

Theorem 6.1.12 *$(L, T_{F(L)})$ is a compact and connected space.*

Proof. First we prove $(L, T_{F(L)})$ is compact. In fact, suppose $\{F_i \mid i \in \Gamma\}$ is any open covering of L and $F_i \in F(L), i \in \Gamma$, then

$$O \in L = \bigcup_{i \in \Gamma} F_i,$$

and therefore, there exists $i_0 \in \Gamma$, such that $O \in F_{i_0}$, hence $F_{i_0} = L$. It can be seen that there exists a finite covering of L in $\{F_i \mid i \in \Gamma\}$, it shows that $(L, \mathcal{T}_{F(L)})$ is compact.

Clearly, the clopen sets in $(L, \mathcal{T}_{F(L)})$ are only $\emptyset$ and L, hence $(L, \mathcal{T}_{F(L)})$ is a connected space. □

Theorem 6.1.13 *If* $|L| \geqslant 2$, *then* $(L, \mathcal{T}_{F(L)})$ *is not a zero-dimensional space.*

Proof. Assume $|L| \geqslant 2$. Since the clopen sets in $(L, \mathcal{T}_{F(L)})$ are only $\emptyset$ and L, obviously, $(L, \mathcal{T}_{F(L)})$ is not a zero-dimensional space. □

6.2 Product Topology and Quotient Topology

The purpose of this section is to investigate the structure of product topology and quotient topology for filter spaces, and the continuity of lattice implication homomorphisms in filter spaces.

In this section, if no speciality, let L_1 and L_2 be lattice implication algebras and $L_1 \times L_2$ denotes their product algebra

$$(L_1 \times L_2, \vee, \wedge, ', \rightarrow, (O_1, O_2), (I_1, I_2)),$$

and p_1 and p_2 be the mappings defined in Lemma 5.1.1. By Lemma 5.1.1, p_1 and p_2 are lattice implication homomorphisms from $L_1 \times L_2$ to L_1 and L_2, respectively.

Let

$$C_F(L_1, L_2) = \{f \mid f \text{ is a continuous mapping from } \mathcal{T}_{F(L_1)} \text{ to } \mathcal{T}_{F(L_2)}\}.$$

Then we have the following theorem.

Theorem 6.2.1 $\text{hom}(L_1, L_2) \subseteq C_F(L_1, L_2)$.

Proof. If $f \in \text{hom}(L_1, L_2)$, then for any $F \in F(L_2)$, by Theorem 3.1.12, $f^{-1}(F) \in F(L_1)$. □

There are two topological structures on lattice implication product algebras $L_1 \times L_2$, the one is the filter topology $\mathcal{T}_{F(L_1 \times L_2)}$ which is a topology generated by the base $F(L_1 \times L_2)$, and the other one is the product topology

$$\mathcal{T} = \mathcal{T}_{F(L_1)} \times \mathcal{T}_{F(L_2)}$$

of the filter spaces $(L_1, \mathcal{T}_{F(L_1)})$ and $(L_2, \mathcal{T}_{F(L_2)})$, which is a topology generated by the base

$$\{U_1 \times U_2 \mid U_1 \in \mathcal{T}_{F(L_1)}, U_2 \in \mathcal{T}_{F(L_2)}\}.$$

Now we discuss the relations between them.

According to Theorem 6.2.1 and Lemma 5.1.1, the filter topology $\mathcal{T}_{F(L_1 \times L_2)}$ for $L_1 \times L_2$ makes two mappings p_1 and p_2 be continuous w.r.t. the topology $\mathcal{T}_{F(L_1)}$ and the topology $\mathcal{T}_{F(L_2)}$, respectively, *i.e.*,

$$p_i \in C_F(L_1 \times L_2, L_i), i = 1, 2.$$

Theorem 6.2.2 *If $\mathcal{T}$ is the product topology of the filter spaces $(L_1, \mathcal{T}_{F(L_1)})$ and $(L_2, \mathcal{T}_{F(L_2)})$, $\mathcal{T}_{F(L_1 \times L_2)}$ is the filter topology generated by the base $F(L_1 \times L_2)$, then $\mathcal{T} = \mathcal{T}_{F(L_1 \times L_2)}$.* □

Theorem 6.2.2 shows that the product topology of two filter spaces for two lattice implication algebras is the filter topology of the lattice implication product algebra of these two lattice implication algebras. By the induction, this conclusion can be generalized to the case for finitely many lattice implication algebras, *i.e.*,

$$\prod_{\lambda \in \Gamma} \mathcal{T}_{F(L_\lambda)} = \mathcal{T}_{F(\prod\limits_{\lambda \in \Gamma} L_\lambda)},$$

where Γ is a finite index set.

Now we begin to discuss the initial topologies and the final topologies obtained from filter topologies and lattice implication homomorphisms. Given any mapping $f\colon\ L_1 \longrightarrow L_2$, the following two topological structures can be obtained:

(1). The initial topology w.r.t. the topology $\mathcal{T}_{F(L_2)}$ and the mapping f:

$$\mathcal{T}_l(\mathcal{T}_{F(L_2)}, f) = \{f^{-1}(U) \mid U \in \mathcal{T}_{F(L_2)}\}.$$

(2). The final topology w.r.t. the topology $\mathcal{T}_{F(L_1)}$ and the mapping f:

$$\mathcal{T}_r(\mathcal{T}_{F(L_1)}, f) = \{U \mid U \subseteq L_2, f^{-1}(U) \in \mathcal{T}_{F(L_1)}\}.$$

Theorem 6.2.3 *For any $f \in \hom(L_1, L_2)$, we have $\mathcal{T}_l(\mathcal{T}_{F(L_2)}, f) \subseteq \mathcal{T}_{F(L_1)}$, $\mathcal{T}_r(\mathcal{T}_{F(L_1)}, f) \supseteq \mathcal{T}_{F(L_2)}$.*

Proof. Since the initial topology $\mathcal{T}_l(\mathcal{T}_{F(L_2)}, f)$ is the coarsest topology which makes f be continuous w.r.t. the topology $\mathcal{T}_{F(L_2)}$, and $\mathcal{T}_r(\mathcal{T}_{F(L_1)}, f)$ is the finest topology which makes f be continuous w.r.t. the topology $\mathcal{T}_{F(L_1)}$, and f is continuous w.r.t. the filter topology whenever f is a lattice implication homomorphism. Hence the conclusion holds. □

Example 6.2.1 *Let L_1 be the lattice implication algebra defined in Example 2.1.6. then*

$$\mathcal{T}_{F(L_1)} = \{\emptyset, \{a, I\}, \{b, I\}, \{a, b, I\}, \{I\}, \{O, a, b, I\}\}.$$

Suppose L_2 is a Boolean algebra $\{O_2, I_2\}$, then

$$\mathcal{T}_{F(L_2)} = \{\emptyset, \{I_2\}, \{O_2, I_2\}\}.$$

Construct a mapping $f\colon\ L_1 \longrightarrow L_2$ *as follows:*

$$f(I) = f(a) = I_2, f(O) = f(b) = O_2.$$

It can be verified that $f \in \hom(L_1, L_2)$, *and*

$$T_l(T_{F(L_2)}, f) = \{\emptyset, \{a, I\}, \{O, a, b, I\}\} \neq T_{F(L_1)},$$

$$T_r(T_{F(L_1)}, f) = \{\emptyset, \{I_2\}, \{O_2, I_2\}\} = T_{F(L_2)}.$$

Example 6.2.2 *Let* L_1 *be the Boolean algebra* $\{O, I\}$ *and* L_2 *the Łukasiewicz implication algebra defined in Example 2.1.2. Then*

$$T_{F(L_1)} = \{\emptyset, \{I\}, \{O, I\}\},$$

$$T_{F(L_2)} = \{\emptyset, \{1\}, [0, 1]\}.$$

A mapping $f\colon\ L_1 \longrightarrow L_2$ *is constructed as follows:*

$$f(O) = 0, f(I) = 1.$$

Obviously, $f \in \hom(L_1, L_2)$. *Since* $f^{-1}([\frac{1}{2}, 1]) = \{I\} \in T_{F(L_1)}$, *we have*

$$[\frac{1}{2}, 1] \in T_r(T_{F(L_1)}, f).$$

However, $[\frac{1}{2}, 1] \notin T_{F(L_2)}$, *so* $T_{F(L_2)} \neq T_r(T_{F(L_1)}, f)$. *It can be verified that* $T_{F(L_1)} = T_l(T_{F(L_2)}, f)$.

Theorem 6.2.4 *For any* $f \in \hom(L_1, L_2)$,

(1). If f *is injective, then* $T_{F(L_1)} = T_l(T_{F(L_2)}, f)$.
(2). If f *is surjective, then* $T_{F(L_2)} = T_r(T_{F(L_1)}, f)$.

Proof. (1). we need only to prove that: for any $F \in F(L_1)$, $F \in T_l(T_{F(L_2)}, f)$. In what follows, we prove that

$$F = \bigcup_{x \in F} f^{-1}([f(x))).$$

In fact, it is trivial that

$$F \subseteq \bigcup_{x \in F} f^{-1}([f(x))).$$

On the other hand, for any $a \in \bigcup_{x \in F} f^{-1}([f(x)))$, there exists $x \in F$, such that $a \in f^{-1}([f(x)))$, *i.e.*, $f(a) \in [f(x))$. By Corollary 3.2.2, there exists $n \in \mathbb{N}$, such that $(f(x))^n \to f(a) = I_2$. Since f is a lattice implication homomorphism, thus $f(x^n \to a) = I_2$. And since f is injective, we have $x^n \to a = I_1$. It shows that $a \in [x) \subseteq F$. And therefore,

$$\bigcup_{x \in F} f^{-1}([f(x))) \subseteq F.$$

(2). It can be proved similarly to (1). □

6.3 Lattice Topology

In this section, we discuss the lattice topology on the lattice implication algebras L.

Definition 6.3.1 *Let C be a complete lattice, $\mathcal{A} \subseteq C$, $\mathcal{A}$ is said to be a lattice topology on C, if*

(1). $O, I \in \mathcal{A}$;
(2). $a, b \in \mathcal{A}$ implies $a \wedge b \in \mathcal{A}$;
(3). for any $t \in \Gamma$, $a_t \in \mathcal{A}$ implies $\bigvee_{t\in\Gamma} a_t \in \mathcal{A}$.

If $\mathcal{A}$ is a lattice topology on C, we also say that $(C, \mathcal{A})$ is a topological lattice.

Now we discuss lattice topologies generated by filters. Here, we always assume L is a complete lattice implication algebra.

Assume $(L, \mathcal{A})$ is a topological lattice, the elements of $\mathcal{A}$ are said to be open elements, $a \in L$ is said to be a closed element if and only if $a' \in \mathcal{A}$. And for any $a \in L$,

$$a^\circ = \vee\{b \mid b \in \mathcal{A}, b \leqslant a\}$$

is called the inner point (or the interior) of a.

$$\bar{a} = \wedge\{b \mid \ b' \in \mathcal{A}, a \leqslant b\}$$

is called the closure of a.

Theorem 6.3.1 *For any $F \in F(L)$, $\mathcal{A} = \{O\} \cup F$ is a lattice topology on L.*

Proof. First, since $F \in F(L)$, then $I \in F$, therefore, $O, I \in \mathcal{A}$.

Second, given $a, b \in \mathcal{A}$, we need to prove $a \wedge b \in \mathcal{A}$. In fact, if $a = O$ or $b = O$, then $a \wedge b = O \in \mathcal{A}$. If $a \neq O$ and $b \neq O$, then $a, b \in F$. By Theorem 3.1.1, F is a dual ideal of L, so $a \wedge b \in F \subseteq \mathcal{A}$.

Finally, for any $t \in \Gamma$, suppose that $a_t \in \mathcal{A}$. If $a_t = O$ holds for any $t \in \Gamma$, then

$$\bigvee_{t\in\Gamma} a_t = O \in \mathcal{A}.$$

If there exists $t_0 \in \Gamma$, $a_{t_0} \neq O$, then $a_{t_0} \in F$. Since F is a filter of L and

$$a_{t_0} \leqslant \bigvee_{t\in\Gamma} a_t,$$

hence,

$$\bigvee_{t\in\Gamma} a_t \in F \subseteq \mathcal{A}.$$

As above indicate that $\mathcal{A} = \{O\} \cup F$ is a lattice topology on L. □

For any $A \subseteq L$, denote $T(A)$ the lattice topology generated by A, which is the least lattice topology containing A. Clearly, for any $F \in F(L)$, $T(F) = \{O\} \cup F$.

Theorem 6.3.2 *For any $F \in F(L)$, $(L, T(F))$ is a compact topological lattice.*

Proof. Since each filter of L contains I, so this statement is true. □

6.4 Prime Spaces

The prime filters of lattice implication algebras have been discussed in Chapter 3. In this section, we discuss the prime spaces of lattice implication algebras.

Let L be a lattice implication algebra. For any $a \in L$, denote

$$X_a = \{F \mid F \in PF(L), a \notin F\},$$

$$X = \{X_a \mid a \in L\},$$

and for any $F \subseteq L$, denote

$$F^{\Delta} = \{P \mid P \in PF(L), F \nsubseteq P\}.$$

By Corollary 3.4.3, if $a \neq I$, then $X_a \neq \emptyset$. Obviously, $X_I = \emptyset, X_O = PF(L)$.

Lemma 6.4.1 $\bigcup_{a \in L} X_a = PF(L)$.

Proof. It is trivial that

$$\bigcup_{a \in L} X_a \subseteq PF(L).$$

On the other hand, for any $F \in PF(L)$, since F is a proper filter, then there exists $a \in L$ such that $a \notin F$, so

$$F \in X_a \subseteq \bigcup_{a \in L} X_a.$$

It follows that

$$PF(L) \subseteq \bigcup_{a \in L} X_a.$$

Hence,

$$\bigcup_{a \in L} X_a = PF(L). \quad \square$$

Lemma 6.4.2 *(1). For any $a \in L$, $[a)^{\Delta} = X_a$.*
(2). For any $M, N \in F(L)$, $M^{\Delta} \cap N^{\Delta} = (M \cap N)^{\Delta}$.

(3). For any $G_\lambda \in F(L)$, $\lambda \in \Gamma$, $(\bigcup_{\lambda\in\Gamma} G_\lambda)^\vartriangle = \bigcup_{\lambda\in\Gamma}(G_\lambda)^\vartriangle$.

Proof. (1). For any $P \in PF(L)$, since $[a) \nsubseteq P$ if and only if $a \notin P$, so $[a)^\vartriangle = X_a$.

(2). For any $P \in M^\vartriangle \cap N^\vartriangle$, *i.e.*, $P \in PF(L)$, $M \nsubseteq P$ and $N \nsubseteq P$. Then there exist $m \in M, n \in N$ such that $m, n \notin P$. Since P is a prime filter and M, N are filters, we have $m \vee n \in M \cap N, m \vee n \notin P$. Hence $M \cap N \nsubseteq P, P \in (M \cap N)^\vartriangle$, *i.e.*, $M^\vartriangle \cap N^\vartriangle \subseteq (M \cap N)^\vartriangle$.

On the other hand, for any $P \in (M\cap N)^\vartriangle$, $P \in PF(L)$ and $M\cap N \nsubseteq P$. It is clear that $M \nsubseteq P$ and $N \nsubseteq P$, then $P \in M^\vartriangle$ and $P \in N^\vartriangle$, so $P \in M^\vartriangle \cap N^\vartriangle$, $(M \cap N)^\vartriangle \subseteq M^\vartriangle \cap N^\vartriangle$. Consequently, $M^\vartriangle \cap N^\vartriangle = (M \cap N)^\vartriangle$.

(3). For any $P \in \bigcup_{\lambda\in\Gamma}(G_\lambda)^\vartriangle$, there exists $\lambda_0 \in \Gamma$, such that $P \in (G_{\lambda_0})^\vartriangle$, *i.e.*, $P \in PF(L)$ and $G_{\lambda_0} \nsubseteq P$. It follows that

$$\bigcup_{\lambda\in\Gamma} G_\lambda \nsubseteq P,$$

i.e.,

$$P \in (\bigcup_{\lambda\in\Gamma} G_\lambda)^\vartriangle.$$

Hence,

$$\bigcup_{\lambda\in\Gamma}(G_\lambda)^\vartriangle \subseteq (\bigcup_{\lambda\in\Gamma} G_\lambda)^\vartriangle.$$

On the other hand, for any $P \in (\bigcup_{\lambda\in\Gamma} G_\lambda)^\vartriangle$, then $P \in PF(L)$ and $\bigcup_{\lambda\in\Gamma} G_\lambda \nsubseteq P$. So there exists $\lambda_0 \in \Gamma$, such that $G_{\lambda_0} \nsubseteq P$, *i.e.*, $P \in (G_{\lambda_0})^\vartriangle$, hence

$$P \in \bigcup_{\lambda\in\Gamma}(G_\lambda)^\vartriangle,$$

therefore,

$$(\bigcup_{\lambda\in\Gamma} G_\lambda)^\vartriangle \subseteq \bigcup_{\lambda\in\Gamma}(G_\lambda)^\vartriangle.$$

Consequently, $(\bigcup_{\lambda\in\Gamma} G_\lambda)^\vartriangle = \bigcup_{\lambda\in\Gamma}(G_\lambda)^\vartriangle$. □

Lemma 6.4.3 *Let $F \in F(L)$, then $a \in F$ if and only if $X_a \subseteq F^\vartriangle$.*

Proof. Assume $a \in F$, for any $P \in X_a$, we have $a \notin P$. It follows that $F \nsubseteq P$, $P \in F^\vartriangle$. So $X_a \subseteq F^\vartriangle$.

Conversely, assume $X_a \subseteq F^\vartriangle$, if $a \notin F$, by Corollary 3.4.3, there exists $P_a \in PF(L)$ such that $a \notin P_a$ and $F \subseteq P_a$, so $P_a \in X_a$ and $P_a \notin F^\vartriangle$, which imply $X_a \nsubseteq F^\vartriangle$, this is a contradiction, so $a \in F$. □

Lemma 6.4.4 *For any $a, b \in L$,*

$$X_a \cap X_b = X_{a\vee b},$$
$$X_a \cup X_b = X_{a\wedge b}.$$

Proof. Since

$$\begin{aligned} F \in X_a \cap X_b \text{ iff } & a \notin F, b \notin F \text{ and } F \in PF(L) \\ \text{iff } & a \vee b \notin F \text{ and } F \in PF(L) \\ \text{iff } & F \in X_{a \vee b}, \end{aligned}$$

hence

$$X_a \cap X_b = X_{a \vee b}.$$

And since

$$\begin{aligned} F \in X_a \cup X_b \text{ iff } & F \in PF(L) \text{ and } a \notin F \text{ or } b \notin F \\ \text{iff } & F \in PF(L) \text{ and } a \wedge b \notin F \\ \text{iff } & F \in X_{a \wedge b}, \end{aligned}$$

hence $X_a \cup X_b = X_{a \wedge b}$. □

By Lemma 6.4.4, we obtain that for any $a_1, a_2, \cdots, a_n \in L$,

$$X_{a_1} \cap X_{a_2} \cap \cdots \cap X_{a_n} = X_{\bigvee\limits_{i=1}^{n} a_i},$$

$$X_{a_1} \cup X_{a_2} \cup \cdots \cup X_{a_n} = X_{\bigwedge\limits_{i=1}^{n} a_i}.$$

By Lemma 6.4.1, $X = \{X_a \mid a \in L\}$ is a covering of $PF(L)$. By Lemma 6.4.4, $X = \{X_a \mid a \in L\}$ is closed w.r.t. finite intersection. Then taking X as a base, we can introduce a topology $\mathcal{T}$ for $PF(L)$ such that $(PF(L), \mathcal{T})$ becomes a topological space. We call it as the prime space of L. In the following, we discuss its structure and some properties.

Theorem 6.4.1 *For any $F \in F(L)$, $F^{\triangle}$ is an open set of the prime space $(PF(L), \mathcal{T})$, i.e., $F^{\triangle} \in \mathcal{T}$. And for any open set T of this prime space, there exists a unique filter $F \in F(L)$, such that $T = F^{\triangle}$.*

Proof. At first, for any $F \in F(L)$, by Lemma 6.1.3,

$$F = \bigcup_{a \in F} [a).$$

And it follows from Lemma 6.4.2 that:

$$F^{\triangle} = (\bigcup_{a \in F} [a))^{\triangle} = \bigcup_{a \in F} [a)^{\triangle} = \bigcup_{a \in F} X_a \in \mathcal{T}.$$

Moreover, for any $T \in \mathcal{T}$, since $\{X_a \mid a \in L\}$ is a base of $\mathcal{T}$, then there exists $A \subseteq L$, such that $T = \bigcup_{a \in A} X_a$. By Lemma 6.4.2,

$$T = \bigcup_{a \in A} X_a = \bigcup_{a \in A} [a)^{\triangle} = (\bigcup_{a \in A} [a))^{\triangle}.$$

Now we prove

$$(\bigcup_{a\in A}[a))^{\Delta} = [A)^{\Delta}.$$

In fact, if $P \in (\bigcup_{a\in A}[a))^{\Delta}$, then there exists $a \in A$, $P \in X_a$, hence $P \in PF(L)$ and $a \notin P$. It follows that $[A) \nsubseteq P$ and $P \in [A)^{\Delta}$.

Conversely, suppose that $P \in [A)^{\Delta}$, then $P \in PF(L)$ and $[A) \nsubseteq P$, it follows that there exists $x \in [A)$ such that $x \notin P$. By $x \in [A)$, there exist $a_1, \cdots, a_n \in A$ such that

$$a_1 \otimes \cdots \otimes a_n \leqslant x.$$

If $\bigcup_{a\in A}[a) \subseteq P$, then $a_1, \cdots, a_n \in P$ and hence

$$a_1 \otimes \cdots \otimes a_n \in P,$$

this implies $x \in P$, this is a contradiction. So we have proved that

$$P \in (\bigcup_{a\in A}[a))^{\Delta}.$$

By $T = [A)^{\Delta}$, we know that every open set is of the form F^{Δ} for some $F \in F(L)$.

Now we prove the representation form of T is unique. In fact, if $T = F_1^{\Delta} = F_2^{\Delta}$, then for any $a \in L$, by Lemma 6.4.3, we have

$$\begin{aligned} a \in F_1 \ \text{ iff } \ & X_a \subseteq F_1^{\Delta} = F_2^{\Delta} \\ \text{ iff } \ & a \in F_2, \end{aligned}$$

it shows that $F_1 = F_2$. □

According to Theorem 6.4.1, the topology for the prime space $(PF(L), \mathcal{T})$ is

$$\mathcal{T} = \{F^{\Delta} \mid F \in F(L)\}.$$

Theorem 6.4.2 *For any S, $\emptyset \neq S \subseteq PF(L)$, then S is a compact and open set of the prime space $(PF(L), \mathcal{T})$ if and only if there exists $a \in L$, such that $S = X_a$.*

Proof. Assume S is a compact and open set of the prime space $(PF(L), \mathcal{T})$. Since $\{X_a \mid a \in L\}$ is a base for this prime space, so there exists $A \subseteq L$, such that

$$S = \bigcup_{a\in A} X_a,$$

then $\{X_a \mid a \in A\}$ is an open covering of S, so by the compactness of S, there exist $a_1, a_2, \cdots, a_n \in A$, such that

$$S = \bigcup_{i=1}^{n} X_{a_i} = X_{\bigwedge_{i=1}^{n} a_i}.$$

Take

$$a = \bigwedge_{i=1}^{n} a_i,$$

we get $S = X_a$.

Conversely, we need to prove that: for any $a \in L$, X_a is compact and open. In fact, by the definition, X_a is an open set. Now we prove its compactness. By Theorem 6.4.1, we can suppose

$$\{(G_\lambda)^\Delta \mid G_\lambda \in F(L), \lambda \in \Gamma\}$$

is any open covering of X_a, *i.e.*, we have

$$X_a \subseteq \bigcup_{\lambda \in \Gamma} (G_\lambda)^\Delta = (\bigcup_{\lambda \in \Gamma} G_\lambda)^\Delta.$$

By Lemma 6.4.3,

$$a \in \bigcup_{\lambda \in \Gamma} G_\lambda.$$

Hence, there exists $\lambda_0 \in \Gamma$ such that $a \in G_{\lambda_0}$, and then $X_a \subseteq (G_{\lambda_0})^\Delta$, therefore, X_a is compact. □

Since $X_O = PF(L)$ and X_O is a compact set, we have the following theorem:

Theorem 6.4.3 *The prime space $(PF(L), \mathcal{T})$ is a compact space.* □

Theorem 6.4.4 *For any $P \in PF(L)$,*

$$\overline{\{P\}} = PF(L) \setminus P^\Delta$$

holds in the prime space $(PF(L), \mathcal{T})$, where $\overline{\{P\}}$ is the closure of the singleton $\{P\}$.

Proof. Note that $Q \in \overline{\{P\}}$ if and only if every neighborhood of Q contains P. And since

$$X = \{X_a \mid a \in L\}$$

is a base of the prime space $(PF(L), \mathcal{T})$, hence, we can only consider the case of X_a, and furthermore,

$$\begin{aligned}\overline{\{P\}} &= \{Q \mid\ Q \in PF(L),\ \text{for any } a \in L, Q \in X_a \text{ implies } P \in X_a\} \\ &= \{Q \mid Q \in PF(L),\ \text{for any } a \in L, a \notin Q \text{ implies } a \notin P\} \\ &= \{Q \mid Q \in PF(L), P \subseteq Q\} \\ &= PF(L) \setminus P^\Delta.\ \square\end{aligned}$$

According to the properties of closures and by Theorem 6.4.4, we can get the following corollary:

Corollary 6.4.1 *For any finite subset* $\{P_1, P_2, \cdots, P_n\} \subseteq PF(L)$, *its closure is*

$$\overline{\{P_1, P_2, \cdots, P_n\}} = PF(L) \setminus \bigcap_{i=1}^{n} P_i^{\Delta}. \quad \square$$

Theorem 6.4.5 *The prime space* $(PF(L), \mathcal{T})$ *is a* T_0*-space.*

Proof. For any $P, Q \in PF(L)$, if $P \neq Q$, then by Theorem 6.4.1, $P^{\Delta} \neq Q^{\Delta}$. It follows from Theorem 6.4.4 that $\overline{\{P\}} \neq \overline{\{Q\}}$. Therefore, $(PF(L), \mathcal{T})$ is a T_0-space. □

Theorem 6.4.6 *If the set* $\{[x) \mid x \in L\}$ *is at most countable, then the prime space* $(PF(L), \mathcal{T})$ *satisfies the second axiom of countability.*

Proof. Assume the set $\{[x) \mid x \in L\}$ is at most countable, then the cardinal number of the set

$$\{[a)^{\Delta} \mid a \in L\} = \{X_a \mid a \in L\}$$

is at most countable, hence $(PF(L), \mathcal{T})$ has a countable base, and $(PF(L), \mathcal{T})$ satisfies the second axiom of countability. □

Note that a topological space is a T_1-space if and only if its every singleton is closed. According to Theorem 6.4.4, in general, the prime space $(PF(L), \mathcal{T})$ is not necessarily a T_1-space. $(PF(L), \mathcal{T})$ is a T_1-space if and only if $P^{\Delta} = PF(L) \setminus \{P\}$ holds for any $P \in PF(L)$.

Theorem 6.4.7 *The prime space* $(PF(L), \mathcal{T})$ *is a* T_1*-space if and only if for any* $P, Q \in PF(L)$, $P \setminus Q \neq \emptyset$ *holds whenever* $P \neq Q$.

Proof. If the prime space $(PF(L), \mathcal{T})$ is a T_1-space, then for any $P \in PF(L)$,

$$P^{\Delta} = PF(L) \setminus \{P\}.$$

Hence, for any $Q \in PF(L)$, if $Q \neq P$, then $Q \in P^{\Delta}$, *i.e.*, $P \nsubseteq Q$. Therefore, $P \setminus Q \neq \emptyset$.

Conversely, if for any $P, Q \in PF(L)$, $P \setminus Q \neq \emptyset$ holds whenever $P \neq Q$, then the closure of every singleton is itself by Theorem 6.4.4, *i.e.*, every singleton is closed. Thus, the prime space $(PF(L), \mathcal{T})$ is a T_1-space. □

Theorem 6.4.7 shows that: if the prime space $(PF(L), \mathcal{T})$ is a T_1-space, then there does not exist an inclusion relation between any two different prime filters.

Theorem 6.4.8 *Suppose* L *is a lattice* H *implication algebra, then its prime space* $(PF(L), \mathcal{T})$ *has the following properties:*

(1). all of clopen sets compose a base for $\mathcal{T}$.
(2). $(PF(L), \mathcal{T})$ is a compact Hausdorff space, i.e., a compact T_2-space.

Proof. (1). Since $X = \{X_a \mid a \in L\}$ is a base of $(PF(L), \mathcal{T})$, we only need to prove every element of X is a clopen set. In fact, for any $a \in L$, since $a \vee a' = I, a \wedge a' = O$, by Lemma 6.4.4, we have

$$X_a \cup X_{a'} = X_{a \wedge a'} = X_O = PF(L),$$

and

$$X_a \cap X_{a'} = X_{a \vee a'} = X_I = \emptyset,$$

i.e., X_a and $X_{a'}$ are supplementary each other in $PF(L)$, hence, X_a is a clopen set. And thus, all of clopen sets compose a base for $\mathcal{T}$.

(2). For any $P, Q \in PF(L)$, $P \neq Q$, we can assume $P \nsubseteq Q$. Take any $a \in P \setminus Q$, then $a' \in Q \setminus P$, and thus, $Q \in X_a, P \in X_{a'}, X_a \cap X_{a'} = \emptyset$. Hence, $(PF(L), \mathcal{T})$ is a Hausdorff space. By Theorem 6.4.3, $(PF(L), \mathcal{T})$ is also compact. □

Corollary 6.4.2 *The prime space for a lattice H implication algebra is zero-dimensional.* □

Theorem 6.4.9 *If L is any lattice H implication algebra with $|L| > 2$, then the prime space for L is not connected.*

Proof. Note that a topology for a set Y is connected if and only if the clopen subsets of Y are only Y and $\emptyset$. By Theorem 6.4.8, we know that for any $a \in L$, X_a is a clopen subset of L, so the prime space for L is not connected.
□

7 Connections with Related Algebras

BCK-algebra [194], MV-algebra [52], FI-algebra [525], residuated lattice [153], R_0-algebra [500], and lattice implication algebra [534] are all related to logic. This chapter is mainly devoted to the discussion of relations between lattice implication algebra and the other algebras.

7.1 Lattice Implication Algebras and BCK-Algebras

In this section, the dual relations between lattice implication algebra and bounded commutative BCK-algebra, lattice H implication algebra and bounded implicative BCK-algebra are pointed out.

Definition 7.1.1 (BCK-Algebra) *[194] A (2,0)-type algebra $(X, *, \theta)$ is called a BCK-algebra, if the following conditions hold for any $x, y, z \in X$,*

(1). $[(x * y) * (x * z)] * (z * y) = \theta$;
(2). $[x * (x * y)] * y = \theta$;
(3). $x * x = \theta$;
(4). $\theta * x = \theta$;
(5). $x * y = y * x = \theta$ *implies* $x = y$.

We can define a partial ordering $\leqslant$ on a BCK-algebra X by

$$x \leqslant y \text{ iff } x * y = \theta.$$

A BCK-algebra X is said to be bounded if there exists $e \in X$ such that $x \leqslant e$ for any $x \in X$, and such element e is said to be the unit of X. In a bounded BCK-algebra, we denote $e * x$ by $N(x)$. A BCK-algebra X is said to be commutative if

$$x \sqcap y = y \sqcap x$$

for any $x, y \in X$, where $x \sqcap y = y * (y * x)$. A BCK-algebra X is said to be implicative if $x * (y * x) = x$ for any $x, y \in X$. In a bounded BCK-algebra X we let $x \sqcup y = N(N(x) \sqcap N(y))$.

Theorem 7.1.1 *Let $(L, \vee, \wedge, ', \rightarrow, O, I)$ be a lattice implication algebra. If define $x * y = y \rightarrow x$ for any $x, y \in L$, then $(L, *, \theta)$ is a bounded commutative BCK-algebra with $e = O, \theta = I$.*

Proof. By properties of lattice implication algebra, it is easy to prove that $(L,*,\theta)$ is a BCK-algebra.

For boundedness, by $x*O=O\to x=I=\theta$, we know that O is the unit element and

$$\begin{aligned}N(x)&=e*x\\&=x\to e\\&=x\to O\\&=x'.\end{aligned}$$

For commutativity, we have

$$\begin{aligned}x\sqcap y&=y*(y*x)\\&=(x\to y)\to y\\&=(y\to x)\to x\\&=x*(x*y)\\&=y\sqcap x.\ \square\end{aligned}$$

Theorem 7.1.2 *Let $(L,*,\theta)$ be a bounded commutative BCK-algebra. If define*

$$\begin{aligned}x\vee y&=x\sqcap y,\\x\wedge y&=x\sqcup y,\\x'&=N(x),\\y\to x&=x*y,\end{aligned}$$

for any $x,y\in L$, then $(L,\vee,\wedge,{}',\to)$ is a lattice implication algebra.

Proof. We know that $(L,\sqcup,\sqcap)$ is a dual lattice (*cf.* [188], p. 58, Theorem 11). Since $x\vee\theta=x\sqcap\theta=\theta$ holds for any $x\in L$, we have $x\leqslant\theta,\theta=I$. And for $x\wedge e=x\sqcup e=e$ holds for any $x\in L$, we have $x\geqslant e$, $e=O$. In $(L,\vee,\wedge)$, if $x\leqslant y$, then $x\sqcup y=x$ and

$$\begin{aligned}x'&=N(x)\\&=N(x)\sqcap N(y)\\&=x'\vee y',\end{aligned}$$

i.e., $y'\leqslant x'$. We know that

$$(x')'=N(N(x))=x$$

holds for any $x\in L$ (*cf.* [188], p. 58, Lemma 3). Then $(L,\vee,\wedge,{}')$ is a bounded lattice with an order-reversing involution.

In the following, we prove that $(L,\vee,\wedge,{}',\to)$ satisfies the definition of a lattice implication algebra. In fact, for every $x,y\in L$,

(1).

$$\begin{aligned} x \to (y \to z) &= (z * y) * x \\ &= (z * x) * y \\ &= y \to (x \to z) \end{aligned}$$

holds (*cf.* [188], p. 39, Theorem 5).

(2). $x \to x = x * x = \theta = I$.

(3).

$$\begin{aligned} y' \to x' &= N(y) \to N(x) \\ &= N(x) * N(y) \\ &= (e * x) * N(y) \\ &= (e * N(y)) * x \\ &= N(N(y)) * x \\ &= y * x \\ &= x \to y. \end{aligned}$$

(4). $x \to y = y \to x = I$ implies $y * x = x * y = \theta$ and hence $x = y$.

(5).

$$\begin{aligned} (x \to y) \to y &= y * (y * x) \\ &= x \sqcap y \\ &= y \sqcap x \\ &= x * (x * y) \\ &= (y \to x) \to x. \end{aligned}$$

By Theorem 2.1.8, we also need only to prove

$$(x \to y) \to y = x \vee y.$$

In fact,

$$\begin{aligned} (x \to y) \to y &= y * (y * x) \\ &= x \sqcap y \\ &= x \vee y. \ \Box \end{aligned}$$

According to Theorem 7.1.1 and Theorem 7.1.2, we have the following corollary.

Corollary 7.1.1 *Let L be a set, $\vee, \wedge, \to, \sqcup, \sqcap$ and $*$ binary operations on L, $'$ and N unary operations on L. If*

$$\begin{aligned} x \wedge y &= x \sqcup y, \\ x \vee y &= x \sqcap y, \\ x' &= N(x), \\ y \to x &= x * y \end{aligned}$$

hold for any $x, y \in L$. Then $(L, \vee, \wedge, ', \to)$ is a lattice implication algebra if and only if $(L, \sqcup, \sqcap, N, *)$ is a bounded commutative BCK-algebra. □

Theorem 7.1.3 *Let L be a set, $\vee, \wedge, \to, \sqcup, \sqcap$ and $*$ binary operations on L, $'$ and N unary operations on L. If*

$$\begin{aligned} x \wedge y &= x \sqcup y, \\ x \vee y &= x \sqcap y, \\ x' &= N(x), \\ y \to x &= x * y \end{aligned}$$

hold for any $x, y \in L$, then $(L, \vee, \wedge, ', \to)$ is a lattice H implication algebra if and only if $(L, \sqcup, \sqcap, N, *)$ is a bounded implicative BCK-algebra.

Proof. Suppose $(L, \vee, \wedge, ', \to)$ is a lattice H implication algebra. We know that $(L, \sqcup, \sqcap, N, *)$ is a bounded commutative BCK-algebra by Corollary 7.1.1, and

$$\begin{aligned} x * (y * x) &= (x \to y) \to x \\ &= (x' \vee y)' \vee x \\ &= (x \wedge y') \vee x \\ &= x \end{aligned}$$

holds for any $x, y \in L$. That is to say, $(L, \sqcup, \sqcap, N, *)$ is a bounded implicative BCK-algebra.

Conversely, suppose $(L, \sqcup, \sqcap, N, *)$ is a bounded implicative BCK-algebra. We know that $(L, \sqcup, \sqcap, N, *)$ is a Boolean algebra (*cf.* [188], p. 87, Theorem 14) and hence $(L, \vee, \wedge, ', \to)$ is a lattice H implication algebra. □

Corollary 7.1.2 *Let L be a set, $\vee, \wedge, \to, \sqcup, \sqcap$ and $*$ binary operations on L, $'$ and N unary operations on L. If*

$$\begin{aligned} x \wedge y &= x \sqcup y, \\ x \vee y &= x \sqcap y, \\ x' &= N(x), \\ y \to x &= x * y \end{aligned}$$

hold for any $x, y \in L$, then $(L, \sqcup, \sqcap, N, *)$ is a Boolean algebra if and only if $(L, \vee, \wedge, ', \to)$ is a Boolean algebra if and only if $(L, \vee, \wedge, ', \to)$ is a lattice H implication algebra if and only if $(L, \sqcup, \sqcap, N, *)$ is a bounded implicative BCK-algebra. □

7.2 Lattice Implication Algebras and MV-Algebras

The theory of MV-algebras, developed by Chang [52], first appeared in 1958. MV-algebras were developed to provide an algebraic proof of the completeness theorem for Łukasiewicz infinite-valued propositional logic. The properties and structure of MV-algebra have been studied by many researchers. In this section, we only discuss the relations between lattice implication algebras and MV-algebras.

Definition 7.2.1 *[52] A* $(2,2,1,0,0)-$*type algebra* $(A,+,\cdot,*,O,I)$ *is called an* MV*-algebra, if* A *is a non-empty set and the following conditions hold for any* $x,y,z \in A$:

$(AX1)$ $x+y=y+x$;
$(AX1')$ $x\cdot y=y\cdot x$;
$(AX2)$ $x+(y+z)=(x+y)+z$;
$(AX2')$ $x\cdot(y\cdot z)=(x\cdot y)\cdot z$;
$(AX3)$ $x+x^*=I$;
$(AX3')$ $x\cdot x^*=O$;
$(AX4)$ $x+I=I$;
$(AX4')$ $x\cdot O=O$;
$(AX5)$ $x+O=x$;
$(AX5')$ $x\cdot I=x$;
$(AX6)$ $(x+y)^*=x^*\cdot y^*$;
$(AX6')$ $(x\cdot y)^*=x^*+y^*$;
$(AX7)$ $x=(x^*)^*$;
$(AX8)$ $O^*=I$;
$(AX9)$ $x\vee y=y\vee x$;
$(AX9')$ $x\wedge y=y\wedge x$;
$(AX10)$ $x\vee(y\vee z)=(x\vee y)\vee z$;
$(AX10')$ $x\wedge(y\wedge z)=(x\wedge y)\wedge z$;
$(AX11)$ $x+(y\wedge z)=(x+y)\wedge(x+z)$;
$(AX11')$ $x\cdot(y\vee z)=(x\cdot y)\vee(x\cdot z)$,

where $x\vee y=(x\cdot y^*)+y$ *and* $x\wedge y=(x+y^*)\cdot y$.

MV-algebras constitute a generalization of Boolean algebras and arise from the many-valued logic of Łukasiewicz in the same manner as Boolean algebras arise from two-valued logics. Pleass see [16, 52, 314–317, 488] for details.

We define the binary relation $\leqslant$ on $(A,+,\cdot,*,O,I)$ by $x\leqslant y$ if $x\vee y=y$ for any $x,y\in A$. We can get the following theorem.

Theorem 7.2.1 *If* $(A,+,\cdot,*,O,I)$ *is an* MV*-algebra, then*

(1). $\leqslant$ *is a partial order on* A; $x\vee y, x\wedge y$ *are the least upper bound and the greatest lower bound of* x *and* y*, respectively, and* $O\leqslant x\leqslant I$ *holds for any* $x\in A$.

(2). $(A, \vee, \wedge, *)$ is a distributive lattice with an order-reversing involution.

Proof. (1). The proof is clear by the reference [52].

(2). The distributivity of A with respect to operations $\vee$ and $\wedge$ was proved by reference [313]. We know that

$$(x \vee y)^* = x^* \wedge y^*$$

holds by reference [52], and

$$x = (x^*)^*$$

holds for any $x \in A$. According to the reference [528], $*$ is an order-reversing involution and A is a distributive lattice with an order-reversing involution. □

Theorem 7.2.2 *Let $A = (A, +, \cdot, *, O, I)$ be an MV-algebra. Define a binary operation $\rightarrow$ and a unary operation $'$ on A as follows: for any $x, y \in A$,*

$$x \rightarrow y = x^* + y,$$
$$x' = x^*.$$

Then $(A, \vee, \wedge, \rightarrow, ')$ is a lattice implication algebra.

Proof. According to Theorem 2.1.8, we only need to prove A is a quasi-lattice implication algebra and

$$(x \rightarrow y) \rightarrow y = x \vee y$$

holds for any $x, y \in A$.

For the condition (I_1) of Definition 2.1.1,

$$\begin{aligned} x \rightarrow (y \rightarrow z) &= x^* + (y^* + z) \\ &= (x^* + y^*) + z \\ &= (y^* + x^*) + z \\ &= y^* + (x^* + z) \\ &= y \rightarrow (x \rightarrow z). \end{aligned}$$

For the condition (I_2) of Definition 2.1.1, $x \rightarrow x = x^* + x = I$.

For the condition (I_3) of Definition 2.1.1,

$$\begin{aligned} x \rightarrow y &= x^* + y \\ &= x^* + (y^*)^* \\ &= (y^*)^* + x^* \\ &= y^* \rightarrow x^* \\ &= y' \rightarrow x'. \end{aligned}$$

For the condition (I_4) of Definition 2.1.1, according to Theorem 1.13 of the reference [52], we have $x \leqslant y$ if and only if $x^* + y = I$ for any $x, y \in A$. If $x \to y = y \to x = I$, then $x^* + y = y^* + x = I$, it follows that $x \leqslant y$, $y \leqslant x$, and hence $x = y$.

For the condition (I_5) of Definition 2.1.1,

$$\begin{aligned}(x \to y) \to y &= (x^* + y)^* + y \\ &= x^{**} \cdot y^* + y \\ &= (x \cdot y^*) + y \\ &= x \vee y \\ &= y \vee x \\ &= y \cdot x^* + x \\ &= (y^* + x^{**})^* + x \\ &= (y^* + x)^* + x \\ &= (y \to x) \to x.\end{aligned}$$

From the proof of (I_5), we know that

$$(x \to y) \to y = x \vee y$$

for any $x, y \in A$. □

An equivalent description of MV-algebra was given in [314] as follows.

Lemma 7.2.1 *(1). Let $B = (B, +, \cdot, *, O, I)$ be a $(2,2,1,0,0)$–type algebra. If the following conditions hold for any $x, y, z \in B$:*

(p1) $(x + y) + z = x + (y + z)$;
(p2) $x + O = x$;
(p3) $x + y = y + x$;
(p4) $x + I = I$;
(p5) $(x^*)^* = x$;
(p6) $O^* = I$;
(p7) $x + x^* = I$;
(p8) $(x^* + y)^* + y = (x + y^*)^* + x$;
(p9) $x \cdot y = (x^* + y^*)^*$,

then B is an MV-algebra.
(2). MV-algebra satisfies (p1)-(p9). □

Theorem 7.2.3 *Let $L = (L, \vee, \wedge, \to, ')$ be a lattice implication algebra. Define operations $*$, $+$ and $\cdot$ on L as follows: for any $x, y \in L$,*

$$(D1)\ \ x^* = x';$$
$$(D2)\ \ x + y = x' \to y;$$
$$(D3)\ \ x \cdot y = (x \to y')',$$

then $(L, +, \cdot, *, O, I)$ is an MV-algebra.

Proof. In order to prove $(A, +, \cdot, *, O, I)$ is an MV-algebra, we only need to prove $(A, +, \cdot, *, O, I)$ satisfies $(p1)$-$(p9)$. For any $x, y, z \in L$,

$$\begin{aligned}(x + y) + z &= (x' \to y)' \to z \\ &= z' \to (x' \to y) \\ &= x' \to (y' \to z) \\ &= x + (y + z),\end{aligned}$$

hence $(p1)$ holds. $(p2)$-$(p9)$ can be proved similarly. □

Wu [525] proposed the concept of fuzzy implication algebra as follows.

Definition 7.2.2 (FI-Algebra) *[525] A (2, 0)-type algebra $(X, \to, O)$ is called a fuzzy implication algebra if for any $x, y, z \in X$,*

(1). $x \to (y \to z) = y \to (x \to z)$;
(2). $(x \to y) \to ((y \to z) \to (x \to z)) = I$;
(3). $x \to x = I$;
(4). *if* $x \to y = y \to x = I$, *then* $x = y$;
(5). $O \to I = I$,

where $I = O \to O$.

An FI-algebra $(X, \to, O)$ is called normal, if $(x')' = x$ holds for any $x \in X$, where $x' = x \to O$.

Liu and Wang [266] proved the following theorem.

Theorem 7.2.4 *[266]*

(1). Let $A = (A, +, \cdot, *, O, I)$ be an MV-algebra. Define a binary operation $\to$ on A as follows: for any $x, y \in A$,

$$x \to y = x^* + y,$$

then $(A, \to, O)$ is an FI-algebra and

$$(x \to y) \to y = (y \to x) \to x \tag{7.1}$$

holds.

(2). Let $(X, \rightarrow, O)$ be a normal FI-algebra and Eq. (7.1) holds for any $x, y \in X$. Define binary operations $+$, $\cdot$ and unary operation $*$ on X as follows:

$$
\begin{aligned}
x + y &= x' \rightarrow y, \\
x \cdot y &= (x \rightarrow y')', \\
x^* &= x',
\end{aligned}
$$

then $(X, +, \cdot, *, O, I)$ is an MV-algebra. □

From Theorems 7.2.2, 7.2.3, and 7.2.4, we can get the relation between lattice implication algebra and FI-algebra as follows.

Corollary 7.2.1 *Let N be a normal FI-algebra. N satisfies Eq. (7.1) if and only if N is a lattice implication algebra.* □

Wang [500] introduced the concept of R_0-algebra and studied the logic system based on it.

Definition 7.2.3 *[500] Let $(M, \vee, \wedge, O, I)$ be a bounded distributive lattice with an order-reversing involution $\neg$ and $\rightarrow$: $M \times M \longrightarrow M$ a mapping. M is called a R_0-algebra if the following conditions hold for any $a, b, c \in M$:*

(R1). $(\neg a) \rightarrow (\neg b) = b \rightarrow a$;
(R2). $I \rightarrow a = a$;
(R3). $b \rightarrow c \leqslant (a \rightarrow b) \rightarrow (a \rightarrow c)$;
(R4). $a \rightarrow (b \rightarrow c) = b \rightarrow (a \rightarrow c)$;
(R5). $a \rightarrow (b \vee c) = (a \rightarrow b) \vee (a \rightarrow c)$;
(R6). $(a \rightarrow b) \vee ((a \rightarrow b) \rightarrow ((\neg a) \vee b)) = I$;
(R7). $a \rightarrow a = I$;
(R8). $a \rightarrow (b \wedge c) = (a \rightarrow b) \wedge (a \rightarrow c)$.

Pei [358] studied the relations between R_0-algebra and FI-algebra, and proved the following theorem.

Theorem 7.2.5 *[358] A R_0-algebra is a normal FI-algebra.* □

From Theorem 7.2.5 and the properties of lattice implication algebra, we can get the following corollary.

Corollary 7.2.2 *(1). If M is a R_0-algebra and Eq. (7.1) holds, then M is a lattice implication algebra;*
(2). If L is a lattice implication algebra and (R6) holds, then L is a R_0-algebra.
□

7.3 Lattice Implication Algebras and Related Algebras

In this section, we study the relations between lattice implication algebras and other related algebras.

Theorem 7.3.1 *Let* $(L, \rightarrow, *, O, I)$ *be a* $(2,1,0,0)$*-type algebra, and satisfies the following conditions: for any* $x, y, z \in L$

(1). $x \rightarrow (y \rightarrow z) = y \rightarrow (x \rightarrow z)$*;*
(2). $x \rightarrow y = y^* \rightarrow x^*$*;*
(3). $(x \rightarrow y) \rightarrow y = (y \rightarrow x) \rightarrow x$*;*
(4). $(x^*)^* = x$*;*
(5). $O^* = I$*;*
(6). $x \rightarrow O = x^*$*;*
(7). $x \rightarrow I = I$*;*
(8). $x \rightarrow x = I$*.*

Let

$$\begin{aligned} x' &= x^*, \\ x \vee y &= (x \rightarrow y) \rightarrow y, \\ x \wedge y &= (x^* \vee y^*)^*, \end{aligned}$$

then $(L, \vee, \wedge, ', \rightarrow, O, I)$ *is a lattice implication algebra. Conversely, every lattice implication algebra satisfies the conditions* (1)-(8).

Proof. It is clear that every lattice implication algebra satisfies the conditions (1)-(8).

Let $(L, \rightarrow, *, O, I)$ be a $(2,1,0,0)$-type algebra satisfying the conditions (1)-(8) and for any $x, y \in L$,

$$\begin{aligned} x \vee y &= (x \rightarrow y) \rightarrow y, \\ x \wedge y &= (x^* \vee y^*)^*. \end{aligned}$$

Then for any $x, y \in L$, we have:

(1).

$$\begin{aligned} x \vee y &= (x \rightarrow y) \rightarrow y \\ &= (y \rightarrow x) \rightarrow x \\ &= y \vee x. \end{aligned}$$

(2).

$$\begin{aligned} x \vee x &= (x \rightarrow x) \rightarrow x \\ &= I \rightarrow x \\ &= x. \end{aligned}$$

(3).

$$\begin{aligned} x \vee (y \vee z) &= (x \to (y \vee z)) \to (y \vee z) \\ &= (x \to ((z \to y) \to y)) \to ((z \to y) \to y) \\ &= (z \to y) \to (((z \to y) \to (x \to y)) \to y) \\ &= y^* \to (((z \to y) \to (x \to y)) \to (z \to y)^*) \\ &= y^* \to (((x \to y)^* \to (z \to y)^*) \to (z \to y)^*) \\ &= y^* \to ((x \to y)^* \vee (z \to y)^*) \\ &= y^* \to (((z \to y)^* \to (x \to y)^*) \to (x \to y)^*) \\ &= y^* \to (((x \to y) \to (z \to y)) \to (x \to y)^*) \\ &= (x \to y) \to (((x \to y) \to (z \to y)) \to y) \\ &= (z \to ((x \to y) \to y) \to ((x \to y) \to y) \\ &= (z \to (x \vee y)) \to (x \vee y) \\ &= z \vee (x \vee y) \\ &= (x \vee y) \vee z. \end{aligned}$$

(4).

$$\begin{aligned} x \wedge (x \vee y) &= (x^* \vee (x \vee y)^*)^* \\ &= (((x \vee y)^* \to x^*) \to x^*)^* \\ &= ((x \to (x \vee y)) \to x^*)^* = (I \to x^*)^* \\ &= (x^*)^* \\ &= x. \end{aligned}$$

Similarly, we can prove the following conclusions:
(5). $x \wedge y = y \wedge x$.
(6). $x \wedge x = x$.
(7). $x \wedge (y \wedge z) = (x \wedge y) \wedge z$.
(8). $x \vee (x \wedge y) = x$.
Then, $(L, \vee, \wedge)$ is a lattice, and the order relation $\leqslant$ is given by:

$$\begin{aligned} x \leqslant y \quad &\text{iff} \quad x \vee y = y \\ &\text{iff} \quad x \wedge y = x \\ &\text{iff} \quad x \to y = I. \end{aligned}$$

In fact, for any $x, y \in L$. If $x \to y = I$, then

$$\begin{aligned} x \vee y &= (x \to y) \to y \\ &= I \to y \\ &= y. \end{aligned}$$

If $x \vee y = y$, then

$$\begin{aligned} x \to y &= x \to (x \vee y) \\ &= x \to ((x \to y) \to y) \\ &= (x \to y) \to (x \to y) \\ &= I. \end{aligned}$$

It is clear that $O \leqslant x \leqslant I$. If $x \leqslant y$, then

$$y^* \to x^* = x \to y = I,$$

i.e.,

$$y^* \leqslant x^*.$$

It follows that $(L, \vee, \wedge, *, O, I)$ is a bounded lattice with order-reversing involution, and for any $x, y \in L$,

(I_1) $x \to (y \to z) = y \to (x \to z)$,
(I_2) $x \to x = I$,
(I_3) $x \to y = y^* \to x^*$,
(I_4) if $x \to y = y \to x = I$, then $x \leqslant y$ and $y \leqslant x$, and then, $x = y$,
(I_5) $(x \to y) \to y = (y \to x) \to x$.

By $x \vee y = (x \to y) \to y$ and Theorem 2.1.8, $(L, \vee, \wedge, ', \to, O, I)$ is a lattice implication algebra. □

Zhu and Tu [609] gave another equivalent description for a $(2,2,2,1,0,0)$-type algebra to be a lattice implication algebra. They proved the following theorem:

Theorem 7.3.2 *[609] A $(2,2,2,1,0,0)$-type algebra $(X, \vee, \wedge, \to, ', O, I)$ is a lattice implication algebra if and only if it satisfies the following conditions: for any $x, y, z \in X$,*

(1). $O \to x = I$;
(2). $(x \to z) \to (x \to y) = (z \to x) \to (z \to y)$;
(3). $x \to y = y' \to x'$;
(4). $(x')' = x$;
(5). $x' = x \to O$;
(6). $x \vee y = (x \to y) \to y$;
(7). $(x \wedge y)' = x' \vee y'$. □

Theorem 7.3.3 *Suppose L is a non-empty set, O and I are two elements of L and $\to$ is a binary operation. Define two binary operations $\wedge$, $\vee$ and a unary operation $'$ on L as follows: for any $x, y \in L$,*

(1). $x' = x \to O$;
(2). $x \vee y = (x \to y) \to y$;
(3). $x \wedge y = ((y \to x) \to y')'$.

If the operation $\rightarrow$ satisfies the following conditions: for any $x, y, z \in L$,

$$
\begin{aligned}
&(I_1)\ \ x \rightarrow (y \rightarrow z) = y \rightarrow (x \rightarrow z);\\
&(I_2)\ \ x \rightarrow x = I;\\
&(I_3)\ \ x \rightarrow y = y' \rightarrow x';\\
&(I_4)\ \ x \rightarrow y = y \rightarrow x = I \text{ implies } x = y;\\
&(I_5)\ \ (x \rightarrow y) \rightarrow y = (y \rightarrow x) \rightarrow x;\\
&(l_1)\ \ (x \vee y) \rightarrow z = (x \rightarrow z) \wedge (y \rightarrow z);\\
&(l_2)\ \ (x \wedge y) \rightarrow z = (x \rightarrow z) \vee (y \rightarrow z),
\end{aligned}
$$

then $(L, \vee, \wedge, ', \rightarrow, O, I)$ is a lattice implication algebra.

In order to prove this theorem, we firstly prove the following lemmas with those conditions of Theorem 7.3.3.

Lemma 7.3.1 *For any $x \in L$, we have $(x')' = x$.*

Proof. For any $x \in L$,

$$
\begin{aligned}
x \rightarrow (x')' &= x \rightarrow ((x \rightarrow O) \rightarrow O)\\
&= (x \rightarrow O) \rightarrow (x \rightarrow O) = I.
\end{aligned}
$$

$$
\begin{aligned}
(x')' \rightarrow x &= x' \rightarrow ((x')')'\\
&= (x \rightarrow O) \rightarrow (((x \rightarrow O) \rightarrow O) \rightarrow O)\\
&= ((x \rightarrow O) \rightarrow O) \rightarrow ((x \rightarrow O) \rightarrow O)\\
&= I.
\end{aligned}
$$

Then $(x')' = x$ holds by (I_4). □

Lemma 7.3.2 *(1). $O' = I$.*
(2). $I' = O$.

Proof. (1). $O' = O \rightarrow O = I$.
(2). $I' = (O')' = O$. □

Lemma 7.3.3 *For any $x \in L$, $I \rightarrow x = x$.*

Proof. According to Lemmas 7.3.1 and 7.3.2 and the definition of operation $'$, we have

$$I \rightarrow x = x' \rightarrow I' = x' \rightarrow O = (x')' = x. \ \square$$

Lemma 7.3.4 *For any $x, y \in L$, $(x \vee y)' = x' \wedge y'$, $(x \wedge y)' = x' \vee y'$.*

Proof.

$$\begin{aligned}(x \vee y)' &= (x \vee y) \to O \\ &= (x \to O) \wedge (y \to O) \\ &= x' \wedge y'.\end{aligned}$$

$$\begin{aligned}(x \wedge y)' &= (x \wedge y) \to O \\ &= (x \to O) \vee (y \to O) \\ &= x' \vee y'. \ \square\end{aligned}$$

Lemma 7.3.5 *For any $x, y \in L$,*

$$x \to (y \vee z) = (x \to y) \vee (x \to z),$$

$$x \to (y \wedge z) = (x \to y) \wedge (x \to z).$$

Proof.

$$\begin{aligned}x \to (y \vee z) &= (y \vee z)' \to x' \\ &= (y' \wedge z') \to x' \\ &= (y' \to x') \vee (z' \to x') \\ &= (x \to y) \vee (x \to z).\end{aligned}$$

$$\begin{aligned}x \to (y \wedge z) &= (y \wedge z)' \to x' \\ &= (y' \vee z') \to x' \\ &= (y' \to x') \wedge (z' \to x') \\ &= (x \to y) \wedge (x \to z). \ \square\end{aligned}$$

Lemma 7.3.6 *For any $x, y \in L$, $x \vee y = y \vee x$, $x \wedge y = y \wedge x$.*

Proof. According to (I_5) and the definitions of $\vee$ and $\wedge$, we have

$$\begin{aligned}x \vee y &= (x \to y) \to y \\ &= (y \to x) \to x \\ &= y \vee x.\end{aligned}$$

$$\begin{aligned}x \wedge y &= ((y \to x) \to y')' \\ &= ((x' \to y') \to y')' \\ &= ((y' \to x') \to x')' \\ &= ((x \to y) \to x')' \\ &= y \wedge x. \ \square\end{aligned}$$

Lemma 7.3.7 *For any $x \in L$, $x \vee I = I \vee x = I$, and $x \wedge I = I \wedge x = x$.*

Proof.

$$\begin{aligned} x \vee I &= I \vee x \\ &= (I \to x) \to x \\ &= x \to x = I. \end{aligned}$$

$$\begin{aligned} x \wedge I &= I \wedge x \\ &= ((I \to x) \to I')' \\ &= (x \to O)' \\ &= (x')' \\ &= x. \ \square \end{aligned}$$

Lemma 7.3.8 *For any $x \in L$, $x \vee O = O \vee x = x$, and $x \wedge O = O \wedge x = O$.*

Proof.

$$\begin{aligned} x \vee O &= O \vee x \\ &= (x \to O) \to O \\ &= (x')' \\ &= x. \end{aligned}$$

$$\begin{aligned} x \wedge O &= O \wedge x \\ &= ((x \to O) \to x')' \\ &= (x' \to x')' \\ &= I' \\ &= O. \ \square \end{aligned}$$

Lemma 7.3.9 *For any $x \in L$, $x \to (x \vee y) = I$, and $(x \wedge y) \to x = I$.*

Proof.

$$\begin{aligned} x \to (x \vee y) &= (x \to x) \vee (x \to y) \\ &= I \vee (x \to y) \\ &= I. \end{aligned}$$

$$\begin{aligned} (x \wedge y) \to x &= (x \to x) \vee (y \to x) \\ &= I \vee (y \to x) \\ &= I. \ \square \end{aligned}$$

Lemma 7.3.10 *For any $x, y \in L$,*

$$(x \vee y) \vee z = x \vee (y \vee z),$$

$$(x \wedge y) \wedge z = x \wedge (y \wedge z).$$

Proof.

$$\begin{aligned}
&((x \vee y) \vee z) \to (x \vee (y \vee z)) \\
&= ((x \vee y) \to (x \vee (y \vee z))) \wedge (z \to (x \vee (y \vee z))) \\
&= ((x \to (x \vee (y \vee z))) \wedge (y \to (x \vee (y \vee z)))) \wedge (z \to (x \vee (y \vee z))) \\
&= (I \wedge (y \to (x \vee (y \vee z)))) \wedge (z \to (x \vee (y \vee z))) \\
&= (y \to (x \vee (y \vee z))) \wedge (z \to (x \vee (y \vee z))) \\
&= (y \vee z) \to (x \vee (y \vee z)) = I.
\end{aligned}$$

$$\begin{aligned}
&(x \vee (y \vee z)) \to ((x \vee y) \vee z) \\
&= (x \to ((x \vee y) \vee z)) \wedge ((y \vee z) \to ((x \vee y) \vee z)) \\
&= (x \to ((x \vee y) \vee z)) \wedge ((y \to ((x \vee y) \wedge z)) \wedge (z \to ((x \vee y) \vee z))) \\
&= (x \to ((x \vee y) \vee z)) \wedge ((y \to ((x \vee y) \vee z)) \wedge I) \\
&= (x \to ((x \vee y) \vee z)) \wedge (y \to ((x \vee y) \vee z)) \\
&= (x \vee y) \to ((x \vee y) \vee z) = I.
\end{aligned}$$

Then we have $(x \vee y) \vee z = x \vee (y \vee z)$.

$$\begin{aligned}
(x \wedge y) \wedge z &= (((x \wedge y) \wedge z)')' \\
&= ((x \wedge y)' \vee z')' \\
&= ((x' \vee y') \vee z')' \\
&= (x' \vee (y \wedge z)')' \\
&= ((x \wedge (y \wedge z))')' \\
&= x \wedge (y \wedge z). \ \square
\end{aligned}$$

Lemma 7.3.11 *For any $x \in L$, $x \vee x = x$, and $x \wedge x = x$.*

Proof.

$$\begin{aligned}
x \vee x &= (x \to x) \to x \\
&= I \to x \\
&= x.
\end{aligned}$$

$$\begin{aligned}
x \wedge x &= ((x \to x) \to x')' \\
&= (I \to x')' \\
&= (x')' \\
&= x. \ \square
\end{aligned}$$

Lemma 7.3.12 *For any $x, y \in L$, $x \wedge (x \vee y) = x$, and $x \vee (x \wedge y) = x$.*

Proof. By Lemmas 7.3.7, 7.3.9 and (I_2), we have $(x \wedge (x \vee y)) \to x = I$ and

$$\begin{aligned} x \to (x \wedge (x \vee y)) &= (x \to x) \wedge (x \to (x \vee y)) \\ &= I \wedge I \\ &= I, \end{aligned}$$

then $x \wedge (x \vee y) = x$.

By Lemmas 7.3.1 and 7.3.4, we have

$$\begin{aligned} x \vee (x \wedge y) &= ((x \vee (x \wedge y))')' \\ &= (x' \wedge (x \wedge y)')' \\ &= (x' \wedge (x' \vee y'))' = (x')' \\ &= x. \ \square \end{aligned}$$

We go back to the proof of Theorem 7.3.3. By Lemmas 7.3.6, 7.3.10, 7.3.11 and 7.3.12 we know that $(L, \vee, \wedge)$ is a lattice. Let $\leqslant$ be the partial order induced by the lattice structure of L, *i.e.*,

$$\begin{aligned} x \leqslant y \ \text{ iff } \ & x \vee y = y \\ \text{iff } \ & x \wedge y = x. \end{aligned}$$

Lemmas 7.3.7 and 7.3.8 tell us that L is a bounded lattice. Lemmas 7.3.1 and 7.3.4 tell us that the operation $'$ is an order-reversing involution of L. In a word, $(L, \vee, \wedge, ', O, I)$ is a bounded lattice with an order-reversing involution and hence $(L, \vee, \wedge, ', \to, O, I)$ is a lattice implication algebra.

8 Related Issues

In this chapter, we discuss some issues related to lattice implication algebras. In Sections 8.1 and 8.2, we establish the categories of lattice implication algebras and fuzzy lattice implication algebras and discuss their basic properties. In Section 8.3 we presented the fuzzy power sets theory based on lattice implication algebras, which is a generalization of Bandler and Kohout's fuzzy power sets theory. In Section 8.4, we discuss the properties of semigroups $(L, \oplus)$, $(L, \otimes)$ and adjoint semigroup $(M(L), \circ, 1_L)$, they are induced by a lattice implication algebra L. In Section 8.5, we formalize lattice implication algebra system in a first-order language with identity and study the logical properties of this algebra.

8.1 Category of Lattice Implication Algebras

In this section we discuss the total properties and relations between lattice implication algebras, and give basic concepts and results of the category of lattice implication algebras. We refer the readers to the reference [415] for details of basic concepts of category theory.

Definition 8.1.1 *The category of lattice implication algebras, denoted by* $\boldsymbol{LC}$*, is the category that has as objects all lattice implication algebras and as arrows all lattice implication homomorphisms between the corresponding lattice implication algebras with the usual composition of functions.*

Let $f: A \longrightarrow B$ be an arrow of $\boldsymbol{LC}$, it is trivial to verify that f is invertible if and only if f is a lattice implication isomorphism from A to B. Moreover, if the lattice implication algebra A has exactly two elements, O and I, then there is exactly one lattice implication homomorphism from A to any lattice implication algebra, and hence A is an initial object of $\boldsymbol{LC}$. Similarly, the lattice implication algebra B is a terminal object of $\boldsymbol{LC}$ if B has exactly one element.

Definition 8.1.2 *If* $S: \boldsymbol{D} \longrightarrow \boldsymbol{C}$ *is a functor from the category* $\boldsymbol{D}$ *to the category* $\boldsymbol{C}$ *and* c *an object of* $\boldsymbol{C}$*, a universal arrow from* c *to* S *is a pair* $\langle r, u\rangle$ *consisting of an object* r *of* $\boldsymbol{D}$ *and an arrow* $u\colon c \longrightarrow Sr$ *of* $\boldsymbol{C}$*, such that to* *every pair $\langle d, f\rangle$ with d an object of $\mathbf{D}$ and $f\colon c \longrightarrow Sd$ an arrow of $\mathbf{C}$, there is a unique arrow $f'\colon r \longrightarrow d$ of $\mathbf{D}$ with $Sf' \circ u = f$.*

The notion of a universal arrow from S to c is dual to that of a universal arrow from c to S. If $\mathbf{D}$ is the category $(j \leftarrow I \rightarrow k)$, $S: \mathbf{D} \longrightarrow \mathbf{LC}$ is a functor, $u\colon \{O, I\} \longrightarrow Si$ is the lattice implication homomorphism from $\{O, I\}$ to Si, for any object d of $\mathbf{D}$ and lattice implication homomorphism f from $\{O, I\}$ to Sd, there exists only one arrow f' of $\mathbf{D}$ from i to d such that

$$Sf' \circ u = f,$$

and hence the universal arrow from $\{O, I\}$ to S is $\langle i, u\rangle$. Similarly, if $T\colon \mathbf{D}^{op} \longrightarrow \mathbf{LC}$ is a functor, $v\colon Ti \longrightarrow \{O, I\}$ is the lattice implication homomorphism, then $\langle i, v\rangle$ is the universal arrow from T to $\{O, I\}$, where $Ti = \{O, I\}$.

Let $\mathbf{F(LC)}$ be the forgetful category of $\boldsymbol{LC}$, $S: \mathbf{LC} \longrightarrow \mathbf{F(LC)}$ the forgetful functor, and i an initial object of $\boldsymbol{LC}$. Let $u\colon \{O, I\} \longrightarrow Si$ be the unique arrow of $\mathbf{F(LC)}$ from $\{O, I\}$ to Si, then $\langle i, u\rangle$ is a universal arrow from $\{O, I\}$ to S. Similarly, a universal arrow from S to $\{O, I\}$ is a pair $\langle j, v\rangle$ with j an initial object of $\boldsymbol{LC}$ and $v\colon Sj \longrightarrow \{O, I\}$ the unique arrow of $\mathbf{F(LC)}$ from Sj to $\{O, I\}$.

Definition 8.1.3 *Let $\boldsymbol{C}$ and $\boldsymbol{J}$ be two categories, $\Delta: \boldsymbol{C} \longrightarrow \mathbf{C}^{\mathbf{J}}$ be the diagonal functor. A limit for a functor $F: \boldsymbol{J} \longrightarrow \boldsymbol{C}$ is a universal arrow from Δ to F.*

In this definition, if $\mathbf{J}$ is a discrete category, or $\mathbf{J} =\downarrow\downarrow$, or $\mathbf{J} = (\rightarrow, \leftarrow)$, a limit for a functor $F: \mathbf{J} \longrightarrow \mathbf{C}$ is called the product, an equalizer, or the pullback of F, respectively.

Theorem 8.1.1 *Let $\{A_\lambda \mid \lambda \in \Gamma\}$ be a family of objects of $\mathbf{LC}$, the product of $\{A_\lambda \mid \lambda \in \Gamma\}$ is the lattice implication product algebra $\prod_{\lambda\in\Gamma} A_\lambda$ as object together with projecting lattice implication homomorphism*

$$P_i\colon \prod_{\lambda\in\Gamma} A_\lambda \longrightarrow A_i,$$

for any $i \in \Gamma$, where for any $f \in \prod_{\lambda\in\Gamma} A_\lambda$, $P_i(f) = f(i)$.

Proof. For any $f, g \in \prod_{\lambda\in\Gamma} A_\lambda$ and $i \in \Gamma$, $P_i(O) = O(i) = O_i$ and

$$\begin{aligned} P_i(f \rightarrow g) &= (f \rightarrow g)(i) \\ &= f(i) \rightarrow g(i) \\ &= P_i(f) \rightarrow P_i(g). \end{aligned}$$

It follows that P_i is an arrow of $\boldsymbol{LC}$ from $\prod_{\lambda\in\Gamma} A_\lambda$ to A_i. For any object C of $\boldsymbol{LC}$ and arrows $h_i\colon C \longrightarrow A_i$ of $\boldsymbol{LC}$, let

$$\delta\colon C \longrightarrow \prod_{\lambda\in\Gamma} A_\lambda$$

with $\delta(x)(i) = h_i(x)$ for any $x \in C$ and $i \in \Gamma$. For any $x, y \in C$ and $i \in \Gamma$, it follows that $\delta(O)(i) = h_i(O) = O_i$ and

$$\begin{aligned}\delta(x \rightarrow y)(i) &= h_i(x \rightarrow y)\\ &= h_i(x) \rightarrow h_i(y)\\ &= \delta(x)(i) \rightarrow \delta(y)(i)\\ &= (\delta(x) \rightarrow \delta(y))(i),\end{aligned}$$

i.e., $\delta(x \rightarrow y) = \delta(x) \rightarrow \delta(y)$. So, δ is an arrow of $\boldsymbol{LC}$ from C to $\prod_{\lambda\in\Gamma} A_\lambda$, and for any $x \in C$ and $i \in \Gamma$,

$$\begin{aligned}P_i \circ \delta(x) &= P_i(\delta(x))\\ &= \delta(x)(i)\\ &= h_i(x),\end{aligned}$$

that is, for any $i \in \Gamma$, $h_i = P_i \circ \delta$.

If δ_1 and δ_2 are arrows of $\boldsymbol{LC}$ from C to $\prod_{\lambda\in\Gamma} A_\lambda$ with $P_i \circ \delta_1 = h_i$ and $P_i \circ \delta_2 = h_i$ for any $i \in \Gamma$, then $\delta_1 = \delta_2$. In fact, if $\delta_1 \neq \delta_2$, then there exists $x \in C$ such that $\delta_1(x) \neq \delta_2(x)$ and hence there exists $i \in \Gamma$ such that $\delta_1(x)(i) \neq \delta_2(x)(i)$, this is a contradiction with $P_i \circ \delta_1 = h_i = P_i \circ \delta_2$. □

Theorem 8.1.2 *Let A and B be objects of* **LC**, *f, g arrows of* **LC** *from A to B,*

$$E = \{x \mid x \in A, f(x) = g(x)\},$$
$$i\colon E \longrightarrow A$$

with $i(x) = x$ for any $x \in E$, then $\langle E, i\rangle$ is an equalizer of f and g.

Proof. If $x \in E$, then

$$\begin{aligned}f(x') &= (f(x))'\\ &= (g(x))'\\ &= g(x'),\end{aligned}$$

and hence $x' \in E$. Similarly, for any $x, y \in A$, $x \vee y$, $x \wedge y$, $x \rightarrow y \in E$ provided $x, y \in E$. So, E is a lattice implication subalgebra of A and hence an object of $\boldsymbol{LC}$. It is trivial to verify that i is an arrow of $\boldsymbol{LC}$ and $f \circ i = g \circ i$.

If C is an object of $\boldsymbol{LC}$ and

$$h\colon C \longrightarrow A$$

is an arrow of $\boldsymbol{LC}$ with $f \circ h = g \circ h$, then $f(h(x)) = g(h(x))$ and hence $h(x) \in E$ for any $x \in C$. Let $k\colon C \longrightarrow E$ with $k(x) = h(x)$ for any $x \in C$, then k is an arrow of $\boldsymbol{LC}$ and $i \circ k = h$.

If $k_1, k_2\colon C \longrightarrow E$ are arrows of $\boldsymbol{LC}$ with $i \circ k_1 = h$ and $i \circ k_2 = h$, then $k_1(x) = h(x) = k_2(x)$ for any $x \in C$, and hence $k_1 = k_2$. □

Let A, B, and X be objects of $\boldsymbol{LC}$, $\varphi\colon\ A \longrightarrow X$ and $\psi\colon\ B \longrightarrow X$ arrows of $\boldsymbol{LC}$, and

$$Y = \{(a,b) \mid a \in A, b \in B, \varphi(a) = \psi(b)\}.$$

Define operations $'$, $\vee$, $\wedge$, and $\rightarrow$ on Y as follows:

$$\begin{aligned}(a,b)' &= (a',b'),\\ (a_1,b_1)\vee(a_2,b_2) &= (a_1\vee a_2, b_1\vee b_2),\\ (a_1,b_1)\wedge(a_2,b_2) &= (a_1\wedge a_2, b_1\wedge b_2),\\ (a_1,b_1)\rightarrow(a_2,b_2) &= (a_1\rightarrow a_2, b_1\rightarrow b_2).\end{aligned}$$

It can be verified that $(Y, \vee, \wedge, ', \rightarrow)$ is a lattice implication algebra.

Let $\alpha\colon\ Y \longrightarrow A$ and $\beta\colon\ Y \longrightarrow B$ with $\alpha((a,b)) = a$, $\beta((a,b)) = b$ for any $(a,b) \in Y$, we have the following theorem.

Theorem 8.1.3 *(Y, α, β) is the pullback of φ and ψ.*

Proof. It is trivial to verify that $(Y, \vee, \wedge, ', \rightarrow)$ is an object of $\boldsymbol{LC}$ and α, β are arrows of $\boldsymbol{LC}$. For any $(a,b) \in Y$,

$$\varphi\circ\alpha((a,b)) = \varphi(a) = \psi(b) = \psi\circ\beta((a,b)),$$

i.e., $\varphi\circ\alpha = \psi\circ\beta$.

If Z is an object of $\boldsymbol{LC}$ and $\gamma\colon\ Z \longrightarrow A$, $\delta\colon\ Z \longrightarrow B$ are arrows of $\boldsymbol{LC}$ such that $\varphi\circ\gamma = \psi\circ\delta$, then $\varphi(\gamma(z)) = \psi(\delta(z))$ and hence $(\gamma(z), \delta(z)) \in Y$ for any $z \in Z$.

Let $h\colon\ Z \longrightarrow Y$ with $h(z) = (\gamma(z), \delta(z))$. For any $z_1, z_2 \in Z$,

$$\begin{aligned}h(z_1\rightarrow z_2) &= (\gamma(z_1\rightarrow z_2), \delta(z_1\rightarrow z_2))\\ &= (\gamma(z_1)\rightarrow\gamma(z_2), \delta(z_1)\rightarrow\delta(z_2))\\ &= (\gamma(z_1),\delta(z_1))\rightarrow(\gamma(z_2),\delta(z_2))\\ &= h(z_1)\rightarrow h(z_2),\end{aligned}$$

and $h(O) = O$, so h is an arrow of $\boldsymbol{LC}$. $\gamma = \alpha\circ h$ and $\delta = \beta\circ h$ are obvious.

If $h_1, h_2\colon\ Z \longrightarrow Y$ are arrows of $\boldsymbol{LC}$ with $\gamma = \alpha\circ h_1$, $\delta = \beta\circ h_1$, $\gamma = \alpha\circ h_2$ and $\delta = \beta\circ h_2$, then $h_1(z) = (\gamma(z), \delta(z)) = h_2(z)$ for each $z \in Z$ and hence $h_1 = h_2$. $\square$

Corollary 8.1.1 *(Y, α, β) is the kernel pair of φ if $A = B$ and $\varphi = \psi$.* $\square$

Definition 8.1.4 *Let $\boldsymbol{C}$ be a category with terminal object t, a subobjects classifier of $\boldsymbol{C}$ is an object Ω of $\boldsymbol{C}$ and an arrow $True\colon\ t \longrightarrow \Omega$ of $\boldsymbol{C}$ with the following Ω-axiom: for any monic arrow $f\colon\ a \longrightarrow d$, there is a unique arrow $g\colon\ d \longrightarrow \Omega$ such that (a, f, h) is the pullback of g and $True$, where h is the unique arrow of $\boldsymbol{C}$ from a to t.*

Theorem 8.1.4 *$\boldsymbol{LC}$ has no subobjects classifier.*

Proof. If $\boldsymbol{LC}$ has subobjects classifier (Ω, f), then $f\colon \{I\} \longrightarrow \Omega$ is an arrow of $\boldsymbol{LC}$, $\{I\}$ is a terminal object of $\boldsymbol{LC}$, and hence $\Omega = \{I'\}$ is also a terminal object of $\boldsymbol{LC}$, because f is a lattice implication homomorphism.

Let $A = \{O, I\}$ and $B = \{O, I\}$ be two objects of $\boldsymbol{LC}$, $g\colon A \longrightarrow B$ with $g(O) = O$ and $g(I) = I$, then g is a monic arrow of $\boldsymbol{LC}$, but (A, f_1, f_2) is not a pullback square, where

$$f_1(O) = f_1(I) = I,$$

$$f_2(O) = f_2(a) = f_2(I) = I'.$$

In fact, for arrows $g_1:\ B \longrightarrow \{I\}$ and $g_2:\ B \longrightarrow B$ with $g_1(x) = I$ and $g_2(x) = x$ for any $x \in B$, $f_1 \circ g_1 = f_2 \circ g_2$, but for any arrow $h\colon\ B \longrightarrow A$, $g \circ h \neq g_2$. □

In the following we discuss the relation between the category of lattice implication algebras and the category of l-groups with order unit. We use the properties of lattice-ordered groups and MV-algebras to describe the lattice implication algebras, and get some important conclusions.

The category of MV-algebras is the category that has as objects all MV-algebras and as arrows all homomorphisms between the corresponding MV-algebras. And the category of l-groups with an order unit is the category that has as objects all l-groups with the order unit and as arrows all unital l-homomorphism between the corresponding l-groups with the order unit.

Let $A = (A, +, \cdot, *, O, I)$ be an MV-algebra, define $\Gamma(A) = (A, \rightarrow, ', O, I)$, where, $x \rightarrow y = x^* + y$ and $x' = x^*$ hold for any $x, y \in A$. For MV-algebras A and $A^\diamond$, and the homomorphism $\psi : A \longrightarrow A^\diamond$, we define

$$\Gamma(\psi) : \Gamma(A) \longrightarrow \Gamma(A^\diamond),$$

such that $\Gamma(\psi) = \psi$.

According to Theorems 7.2.2 and 7.2.3, we know that $\Gamma(A)$ is a lattice implication algebra and Γ is a faithful and full functor from the category of MV-algebras to the category of lattice implication algebras.

Let $G = (G, +, -, O_G, \vee_G, \wedge_G)$ be a lattice-ordered Abelian group with identity O_G about operation $+$ (we call it l-group simply). The inverse of x about operation $+$ is written as $-x$, and $a + a + \cdots + a$ (to n summands) as na. Let $\leqslant_G$ be the order relation on G induced by a lattice operation and $G^+ = \{g \in G \mid O_G \leqslant_G g\}$. An element $u \in G$ is called an order unit if for any $g \in G$, there exists a number n such that $g \leqslant_G n \cdot u$. In what follows, the l-group $G = (G, +, -, O_G, \vee_G, \wedge_G)$ is denoted simply by G whenever this may cause no confusion.

Suppose that G and $G^\triangle$ are l-groups with order units u and $u^\triangle$ respectively. The mapping

$$\lambda\colon\ G \longrightarrow G^\triangle$$

is called a unital l-homomorphism if and only if λ is a group homomorphism and a lattice homomorphism such that $\lambda(u) = u'$. A unital l-homomorphism is called a unital l-isomorphism if it is bijective. We write

$$(G, u) \cong (G^{\Delta}, u^{\Delta})$$

if there exists a unital l-isomorphism from G to G^{Δ}.

Definition 8.1.5 *Let G be an l-group with an order unit u, and define $\Sigma(G, u) = (A, \boxplus, \cdot, *, O, I)$ as follows:*

$$A = [O_G, u] = \{g \in G \mid O_G \leqslant_G g \leqslant_G u\},$$

and for any $x, y \in A$,

$$\begin{aligned} x \boxplus y &= u \wedge_G (x + y); \\ x^* &= u - x; \\ x \cdot y &= (x^* \boxplus y^*)^*; \\ O &= O_G; \\ I &= u. \end{aligned}$$

Furthermore, given a unital l-homomorphism $\theta\colon\ (G, u) \longrightarrow (G^{\nabla}, u^{\nabla})$, define

$$\Sigma(\theta)\colon\ \Sigma(G, u) \longrightarrow \Sigma(G^{\nabla}, u^{\nabla})$$

by $\Sigma(\theta) = \theta|_A$, i.e., $\Sigma(\theta)$ is the restriction of θ to A.

Lemma 8.1.1 *[315]*

(1). For any l-group G with an order unit u, $\Sigma(G, u)$ is an MV-algebra.
(2). Σ is a faithful and full functor from the category of l-group with an order unit to MV-algebra category. For any such group (G, u), the lattice operations on the unit interval $[O_G, u]$ of G agree with the lattice operations on the MV-algebra $\Sigma(G, u)$ as given by Definition 7.2.1. □

Lemma 8.1.2 *[315] Let $A = (A, \boxplus, \cdot, *, O, I)$ be an MV-algebra. Then there exists an l-group G with an order unit u such that*

$$A \cong \Sigma(G, u). \ \square$$

Lemma 8.1.1 tells us that the category of MV-algebras and the category of l-groups with the order unit are equivalent categories. We can use the properties of l-groups with the order unit to describe MV-algebras and lattice implication algebras.

Let Γ be the faithful and full functor from the category of MV-algebras to the category of the lattice implication algebras, and Σ the faithful and full functor from the category of l-groups with an order unit to the category of MV-algebras which are defined as above.

Theorem 8.1.5 *The composition of functor Γ and Σ is a faithful and full functor from the category of l-groups with an order unit to the category of the lattice implication algebras.* □

Corollary 8.1.2 *For any lattice implication algebra L, there exists an l-group G with an order unit u such that $L \cong \Psi(G, u)$. Here,*

$$\begin{aligned} \Psi(G,u) &= (G(u), \rightarrow, ', O, I), \\ G(u) &= [O_G, u] = \{g \in G \mid O_G \leqslant_G g \leqslant_G u\}, \\ O &= O_G, \\ I &= u, \end{aligned}$$

and for any $x, y \in L$,

$$\begin{aligned} x \rightarrow y &= u \wedge (u - x + y), \\ x' &= u - x. \end{aligned}$$

Proof. Suppose $x + y = u \wedge (x + y)$, $x^* = u - x$, then

$$x \rightarrow y = x^* + y = u \wedge (u - x + y).$$

We can get $\Psi = \Gamma \circ \Sigma$ by Theorem 8.1.5. □

Corollary 8.1.2 tells us that the category of the lattice implication algebras and the category of l-groups with an order unit are equivalent categories.

Theorem 8.1.6 *If L is complete lattice implication algebra with the infinite distributive law, $\{x_j \mid j \in J\} \subseteq L$, $\{y_k \mid k \in K\} \subseteq L$, $x \in L$, then*

(1). $(\bigvee_{j \in J} x_j)' = \bigwedge_{j \in J} x_j'$;

(2). $(\bigwedge_{j \in J} x_j)' = \bigvee_{j \in J} x_j'$;

(3). $x \rightarrow (\bigvee_{j \in J} x_j) = \bigvee_{j \in J} (x \rightarrow x_j)$;

(4). $x \rightarrow (\bigwedge_{j \in J} x_j) = \bigwedge_{j \in J} (x \rightarrow x_j)$;

(5). $(\bigvee_{j \in J} x_j) \rightarrow x = \bigwedge_{j \in J} (x_j \rightarrow x)$;

(6). $(\bigwedge_{j \in J} x_j) \rightarrow x = \bigvee_{j \in J} (x_j \rightarrow x)$;

(7). $(\bigvee_{j \in J} x_j) \otimes (\bigvee_{k \in K} y_k) = \bigvee_{j \in J} \bigvee_{k \in K} (x_j \otimes y_k)$;

(8). $(\bigwedge_{j \in J} x_j) \otimes (\bigwedge_{k \in K} y_k) = \bigwedge_{j \in J} \bigwedge_{k \in K} (x_j \otimes y_k)$.

Proof. We only prove (1), (3) and (8), others can be proved similarly.

For (1). By $\vee_{j \in J} x_j \geqslant x_j$ for any $j \in J$, it follows that $(\vee_{j \in J} x_j)' \leqslant x_j'$ and hence $(\vee_{j \in J} x_j)' \leqslant \wedge_{j \in J} x_j'$. On the other hand, $\wedge_{j \in J} x_j' \leqslant x_j'$ for any

$j \in J$, it follows that $(\wedge_{j\in J} x_j')' \geqslant x_j$ and hence $(\wedge_{j\in J} x_j')' \geqslant \vee_{j\in J} x_j$, that is, $\wedge_{j\in J} x_j' \leqslant (\vee_{j\in J} x_j)'$.

For (3). We need only to prove that

$$x \to (\bigvee_{i\in J} x_i) = \bigvee_{i\in J} (x \to x_i)$$

holds in $G(u)$. In fact, by the properties of l-group,

$$\begin{aligned} x \to (\bigvee_{i\in J} x_i) &= u \wedge (u - x + \bigvee_{i\in J} x_i) \\ &= u \wedge (\bigvee_{i\in J} (u - x + x_i)) \\ &= \bigvee_{i\in J} (u \wedge (u - x + x_i)) \\ &= \bigvee_{i\in J} (x \to x_i). \end{aligned}$$

For (8). We have

$$\begin{aligned} (\bigwedge_{j\in J} x_j) \otimes (\bigwedge_{k\in K} y_k) &= \left(\left(\bigwedge_{j\in J} x_j\right) \to \left(\bigwedge_{k\in K} y_k\right)'\right)' \\ &= \left(\left(\bigwedge_{j\in J} x_j\right) \to \left(\bigwedge_{k\in K} y_k'\right)\right)' \\ &= \left(\bigvee_{j\in J} \bigvee_{k\in K} (x_j \to y_k')\right)' \\ &= \bigwedge_{j\in J} \bigwedge_{k\in K} (x_j \to y_k')' \\ &= \bigwedge_{j\in J} \bigwedge_{k\in K} (x_j \otimes y_k). \ \square \end{aligned}$$

8.2 Category of Fuzzy Lattice Implication Algebras

In this section, we establish the category of fuzzy lattice implication algebras ***FLC*** and discuss its basic properties and the relations between ***FLC*** and ***LC***.

Definition 8.2.1 *The category of fuzzy lattice implication algebras, denoted by* ***FLC****, is defined as:*

$$\begin{aligned} Ob(\boldsymbol{FLC}) &= \{(L, A) \mid L \textit{ is a lattice implication algebra}, \\ &\quad A \textit{ is a fuzzy lattice implication subalgebra of } L\}; \\ \hom((L_1, A_1), (L_2, A_2)) &= \{f \mid f \colon L_1 \longrightarrow L_2 \textit{ is a lattice implication} \\ &\quad \textit{homomorphism, satisfies } A_1 \subseteq A_2 \circ f\}. \end{aligned}$$

The composition of arrows is the usual composition of functions. In this category, the identity of object (L, A) is the identity function $i\colon L \longrightarrow L$ with $i(x) = x$ for any $x \in L$. If $L = \{I\}$ and $A\colon L \longrightarrow [0, 1]$ with $A(I) = 1$, then (L, A) is a terminal object of ***FLC****.*

Theorem 8.2.1 *Let $\{(L_\lambda, A_\lambda) \mid \lambda \in \Gamma\}$ be a family of objects of* ***FLC*** *and $\prod_{\lambda\in\Gamma} L_\lambda$ the lattice implication product algebra of $\{L_\lambda \mid \lambda \in \Gamma\}$. $A : \prod_{\lambda\in\Gamma} L_\lambda \longrightarrow [0, 1]$ and $P_i : \prod_{\lambda\in\Gamma} L_\lambda \longrightarrow L_i (i \in \Gamma)$ are defined respectively as: for any $f \in \prod_{\lambda\in\Gamma} L_\lambda$,*

$$A(f) = \bigwedge_{\lambda\in\Gamma} A_\lambda(f(\lambda)),$$

$$P_i(f) = f(i),$$

then the product of $\{(L_\lambda, A_\lambda) \mid \lambda \in \Gamma\}$ is $(\prod_{\lambda\in\Gamma} L_\lambda, A)$ as object together with $\{P_\lambda \mid \lambda \in \Gamma\}$.

Proof. (1). A is a fuzzy lattice implication subalgebra of $\prod_{\lambda\in\Gamma} L_\lambda$. In fact,

$$\begin{aligned} A(I) &= \bigwedge_{\lambda\in\Gamma} A_\lambda(I(\lambda)) \\ &= \bigwedge_{\lambda\in\Gamma} A_\lambda(I_\lambda) \\ &= \bigwedge_{\lambda\in\Gamma} A_\lambda(O_\lambda) \\ &= \bigwedge_{\lambda\in\Gamma} A_\lambda(O(\lambda)) \\ &= A(O). \end{aligned}$$

For any $f, g \in \prod_{\lambda\in\Gamma} L_\lambda$ and $\lambda \in \Gamma$, by

$$A_\lambda(f(\lambda) \to g(\lambda)) \geqslant \min\{A_\lambda(f(\lambda)), A_\lambda(g(\lambda))\},$$

it follows that

$$\begin{aligned}A(f \to g) &= \bigwedge_{\lambda\in\Gamma} A_\lambda((f \to g)(\lambda))\\ &= \bigwedge_{\lambda\in\Gamma} A_\lambda(f(\lambda) \to g(\lambda))\\ &\geqslant \bigwedge_{\lambda\in\Gamma} \min\{A_\lambda(f(\lambda)), A_\lambda(g(\lambda))\}\\ &= \min\{\bigwedge_{\lambda\in\Gamma} A_\lambda(f(\lambda)), \bigwedge_{\lambda\in\Gamma} A_\lambda(g(\lambda))\}\\ &= \min\{A(f), A(g)\}.\end{aligned}$$

(2). For any $i \in \Gamma$, P_i is an arrow of ***FLC***. In fact, for any $f \in \prod_{\lambda\in\Gamma} L_\lambda$,

$$\begin{aligned}(A_i \circ P_i)(f) &= A_i(P_i(f))\\ &= A_i(f(i))\\ &\geqslant \bigwedge_{\lambda\in\Gamma} A_\lambda(f(\lambda))\\ &= A(f),\end{aligned}$$

i.e., $A \subseteq A_i \circ P_i$.

(3). For any object (L, B) of ***FLC*** and arrows

$$\zeta_j\colon\ (L, B) \longrightarrow (L_j, A_j)(\text{for any } j \in \Gamma)$$

of ***FLC***, let

$$\zeta\colon\ (L, B) \longrightarrow (\prod_{\lambda\in\Gamma} L_\lambda, A)$$

with $\zeta(x)(j) = \zeta_j(x)$ for any $x \in L$ and $j \in \Gamma$, then ζ is an arrow of ***FLC*** and $P_j \circ \zeta = \zeta_j$. In fact, for any $x \in L$,

$$\begin{aligned}(A \circ \zeta)(x) &= A(\zeta(x))\\ &= \bigwedge_{\lambda\in\Gamma} A_\lambda(\zeta(x)(\lambda))\\ &= \bigwedge_{\lambda\in\Gamma} A_\lambda(\zeta_\lambda(x)).\end{aligned}$$

By $B \subseteq A_j \circ \zeta_j$, it follows that $B(x) \leqslant A_j(\zeta_j(x))$ and hence

$$B(x) \leqslant \bigwedge_{j\in\Gamma} A_j(\zeta_j(x)) = (A \circ \zeta)(x),$$

i.e., $B \subseteq A \circ \zeta$ and hence ζ is an arrow of ***FLC***. For any $x \in L$ and $j \in \Gamma$,

$$\begin{aligned}(P_j \circ \zeta)(x) &= P_j(\zeta(x))\\ &= \zeta(x)(j)\\ &= \zeta_j(x),\end{aligned}$$

it follows that $P_j \circ \zeta = \zeta_j$.

(4) If the arrow

$$\eta\colon\ (L, B) \longrightarrow (\prod_{\lambda\in\Gamma} L_\lambda, A)$$

of ***FLC*** satisfies $P_j \circ \eta = \zeta_j$ for any $j \in \Gamma$, then $\eta = \zeta$. Otherwise, there exists $x \in L$ such that $\eta(x) \neq \zeta(x)$ and hence there exists $j \in \Gamma$ with

$$\eta(x)(j) \neq \zeta(x)(j),$$

by $\eta(x)(j) = P_j(\eta(x)) = (P_j \circ \eta)(x)$ and $\zeta(x)(j) = P_j(\zeta(x)) = (P_j \circ \zeta)(x)$, it follows that $(P_j \circ \eta)(x) \neq (P_j \circ \eta)(x)$. This is a contradiction with $P_j \circ \eta = \zeta_j = P_j \circ \zeta$. □

Theorem 8.2.2 *Let (L_1, A_1) and (L_2, A_2) be objects of **FLC**, f, g arrows of **FLC** from (L_1, A_1) to (L_2, A_2), $E = \{x \mid x \in L_1, f(x) = g(x)\}$, $B\colon\ E \longrightarrow [0,1]$ with $B(x) = A_1(x)$ and $i\colon\ (E, A) \longrightarrow (L_1, A_1)$ with $i(x) = x$ for any $x \in E$, then E is a lattice implication subalgebra of L_1 and $\langle (E, B), i\rangle$ is an equalizer of f and g.*

Proof. (1). By Theorem 8.1.2, E is a lattice implication subalgebra of L_1. For any $x, y \in E$,

$$B(I) = A_1(I) = A_1(O) = B(O),$$

$$\begin{aligned} B(x \to y) &= A_1(x \to y) \\ &\geqslant \min\{A_1(x), A_1(y)\} \\ &= \min\{B(x), B(y)\}, \end{aligned}$$

it follows that (E, B) is an object of ***FLC***. By

$$\begin{aligned} (A_1 \circ i)(x) &= A_1(i(x)) \\ &= A_1(x) \\ &= B(x), \end{aligned}$$

it follows that $B \subseteq A_1 \circ i$ and hence i is an arrow of ***FLC***. $f \circ i = g \circ i$ is trivial.

(2). If there exist some object (L_3, A_3) and arrow $\zeta\colon\ (L_3, A_3) \longrightarrow (L_1, A_1)$ of ***FLC*** with $f \circ \zeta = g \circ \zeta$, then there exists a unique arrow $\sigma\colon\ (L_3, A_3) \longrightarrow (E, B)$ such that $i \circ \sigma = \zeta$. In fact, for any $x \in L_3$,

$$\begin{aligned} f(\zeta(x)) &= (f \circ \zeta)(x) \\ &= (g \circ \zeta)(x) \\ &= g(\zeta(x)), \end{aligned}$$

i.e., $\zeta(x) \in E$. Let $\sigma\colon\ L_3 \longrightarrow E$ with $\sigma(x) = \zeta(x)$ for any $x \in L_3$, then σ is a lattice implication homomorphism and

$$
\begin{aligned}
(B \circ \sigma)(x) &= B(\sigma(x)) \\
&= B(\zeta(x)) \\
&= A_1(\zeta(x)) \\
&= (A_1 \circ \zeta)(x) \\
&\geqslant A_3(x),
\end{aligned}
$$

i.e., $A_3 \subseteq B \circ \sigma$ and σ is an arrow of ***FLC***. $i \circ \sigma = \zeta$ is trivial. If the arrow $\eta\colon\ (L_3, A_3) \longrightarrow (E, B)$ of ***FLC*** satisfies $i \circ \eta = \zeta$, then

$$
\begin{aligned}
\eta(x) &= i(\eta(x)) \\
&= \zeta(x) \\
&= \sigma(x)
\end{aligned}
$$

for any $x \in L_3$, *i.e.*, $\eta = \sigma$. $\square$

Let (L_1, A_1), (L_2, A_2), and (L_3, A_3) be objects of ***FLC***, $f\colon\ (L_1, A_1) \longrightarrow (L_3, A_3)$ and $g\colon\ (L_2, A_2) \longrightarrow (L_3, A_3)$ arrows of ***FLC***, and $L_4 = \{(a, b) \mid a \in L_1, b \in L_2, f(a) = g(b)\}$. Define operations $'$, $\vee$, $\wedge$ and $\rightarrow$ on L_4 as follows:

$$
\begin{aligned}
(a, b)' &= (a', b'), \\
(a_1, b_1) \vee (a_2, b_2) &= (a_1 \vee a_2, b_1 \vee b_2), \\
(a_1, b_1) \wedge (a_2, b_2) &= (a_1 \wedge a_2, b_1 \wedge b_2), \\
(a_1, b_1) \rightarrow (a_2, b_2) &= (a_1 \rightarrow a_2, b_1 \rightarrow b_2).
\end{aligned}
$$

By Theorem 8.1.3, $(L_4, \vee, \wedge, ', \rightarrow)$ is a lattice implication algebra. Let $A_4:\ L_4 \longrightarrow [0, 1]$, $\alpha:\ L_4 \longrightarrow L_1$ and $\beta:\ L_4 \longrightarrow L_2$ with $A_4((a, b)) = A_1(a) \wedge A_2(b)$, $\alpha((a, b)) = a$ and $\beta((a, b)) = b$ for any $(a, b) \in L_4$.

Theorem 8.2.3 *(L_4, α, β) is the pullback of f and g.*

Proof. (1). (L_4, A_4) is an object of ***FLC***. In fact, (I_1, I_2) and (O_1, O_2) are the greatest and the smallest elements of L_4, respectively, where I_i and O_i are the greatest and the smallest elements of L_i respectively, $i = 1, 2$. For any $(a_1, b_1), (a_2, b_2) \in L_4$,

$$
\begin{aligned}
A_4(I_1, I_2) &= A_1(I_1) \wedge A_2(I_2) \\
&= A_1(O_1) \wedge A_2(O_2) \\
&= A_4(O_1, O_2),
\end{aligned}
$$

$$
\begin{aligned}
A_4((a_1, b_1) \rightarrow (a_2, b_2)) &= A_4(a_1 \rightarrow a_2, b_1 \rightarrow b_2) \\
&= A_1(a_1 \rightarrow a_2) \wedge A_2(b_1 \rightarrow b_2) \\
&\geqslant \min\{A_1(a_1), A_1(a_2)\} \wedge \min\{A_2(b_1), A_2(b_2)\} \\
&= \min\{A_1(a_1) \wedge A_2(b_1), A_1(a_2) \wedge A_2(b_2)\} \\
&= \min\{A_4(a_1, b_1), A_4(a_2, b_2)\},
\end{aligned}
$$

it follows that A_4 is a fuzzy subalgebra of L_4 and (L_4, A_4) is an object of $\boldsymbol{FLC}$.

(2). α and β are arrows of $\boldsymbol{FLC}$ and $f \circ \alpha = g \circ \beta$. In fact, it is routine to check that α and β are lattice implication homomorphisms. For any $(a, b) \in L_4$,

$$\begin{aligned}(A_1 \circ \alpha)(a, b) &= A_1(\alpha(a, b)) \\ &= A_1(a) \geqslant A_4(a, b), \\ (A_2 \circ \beta)(a, b) &= A_2(\beta(a, b)) \\ &= A_2(b) \geqslant A_4(a, b), \\ (f \circ \alpha)(a, b) &= f(\alpha(a, b)) \\ &= f(a) \\ &= g(b) \\ &= (g \circ \beta)(a, b),\end{aligned}$$

i.e., $f \circ \alpha = g \circ \beta$, α and β are arrows of $\boldsymbol{FLC}$.

(3). Suppose that (L_5, A_5) is an object of $\boldsymbol{FLC}$, $h_1 \colon (L_5, A_5) \longrightarrow (L_1, A_1)$ and $h_2 \colon (L_5, A_5) \longrightarrow (L_2, A_2)$ are arrows of $\boldsymbol{FLC}$ with $f \circ h_1 = g \circ h_2$, then for any $x \in L_5$,

$$f(h_1(x)) = g(h_2(x)),$$

i.e., $(h_1(x), h_2(x)) \in L_4$. Let $\zeta \colon (L_5, A_5) \longrightarrow (L_4, A_4)$ satisfies that $\zeta(x) = (h_1(x), h_2(x))$ for any $x \in L_5$, then ζ is an arrow of $\boldsymbol{FLC}$ and $\alpha \circ \zeta = h_1$, $\beta \circ \zeta = h_2$. In fact, by Theorem 8.1.3, ζ is a lattice implication homomorphism. Moreover, for any $x \in L_5$,

$$A_5(x) \leqslant A_1(h_1(x)),$$

$$A_5(x) \leqslant A_2(h_2(x)).$$

These imply

$$\begin{aligned}(A_4 \circ \zeta)(x) &= A_4(\zeta(x)) \\ &= A_4(h_1(x), h_2(x)) \\ &= A_1(h_1(x)) \wedge A_2(h_2(x)) \\ &\geqslant A_5(x),\end{aligned}$$

i.e., $A_5 \subseteq A_4 \circ \zeta$ and ζ is an arrow of $\boldsymbol{FLC}$. $\alpha \circ \zeta = h_1$ and $\beta \circ \zeta = h_2$ are trivial.

Let $\eta : (L_5, A_5) \longrightarrow (L_4, A_4)$ be an arrow of $\boldsymbol{FLC}$ with $\alpha \circ \eta = h_1$, $\beta \circ \eta = h_2$, then for any $x \in L_5$,

$$\alpha(\eta(x)) = h_1(x),$$

$$\beta(\eta(x)) = h_2(x),$$

it follows that

$$\eta(x) = (h_1(x), h_2(x)) = \zeta(x),$$

and hence $\eta = \zeta$. □

Theorem 8.2.4 *The category* **LC** *is isomorphic to a full subcategory of* **FLC**. □

Definition 8.2.2 *The category* $\mathbf{FLC}_R$ *is defined as:*

$$Ob(\mathbf{FLC}_R) = Ob(\mathbf{FLC}),$$

$$\begin{aligned}\hom((L_1, A_1), (L_2, A_2)) = \{R \mid R\colon\ & L_1 \times L_2 \longrightarrow [0,1], \\ & R(u,v) \leqslant A_1(u) \wedge A_2(v) \text{ for any } u \in L_1, v \in L_2\}.\end{aligned}$$

If $R_1 \in \hom((L_1, A_1), (L_2, A_2))$, $R_2 \in \hom((L_2, A_2), (L_3, A_3))$, *then the composition* $R_2 \circ R_1$ *is defined as: for any* $u \in L_1$ *and* $v \in L_3$,

$$(R_2 \circ R_1)(u, v) = \bigvee_{x \in L_2} (R_1(u, x) \wedge R_2(x, v)).$$

In this category, the identity arrow R_{id} *of an object* (L, A) *satisfies: for any* $u, v \in L$,

$$R_{id}(u,v) = \begin{cases} A(u); & u = v, \\ 0; & u \neq v. \end{cases}$$

Theorem 8.2.5 H : **FLC** $\longrightarrow$ $\mathbf{FLC}_R$ *is a functor, where the object function and arrow function of* H *are defined respectively as follows: for each* $(L_1, A_1), (L_2, A_2) \in Ob(\mathbf{FLC})$, $f \in \hom((L_1, A_1), (L_2, A_2))$ *and* $u \in L_1$, $v \in L_2$,

$$H(L_1, A_1) = (L_1, A_1),$$

$$H(f)(u,v) = \begin{cases} A_1(u), & f(u) = v, \\ 0, & f(u) \neq v. \end{cases}$$

Proof. (1). $H(f) \in \hom_R((L_1, A_1), (L_2, A_2))$. In fact, for any $u \in L_1$, $v \in L_2$, if $f(u) \neq v$, then $H(f)(u,v) = 0 \leqslant A_1(u) \wedge A_2(v)$; if $f(u) = v$, then $A_1 \subseteq A_2 \circ f$ by $f \in \hom((L_1, A_1), (L_2, A_2))$. It follows that

$$\begin{aligned} A_1(u) &\leqslant (A_2 \circ f)(u) \\ &= A_2(f(u)) \\ &= A_2(v), \end{aligned}$$

and hence

$$\begin{aligned} H(f)(u,v) &= A_1(u) \\ &\leqslant A_1(u) \wedge A_2(v). \end{aligned}$$

(2). In category ***FLC***, for identity arrow $i\colon\ L \longrightarrow L$ of (L, A), $i(u) = u$ for any $u \in L$. This implies

$$H(i)(u,v) = \begin{cases} A(u), & i(u) = v, \\ 0, & i(u) \neq v. \end{cases}$$

$$= \begin{cases} A(u), & u = v, \\ 0, & u \neq v, \end{cases}$$

and $H(i)$ is the identity arrow of (L, A) as for the category $\boldsymbol{FLC}_R$.

(3). $H(g \circ f) = H(g) \circ H(f)$ for any $f \in \hom((L_1, A_1), (L_2, A_2))$ and $g \in \hom((L_2, A_2), (L_3, A_3))$. In fact, for any $u \in L_1$ and $v \in L_3$, if $(g \circ f)(u) \neq v$, then $g(f(u)) \neq v$ and $H(g)(f(u), v) = 0$, this implies

$$\begin{aligned}(H(g) \circ H(f))(u,v) &= \bigvee_{x \in L_2} (H(f)(u,x) \wedge H(g)(x,v)) \\ &= H(f)(u, f(u)) \wedge H(g)(f(u), v) \\ &= 0,\end{aligned}$$

and hence $H(g \circ f)(u,v) = (H(g) \circ H(f))(u,v)$; if $(g \circ f)(u) = v$, then $g(f(u)) = v$, it follows that

$$H(g \circ f)(u,v) = A_1(u),$$

and

$$\begin{aligned}&(H(g) \circ H(f))(u,v) \\ &= \bigvee_{x \in L_2} (H(f)(u,x) \wedge H(g)(x,v)) \\ &= H(f)(u, f(u)) \wedge H(g)(f(u), v) \\ &= A_1(u) \wedge A_2(f(u)) \\ &= A_1(u)\end{aligned}$$

by $A_1(u) \leqslant (A_2 \circ f)(u) = A_2(f(u))$. □

Corollary 8.2.1 *The category* ***LC*** *is isomorphic to a subcategory of* $\boldsymbol{FLC}_R$. □

8.3 Fuzzy Power Sets

In Zadeh's fuzzy sets theory, if A and B are fuzzy subsets of a crisp universe U, then A is a subset of B if and only if for any $u \in U$, $A(u) \leqslant B(u)$. According to this definition, A is either utterly or not at all a subset of B, whereas far more in accord with the spirit of fuzzy sets theory would be the assignment of a degree to A's substances in B. In order to extend this definition, in [12], Bandler and Kohout established fuzzy power set theory based on the fuzzy implication operator and discussed the relative properties. In this section, we establish fuzzy power sets theory based on the lattice implication algebras.

Definition 8.3.1 *[12] Given a fuzzy implication operator $\rightarrow$, and a fuzzy subset B of a crisp universe U, the fuzzy power set $P(B)$ of B is given by the membership function, also denoted by $P(B)$, with*

$$P(B)(A) = \bigwedge_{x \in U} (A(x) \rightarrow B(x)).$$

The degree to which A is a subset of B is

$$\pi(A \subseteq B) = P(B)(A).$$

We extend this idea to L-type fuzzy subsets by using a lattice implication algebra. In what follows, we suppose that L is a complete lattice implication algebra.

Definition 8.3.2 *Let $B \in \mathscr{F}_L(U)$. $P(B) \in \mathscr{F}_L(\mathscr{F}_L(U))$ is called the fuzzy power set of B defined as follows:*

$$P(B)(A) = \bigwedge_{x \in U} (A(x) \rightarrow B(x)), \textit{for any } A \in \mathscr{F}_L(U).$$

Definition 8.3.3 *Let $A, B \in \mathscr{F}_L(U)$. $P(B)(A)$ is called the degree of A to be the subset of B and denoted by $\pi(A \subseteq B)$. The degree of A equal to B, denoted by $\pi(A = B)$, is defined as*

$$\pi(A = B) = \pi(A \subseteq B) \wedge \pi(B \subseteq A).$$

If $L = \{0, 1\}$, then $\mathscr{F}_L(U)$ and $P(B)$ are the classical power sets of U and B, respectively. In this case, $\pi(A \subseteq B) = 1$ if and only if A is a classical subset of B.

Theorem 8.3.1 *For any $A, A_1, A_2, B, B_1, B_2 \in \mathscr{F}_L(U)$,*

(1). $\pi(A \subseteq B) = \pi(B' \subseteq A')$, where for any $u \in U$, $A'(u) = (A(u))'$ and $B'(u) = (B(u))'$;
(2). $\pi(A_1 \cup A_2 \subseteq B) = \pi(A_1 \subseteq B) \wedge \pi(A_2 \subseteq B)$;
(3). $\pi(A_1 \cap A_2 \subseteq B) \geqslant \pi(A_1 \subseteq B) \vee \pi(A_2 \subseteq B)$;
(4). $\pi(A \subseteq B_1 \cup B_2) \geqslant \pi(A \subseteq B_1) \vee \pi(A \subseteq B_2)$;
(5). $\pi(A \subseteq B_1 \cap B_2) = \pi(A \subseteq B_1) \wedge \pi(A \subseteq B_2)$.

Proof. We only prove (2).

$$\begin{aligned}\pi(A_1 \cup A_2 \subseteq B) &= \bigwedge_{x\in U} ((A_1 \cup A_2)(x) \to B(x)) \\ &= \bigwedge_{x\in U} ((A_1(x) \vee A_2(x)) \to B(x)) \\ &= \bigwedge_{x\in U} ((A_1(x) \to B(x)) \wedge (A_2(x) \to B(x))) \\ &= \bigwedge_{x\in U} (A_1(x) \to B(x)) \wedge \bigwedge_{x\in U} (A_2(x) \to B(x)) \\ &= \pi(A_1 \subseteq B) \wedge \pi(A_2 \subseteq B). \quad \square\end{aligned}$$

Definition 8.3.4 *Let $A, B \in \mathscr{F}_L(U)$, $\alpha \in L$.*

(1). A is said α-strongly included in B, denoted by $A \prec_\alpha B$, if $x \in (A' \cup B)_\alpha$ for any $x \in U$;
(2). A is said α-included in B, denoted by $A \subseteq_\alpha B$, if $\pi(A \subseteq B) \geqslant \alpha$;
(3). A is said α-equal to B, denoted by $A =_\alpha B$, if $A \subseteq_\alpha B$ and $B \subseteq_\alpha A$.

Corollary 8.3.1 *Let $A, B \in \mathscr{F}_L(U)$.*

(1). $A \subseteq B$ (in Zadeh's inclusion sense) if and only if $A \subseteq_I B$;
(2). $A = B$ (in Zadeh's equality sense) if and only if $A =_I B$. $\square$

Corollary 8.3.2 *Let $A, B \in \mathscr{F}_L(U)$.*

(1). If there exists $x \in U$ such that $A(x) = I$ and $B(x) = O$, then $\pi(A \subseteq B) = O$;
(2). If A is normal, i.e., there exists $x \in U$ such that $A(x) = I$, then $\pi(A \subseteq A') = O$;
(3). $\pi(A = \emptyset) = \pi(A \subseteq \emptyset) = \bigwedge_{x\in U} A'(x)$. $\square$

Theorem 8.3.2 *Let $A, B \in \mathscr{F}_L(U)$ and $\alpha \in L$. If $A \prec_\alpha B$, then $A \subseteq_\alpha B$.*

Proof. By $A \prec_\alpha B$, it follows that $x \in (A' \cup B)_\alpha$ for any $x \in U$, *i.e.*,

$$A'(x) \vee B(x) = (A' \cup B)(x) \geqslant \alpha$$

and hence

$$A(x) \to B(x) \geqslant A'(x) \vee B(x) \geqslant \alpha$$

$$\pi(A \subseteq B) = \bigwedge_{x\in U} (A(x) \to B(x)) \geqslant \alpha. \quad \square$$

Definition 8.3.5 *Let $A \in \mathscr{F}_L(U)$. The crispness $K(A)$ and fuzziness $F(A)$ of A are defined respectively as*

$$K(A) = \bigwedge_{x\in U} (A(x) \vee A'(x)),$$
$$F(A) = (K(A))'.$$

It is clear that $K(A) = K(A')$ and $F(A) = F(A')$.

Theorem 8.3.3 *Let $A \in \mathscr{F}_L(U)$, then $K(A) = (h(A \cap A'))'$, where*

$$h(B) = \bigvee_{x \in U} B(x)$$

is the height of B. □

Theorem 8.3.4 *Let $A \in \mathscr{F}_L(U)$. If A is a crisp subset of U, then $K(A) = I$. If $K(A) = I$ and I is $\vee$-irreducible, then A is a crisp subset of U.*

Proof. If A is a crisp subset of U, then $A(x) = O$ or $A(x) = I$ for any $x \in U$, it follows that

$$K(A) = \bigwedge_{x \in U} (A(x) \vee A'(x)) = I.$$

Conversely, If $K(A) = I$ and I is $\vee$-irreducible, then for any $x \in U$,

$$A(x) \vee A'(x) = I,$$

it follows that $A(x) = O$ or $A(x) = I$. □

According to [12], there seems to be plausible ways to define the degree to which sets A and B may be said to be disjoint. One is the degree to which each is a subset of the other's complement. The second is the degree to which their intersection is empty. We give the definition as follows.

Definition 8.3.6 *Let $A, B \in \mathscr{F}_L(U)$. The degree of disjointness $\pi(A \pitchfork B)$ of A and B can be defined in two ways as*

(1). $\pi(A \pitchfork_1 B) = \pi(A \subseteq B')$;
(2). $\pi(A \pitchfork_2 B) = \pi(A \cap B = \emptyset)$.

According to Definition 8.3.6, if A and B are crisp subsets of U, then $\pi(A \pitchfork B) = I$ if and only if $A \cap B = \emptyset$ and $\pi(A \pitchfork B) = O$ if and only if $A \cap B \neq \emptyset$. That is to say, the concept of the degree of disjointness for fuzzy subset is an extension of the corresponding concept for crisp subsets. By the properties of lattice implication algebra, we know that

$$\pi(A \subseteq B') = \pi((B')' \subseteq A') = \pi(B \subseteq A'),$$

i.e., $\pi(A \pitchfork_1 B) = \pi(B \pitchfork_1 A)$. $\pi(A \pitchfork_2 B) = \pi(B \pitchfork_2 A)$ is trivial.

Theorem 8.3.5 *Let $A, B \in \mathscr{F}_L(U)$.*

(1). $\pi(A \pitchfork_1 B) = \pi(A \cap B' = A)$;
(2). $\pi(A \pitchfork_2 B) \leqslant \pi(A \pitchfork_1 B)$.

Proof. (1).

$$\begin{aligned}\pi(A\cap B'=A) &= \pi(A\subseteq A\cap B')\\ &= \bigwedge_{x\in U}(A(x)\to(A(x)\wedge B'(x)))\\ &= \bigwedge_{x\in U}(A(x)\to B'(x))\\ &= \pi(A\subseteq B')\\ &= \pi(A\pitchfork_1 B).\end{aligned}$$

(2).

$$\begin{aligned}\pi(A\pitchfork_2 B) &= \pi(A\cap B=\emptyset)\\ &= \pi(A\cap B\subseteq\emptyset)\\ &= \bigwedge_{x\in U}((A(x)\wedge B(x))\to O)\\ &= \bigwedge_{x\in U}(A'(x)\vee B'(x))\\ &\leqslant \bigwedge_{x\in U}(A(x)\to B'(x))\\ &= \pi(A\subseteq B')\\ &= \pi(A\pitchfork_1 B).\ \square\end{aligned}$$

Corollary 8.3.3 *Let $A,B\in\mathscr{F}_L(U)$, then*

(1). $\pi(A\pitchfork_1 B)=\begin{cases}O, & \text{there exists } x\in U, A(x)=B(x)=I,\\ I, & \text{for any } x\in U, A(x)\leqslant B'(x).\end{cases}$

(2). $\pi(A\pitchfork_2 B)=\begin{cases}O, & \text{there exists } x\in U, A(x)=B(x)=I,\\ I, & \text{for any } x\in U, A(x)=O \text{ or } B(x)=O.\end{cases}$ $\square$

Corollary 8.3.4 *Let $A\in\mathscr{F}_L(U)$, then*

(1). $\pi(A\pitchfork_1 A')=I$;
(2). $\pi(A\pitchfork_2 A')=K(A)$;
(3). $\pi(A\pitchfork_1\emptyset)=I$;
(4). $\pi(A\pitchfork_2\emptyset)=I$;
(5). $\pi(A\cup A'=U)=\pi(A\pitchfork_2 A')$. $\square$

In the lattice implication algebra defined in Example 2.1.7, $\bar{\alpha}$ is defined as $\bar{\alpha}(x)=\alpha$ for any $\alpha\in L$ and $x\in U$.

By the properties of lattice implication algebra, it is easy to prove the following corollaries.

Corollary 8.3.5 *Let $A,B,C\in\mathscr{F}_L(U)$, $\alpha\in L$.*

(1). $A\to B=\bar{I}$ *if and only if* $A\subseteq_I B$;
(2). $A\to B\geqslant\bar{\alpha}$ *if and only if* $A\subseteq_\alpha B$;
(3). $\pi(A\subseteq B\to C)=\pi(B\subseteq A\to C)$;
(4). $\pi(A\to B\subseteq B)=\pi(B\to A\subseteq A)$;
(5). $\pi(A\subseteq B)\leqslant\pi(B\to C\subseteq A\to C)$;
(6). $\pi(A\subseteq B)\leqslant\pi(C\to A\subseteq C\to B)$. $\square$

Corollary 8.3.6 *$(\mathscr{F}_L(U), \cup, \cap, ', \rightarrow)$ is a lattice H implication algebra if and only if L is a lattice H implication algebra.* □

Theorem 8.3.6 *The function $f: L \longrightarrow \mathscr{F}_L(U)$, defined by $f(\alpha) = \bar{\alpha}$, $\alpha \in L$, is a lattice implication injective homomorphism.*

Proof. It is trivial that f is injective. For any $\alpha, \beta \in L$ and $x \in U$,

$$f(O) = \bar{O},$$

and

$$\begin{aligned} f(\alpha \rightarrow \beta)(x) &= \alpha \rightarrow \beta \\ &= f(\alpha)(x) \rightarrow f(\beta)(x) \\ &= (f(\alpha) \rightarrow f(\beta))(x), \end{aligned}$$

i.e., $f(\alpha \rightarrow \beta) = f(\alpha) \rightarrow f(\beta)$. So, f is a lattice implication homomorphism. □

By Theorem 8.3.1, the function

$$f\colon \mathscr{F}_L(U) \longrightarrow L$$

defined by $f(A) = \pi(B \subseteq A)$, is a $\cap$-homomorphism for any $B \in \mathscr{F}_L(U)$.

Definition 8.3.7 *Let $\alpha \in L$, $J \subseteq L$.*

(1). J is called an α-ideal of L if it satisfies
a) for any $a, b \in L$, $a \in J$ and $b \in J$ imply $a \vee b \in J$,
b) for any $a, b \in L$, $a \rightarrow b \geqslant \alpha$ and $b \in J$ imply $a \in J$.

(2). J is called a dual α-ideal of L if it satisfies
a) for any $a, b \in L$, $a \in J$ and $b \in J$ imply $a \wedge b \in J$,
b) for any $a, b \in L$, $a \rightarrow b \geqslant \alpha$ and $a \in J$ imply $b \in J$.

Theorem 8.3.7 *Let $\alpha \in L$, $J \subseteq L$ and*

$$J_1 = \{A \in \mathscr{F}_L(U) \mid \textit{for any } x \in U, A(x) \in J\}.$$

(1). J is an α-ideal (or a dual α-ideal) of L if and only if J_1 is an $\bar{\alpha}$-ideal (or a dual $\bar{\alpha}$-ideal) of $\mathscr{F}_L(U)$.

(2). J is a filter (or an implicative filter) of L if and only if J_1 is a filter (or an implicative filter) of $\mathscr{F}_L(U)$.

Proof. (1). Suppose J is an α-ideal of L. For any $A, B \in J_1$ and $x \in U$, $A(x) \in J$ and $B(x) \in J$, it follows that $A(x) \vee B(x) \in J$ and hence $A \cup B \in J_1$. If $B \in J_1$ and $A \rightarrow B \geqslant \bar{\alpha}$, then for any $x \in U$,

$$A(x) \rightarrow B(x) = (A \rightarrow B)(x) \geqslant \alpha,$$

i.e., $A(x) \in J$ and hence $A \in J_1$. So, we proved that J_1 is an $\bar{\alpha}$-ideal of $\mathscr{F}_L(U)$.

Conversely, if J_1 is an $\bar{\alpha}$-ideal of $\mathscr{F}_L(U)$ and $a, b \in J$, then $\bar{a}, \bar{b} \in J_1$ and hence $\bar{a} \cup \bar{b} \in J_1$, *i.e.*,

$$a \vee b = \bar{a}(x) \vee \bar{b}(x) = (\bar{a} \cup \bar{b})(x) \in J.$$

If $b \in J$ and $a \to b \geqslant \alpha$, then $\bar{b} \in J_1$ and $\bar{a} \to \bar{b} \geqslant \bar{\alpha}$, it follows that $\bar{a} \in J_1$ and $a \in J$. This proved that J is an α-ideal of L.

For the case of dual α-ideal, the proof is similar.

(2). Suppose J is a filter of L and $A, A \to B \in J_1$. For any $x \in U$, $A(x) \in J$ and

$$A(x) \to B(x) = (A \to B)(x) \in J,$$

it follows that $B(x) \in J$ and hence $B \in J_1$. $I \in J$ implies $\bar{I} \in J_1$. This proved that J_1 is a filter of $\mathscr{F}_L(U)$.

Conversely, if J_1 is a filter of $\mathscr{F}_L(U)$, then $\bar{I} \in J_1$ implies $I \in J$. Furthermore, if $a, a \to b \in J$, then $\bar{a} \in J_1$,

$$\bar{a} \to \bar{b} = \overline{a \to b} \in J_1,$$

it follows that $\bar{b} \in J_1$ and hence $b \in J$.

For the case of implicative filter, the proof is similar. □

Theorem 8.3.8 *Suppose $H \subseteq \mathscr{F}_L(U)$ and $x \in U$. Let*

$$H_x = \{a \in L \mid \text{ there exists } A \in H \text{ s.t., } A(x) = a\}.$$

(1). If H is an $\bar{\alpha}$-ideal (or a dual $\bar{\alpha}$-ideal) of $\mathscr{F}_L(U)$, then H_x is an α-ideal (or a dual α-ideal) of L for any $x \in U$.
(2). If H is a filter (or an implicative filter) of $\mathscr{F}_L(U)$, then H_x is a filter (or an implicative filter) of L for any $x \in U$.

Proof. (1). Suppose H is an $\bar{\alpha}$-ideal of $\mathscr{F}_L(U)$, $x \in U$, $a, b \in H_x$, then there exist $A \in H$ and $B \in H$ such that $A(x) = a$ and $B(x) = b$. By $A \cup B \in H$, it follows that

$$a \vee b = A(x) \vee B(x) = (A \cup B)(x) \in H_x.$$

If $b \in H_x$ and $a \to b \geqslant \alpha$, then there exists $B \in H$ such that $B(x) = b$. Let $A \in \mathscr{F}_L(U)$ satisfy

$$A(y) = \begin{cases} a, & y = x, \\ B(y), & y \neq x, \end{cases}$$

then $A \to B \geqslant \bar{\alpha}$, it follows that $A \in H$ and hence $a \in H_x$.

For the case of dual α-ideal, the proof is similar.

(2). Suppose H is a filter of $\mathscr{F}_L(U)$, $x \in U$. $\bar{I} \in H$ implies that $I \in H_x$. If $a \in H_x$, $a \to b \in H_x$, then there exist $A \in H$ and $B \in H$ such that $A(x) = a$ and $B(x) = a \to b$. Let $C, D \in \mathscr{F}_L(U)$ satisfy

$$C(y) = \begin{cases} a, \ y = x, \\ I, \ y \neq x, \end{cases}$$

$$D(y) = \begin{cases} b, \quad\quad y = x, \\ B(y), \ y \neq x, \end{cases}$$

for any $y \in U$, then $C \in H$ by $A \leqslant C$, and

$$\begin{aligned}(C \to D)(y) &= C(y) \to D(y) \\ &= \begin{cases} a \to b, \ y = x, \\ B(y), \ \ y \neq x, \end{cases}\end{aligned}$$

i.e., $C \to D = B \in H$. This implies that $D \in H$ and hence $b \in H_x$ by $D(x) = b$.

For the case of implicative filter, the proof is similar. □

Theorem 8.3.9 *Let $A \in \mathscr{F}_L(U)$, $\alpha \in L$ and*

$$A_1 = \{B \in \mathscr{F}_L(U) \mid \pi(A \subseteq B) \geqslant \alpha\},$$
$$A_2 = \{B \in \mathscr{F}_L(U) \mid \pi(B \subseteq A) \geqslant \alpha\},$$

then

(1). A_1 is a dual ideal of $\mathscr{F}_L(U)$;
(2). A_2 is an ideal of $\mathscr{F}_L(U)$;
(3). if $[\alpha, I]$ is an implicative filter of L, then A_1 is a filter of $\mathscr{F}_L(U)$. □

8.4 Adjoint Semigroups

The semigroup is an important and interesting algebraic structure. Its theory has close relations to the formal language theory, the automatic machine theory and so on. In this section, we discuss the properties of the semigroups $(L, \otimes)$, $(L, \oplus)$ and the adjoint semigroups $(M(L), \circ, 1_L)$, induced by a lattice implication algebra L.

Definition 8.4.1 *Let L be a lattice implication algebra. For $a \in L$, denote the mapping*

$$\begin{aligned}\theta_l\colon\ L &\longrightarrow L, \\ x &\longmapsto a \to x,\end{aligned}$$

as l_a.

Theorem 8.4.1 *Let L be a lattice implication algebra, $a \in L$. If $l_a^2 = l_a$, then the following conclusions hold:*

(1). $a \vee a' = I$;

(2). $a \to a' = a'$;
(3). $a' \to a = a$;
(4). l_a *is an implication homomorphism.*

where, $l_a^2 = l_a \circ l_a$, *i.e.,* $l_a^2(x) = a \to (a \to x)$. □

Theorem 8.4.2 *Let* L *be a lattice implication algebra,* $a \in L$. *If* $l_a^2 = l_a$, *then*

(1). $\{l_a(x) \mid x \in L\} = A(a')$;
(2). $\{x \mid x \in L, l_a(x) = I\} = A(a)$.

Proof. The proof of (2) is clear, we only give the proof of (1). For any $x \in L$, we have $a' = a \to O \leqslant a \to x$, *i.e.*,

$$\{l_a(x) \mid x \in L\} \subseteq A(a').$$

According to Theorem 8.4.1, for any $x \in A(a')$, then $a' \to x = I$, and

$$\begin{aligned} x &= I \to x \\ &= (a \vee a') \to x \\ &= (a' \to x) \wedge (a \to x) \\ &= a \to x, \end{aligned}$$

i.e., $A(a') \subseteq \{l_a(x) \mid x \in L\}$. □

Lemma 8.4.1 *Let* L *be a lattice implication algebra,* $a \in L$. *If* $l_a^2 = l_a$, *then* $l_a(x) = x \vee a'$ *holds for any* $x \in L$. □

Theorem 8.4.3 *Let* L *be a lattice implication algebra,*

$$M = \{a \mid a \in L, l_a^2 = l_a\}.$$

Then $(M, \vee, \wedge, \to, ')$ *is a lattice implication subalgebra of* L.

Proof. $O \in M$ is clear. According to Theorem 2.3.5, we only need to prove $a \to b \in M$ holds for any $a, b \in M$. By Lemma 8.4.1, for any $a, b \in M$, and for any $x \in L$, we have

$$\begin{aligned} &(a \to b) \to ((a \to b) \to x) \\ =& (a' \vee b) \to ((a' \vee b) \to x) \\ =& (a' \vee b) \to ((a' \to x) \wedge (b \to x)) \\ =& (a' \to (a' \to x)) \wedge (a' \to (b \to x)) \wedge (b \to (a' \to x)) \wedge (b \to (b \to x)) \\ =& (a' \to x) \wedge (b \to x) \\ =& (a' \vee b) \to x \\ =& (a \to b) \to x, \end{aligned}$$

i.e., $a \to b \in M$, and $(M, \vee, \wedge, \to, ', O, I)$ is a lattice implication subalgebra of L. □

Definition 8.4.2 *Let L be a lattice implication algebra. The self-mapping $\alpha\colon L \longrightarrow L$ is called regular, if $\alpha(x \to y) = x \to \alpha(y)$ holds for any $x, y \in L$.*

Let

$$M(L) = \{\alpha \mid \alpha = l_{a_1} \circ l_{a_2} \circ \cdots \circ l_{a_n}, a_1, a_2, \cdots, a_n \in L\}$$

be the set of all finite times composition of left-mapping of lattice implication algebra, then $(M(L), \circ, 1_L)$ is a commutative semigroup w.r.t. the operation $\circ$ of mapping composition with a unit element. Here, the unit element $1_L = l_I$. If define a partial order relation $\leqslant$ on $M(L)$:

$$\alpha \leqslant \beta \quad \text{iff} \quad \alpha(x) \to \beta(x) = I$$

holds for any $x \in L$, then $(M(L), \circ, 1_L)$ is a commutative partially ordered semigroup, called an adjoint semigroup of L. It is clear that for any subset S of L,

$$M(S) = \{\alpha = l_{a_1} \circ l_{a_2} \circ \cdots \circ l_{a_n} \mid a_1, a_2, \cdots, a_n \in S\}$$

is a subsemigroup of $(M(L), \circ, 1_L)$. In the following, we write $(M(L), \circ, 1_L)$ as $M(L)$ briefly.

We can get the following theorem.

Theorem 8.4.4 *Let L be a lattice implication algebra. If $\alpha \in M(L)$, then α is regular.*

Definition 8.4.3 *Let L be a lattice implication algebra, $\alpha \in M(L)$. The following subsets of L,*

$$S(\alpha) = \{x \in L \mid \alpha(x) = x\},$$

$$Im(\alpha) = \{\alpha(x) \mid x \in L\},$$

$$D\text{-}\ker(\alpha) = \{x \in L \mid \alpha(x) = I\}$$

are called stabilizer, image and dual kernel of α, respectively.

For any $\alpha \in M(L)$, since $\alpha(I) = I$, we have $I \in S(\alpha)$ and $S(\alpha) \neq \emptyset$. By Theorem 8.4.4, we can get the following theorem.

Theorem 8.4.5 *Let L be a lattice implication algebra. For any $\alpha \in M(L)$, $y \in L$, if $x \in Im(\alpha)$, then $y \to x \in Im(\alpha)$; if $x \in S(\alpha)$, then $y \to x \in S(\alpha)$.* □

Lemma 8.4.2 *Let L be a lattice implication algebra. For any $\alpha \in M(L)$, we have $S(\alpha) \subseteq Im(\alpha)$ and $S(\alpha) \cap D\text{-}\ker(\alpha) = \{I\}$.* □

Lemma 8.4.3 *Let L be a lattice implication algebra. For any $\alpha, \beta \in M(L)$, we have*

$$S(\alpha) \cap S(\beta) \subseteq S(\alpha \circ \beta) \subseteq Im(\alpha \circ \beta) \subseteq Im(\alpha) \cap Im(\beta).$$

Proof. According to Lemma 8.4.2, we only need to prove

$$S(\alpha) \cap S(\beta) \subseteq S(\alpha \circ \beta)$$

and

$$Im(\alpha \circ \beta) \subseteq Im(\alpha) \cap Im(\beta).$$

If $x \in S(\alpha) \cap S(\beta)$, then $\alpha(x) = \beta(x) = x$, it follows that

$$\alpha \circ \beta(x) = \alpha(\beta(x)) = \alpha(x) = x,$$

i.e., $x \in S(\alpha \circ \beta)$, and $S(\alpha) \cap S(\beta) \subseteq S(\alpha \circ \beta)$.

If $x \in Im(\alpha \circ \beta)$, then there exists $y \in L$ such that

$$(\alpha \circ \beta)(y) = x,$$

$$x = \alpha(\beta(y)) \in Im(\alpha).$$

Since $M(L)$ satisfies the commutative law, it follows that

$$x = \alpha(\beta(y)) = \beta(\alpha(y)) \in Im(\beta),$$

i.e., $x \in Im(\alpha) \cap Im(\beta)$, and $Im(\alpha \circ \beta) \subseteq Im(\alpha) \cap Im(\beta)$. □

Lemma 8.4.4 *Let L be a lattice implication algebra. For any $\alpha \in M(L)$, we have $D\text{-}\ker(\alpha) = \{\alpha(x) \to x \mid x \in L\}$.*

Proof. Firstly, for any $x \in L$, we know

$$\alpha(\alpha(x) \to x) = \alpha(x) \to \alpha(x) = I,$$

i.e., $\alpha(x) \to x \in \text{D-}\ker(\alpha)$ and $\{\alpha(x) \to x \mid x \in L\} \subseteq \text{D-}\ker(\alpha)$.

Secondly, for any $x \in \text{D-}\ker(\alpha)$, $\alpha(x) = I$,

$$x = I \to x = \alpha(x) \to x,$$

i.e.,

$$x \in \{\alpha(x) \to x \mid x \in L\},$$

$$\text{D-}\ker(\alpha) \subseteq \{\alpha(x) \to x \mid x \in L\}.$$

That is to say $\text{D-}\ker(\alpha) = \{\alpha(x) \to x \mid x \in L\}$. □

Theorem 8.4.6 *Let L be a lattice implication algebra. For any $\alpha \in M(L)$, the following statements are equivalent:*

(1). $\alpha^2 = \alpha$;
(2). $S(\alpha) = Im(\alpha)$;
(3). $D\text{-}\ker(\alpha) \cap Im(\alpha) = \{I\}$;
(4). $D\text{-}\ker(\alpha)$ *is a filter of* L;
(5). $\alpha(x \to y) = \alpha(x) \to \alpha(y)$ *holds for any* $x, y \in L$.

Proof. ((1)$\Longrightarrow$(2)). On one hand, we have $S(\alpha) \subseteq Im(\alpha)$ by Lemma 8.4.2. And on the other hand, $\alpha^2 = \alpha$ and $\alpha(\alpha(x)) = \alpha^2(x) = \alpha(x)$ hold for any $x \in L$, *i.e.*, $\alpha(x) \in S(\alpha)$, $S(\alpha) \supseteq Im(\alpha)$. In a word, $Im(\alpha) = S(\alpha)$.

((2)$\Longrightarrow$(3)). The proof is clear by $S(\alpha) \cap \text{D-}\ker(\alpha) = \{I\}$.

((3)$\Longrightarrow$(4)). Firstly, $I \in \text{D-}\ker(\alpha)$ is clear. Secondly, if x, $x \to y \in \text{D-}\ker(\alpha)$, then

$$\begin{aligned} \alpha(x) &= \alpha(x \to y) = I, \\ x \to \alpha(y) &= \alpha(x \to y) = I, \end{aligned}$$

i.e., $x \leqslant \alpha(y)$. Because α is an order-preserving mapping,

$$\begin{aligned} \alpha(x) &\leqslant \alpha(\alpha(y)), \\ \alpha(\alpha(y)) &= I, \\ \alpha(y) &\in \text{D-}\ker(\alpha). \end{aligned}$$

Since $\alpha(y) \in Im(\alpha)$, it follows that

$$\alpha(y) \in \text{D-}\ker(\alpha) \cap Im(\alpha),$$

and hence $\alpha(y) = I$, $y \in \text{D-}\ker(\alpha)$, *i.e.*, $\text{D-}\ker(\alpha)$ is a filter of L.

((4)$\Longrightarrow$(5)). For any $x, y \in L$, on one hand, by $x \leqslant \alpha(x)$, it follows that

$$\alpha(x) \to \alpha(y) \leqslant x \to \alpha(y) = \alpha(x \to y).$$

On the other hand,

$$\begin{aligned} &\alpha((\alpha(x) \to x) \to ((x \to \alpha(y)) \to (\alpha(x) \to y))) \\ &= \alpha((x \to \alpha(y)) \to ((\alpha(x) \to x) \to (\alpha(x) \to y))) \\ &= \alpha((x \to \alpha(y)) \to ((x \wedge \alpha(x)) \to y)) \\ &= \alpha((x \to \alpha(y)) \to (x \to y)) \\ &= (x \to \alpha(y)) \to \alpha(x \to y) \\ &= (x \to \alpha(y)) \to (x \to \alpha(y)) \\ &= I, \end{aligned}$$

i.e., $(\alpha(x) \to x) \to ((x \to \alpha(y)) \to (\alpha(x) \to y)) \in \text{D-}\ker(\alpha)$. By $\alpha(x) \to x \in \text{D-}\ker(\alpha)$, and $\text{D-}\ker(\alpha)$ is a filter of L, it follows that

$$(x \to \alpha(y)) \to (\alpha(x) \to y) \in \text{D-}\ker(\alpha),$$

$$\alpha((x \to \alpha(y)) \to (\alpha(x) \to y)) = I,$$

$$\alpha(x \to y) \to (\alpha(x) \to \alpha(y)) = I,$$

$$\alpha(x \to y) \leqslant \alpha(x) \to \alpha(y),$$

i.e., $\alpha(x \to y) = \alpha(x) \to \alpha(y)$ holds for every $x, y \in L$.

((5)$\Longrightarrow$(1)). For any $x \in L$,

$$\begin{aligned}\alpha^2(x) \to \alpha(x) &= \alpha(\alpha(x) \to x)\\ &= \alpha(x) \to \alpha(x)\\ &= I,\\ \alpha(x) \to \alpha^2(x) &= \alpha(x \to \alpha(x))\\ &= \alpha(I)\\ &= I,\end{aligned}$$

i.e., $\alpha^2 = \alpha$. □

Theorem 8.4.7 *Let L be a lattice implication algebra. If $\alpha \in M(L)$, then $S(\alpha)$ is a filter of L.*

Proof. It is clear that $I \in S(\alpha)$. If x, $x \to y \in S(\alpha)$, then

$$\alpha(x) = x,$$

$$\alpha(x \to y) = x \to y.$$

Since $y \leqslant \alpha(y)$, we only need to prove $\alpha(y) \leqslant y$.

$$\begin{aligned}x \to (\alpha(y) \to y) &= \alpha(y) \to (x \to y)\\ &= \alpha(y) \to \alpha(x \to y)\\ &= \alpha(y) \to (x \to \alpha(y))\\ &= x \to (\alpha(y) \to \alpha(y))\\ &= x \to I\\ &= I,\end{aligned}$$

i.e., $x \leqslant \alpha(y) \to y$. And

$$\begin{aligned}\alpha(y) \to y &= x \vee (\alpha(y) \to y)\\ &= ((\alpha(y) \to y) \to x) \to x\\ &= ((\alpha(y) \to y) \to x) \to \alpha(x)\\ &= \alpha(((\alpha(y) \to y) \to x) \to x)\\ &= \alpha(x \vee (\alpha(y) \to y))\\ &= \alpha(\alpha(y) \to y)\\ &= \alpha(y) \to \alpha(y)\\ &= I,\end{aligned}$$

i.e., $\alpha(y) \leqslant y$. □

For any $\alpha \in M(L), x, y \in L$,

$$\alpha(x \vee y) = \alpha(x) \vee \alpha(y),$$

$$\alpha(x \wedge y) = \alpha(x) \wedge \alpha(y),$$

that is to say, α is a lattice endomorphism of L. By Theorem 8.4.6, if $\alpha^2 = \alpha$,

$$\alpha(x \to y) = \alpha(x) \to \alpha(y),$$

α is an implication homomorphism of L.

Definition 8.4.4 *Let L be a lattice implication algebra, A, B two non-empty sets of L. Define*

$$A \otimes B = \{a \otimes b \mid a \in A, b \in B\}.$$

We call $A \otimes B$ the direct sum of A and B, write as $A \dot{\otimes} B$ if $A \cap B = \{I\}$.

Theorem 8.4.8 *Let L be a lattice implication algebra, $\alpha \in M(L)$.*

$$L = Im(\alpha) \dot{\otimes} D\text{-}\ker(\alpha) \text{ iff } \alpha^2 = \alpha.$$

Proof. If $\alpha^2 = \alpha$, by Theorems 8.4.6 and 2.1.21, we have $Im(\alpha) \cap \text{D-}\ker(\alpha) = \{I\}$ and for any $x \in L$

$$\alpha(x) \otimes (\alpha(x) \to x) = x \wedge \alpha(x) = x.$$

Since $\alpha(x) \in Im(\alpha)$, $\alpha(x) \to x \in \text{D-}\ker(\alpha)$, we know $L = Im(\alpha) \dot{\otimes} \text{D-}\ker(\alpha)$.

If $L = Im(\alpha) \dot{\otimes} \text{D-}\ker(\alpha)$, then $Im(\alpha) \cap \text{D-}\ker(\alpha) = \{I\}$. We have $\alpha^2 = \alpha$ by Theorem 8.4.6. □

If $\alpha, \beta \in M(L)$ and α, β are idempotent elements of $M(L)$, then $\alpha \circ \beta$ is an idempotent element of $M(L)$.

Theorem 8.4.9 *Let L be a lattice implication algebra, $a \in L$. The following conclusions are equivalent:*

(1). l_a is an idempotent element of the semigroup $(M(L), \circ)$;
(2). a is an idempotent element of the semigroup $(L, \otimes)$;
(3). a' is an idempotent element of the semigroup $(L, \oplus)$.

Proof. ((1)$\Longrightarrow$(2)). If l_a is an idempotent element of the semigroup $(M(L), \circ)$, then $l_a^2 = l_a$, *i.e.*,

$$a \to (a \to x) = a \to x$$

holds for any $x \in L$. Let $x = O$, we know that

$$a \to a' = a',$$
$$(a \to a')' = a,$$
$$a \otimes a = a,$$

i.e., a is an idempotent element of the semigroup $(L, \otimes)$.

((2)$\Longrightarrow$(1)). If a is an idempotent element of the semigroup $(L, \otimes)$,

$$a \otimes a = a,$$
$$(a \to a')' = a,$$

then $(a \to a')' \to x = a \to x$ holds for any $x \in L$. Since

$$\begin{aligned} a \to x &= (a \to a')' \to x \\ &= x' \to (a \to a') \\ &= a \to (x' \to a') \\ &= a \to (a \to x), \end{aligned}$$

i.e., $l_a^2 = l_a$ and l_a is an idempotent element of the semigroup $(M(L), \circ)$.

Since the operation $\otimes$ is a dual operation of $\oplus$, we can get the equivalence of (2) and (3) easily. □

Theorem 8.4.10 *Let L be a lattice implication algebra. The following statements are equivalent:*

(1). L is a lattice H implication algebra;
(2). $(M(L), \circ)$ is an idempotent semigroup;
(3). $(L, \otimes)$ is an idempotent semigroup;
(4). $(L, \oplus)$ is an idempotent semigroup.

Proof. ((1)$\Longrightarrow$(2)). Let L be a lattice H implication algebra. We know

$$x \to (x \to y) = x \to y$$

holds for any $x, y \in L$, *i.e.*, $l_x^2 = l_x$ holds for any $x \in L$, and l_x is an idempotent element of $(M(L), \circ)$. For any $\alpha \in M(L)$, there exist $a_1, a_2, \cdots, a_n \in L$, such that

$$\alpha = l_{a_1} \circ l_{a_2} \circ \cdots \circ l_{a_n}.$$

Since $l_{a_1}, l_{a_2}, \cdots, l_{a_n}$ are idempotent elements of $(M(L), \circ)$, we know that α is an idempotent element of $(M(L), \circ)$ and $(M(L), \circ)$ is an idempotent semigroup.

((2)$\Longrightarrow$(1)). If $(M(L), \circ)$ is an idempotent semigroup, then for any $x \in L$, l_x is an idempotent element of $(M(L), \circ)$, $l_x^2 = l_x$. We have

$$x \to (x \to y) = x \to y$$

holds for any $y \in L$, and L is a lattice H implication algebra.

The equivalences of (1), (3) and (4) are clear. □

8.5 Logical Properties

Because lattice implication algebra is presented to study lattice-valued logic, it seems appropriate and interesting to take a further look at the model properties of lattice implication algebras from a logical point of view. In this section, we formalize the lattice implication algebra in a first-order language with identity and study the logical properties of this algebra. We prove that the formalized lattice implication algebra theory is preserved under submodels, homomorphisms, and unions of chains; we also prove that this formalized theory is neither complete nor model complete; we establish the concept of ultra-product of lattice implication algebras and discuss its basic properties.

Suppose that $\mathcal{L}$ is a first-order language with identity, F_1, F_2 and F_3 are its 2-placed function symbols, F_4 is its 1-placed function symbol, O and I are its constant symbols. F_1, F_2, F_3, and F_4 are intended to represent $\rightarrow, \wedge, \vee$ and $'$ of lattice implication algebra respectively. We formalize lattice implication algebra theory in this language and denote it by T. T is composed of the following formulae:

(1). $\forall x(F_2(x,x)=x \wedge F_3(x,x)=x)$;
(2). $\forall x \forall y(F_2(x,y)=F_2(y,x) \wedge F_3(x,y)=F_3(y,x))$;
(3). $\forall x \forall y \forall z$

$$(F_2(F_2(x,y),z)=F_2(x,F_2(y,z)) \wedge F_3(F_3(x,y),z)=F_3(x,F_3(y,z)));$$

(4). $\forall x \forall y(F_2(x,F_3(x,y))=F_3(x,F_2(x,y))=x)$;
(5). $\forall x(F_2(x,I)=x \wedge F_2(O,x)=O)$;
(6). $\forall x \forall y(F_2(x,y)=x \rightarrow F_2(F_4(y),F_4(x))=F_4(y))$;
(7). $\forall x(F_4(F_4(x))=x)$;
(8). $\forall x \forall y \forall z(F_1(x,F_1(y,z))=F_1(y,F_1(x,z)))$;
(9). $\forall x(F_1(x,x)=I)$;
(10). $\forall x \forall y(F_1(x,y)=F_1(F_4(y),F_4(x)))$;
(11). $\forall x \forall y((F_1(x,y)=I \wedge F_1(y,x)=I) \rightarrow x=y)$;
(12). $\forall x \forall y(F_1(F_1(x,y),y)=F_1(F_1(y,x),x))$;
(13). $\forall x \forall y \forall z(F_1(x,F_2(y,z))=F_2(F_1(x,y),F_1(x,z)))$;
(14). $\forall x \forall y \forall z(F_1(x,F_3(y,z))=F_3(F_1(x,y),F_1(x,z)))$.

Remark 10. The symbols of the first-order language or of the lattice implication algebra can be understood from the context.

It is trivial that a model for language $\mathcal{L}$ is a model of T if and only if it is a lattice implication algebra.

We say that a theory S is preserved under submodels if and only if any submodel of a model of S is a model of S. S is said to be preserved under unions of chains if and only if the union of any chain of models of S is a model of S. S is preserved under homomorphisms if and only if any homomorphic image of a model of S is a model of S. We refer to [55] for details of these concepts.

Theorem 8.5.1 *T is preserved under submodels and unions of chains.*

Proof. Suppose that

$$A_1 \subseteq A_2 \subseteq \cdots \subseteq A_\alpha \subseteq \cdots, \alpha < \xi$$

is a chain of a lattice implication algebras and A_α is a lattice implication subalgebra of A_β when $\alpha < \beta < \xi$. Let

$$A = \bigcup_{\alpha<\xi} A_\alpha.$$

For any $x, y \in A$, there exists $\alpha < \xi$, such that $x, y \in A_\alpha$. Define

$$\begin{aligned} x \vee y &= x \vee_\alpha y, \\ x \wedge y &= x \wedge_\alpha y, \\ x' &= x'^{\alpha}, \\ x \to y &= x \to_\alpha y. \end{aligned}$$

It is trivial to prove that these operations are well defined and they make A be a lattice implication algebra. This means that the union of a chain of models of T is also a model of T. Because T is a set of universal formulae, by Theorem 2.2.2 of [55], it follows that T is preserved under submodels. □

Theorem 8.5.2 *T is preserved under homomorphisms.*

Proof. Suppose that A is a model of T, B is a model of $\mathcal{L}$ and $f\colon\ A \longrightarrow B$ is a model homomorphism. It follows that A is a lattice implication algebra. For any $x, y \in A$, let

$$x^* = x', \quad x + y = x' \to y, \quad x \cdot y = (x \to y')',$$

we know from Theorem 7.2.3 that $(A, +, \cdot, *)$ is an MV-algebra. If we formalize MV-algebra theory in the first-order language $\mathcal{L}$, it follows that MV-algebra theory is preserved under homomorphisms because each axiom of this theory is positive. That is to say $f(A)$ is an MV-algebra and hence a lattice implication algebra by Theorem 7.2.2. This implies that $f(A)$ is a model of T. □

Theorem 8.5.3 *(1). T is not model complete;*
(2). T is not complete.

Proof. Consider the Łukasiewicz implication algebra $\mathcal{A} = ([0,1], \vee, \wedge, ', \to)$ defined in Example 2.1.2. Let $B = \{0, 0.5, 1\}$. It is easy to check that $\vee$, $\wedge$, $'$ and $\to$ are closed in B and $\mathcal{B} = (B, \vee, \wedge, ', \to)$ is a lattice implication subalgebra of $\mathcal{A}$. That is to say, as models of T, $\mathcal{B}$ is a submodel of $\mathcal{A}$. It is clear that

$$\varphi = \forall x(x = c_0 \vee x = c_{0.5} \vee x = c_1)$$

is satisfiable in $\mathcal{B}_B$ and $\varphi \in Th(\mathcal{B}_B)$. By Proposition 3.1.3 in [55] and φ is not true in $(\mathcal{A}, b)_{b\in B}$, $\mathcal{B}$ is not an elementary submodel of $\mathcal{A}$ and hence T is not model complete. Let

$$\begin{aligned} al(n) = \exists x_1 \cdots \exists x_n(\neg x_1 = x_2 \wedge \cdots \wedge \neg x_1 = x_n \wedge \\ \neg x_2 = x_3 \wedge \cdots \wedge \neg x_2 = x_n \wedge \cdots \wedge \neg x_{n-1} = x_n), \end{aligned}$$

it follows that $\mathcal{B} \models al(3) \wedge \neg al(4)$ and $\mathcal{A} \not\models al(3) \wedge \neg al(4)$, *i.e.*, T is not complete. $\square$

Suppose J is a non-empty index set, L_j is a lattice implication algebra for any $j \in J$, $D \subseteq P(J)$ is an ultra-filter over J and

$$L = \prod\{L_j \mid j \in J\} = \{f \mid f\colon\ J \longrightarrow \bigcup_{j\in J} L_j,\ s.t., \text{ for any } \ j \in J, f(j) \in L_j\}$$

is the lattice implication product algebra of $\{L_j \mid j \in J\}$. Define a binary relation $=_D$ on L by D: for any $f, g \in L$,

$$f =_D g \qquad \text{iff} \qquad \{j \in J \mid f(j) = g(j)\} \in D,$$

then $=_D$ is an equivalent relation on L, we denote by L_D the family of all the equivalent classes, *i.e.*,

$$L_D = \{f_D \mid f \in L\},$$

where $f_D = \{g \in L \mid f =_D g\}$. Define operations $\vee$, $\wedge$, $\rightarrow$ and $'$ on L_D as follows: for any $f_D, g_D \in L_D$,

$$\begin{aligned} f_D \vee g_D &= (f \vee g)_D, \\ f_D \wedge g_D &= (f \wedge g)_D, \\ f_D \rightarrow g_D &= (f \rightarrow g)_D, \\ (f_D)' &= (f')_D. \end{aligned}$$

It is trivial to verify that these definitions do not depend on the choice of representation and they make L_D be a lattice implication algebra. We call this lattice implication algebra the ultra-product of $\{L_j \mid j \in J\}$ w.r.t. the ultra-filter D.

Let

$$F = \{f \in L \mid \{j \in J \mid f(j) = I_j\} \in D\},$$

where I_j is the greatest element of L_j for any $j \in J$.

Theorem 8.5.4 *F is a filter of lattice implication algebra L and $L/F = L_D$.*

Proof. (1). F is a filter of L. In fact, by $\{j \in J \mid I(j) = I_j\} = J \in D$, it follows that $I \in F$. If $f \in F$ and $f \to g \in F$, then $\{j \in J \mid f(j) = I_j\} \in D$ and $\{j \in J \mid f(j) \to g(j) = I_j\} \in D$, it follows by

$$\{j \in J \mid f(j) = I_j\} \cap \{j \in J \mid f(j) \to g(j) = I_j\} \subseteq \{j \in J \mid g(j) = I_j\}$$

that $\{j \in J \mid g(j) = I_j\} \in D$ and hence $g \in F$.

(2). In order to prove $L/F = L_D$, we need only to prove that $f \to g \in F$ and $g \to f \in F$ if and only if $f_D = g_D$ for any $f, g \in L$. Suppose that $f \to g \in F$ and $g \to f \in F$, it follows that $\{j \in J \mid f(j) \to g(j) = I_j\} \in D$ and $\{j \in J \mid g(j) \to f(j) = I_j\} \in D$, *i.e.*, $\{j \in J \mid f(j) \leqslant g(j)\} \in D$ and $\{j \in J \mid g(j) \leqslant f(j)\} \in D$, this implies that

$$\{j \in J \mid f(j) \leqslant g(j)\} \cap \{j \in J \mid g(j) \leqslant f(j)\} \in D,$$

and hence $\{j \in J \mid f(j) = g(j)\} \in D$, *i.e.*, $f_D = g_D$. The converse can be proved similarly. □

Let J, D, $\{L_j \mid j \in J\}$ and L_D be defined as above. Suppose that R_j is a fuzzy subset of L_j for any $j \in J$. Define a fuzzy subset R of L_D as follows: for any $f_D \in L_D$,

$$R(f_D) = \sup\{\xi \mid \{j \in J \mid R_j(f(j)) \geqslant \xi\} \in D\}.$$

If $f_D = g_D$ and $\xi \in [0,1]$ such that $\{j \in J \mid R_j(f(j)) \geqslant \xi\} \in D$, then by $\{j \in J \mid f(j) = g(j)\} \in D$ and

$$\{j \in J \mid R_j(f(j)) \geqslant \xi\} \cap \{j \in J \mid f(j) = g(j)\} \subseteq \{j \in J \mid R_j(g(j)) \geqslant \xi\},$$

it follows that $\{j \in J \mid R_j(g(j)) \geqslant \xi\} \in D$ and hence $R(f_D) \leqslant R(g_D)$. $R(g_D) \leqslant R(f_D)$ can be proved similarly. This means that R is well defined. R is called the fuzzy ultra-product of $\{R_j \mid j \in J\}$ w.r.t. D and denoted by $\prod_D R_j$. If $L_j = A$ and $R_j = B$ for any $j \in J$, then $\prod_D R_j$ is called the fuzzy ultra-power of B w.r.t. D.

Theorem 8.5.5 *[601] If $\{j \in J \mid R_j(f(j)) = \xi\} \in D$, then $R(f_D) = \xi$.* □

Theorem 8.5.6 *[601] $R(f_D) = \inf\{\xi \mid \{j \in J \mid R_j(f(j)) \leqslant \xi\} \in D\}$.* □

Theorem 8.5.7 *[601] Suppose that R_j and H_j are fuzzy subsets of L_j for any $j \in J$, then*

(1). $\prod_D(R_j \cup H_j) = \prod_D R_j \cup \prod_D H_j$;
(2). $\prod_D(R_j \cap H_j) = \prod_D R_j \cap \prod_D H_j$;
(3). $\prod_D(R_j)' = (\prod_D R_j)'$. □

Theorem 8.5.8 *If R_j is a fuzzy filter of L_j for any $j \in J$, then $R = \prod_D R_j$ is a fuzzy filter of L_D.*

Proof. (1). $R(f_D) \leqslant R(I_D)$ for any $f_D \in L_D$. In fact, for any $f_D \in L_D$ and $j \in J$, it follows that $R_j(f(j)) \leqslant R_j(I_j)$. If

$$\{j \in J \mid R_j(f(j)) \geqslant \xi\} \in D,$$

then $\{j \in J \mid R_j(I_j) \geqslant \xi\} \in D$, and hence

$$\begin{aligned} R(f_D) &= \sup\{\xi \mid \{j \in J \mid R_j(f(j)) \geqslant \xi\} \in D\} \\ &\leqslant \sup\{\xi \mid \{j \in J \mid R_j(I_j) \geqslant \xi\} \in D\} \\ &= R(I_D). \end{aligned}$$

(2). $R(g_D) \geqslant \min\{R(f_D \to g_D), R(f_D)\}$ for any $f_D, g_D \in L_D$. In fact, if there exist $f_D, g_D \in L_D$ such that

$$R(g_D) < \min\{R(f_D \to g_D), R(f_D)\},$$

then

$$R(g_D) < R(f_D \to g_D) = \sup\{\xi \mid \{j \in J \mid R_j(f(j) \to g(j)) \geqslant \xi\} \in D\},$$

and

$$R(g_D) < R(f_D) = \sup\{\xi \mid \{j \in J \mid R_j(f(j)) \geqslant \xi\} \in D\},$$

and hence there exist $\xi_1, \xi_2 \in [0, 1]$ such that

$$\{j \in J \mid R_j(f(j) \to g(j)) \geqslant \xi_1\} \in D,$$

$$\{j \in J \mid R_j(f(j)) \geqslant \xi_2\} \in D,$$

and $R(g_D) < \xi_1$, $R(g_D) < \xi_2$. By $R_j(g(j)) \geqslant \min\{R_j(f(j) \to g(j)), R_j(f(j))\}$, it follows that

$$\begin{aligned} \{j \in J \mid R_j(g(j)) \geqslant \xi_1 \wedge \xi_2\} \supseteq \{j \in J \mid R_j(f(j)) \geqslant \xi_1 \wedge \xi_2\} \\ \cap\{j \in J \mid R_j(f(j) \to g(j)) \geqslant \xi_1 \wedge \xi_2\}, \end{aligned}$$

and hence $\{j \in J \mid R_j(g(j)) \geqslant \xi_1 \wedge \xi_2\} \in D$, *i.e.*, a contradiction with the fact that $R(g_D) < \xi_1 \wedge \xi_2$. □

Theorem 8.5.9 *(1). If R_j is a fuzzy implicative filter of L_j for any $j \in J$, then $R = \prod_D R_j$ is a fuzzy implicative filter of L_D.*

(2). If R_j is a fuzzy lattice implication subalgebra of L_j for any $j \in J$, then $R = \prod_D R_j$ is a fuzzy lattice implication subalgebra of L_D. □

Part III

Lattice-Valued Logic Systems

9 Lattice-Valued Propositional Logics

In Chapter 2, a logical algebra – lattice implication algebra has been established, and its properties have been discussed in Chapters 2 – 8. In this chapter, we establish lattice-valued propositional logic $LP(X)$ and gradational lattice-valued propositional logic L_{vpl} based on lattice implication algebra.

In this chapter and the following chapters, L is an infinite distributive lattice implication algebra and all concepts except specially defined are the same as those in classical logic.

9.1 Lattice-Valued Propositional Logic $LP(X)$

9.1.1 Language

The symbols in $LP(X)$ are

(1). the set of propositional variable: $X = \{p, q, r, \cdots\}$;
(2). the set of constants: L;
(3). logical connectives: $\rightarrow$, $'$;
(4). auxiliary symbols:) , (.

The set $\mathcal{F}$ of formulae of $LP(X)$ is the least set Y satisfying the following conditions:

(1). $X \subseteq Y$;
(2). $L \subseteq Y$;
(3). if $p, q \in Y$, then $'(p), \rightarrow (p, q) \in Y$.

In the following, we denote $'(p)$ as p' and $\rightarrow (p, q)$ as $p \rightarrow q$.

Remark 11. From the viewpoint of universal algebra, $LP(X)$ is the free algebra on X w.r.t. the type $T = L \cup \{', \rightarrow\}$, where $\alpha \in L$ is a 0-ary operation.

9.1.2 Semantics

In Chapter 2, we showed in the lattice implication algebra L

$$\alpha \vee \beta = (\alpha \rightarrow \beta) \rightarrow \beta,$$
$$\alpha \wedge \beta = (\alpha' \vee \beta')',$$

hold for any $\alpha, \beta \in L$. Hence, L and $LP(X)$ can be looked as algebras with the same type $T = L \cup \{', \rightarrow\}$ and for any $p, q \in \mathcal{F}$,

$$p \vee q \triangleq (p \rightarrow q) \rightarrow q,$$
$$p \wedge q \triangleq (p' \vee q')'.$$

Definition 9.1.1 *A mapping $v: LP(X) \longrightarrow L$ is called a valuation of $LP(X)$, if it is a T-homomorphism.*

Because $LP(X)$ is a free T-algebra on X, there exists exactly one valuation v such that

$$v|_X = \varphi$$

for any mapping $\varphi: X \longrightarrow L$.

From Definition 9.1.1, we know that if v is a valuation of $LP(X)$, then $v(\alpha) = \alpha$ for any $\alpha \in L$ and the following conclusions hold.

Corollary 9.1.1 *Let $f: LP(X) \longrightarrow L$ be a mapping, then f is a valuation of $LP(X)$ if and only if it satisfies*

(1). $f(\alpha) = \alpha$ for any $\alpha \in L$;
(2). $f(p') = (f(p))'$ for any $p \in \mathcal{F}$;
(3). $f(p \rightarrow q) = f(p) \rightarrow f(q)$ for any $p, q \in \mathcal{F}$. □

Definition 9.1.2 *Let $A \in \mathscr{F}_L(\mathcal{F})$, v be a valuation of $LP(X)$. It is called that v satisfies A if*

$$A(p) \leqslant v(p)$$

for any $p \in \mathcal{F}$. A is called satisfiable if there exists a valuation v, which satisfies A.

Definition 9.1.3 *Let $A \in \mathscr{F}_L(\mathcal{F})$, $p \in \mathcal{F}$, $\alpha \in L$. p is called semantically implied from A with truth value level α if $v(p) \geqslant \alpha$ for any valuation v, which satisfies A, of $LP(X)$. We write this $A \models_\alpha p$. p is called valid with truth value level α (shortly for α-valid) and denoted by $\models_\alpha p$, if $v(p) \geqslant \alpha$ for any valuation v of $LP(X)$. p is called a valid formula, if $\models_I p$. p is called α-valid in A, if*

$$\alpha = \bigwedge\{v(p) \mid v \text{ is a valuation and satisfies } A\}.$$

We shall write this $A \mid\models_\alpha p$.

It is obvious that p is α-valid for any $\alpha \in L$ if p is valid. In the following, we denote $A \mid\models_I p$ by $A \mid\models p$ and $\emptyset \mid\models_\alpha p$ by $\mid\models_\alpha p$ for short.

Example 9.1.1 *For any $p, q \in \mathcal{F}$, the following formulae are all valid.*

(1). $p \rightarrow (q \rightarrow p)$;
(2). $(p \wedge q) \rightarrow p$;

(3). $p \to p$;
(4). $p \to (p \vee q)$;
(5). $p \to (q \vee p)$;
(6). $p \to (q \to (p \wedge q))$;
(7). $(p \to q) \to ((r \to q) \to ((p \vee r) \to q))$,

and the following formulae are all τ_0-valid.

(1). $(p \to (q \to r)) \to ((p \to q) \to (p \to r))$;
(2). $(p \to q) \to ((p \to q') \to p')$,

where $\tau_0 = \bigwedge_{\alpha \in L}(\alpha \vee \alpha')$.

Definition 9.1.4 *Let $p, q \in \mathcal{F}$. p, q are called equivalent if*

$$v(p) = v(q)$$

for any valuation v of $LP(X)$. We write this $p = q$.

It is easy to prove the following conclusions.

Theorem 9.1.1 *For any $p, q \in \mathcal{F}$,*

$$p = q \qquad \text{iff} \qquad p \leftrightarrow q \text{ is a valid formula,}$$

where $p \leftrightarrow q$ is shortly for $(p \to q) \wedge (q \to p)$. □

Theorem 9.1.2 *For any $p, q, r \in \mathcal{F}$, the following conclusions hold.*

(1). $p \to (q \to r) = q \to (p \to r)$;
(2). $p \to q = q' \to p'$;
(3). $(p \to q) \to q = (q \to p) \to p$;
(4). $p \to p = I$;
(5). $(p \vee q) \to r = (p \to r) \wedge (q \to r)$;
(6). $(p \wedge q) \to r = (p \to r) \vee (q \to r)$;
(7). $p \to (q \vee r) = (p \to q) \vee (p \to r)$;
(8). $p \to (q \wedge r) = (p \to q) \wedge (p \to r)$;
(9). $p \to O = p'$;
(10). $I \to p = p$;
(11). $O \to p = p \to I = I$;
(12). $(p \to q) \to (r \to q) = r \to (p \vee q) = (q \to p) \to (r \to p)$;
(13). $(p \to q) \to (p \to r) = (p \wedge q) \to r = (q \to p) \to (q \to r)$. □

Theorem 9.1.3 *For any $A \in \mathscr{F}_L(\mathcal{F})$, $\alpha \in L$, $p, q \in \mathcal{F}$, there holds*

$$A \models_\alpha p \to q \qquad \text{iff} \qquad \alpha \otimes v(p) \leqslant v(q)$$

for any valuation v which satisfies A.

Proof. For any valuation v of $LP(X)$, which satisfies A, we have

$$\begin{aligned} v(p \to q) \geqslant \alpha \ &\text{iff} \ \alpha \to v(p \to q) = I \\ &\text{iff} \ \alpha \to (v(p) \to v(q)) = I \\ &\text{iff} \ \alpha \otimes v(p) \to v(q) = I \\ &\text{iff} \ \alpha \otimes v(p) \leqslant v(q). \end{aligned}$$

The conclusion is obvious from Definition 9.1.3. □

Definition 9.1.5 *Let $A \in \mathscr{F}_L(\mathcal{F})$, A is called closed if*

(1). $A(p \to q) \otimes A(p) \leqslant A(q)$,
(2). $\alpha \to A(p) \leqslant A(\alpha \to p)$,

for any $p, q \in \mathcal{F}$ and $\alpha \in L$.

Theorem 9.1.4 *Each valuation of $LP(X)$ is closed.* □

Corollary 9.1.2 *If $A \mid\models_\alpha p$, then $A \models_\beta p$ for any $\beta \leqslant \alpha$.* □

Corollary 9.1.3 *$\mid\models_\alpha p$ if and only if $\models_\alpha p$ holds, but $A \models_\beta p$ does not hold for any $\beta > \alpha$.* □

Corollary 9.1.4 *$\mid\models p$ if and only if $\models p$.* □

Definition 9.1.6 *A mapping $f : \mathscr{F}_L(Y) \longrightarrow \mathscr{F}_L(Y)$ is called a closure operation, if for any $A, B \in \mathscr{F}_L(Y)$,*

(1). $A \subseteq f(A)$,
(2). $f(A) \subseteq f(B)$ *when* $A \subseteq B$,
(3). $f(f(A)) = f(A)$.

Theorem 9.1.5 *Let* Con: $\mathscr{F}_L(\mathcal{F}) \longrightarrow \mathscr{F}_L(\mathcal{F})$ *be a mapping satisfying*

$$\mathrm{Con}(A) = \bigcap \{v \mid v \textit{ is a valuation of } LP(X) \textit{ and satisfies } A\}$$

for any $A \in \mathscr{F}_L(\mathcal{F})$, then Con *is a closure operation.*

Proof. For any $A, B \in \mathscr{F}_L(\mathcal{F})$.

(1). It is obvious that $A \subseteq \mathrm{Con}(A)$.
(2). If $A \subseteq B$, v is a valuation and satisfies B, then v satisfies A. Hence,

$$\{v \mid v \text{ satisfies } A\} \supseteq \{v \mid v \text{ satisfies } B\}.$$

So

$$\begin{aligned} \mathrm{Con}(A) &= \bigcap \{v \mid v \text{ is a valuation and satisfies } A\} \\ &\subseteq \bigcap \{v \mid v \text{ is a valuation and satisfies } B\} \\ &= \mathrm{Con}(B). \end{aligned}$$

(3). Suppose v is a valuation and satisfies A, then

$$v \supseteq \bigcap\{v' \mid v' \text{ is a valuation and satisfies } A\} = \mathrm{Con}(A).$$

Hence v satisfies $\mathrm{Con}(A)$ and

$$\begin{aligned}&\{v \mid v \text{ is a valuation and satisfies } A\}\\ &\qquad\subseteq \{v \mid v \text{ is a valuation and satisfies } \mathrm{Con}(A)\}.\end{aligned}$$

So we have

$$\begin{aligned}\mathrm{Con}(\mathrm{Con}(A)) &= \bigcap\{v \mid v \text{ is a valuation and satisfies } \mathrm{Con}(A)\}\\ &\subseteq \bigcap\{v \mid v \text{ is a valuation and satisfies } A\}\\ &= \mathrm{Con}(A).\end{aligned}$$

Moreover, $\mathrm{Con}(A) \subseteq \mathrm{Con}(\mathrm{Con}(A))$ by (1). □

Corollary 9.1.5 *For any $A \in \mathscr{F}_L(\mathcal{F})$, $p \in \mathcal{F}$, $\alpha \in L$, $A \,\|\!\!\models_\alpha p$ if and only if* $\mathrm{Con}(A)(p) = \alpha$. □

Let

$$\bar{\mathcal{F}} = {}^{\mathcal{F}}\!/_{=} = \{\overline{p} \mid p \in \mathcal{F}\},$$

where $\overline{p} = \{q \mid q \in \mathcal{F}, p = q\}$. It can be proved that $=$ is a congruence relation on $\mathcal{F}$. So, define operations $\vee$, $\wedge$, $\rightarrow$, $'$ on $\mathcal{F}$ as follows:

$$\begin{aligned}(\overline{p})' &= \overline{p'},\\ \overline{p} \rightarrow \overline{q} &= \overline{p \rightarrow q},\\ \overline{p} \vee \overline{q} &= \overline{p \vee q},\\ \overline{p} \wedge \overline{q} &= \overline{p \wedge q}.\end{aligned}$$

Let $\overline{LP(X)} = (\bar{\mathcal{F}}, \vee, \wedge, ', \rightarrow)$. Then we have the following theorem.

Theorem 9.1.6 *$\overline{LP(X)}$ is a lattice implication algebra.*

Proof. There are two steps for the proof.

First, we prove that $\overline{LP(X)}$ is a complemented lattice with universal bounds $\overline{O}, \overline{I}$. In fact, for the operation $\vee$, there has

$$\begin{aligned}\overline{x} \vee \overline{y} &= \overline{x \vee y}\\ &= \overline{(x \rightarrow y) \rightarrow y}\\ &= \overline{(y \rightarrow x) \rightarrow x}\\ &= \overline{y \vee x}\\ &= \overline{y} \vee \overline{x}.\end{aligned}$$

That means the law of commutativity for $\vee$ holds. Similarly, the associative and idempotent laws hold. Furthermore, we can get the commutative, associative and idempotent laws for $\wedge$ and the absorptive law for $\vee$ and $\wedge$, *i.e.*, $\bar{x} \vee (\bar{x} \wedge \bar{y}) = \bar{x}$. Hence, $(\bar{\mathcal{F}}, \vee, \wedge)$ is a lattice with the partial order $\leqslant$:

$$\begin{aligned} \overline{x} \leqslant \overline{y} \ \text{ iff } & \ \overline{x} \vee \overline{y} = \overline{y} \\ \text{iff } & \ v(x \vee y) = v(y) \text{ for any valuation } v \text{ of } LP(X) \\ \text{iff } & \ v(x) \leqslant v(y) \text{ for any valuation } v \text{ of } LP(X). \end{aligned}$$

If $\overline{x} \leqslant \overline{y}$, then $v(x) \leqslant v(y)$ for any valuation v of $LP(X)$. Hence $(v(y))' \leqslant (v(x))'$, *i.e.*, $v(y') \leqslant v(x')$, $(\overline{y})' \leqslant (\bar{x})'$. Moreover, because $(\overline{x}')' = \overline{x}$ is obvious, then $'$ is an order-reversing involution on $\overline{LP(X)}$.

Sum up above, $(\bar{\mathcal{F}}, \vee, \wedge, ')$ is a complemented lattice.

Second, we need to prove (I_1)–(I_5), (l_1), (l_2) hold for $\overline{LP(X)}$. Here we only prove (I_1), the others can be proved similarly. In fact, there has

$$\begin{aligned} \overline{x} \to (\overline{y} \to \overline{z}) &= \overline{x} \to \overline{y \to z} \\ &= \overline{x \to (y \to z)} \\ &= \overline{y \to (x \to z)} \\ &= \overline{y} \to \overline{x \to z} \\ &= \overline{y} \to (\overline{x} \to \overline{z}) \end{aligned}$$

for any $\overline{x}, \overline{y}, \overline{z} \in \bar{\mathcal{F}}$. This means (I_1) holds. □

Theorem 9.1.7 *(1). If v is a valuation, $\overline{v}\colon \overline{LP(X)} \longrightarrow L$ satisfies $\overline{v}(\overline{x}) = v(x)$ for any $\overline{x} \in \overline{LP(X)}$, then $\overline{v}$ is a lattice implication epimorphism.*

(2). Suppose $\overline{v}\colon \overline{LP(X)} \longrightarrow L$ is a lattice implication homomorphism and satisfies $\overline{v}(\overline{\alpha}) = \alpha$ for any $\alpha \in L$. If $v_1\colon LP(X) \longrightarrow L$ satisfies $v_1(x) = \overline{v}(\overline{x})$ for any $x \in L$, then v_1 is a valuation of $LP(X)$.

Proof. (1). If $\overline{x} = \overline{y}$, then $v(x) = v(y)$ for v being a valuation. Hence, the definition of $\overline{v}$ is rational. Because of $\overline{v}(\overline{\alpha}) = v(\alpha) = \alpha$ for any $\alpha \in L$ and

$$\begin{aligned} \overline{v}(\bar{x} \to \bar{y}) &= \overline{v}(\overline{x \to y}) \\ &= v(x \to y) \\ &= v(x) \to v(y) \\ &= \overline{v}(\overline{x}) \to \overline{v}(\overline{y}) \end{aligned}$$

for any $\overline{x}, \overline{y} \in \overline{LP(X)}$, $\overline{v}$ is a surjection, then $\overline{v}$ is a lattice implication epimorphism.

(2). Similar to (1). □

For any element $\overline{\omega}(x_1, \cdots, x_n)$ of $\overline{LP(X_n)}$, suppose that

$$\overline{\omega(x_1, \cdots, x_n)} = \overline{\omega}(x_1, \cdots, x_n)$$

where $X_n = \{x_1, \cdots, x_n\}$ and $\omega(x_1, \cdots, x_n) \in LP(X_n)$. Define a mapping $f_{\overline{\omega}}\colon L^n \longrightarrow L$ as follows:

for any $(\alpha_1, \cdots, \alpha_n) \in L^n$, there exists exactly one valuation v of $LP(X_n)$, such that

$$v(x_i) = \alpha_i, \quad 1 \leqslant i \leqslant n.$$

then let

$$f_{\overline{\omega}}(\alpha_1, \cdots, \alpha_n) = v(\omega(x_1, \cdots, x_n)).$$

It is easy to validate the rationality of the above definition. We call $f_{\bar{\omega}}$ a truth-value function of L^n.

Theorem 9.1.8 *For any $\overline{\omega}, \overline{\omega_1} \in \overline{LP(X_n)}$, if $f_{\overline{\omega}} = f_{\overline{\omega_1}}$, then $\overline{\omega} = \overline{\omega_1}$.*

Proof. Suppose $\omega(x_1, \cdots, x_n), \omega_1(x_1, \cdots, x_n) \in LP(X_n)$ and satisfy

$$\overline{\omega(x_1, \cdots, x_n)} = \overline{\omega}(x_1, \cdots, x_n),$$

$$\overline{\omega_1(x_1, \cdots, x_n)} = \overline{\omega_1}(x_1, \cdots, x_n).$$

For any valuation v of $LP(X_n)$, suppose

$$v(x_i) = \alpha_i, \qquad 1 \leqslant i \leqslant n,$$

then

$$\begin{aligned} v(\omega(x_1, \cdots, x_n)) &= f_{\overline{\omega}}(\alpha_1, \cdots, \alpha_n) \\ &= f_{\overline{\omega_1}}(\alpha_1, \cdots, \alpha_n) \\ &= v(\omega_1(x_1, \cdots, x_n)). \end{aligned}$$

So

$$\begin{aligned} \overline{\omega}(x_1, \cdots, x_n) &= \overline{\omega(x_1, \cdots, x_n)} \\ &= \overline{\omega_1(x_1, \cdots, x_n)} \\ &= \overline{\omega_1}(x_1, \cdots, x_n) \end{aligned}$$

for arbitrariness of v. □

Theorem 9.1.9 *Suppose $\omega = \omega(x_1, \cdots, x_n) \in \mathcal{F}$, $\tau \in L$, then $\models_\tau \omega$ if and only if $f_{\overline{\omega}}(\alpha_1, \cdots, \alpha_n) \geqslant \tau$ for any $(\alpha_1, \cdots, \alpha_n) \in L^n$.*

Proof. Suppose that $f_{\overline{\omega}}(\alpha_1, \cdots, \alpha_n) \geqslant \tau$ for any $(\alpha_1, \cdots, \alpha_n) \in L^n$. Then $v|_{LP(X_n)}$ is a valuation of $LP(X_n)$ for any valuation v of $LP(X)$. Let

$$v(x_i) = \alpha_i, \qquad 1 \leqslant i \leqslant n.$$

So we have $v(\omega) = f_{\overline{\omega}}(\alpha_1, \cdots, \alpha_n) \geqslant \tau$ and $\models_\tau \omega$.

Conversely, suppose $\models_\tau \omega$. For any $(\alpha_1, \cdots, \alpha_n) \in L^n$, there exists a valuation v of $LP(X)$ such that

$$v(x_i) = \alpha_i, \qquad 1 \leqslant i \leqslant n.$$

Hence $f_{\overline{\omega}}(\alpha_1, \cdots, \alpha_n) = v(\omega) \geqslant \tau$. □

Corollary 9.1.6 *$\omega = \omega(x_1, \cdots, x_n) \in \mathcal{F}$ is valid if and only if $f_{\overline{\omega}}(\alpha_1, \cdots, \alpha_n)$*

$= I$ for any $(\alpha_1, \cdots, \alpha_n) \in L^n$. □

Theorem 9.1.10 *If $|L| < \aleph_0$, $\tau \in L$, then the validity with truth value level τ in $LP(X)$ is decidable.*

Proof. For any $\omega = \omega(x_1, \cdots, x_m) \in \mathcal{F}$, we know that a formula ω is valid with truth value level τ if and only if $f_{\overline{\omega}}(\alpha_1, \cdots, \alpha_m) \geqslant \tau$ for any $(\alpha_1, \cdots, \alpha_m) \in L^m$. By calculating the value of $f_{\overline{\omega}}$, we can decide whether ω is valid with truth value level τ or not. □

Corollary 9.1.7 *If $|L| < \aleph_0$, then the validity in $LP(X)$ is decidable.* □

9.1.3 Syntax

By properties of the lattice implication algebra, it is easy to prove the following conclusion.

Theorem 9.1.11 *For any $p, q, r \in \mathcal{F}$, and $m, n \in \mathbb{N}^+$, the following formulae are valid.*

$$p \to I \tag{9.1}$$
$$p \to p \tag{9.2}$$
$$(p \to q) \to ((q \to r) \to (p \to r)) \tag{9.3}$$
$$(p \to (q \to r)) \to (q \to (p \to r)) \tag{9.4}$$
$$(p \wedge q) \to p \tag{9.5}$$
$$(p \wedge q) \to q \tag{9.6}$$
$$(p \to r) \to ((p \to q) \to (p \to (r \wedge q))) \tag{9.7}$$
$$p \to (p \vee q) \tag{9.8}$$
$$q \to (p \vee q) \tag{9.9}$$
$$(p \to r) \to ((q \to r) \to ((p \vee q) \to r)) \tag{9.10}$$
$$p \to (q \to (p \otimes q)) \tag{9.11}$$
$$p \to (q \to (p \wedge q)) \tag{9.12}$$
$$p \to (q \to p) \tag{9.13}$$
$$(p^m \to (q \to r)) \to ((p^n \to q) \to (p^{m+n} \to r)) \tag{9.14}$$
$$(p \to (q \to r)) \to ((p \otimes q) \to r) \tag{9.15}$$
$$(p \to q) \to (q' \to p') \tag{9.16}$$
$$(p')' \to p \tag{9.17}$$

where $p^2 = p \otimes p$, $p^n = p^{n-1} \otimes p$. □

Definition 9.1.7 *The axiom of lattice-valued propositional logic* $LP(X)$ *is such an* L*-fuzzy subset* $A_L \in \mathscr{F}_L(\mathcal{F})$,

$$A_L(p) = \begin{cases} I, & p \text{ with form of Eqs.}(9.1) \sim (9.17) \text{ in Theorem 9.1.11}, \\ \alpha, & p = \alpha \text{ and } \alpha \in L, \\ O, & \text{otherwise.} \end{cases}$$

Definition 9.1.8 *Let* $A \in \mathscr{F}_L(\mathcal{F})$, $p \in \mathcal{F}$. *A formal proof* ω *of* p *from* A *is a finite sequence as follows:*

$$(p_1, \alpha_1), \cdots, (p_n, \alpha_n),$$

where $p_n = p$, *and for any* i, $1 \leqslant i \leqslant n$, $p_i \in \mathcal{F}$, $\alpha_i \in L$ *and*

(1). $A_L(p_i) = \alpha_i$, *or*
(2). $A(p_i) = \alpha_i$, *or*
(3). there exist $j, k < i$, *such that* $p_j = p_k \to p_i$ *and* $\alpha_i = \alpha_j \otimes \alpha_k$, *or*
(4). there exists $j < i$ *and* $\alpha \in L$, *such that* $p_i = \alpha \to p_j$ *and* $\alpha_i = \alpha \to \alpha_j$.

In the formal proof ω, n *is called the length of* ω *and denoted by* $l(\omega)$, α_n *is called the value of* ω *and denoted by* $\mathrm{val}(\omega)$.

In the following, a formal proof is called a proof for short.

Definition 9.1.9 *Let $A \in \mathscr{F}_L(\mathcal{F})$, $p \in \mathcal{F}$, $\alpha \in L$. p is called an α-theorem of A and written as $A \vdash_\alpha p$, if*

$$\alpha \leqslant \bigvee\{\text{val}(\omega) \mid \omega \text{ is a proof of } p \text{ from } A\}.$$

If

$$\alpha = \bigvee\{\text{val}(\omega) \mid \omega \text{ is a proof of } p \text{ from } A\},$$

then it is written as $A \Vdash_\alpha p$.

Corollary 9.1.8 *If $A \vdash_\alpha p$, then $A \vdash_\beta p$ for any $\beta \leqslant \alpha$.* □

Corollary 9.1.9 $A_L \subseteq \text{Con}(\emptyset) = \bigcap\{v \mid v \text{ is a valuation of } LP(X)\}$. □

Corollary 9.1.10 *If $A \Vdash_\alpha p$, then $A(p) \leqslant \alpha$.* □

Theorem 9.1.12 *Let $A \in \mathscr{F}_L(\mathcal{F})$, $p \in \mathcal{F}$, $\alpha \in L$, and v a valuation of $LP(X)$. If $A \vdash_\alpha p$ and v satisfies A, then $v(p) \geqslant \alpha$.*

Proof. For any proof ω of p from A, we prove that $v(p) \geqslant \text{val}(\omega)$ by induction over the length $l(\omega)$ of the proof.

If $l(\omega) = 1$, then let ω be (p, β). Hence $A_L(p) = \beta$ or $A(p) = \beta$. When $A_L(p) = \beta$, we have $v(p) \geqslant A_L(p) = \beta = \text{val}(\omega)$ for $v \supseteq A_L$. When $A(p) = \beta$, we have $v(p) \geqslant A(p) = \beta = \text{val}(\omega)$ for v satisfying A.

Now suppose $l(\omega) = n$ and the proof ω is

$$(p_1, \alpha_1), (p_2, \alpha_2), \cdots, (p_n, \alpha_n).$$

If $A_L(p_n) = \alpha_n$ or $A(p_n) = \alpha_n$, the proof is similar to that of $l(\omega) = 1$. If there exist $i, j < n$, such that $p_i = p_j \to p_n$ and $\alpha_n = \alpha_i \otimes \alpha_j$, then

$$v(p_i) \geqslant \alpha_i, \qquad v(p_j) \geqslant \alpha_j$$

by induction. Because v is closed, then

$$\begin{aligned} v(p) &= v(p_n) \\ &\geqslant v(p_j \to p_n) \otimes v(p_j) \\ &= v(p_i) \otimes v(p_j) \\ &\geqslant \alpha_i \otimes \alpha_j \\ &= \alpha_n \\ &= \text{val}(\omega). \end{aligned}$$

If there exist $i < n$, $\beta \in L$, such that $p_n = \beta \to p_i$ and $\alpha_n = \beta \to \alpha_i$, then we have

$$\begin{aligned} v(p) &= v(p_n) \\ &= v(\beta \to p_i) \\ &= \beta \to v(p_i) \\ &\geqslant \beta \to \alpha_i \\ &= \alpha_n \\ &= \mathrm{val}(\omega). \end{aligned}$$

We have shown that $v(p) \geqslant \mathrm{val}(\omega)$ for any proof ω of p from A, hence $v(p) \geqslant \bigvee\{\mathrm{val}(\omega) \mid \omega \text{ is a proof of } p \text{ from } A\} \geqslant \alpha$. □

By Theorem 9.1.12, we can get the soundness theorem.

Theorem 9.1.13 (Soundness) *Let $A \in \mathscr{F}_L(\mathcal{F})$, $p \in \mathcal{F}$, $\alpha \in L$. If $A \vdash_\alpha p$, then $A \models_\alpha p$.* □

Theorem 9.1.14 *Let $A \in \mathscr{F}_L(\mathcal{F})$, $p \in \mathcal{F}$, and $\alpha \in L$.*

(1). If $A \| \!\vdash_\alpha p$ and v satisfies A, then $v(p) \geqslant \alpha$.
(2). If $A \| \!\vdash_\alpha p$, then there exists $\beta \in L$ such that $\beta \geqslant \alpha$ and $A \models_\beta p$.

Proof. (1). Suppose v is a valuation which satisfies A. If $A \| \!\vdash_\alpha p$, then $A \vdash_\alpha p$, hence $v(p) \geqslant \alpha$.

(2). If A is unsatisfiable, then $A \| \!\models_I p$. The conclusion is obvious. Let A is satisfiable, then for any valuation v which satisfies A, then $v(p) \geqslant \alpha$ from (1). So let

$$\beta = \bigwedge\{v(p) \mid v \text{ is a valuation and satisfies } A\},$$

then $\beta \geqslant \alpha$ and $A \models_\beta p$. □

Theorem 9.1.15 *Let $A \in \mathscr{F}_L(\mathcal{F})$, $p \in \mathcal{F}$, $\alpha \in L$. If*

$$\alpha = \bigwedge\{B(p) \mid B \in \mathscr{F}_L(\mathcal{F}), B \textit{ is closed, } B \supseteq A \cup A_L\},$$

then $A \| \!\vdash_\alpha p$.

Proof. Suppose $A \| \!\vdash_\beta p$, *i.e.*,

$$\beta = \bigvee\{\mathrm{val}(\omega) \mid \omega \text{ is a proof of } p \text{ from } A\}.$$

It needs to prove $\alpha = \beta$.

(1). For any proof ω of p from A. If Y is closed and $Y \supseteq A \cup A_L$, then

$$\mathrm{val}(\omega) \leqslant Y(p). \tag{9.18}$$

In fact, if $l(\omega) = 1$, then $\mathrm{val}(\omega) = A(p)$ or $\mathrm{val}(\omega) = A_L(p)$, and Eq. (9.18) holds for $Y \supseteq A \cup A_L$.

Suppose Eq. (9.18) holds for any proof ω of p with $l(\omega) < m$. Now, let ω be a proof of p with $l(\omega) = m$:

$$(p_1, \alpha_1), (p_2, \alpha_2), \cdots, (p_m, \alpha_m).$$

If $A(p_m) = \alpha_m$ or $A_L(p_m) = \alpha_m$, then Eq. (9.18) holds.
If there exist $i, j < m$, such that $p_i = p_j \to p_m$ and $\alpha_m = \alpha_i \otimes \alpha_j$. By inductive assumption,

$$\alpha_i \leqslant Y(p_i), \alpha_j \leqslant Y(p_j),$$

hence Eq. (9.18) holds by

$$\begin{aligned} \mathrm{val}(\omega) &= \alpha_m \\ &= \alpha_i \otimes \alpha_j \\ &\leqslant Y(p_i) \otimes Y(p_j) \\ &= Y(p_j \to p_m) \otimes Y(p_j) \\ &\leqslant Y(p_m) \\ &= Y(p). \end{aligned}$$

If there exist $i < m$ and $\alpha_0 \in L$, such that $p_m = \alpha_0 \to p_i$ and $\alpha_m = \alpha_0 \to \alpha_i$, then $\alpha_i \leqslant Y(p_i)$ by inductive assumption. Hence Eq. (9.18) holds by

$$\begin{aligned} \mathrm{val}(\omega) &= \alpha_m \\ &= \alpha_0 \to \alpha_i \\ &\leqslant \alpha_0 \to Y(p_i) \\ &\leqslant Y(\alpha_0 \to p_i) \\ &= Y(p_m) \\ &= Y(p). \end{aligned}$$

Then we know that $\beta \leqslant \alpha$ for Eq. (9.18) holds.

(2). Let $B \in \mathscr{F}_L(\mathcal{F})$ and satisfy

$$B(q) = \bigvee \{\mathrm{val}(\omega) \mid \omega \text{ is a proof of } q \text{ from } A\}$$

for any $q \in \mathcal{F}$. Because both $(q, A(q))$ and $(q, A_L(q))$ are proofs of q from A, hence

$$A(q) \leqslant B(q), \qquad A_L(q) \leqslant B(q),$$

i.e., $A \cup A_L \subseteq B$. Moreover, for any $p, q \in \mathcal{F}$, $\alpha_0 \in L$. If

$$\omega_1 : \ (p_1, \alpha_1), \cdots, (p_m, \alpha_m),$$

$$\omega_2 : \ (q_1, \beta_1), \cdots, (q_n, \beta_n)$$

are two proofs of $p \to q$ and p from A, respectively, then

$$(p_1, \alpha_1), (p_2, \alpha_2), \cdots, (p_m, \alpha_m), (q_1, \beta_1), (q_2, \beta_2), \cdots, (q_n, \beta_n), (q, \alpha_m \otimes \beta_n)$$

is a proof of q from A, and

$$(q_1, \beta_1), (q_2, \beta_2), \cdots, (q_n, \beta_n), (\alpha_0 \to p, \alpha_0 \to \beta_n)$$

is a proof of $\alpha_0 \to p$ from A. Hence,

$$\begin{aligned} & B(p \to q) \otimes B(p) \\ = & (\bigvee\{\text{val}(\omega_1) \mid \omega_1 \text{ is a proof of } p \to q \text{ from } A\}) \\ & \qquad \otimes(\bigvee\{\text{val}(\omega_2) \mid \omega_2 \text{ is a proof of } p \text{ from } A\}) \\ = & \bigvee\{\text{val}(\omega_1) \otimes \text{val}(\omega_2) \mid \omega_1, \omega_2 \text{ are proofs of } p \to q, p \text{ from } A, \text{ resp.}\} \\ \leqslant & \bigvee\{\text{val}(\omega) \mid \omega \text{ is a proof of } q \text{ from } A\} \\ = & B(q) \end{aligned}$$

and

$$\begin{aligned} \alpha_0 \to B(p) &= \alpha_0 \to \bigvee\{\text{val}(\omega_2) \mid \omega_2 \text{ is a proof of } p \text{ from } A\} \\ &= \bigvee\{\alpha_0 \to \text{val}(\omega_2) \mid \omega_2 \text{ is a proof of } p \text{ from } A\} \\ &\leqslant \bigvee\{\text{val}(\omega) \mid \omega \text{ is a proof of } \alpha_0 \to p \text{ from } A\} \\ &= B(\alpha_0 \to p). \end{aligned}$$

So B is closed and $\alpha \leqslant B(p)$, *i.e.*, $\alpha \leqslant \beta$.

Sum up, $\alpha = \beta$ holds. □

Corollary 9.1.11 *If* $A \vdash_\alpha p$, $A \vdash_\beta p \to q$, *then*

(1). $A \vdash_{\alpha \otimes \beta} q$,
(2). $A \vdash_{\alpha_0 \to \alpha} \alpha_0 \to p$ *for any* $\alpha_0 \in L$.

Proof. (1). Because $A \vdash_\alpha p$ and $A \vdash_\beta p \to q$,

$$\alpha \leqslant \bigwedge\{B_1(p) \mid B_1 \supseteq A \cup A_L \text{ and } B_1 \text{ is closed}\},$$
$$\beta \leqslant \bigwedge\{B_2(p \to q) \mid B_2 \supseteq A \cup A_L \text{ and } B_2 \text{ is closed}\}.$$

Moreover,

$$\begin{aligned} &\{B_1(p) \otimes B_2(p \to q) \mid B_1, B_2 \supseteq A \cup A_L, B_1, B_2 \text{ are closed}\} \\ &\qquad \supseteq \{B(p) \otimes B(p \to q) \mid B \supseteq A \cup A_L, B \text{ is closed}\}. \end{aligned}$$

Hence we have

$$\begin{aligned}
&\alpha \otimes \beta \\
&\leqslant (\bigwedge\{B_1(p) \mid B_1 \supseteq A \cup A_L \text{ and } B_1 \text{ is closed}\}) \\
&\qquad\qquad \otimes(\bigwedge\{B_2(p \to q) \mid B_2 \supseteq A \cup A_L \text{ and } B_2 \text{ is closed}\} \\
&= \bigwedge\{B_1(p) \otimes B_2(p \to q) \mid B_1, B_2 \supseteq A \cup A_L \text{ and } B_1, B_2 \text{ are closed}\} \\
&\leqslant \bigwedge\{B(p) \otimes B(p \to q) \mid B \supseteq A \cup A_L \text{ and } B \text{ is closed}\} \\
&\leqslant \bigwedge\{B(q) \mid B \supseteq A \cup A_L \text{ and } B \text{ is closed}\},
\end{aligned}$$

i.e., $A \vdash_{\alpha\otimes\beta} q$.

(2). For any $\alpha_0 \in L$, we have

$$\begin{aligned}
\alpha_0 \to \alpha \leqslant \alpha_0 &\to \bigwedge\{B_1(p) \mid B_1 \supseteq A \cup A_L, B_1 \text{ is closed}\} \\
&= \bigwedge\{\alpha_0 \to B_1(p) \mid B_1 \supseteq A \cup A_L, B_1 \text{ is closed}\} \\
&\leqslant \bigwedge\{B_1(\alpha_0 \to p) \mid B_1 \supseteq A \cup A_L, B_1 \text{ is closed}\},
\end{aligned}$$

then $A \vdash_{\alpha_0 \to \alpha} \alpha_0 \to p$. □

Let Ded: $\mathscr{F}_L(\mathcal{F}) \longrightarrow \mathscr{F}_L(\mathcal{F})$ satisfy

$$\mathrm{Ded}(A)(p) = \bigvee\{\mathrm{val}(\omega) \mid \omega \text{ is a proof of } p \text{ from } A\}$$

for any $p \in \mathcal{F}$ and $A \in \mathscr{F}_L(\mathcal{F})$.

Theorem 9.1.16 Ded *is a closure operation.*

Proof. By Theorem 9.1.15, we have

$$\mathrm{Ded}(A)(p) = \bigwedge\{B(p)|B \supseteq A \cup A_L, B \text{ is closed}\}.$$

(1). It is obvious that $\mathrm{Ded}(A) \supseteq A$.

(2). If $A_1 \subseteq A_2$, then

$$\{B(p)|B \supseteq A_1 \cup A_L, B \text{ is closed}\} \supseteq \{B(p)|B \supseteq A_2 \cup A_L, B \text{ is closed}\}$$

for any $p \in LP(X)$.

So we have

$$\begin{aligned}
\mathrm{Ded}(A_1)(p) &= \bigwedge\{B(p) \mid B \supseteq A_1 \cup A_L, B \text{ is closed}\} \\
&\leqslant \bigwedge\{B(p) \mid B \supseteq A_2 \cup A_L, B \text{ is closed}\} \\
&= \mathrm{Ded}(A_2)(p),
\end{aligned}$$

i.e., $\mathrm{Ded}(A_1) \subseteq \mathrm{Ded}(A_2)$.

(3). By (1), it is known that $\mathrm{Ded}(A) \subseteq \mathrm{Ded}(\mathrm{Ded}(A))$.

For any $B \in \mathscr{F}_L(\mathcal{F})$, if $B \supseteq A \cup A_L$ and B is closed, then $B \supseteq \mathrm{Ded}(A)$. Hence,

$$\{B \mid B \supseteq A \cup A_L, B \text{ is closed}\} \subseteq \{B \mid B \supseteq \mathrm{Ded}(A) \cup A_L, B \text{ is closed}\}.$$

So,

$$\begin{aligned}\mathrm{Ded}(\mathrm{Ded}(A))(p) &= \bigwedge\{B(p) \mid B \supseteq \mathrm{Ded}(A) \cup A_L, B \text{ is closed}\}\\ &\leqslant \bigwedge\{B(p) \mid B \supseteq A \cup A_L, B \text{ is closed}\}\\ &= \mathrm{Ded}(A)(p),\end{aligned}$$

i.e., $\mathrm{Con}(\mathrm{Ded}(A)) \subseteq \mathrm{Ded}(A)$. □

Corollary 9.1.12 Ded $\subseteq$ Con. □

Theorem 9.1.17 (Deduction) *Let $A, A^* \in \mathscr{F}_L(\mathcal{F})$, $p, q \in \mathcal{F}$, $A^* = A \cup \{I/_p\}$ and $\alpha \in L$, then*

(1). if $A \vdash_\alpha p \to q$, then $A^* \vdash_\alpha q$;
(2). if ω_q is a proof of q from A^*, then there exist $n \in \mathbb{N}^+$ and a proof $\omega_{p^n \to q}$ of $p^n \to q$ from A, such that

$$\mathrm{val}(\omega_q) \leqslant \mathrm{val}(\omega_{p^n \to q}).$$

Proof. (1). Since $A \vdash_\alpha p \to q$ and $A \subseteq A^*$, then

$$\alpha \leqslant \mathrm{Ded}(A)(p \to q) \leqslant \mathrm{Ded}(A^*)(p \to q),$$

i.e., $A^* \vdash_\alpha p \to q$. For $A^* \vdash_I p$ and Corollary 9.1.11, then $A^* \vdash_\alpha q$.

(2). Suppose ω_q : $(q_1, \alpha_1), \cdots, (q_m, \alpha_m)$ is a proof of q from A^*, where $q_m = q$, $\alpha_m = \mathrm{val}(\omega_q)$. We prove the conclusion by induction over $l(\omega)$.

If $l(\omega_q) = 1$, then $A_L(q_m) = \alpha_m$ or $A^*(q_m) = \alpha_m$. When $A_L(q_m) = \alpha_m$, the sequence:

$$\begin{aligned}&(q_m, \alpha_m),\\ &(q_m \to (p \to q_m), I),\\ &(p \to q_m, \alpha_m),\end{aligned}$$

is a proof of $p \to q$ from A (denoted by $\omega_{p\to q}$), and $\mathrm{val}(\omega_{p\to q}) = \mathrm{val}(\omega_q)$. When $A^*(q_m) = \alpha_m$, if $p \neq q_m$, then $A(q_m) = \alpha_m$ and

$$\begin{aligned}&(q_m, \alpha_m),\\ &(q_m \to (p \to q_m), I),\\ &(p \to q_m, \alpha_m),\end{aligned}$$

is also the needed proof. If $p = q_m$, then $(p \to q_m, I)$ is a proof of $p \to q$ from A and $\mathrm{val}(\omega_{p\to q}) = I \geqslant \mathrm{val}(\omega_q)$.

Now suppose the conclusion holds for $l(\omega_q) < m$. When $l(\omega_q) = m$, write the proof ω_q as

$$\omega_q\colon\ (q_1,\alpha_1),\cdots,(q_m,\alpha_m).$$

If $A_L(q_m)=\alpha_m$ or $A^*(q_m)=\alpha_m$, similar to $l(\omega_q)=1$, the conclusion holds. If there exist $i,j<m$, such that $p_i=p_j\to p_m$ and $\alpha_m=\alpha_i\otimes\alpha_j$, then there exist $x\in\mathbb{N}^+$ and a proof of $p^x\to p_i$ from A:

$$(p_1,a_1),\cdots,(p_s,a_s)$$

such that $\alpha_i\leqslant a_s$. Also there exist $y\in\mathbb{N}^+$ and a proof of $p^y\to p_j$ from A:

$$(q_1,b_1),\cdots,(q_t,b_t)$$

such that $\alpha_j\leqslant b_t$. Then the sequence

$$\begin{aligned}
&(p_1,a_1),\cdots,(p_s,a_s),(q_1,b_1),\cdots,(q_t,b_t),\\
&((p^x\to(p_j\to p_m))\to((p^y\to p_j)\to(p^{x+y}\to p_m)),I),\\
&((p^y\to p_j)\to(p^{x+y}\to p_m),a_s),\\
&(p^{x+y}\to p_m,a_s\otimes b_t)
\end{aligned}$$

is a proof of $p^{x+y}\to q$ from A and

$$\begin{aligned}
\mathrm{val}(\omega_q)&=\alpha_m\\
&=\alpha_i\otimes\alpha_j\\
&\leqslant a_s\otimes b_t\\
&=\mathrm{val}(\omega_{p^{x+y}\to q}).
\end{aligned}$$

If there exist $i<m$, $\alpha_0\in L$ such that $q_m=\alpha_0\to q_i$ and $\alpha_m=\alpha_0\to\alpha_i$, suppose that $(p_1,a_1),\cdots,(p_s,a_s)$ is a proof of $p^x\to q_i$ from A and $a_s\geqslant\alpha_i$, where $x\in\mathbb{N}^+$. Hence, the sequence

$$\begin{aligned}
&(p_1,a_1),\cdots,(p_s,a_s),\\
&(\alpha_0\to(p^x\to q_i),\alpha_0\to a_s),\\
&((\alpha_0\to(p^x\to q_i))\to(p^x\to(\alpha_0\to q_i)),I),\\
&(p^x\to(\alpha_0\to q_i),\alpha_0\to a_s)
\end{aligned}$$

is a proof of $p^x\to q$ from A and

$$\begin{aligned}
\mathrm{val}(\omega_q)&=\alpha_m\\
&=\alpha_0\to\alpha_i\\
&\leqslant\alpha_0\to a_s\\
&=\mathrm{val}(\omega_{p^x\to q}).\ \square
\end{aligned}$$

Corollary 9.1.13 *Let $A,A^*\in\mathscr{F}_L(\mathcal{F})$, $p,q\in\mathcal{F}$, $A^*=A\cup\{I/_p\}$ and $\alpha\in L$, if $A\Vdash_\alpha p\to q$, then there exists $\beta\in L$, $\beta\geqslant\alpha$ such that $A^*\Vdash_\beta q$.* $\square$

Definition 9.1.10 *For $A\in\mathscr{F}_L(\mathcal{F})$, if*

$$\beta=\bigvee\{\alpha\mid\ \text{there exist } \alpha_1,\alpha_2\in L,p\in\mathcal{F},\ \text{s.t., } A\vdash_{\alpha_1}p, A\vdash_{\alpha_2}p' \text{ and } \alpha=\alpha_1\otimes\alpha_2\},$$

then A is called β'-consistent. A is called consistent if A is I-consistent.

Corollary 9.1.14 *If A is β'-consistent, then*

$$\beta = \bigvee\{\alpha_1 \otimes \alpha_2 \mid A \Vdash_{\alpha_1} p, A \Vdash_{\alpha_2} p', p \in \mathcal{F}\}. \ \square$$

Corollary 9.1.15 *If A is β-consistent, then $\beta \leqslant A(p') \to (A(p))'$ for any $p \in \mathcal{F}$.*

Proof. For any $p \in \mathcal{F}$, because $A \vdash_{A(p)} p$, $A \vdash_{A(p')} p'$ and A is β-consistent, then $A(p) \otimes A(p') \leqslant \beta'$. Hence,

$$\begin{aligned} \beta &\leqslant (A(p) \otimes A(p'))' \\ &= A(p) \to (A(p'))' \\ &= A(p') \to (A(p))'. \ \square \end{aligned}$$

Corollary 9.1.16 *If A is β-consistent, then $\beta \leqslant \mathrm{val}(\omega_{p'}) \to (\mathrm{val}(\omega_p))'$ for any $p \in \mathcal{F}$, proof ω_p of p and proof $\omega_{p'}$ of p' from A.* $\square$

Theorem 9.1.18 *If $A \in \mathscr{F}_L(\mathcal{F})$ is β'-consistent, then*

$$\beta \leqslant \bigvee\{\alpha \mid A \Vdash_\alpha p \otimes p', p \in \mathcal{F}\}.$$

Proof. Since A is β'-consistent, then

$$\beta = \bigvee\{\alpha \mid \alpha = \alpha_1 \otimes \alpha_2, A \Vdash_{\alpha_1} p, A \Vdash_{\alpha_2} p', p \in \mathcal{F}\}.$$

For any $p \in \mathcal{F}$, suppose $A \Vdash_{\alpha_1} p$ and $A \Vdash_{\alpha_2} p'$, then for any proof ω_p of p from A:

$$(p_1, a_1), \cdots, (p_s, a_s)$$

and any proof $\omega_{p'}$ of p' from A:

$$(q_1, b_1), \cdots, (q_t, b_t),$$

the following sequence

$$\begin{aligned} &(p_1, a_1), \cdots, (p_s, a_s), (q_1, b_1), \cdots, (q_t, b_t), \\ &(p \to (p' \to p \otimes p'), I), \\ &(p' \to p \otimes p', a_s), \\ &(p \otimes p', a_s \otimes b_t) \end{aligned}$$

is a proof of $p \otimes p'$ from A. Hence,

$$\mathrm{val}(\omega_p) \otimes \mathrm{val}(\omega_{p'}) = a_s \otimes b_t \leqslant \alpha,$$

where, α satisfies $A \Vdash_\alpha p \otimes p'$. So, we have

$$\begin{aligned}\alpha_1 \otimes \alpha_2 &= (\bigvee\{\mathrm{val}(\omega_p) \mid \omega_p \text{ is a proof of } p \text{ from } A\}) \\ &\quad \otimes(\bigvee\{\mathrm{val}(\omega_{p'}) \mid \omega_{p'} \text{ is a proof of } p' \text{ from } A\}) \\ &= \bigvee\{\mathrm{val}(\omega_p) \otimes \mathrm{val}(\omega_{p'}) \mid \omega_p, \omega_{p'} \text{ are proofs of } p, \\ &\qquad p' \text{ from } A \text{ respectively}\} \\ &\leqslant \alpha,\end{aligned}$$

and

$$\begin{aligned}\beta &= \bigvee\{\alpha_1 \otimes \alpha_2 \mid A\|\!\!-_{\alpha_1} p, A\|\!\!-_{\alpha_2} p', p \in \mathcal{F}\} \\ &\leqslant \bigvee\{\alpha \mid A\|\!\!-_{\alpha} p \otimes \alpha', p \in \mathcal{F}\}. \ \square\end{aligned}$$

Theorem 9.1.19 (Consistent) *If A is satisfiable, then A is consistent.*

Proof. Suppose a valuation v satisfies A, for any $p \in \mathcal{F}$, if $A\|\!\!-_{\alpha_1} p$, $A\|\!\!-_{\alpha_2} p'$, then $v(p) \geqslant \alpha_1$, $v(p') \geqslant \alpha_2$. Hence, $\alpha_1 \otimes \alpha_2 \leqslant v(p) \otimes v(p') = O$, and so, $\beta = \bigvee\{\alpha_1 \otimes \alpha_2 \mid A\|\!\!-_{\alpha_1} p, A\|\!\!-_{\alpha_2} p', p \in \mathcal{F}\} = O$, *i.e.*, A is consistent. □

Theorem 9.1.20 *Let $A \in \mathscr{F}_L(\mathcal{F})$.*

(1). Define a relation $\lesssim$ on $LP(X)$: for any $p, q \in \mathcal{F}$,

$$p \lesssim q \qquad \textit{iff} \qquad A \vdash_I p \to q,$$

then $\lesssim$ is a pre-order, and satisfies:

$$\alpha \lesssim p \qquad \textit{iff} \qquad A \vdash_\alpha p$$

for any $\alpha \in L$ and $p \in \mathcal{F}$.

(2). Define a relation $\approx$ on $LP(X)$: for any $p, q \in \mathcal{F}$,

$$p \approx q \qquad \textit{iff} \qquad p \lesssim q,\ q \lesssim p,$$

then $\approx$ is a congruence relation.

(3). Let $\mathcal{F}_A = \mathcal{F}/_\approx$, define operations $\vee$, $\wedge$, $\to$ and $'$ on it: for any $\overline{p}$, $\overline{q} \in \mathcal{F}_A$,

$$\begin{aligned}(\overline{p})' &= \overline{p'}, \\ \overline{p} \to \overline{q} &= \overline{p \to q}, \\ \overline{p} \vee \overline{q} &= \overline{p \vee q}, \\ \overline{p} \wedge \overline{q} &= \overline{p \wedge q}.\end{aligned}$$

Let $LR_A(X) = (\mathcal{F}_A, \vee, \wedge, ', \to)$, then $LR_A(X)$ is a lattice implication algebra.

(4). Let $f\colon L \longrightarrow LR_A(X)$ satisfy:

$$f(\alpha) = \overline{\alpha}$$

for any $\alpha \in L$, then f is a lattice implication homomorphism.

Proof. (1). For any $p, q, r \in \mathcal{F}$, $\alpha \in L$, we have

a) $p \lesssim p$ is obvious.

b) If $p \lesssim q$ and $q \lesssim r$, then $A \vdash_I p \to q$ and $A \vdash_I q \to r$. Hence $A \vdash_I p \to r$ for $A \vdash_I (p \to q) \to ((q \to r) \to (p \to r))$ and Corollary 9.1.11, *i.e.*, $p \lesssim r$.

c) If $\alpha \lesssim p$, then $A \vdash_I \alpha \to p$ and hence $A \vdash_\alpha p$ by $A \vdash_\alpha \alpha$.

If $A \vdash_\alpha p$, then $A \vdash_{\alpha \to \alpha} \alpha \to p$, *i.e.*, $A \vdash_I \alpha \to p$ and $\alpha \lesssim p$.

(2). It is obvious that $\approx$ is an equivalence relation.

a) For any $p, q \in \mathcal{F}$, because $A \vdash_I (p \wedge q) \to p$ and $A \vdash_I (p \wedge q) \to q$, then $p \wedge q \lesssim p$ and $p \wedge q \lesssim q$, *i.e.*, $p \wedge q$ is a lower bound of p, q w.r.t. $\lesssim$. If $r \lesssim p$ and $r \lesssim q$, then $A \vdash_I r \to p$ and $A \vdash_I r \to q$. Hence $A \vdash_I r \to (p \wedge q)$ for $A \vdash_I (r \to p) \to ((r \to q) \to (r \to (p \wedge q)))$, *i.e.*, $r \lesssim p \wedge q$. So $p \wedge q$ is the infimum of p, q w.r.t. $\lesssim$. Similarly, we can prove that $p \vee q$ is the supremum of p, q w.r.t. $\lesssim$.

b) For any $p, q, r \in \mathcal{F}$. Suppose $p_1 \approx q_1$ and $p_2 \approx q_2$.

i. Because

$$p_1 \wedge p_2 \lesssim p_1 \lesssim q_1 \ ,$$
$$p_1 \wedge p_2 \lesssim p_2 \lesssim q_2 \ ,$$

hence $p_1 \wedge p_2 \lesssim q_1 \wedge q_2$. Similarly, we can prove that $q_1 \wedge q_2 \lesssim p_1 \wedge p_2$. So, $p_1 \wedge p_2 \approx q_1 \wedge q_2$.

ii. Similarly, we can prove that $p_1 \vee p_2 \approx q_1 \vee q_2$.

iii. Because $q_1 \lesssim p_1$, then $A \vdash_I q_1 \to p_1$. Moreover, $A \vdash_I (q_1 \to p_1) \to ((p_1 \to p_2) \to (q_1 \to p_2))$, so we have $A \vdash_I (p_1 \to p_2) \to (q_1 \to p_2)$. Hence, $p_1 \to p_2 \lesssim q_1 \to p_2$. Similarly, $q_1 \to p_2 \lesssim q_1 \to q_2$ and $p_1 \to p_2 \lesssim q_1 \to q_2$. Also we can prove that $q_1 \to q_2 \lesssim p_1 \to p_2$ and $p_1 \to p_2 \approx q_1 \to q_2$.

iv. Because $p_1 \lesssim q_1$, then we have $A \vdash_I p_1 \to q_1$. Moreover,

$$A \vdash_I (p_1 \to q_1) \to (q_1' \to p_1'),$$

so $A \vdash_I q_1' \to p_1'$, *i.e.*, $q_1' \lesssim p_1'$. Similarly, $p_1' \lesssim q_1'$.

Then $\approx$ is a congruence relation.

(3). By the proof of (2), the definitions of $\vee$, $\wedge$, $\to$ and $'$ are rational.

a) It is easy to prove that $\overline{p} \wedge \overline{q}$ and $\overline{p} \vee \overline{q}$ are the infimum and the supremum of $\overline{p}$ and $\overline{q}$ in $LR_A(X)$ with the order $\leqslant$

$$\overline{p} \leqslant \overline{q} \qquad \text{iff} \qquad p \lesssim q,$$

and $LR_A(X)$ is a lattice. Moreover, we have $O \lesssim p \lesssim I$ because $A \vdash_O p$ and $A \vdash_I p \to I$ for any $p \in \mathcal{F}$. Hence, $\overline{O}$ and $\overline{I}$ are the smallest and the greatest element in $LR_A(X)$ respectively.

b) For any $p, q \in \mathcal{F}$, if $\overline{p} \leqslant \overline{q}$, then $p \lesssim q$ and $q' \lesssim p'$, *i.e.*, $(\overline{q})' \leqslant (\overline{p})'$. Because

$$((\overline{p})')' = \overline{(p')'} = \overline{p},$$

thus $'$ is an order-reversing involution on $LR_A(X)$.

c) For any $p, q, r \in \mathcal{F}$, we have

$$p \to (q \to r) \approx q \to (p \to r),$$

for

$$A \vdash_I (p \to (q \to r)) \to (q \to (p \to r)),$$
$$A \vdash_I (q \to (p \to r)) \to (p \to (q \to r)).$$

Hence, $\overline{p} \to (\overline{q} \to \overline{r}) = \overline{q} \to (\overline{p} \to \overline{r})$. Similarly, we can get the other conditions of the lattice implication algebra.

So LR_A is a lattice implication algebra.

(4). It is obvious. □

In $LP(X)$, let

$$\overline{p} = \{q \mid q \in \mathcal{F}, p \approx q\},$$
$$\overline{\overline{p}} = \{q \mid q \in \mathcal{F}, p = q\}.$$

Then we have the following result.

Theorem 9.1.21 *Let* $g\colon\ LR_{\emptyset}(X) \longrightarrow \overline{LP(X)}$ *satisfy: for any* $p \in LP(X)$,

$$g(\overline{p}) = \overline{\overline{p}},$$

then g *is a lattice implication epimorphism.*

Proof. For any $p, q \in \mathcal{F}$. If $p \approx q$, then $\emptyset \mid\!-_I p \to q$, $\emptyset \mid\!-_I q \to p$, *i.e.*, $\mathrm{Ded}(\emptyset)(p \to q) = \mathrm{Ded}(\emptyset)(q \to p) = I$. Because $\mathrm{Ded} \subseteq \mathrm{Con}$, then

$$\mathrm{Con}(\emptyset)(p \to q) = \mathrm{Con}(\emptyset)(q \to p) = I.$$

Hence, $v(p) = v(q)$ for any valuation v of $LP(X)$, *i.e.*, $p = q$. So the definition of g is rational and g is a surjection.

The proof of lattice implication epimorphism is trivial. □

9.1.4 Examples

In this section, we discuss a four-valued propositional logic $Q(X)$ with two incomparable truth values and a fuzzy-valued propositional logic $FP(X)$.

Suppose L is the lattice implication algebra defined in Example 2.1.6, then the corresponding lattice-valued propositional logic based on L is called four-valued propositional logic $Q(X)$. In $Q(X)$, the set of formulae is denoted by $\mathcal{Q}$. We can get the following conclusions.

Theorem 9.1.22 *Let* $A \in \mathscr{F}_L(\mathcal{Q})$, $p \in \mathcal{Q}$, $\alpha \in L$. *Then* $A \models_\alpha p$ *if* $A \vdash_\alpha p$.
□

Theorem 9.1.23 *Let* $A, A^* \in \mathscr{F}_L(\mathscr{Q})$, $p, q \in \mathscr{Q}$, $A^* = A \cup \{I/p\}$. *Then* $A \vdash_\alpha p \to q$ *if and only if* $A^* \vdash_\alpha q$. $\square$

Theorem 9.1.24 *Let* $p \in \mathscr{Q}$, *and there is no constant in* p, *then* $\vdash p$ *if* $\models p$. $\square$

Theorem 9.1.25 *Let* $p \in \mathscr{Q}$, *and there is no constant in* p, *then for any* $\beta \in L$,

(1). $\vdash \beta \vee p$ *if* $\models \beta \vee p$;
(2). $\vdash \beta \to p$ *if* $\models \beta \to p$;
(3). $\vdash p \to \beta$ *if* $\models p \to \beta$. $\square$

In lattice-valued propositional logic system $LP(X)$, if L is the Łukasiewicz implication algebra defined in Example 2.1.2 or the lattice implication algebra $(\mathscr{F}([0,1]), \vee, \wedge, {}', \to)$ defined in Example 2.1.7, we can get two different logic systems denoted by $\mathscr{L}_u$ and $\mathscr{L}_u^*$, respectively. They are widely applied to fuzzy reasoning.

9.2 Gradational Lattice-Valued Propositional Logic L_{vpl}

In this section, we discuss a gradational lattice-valued propositional logic L_{vpl}.

9.2.1 Language

The alphabets of L_{vpl} consists of

(1). propositional variables: $x, y, \cdots$;
(2). symbols for truth value: α, $\alpha \in L$;
(3). logical connectives: $\vee$, $\wedge$, $'$, $\to$;
(4). auxiliary symbols:), (.

The formulae of L_{vpl} is recursively defined as:

(1). a variable is a formula;
(2). a symbol α for truth value is a formula;
(3). if p, q are formulae, then $p \vee q$, $p \wedge q$, p', $p \to q$ are formulae;
(4). there is no formula other than those constructed from (1), (2), and (3).

For convenience, we also use the following notations:

$$p \leftrightarrow q \triangleq (p \to q) \wedge (q \to p) \tag{9.19}$$

$$p \otimes q \triangleq (p \to q')' \tag{9.20}$$

Let $\mathscr{F}_p$ be the set of all formulae of L_{vpl}. Obviously, $(\mathscr{F}_p, \vee, \wedge, {}', \to)$ is an algebra with the same type as $(L, \vee, \wedge, {}', \to)$.

In the following, we let

$$\mathscr{T} \subseteq \mathscr{F}_L(\mathscr{F}_p),$$
$$\mathscr{T}_h \subseteq \{T \mid T\colon\ \mathscr{F}_p \longrightarrow L \text{ is a homomorphic mapping }\} \triangleq \mathscr{T}_H.$$

9.2.2 Rules of Inference

Definition 9.2.1 *Let $D_n \subseteq \mathscr{F}_p^n$. A mapping*

$$r_n\colon\ D_n \longrightarrow \mathscr{F}_p$$

is said to be an n-ary partial operation on $\mathscr{F}_p$. D_n is called the domain of r_n and denoted as $D_n(r_n)$.

For any $n \in \mathbb{N}$, let

$$R_n \subseteq \{r_n \mid r_n \text{ is an } n\text{-ary partial operation on } \mathscr{F}_p\}.$$

Definition 9.2.2 *A mapping*

$$t_n\colon\ L^n \longrightarrow L$$

is said to be an n-ary truth-valued operation on L, if

(1).
$$\alpha \to t_n(\alpha_1, \cdots, \alpha_n) \geqslant t_n(\alpha \to \alpha_1, \cdots, \alpha \to \alpha_n) \tag{9.21}$$
holds for any $\alpha \in L$ and $(\alpha_1, \cdots, \alpha_n) \in L^n$.

(2). t_n is isotone in each argument.

For any $n \in \mathbb{N}$, let

$$T_n \subseteq \{t_n \mid t_n \text{ is an } n\text{-ary truth-valued operation on } L\}.$$

In the following, we write

$$\mathscr{R}_n \subseteq R_n \times T_n,$$
$$\mathscr{R} \subseteq \bigcup_{n=0}^{+\infty} \mathscr{R}_n.$$

If $(r, t) \in \mathscr{R}_n$, $n \in \mathbb{N}$, then (r, t) is said to be an n-ary rule of inference in L_{vpl}. Specially, we write $r_2^0\colon\ D_2^0 \longrightarrow \mathscr{F}_p$

$$(p, p \to q) \longmapsto q, \tag{9.22}$$

where $D_2^0 = \{(p, p \to q) \mid p, q \in \mathscr{F}_p\}$.

Definition 9.2.3 *Let $X \in \mathscr{F}_L(\mathscr{F}_p)$, $(r,t) \in \mathscr{R}_n$, $n \in \mathbb{N}$, $\alpha \in L$. If*

$$X \circ r \supseteq \alpha \otimes \left(t \circ \prod^n X \right) \tag{9.23}$$

in $D_n(r)$, then X is said to be α-I type closed w.r.t. (r,t). If

$$X \circ r \supseteq t \circ \prod^n (\alpha \otimes X) \tag{9.24}$$

in $D_n(r)$, then X is said to be α-II type closed w.r.t. (r,t).

If X is α-I (α-II) type closed w.r.t. (r,t) for any $(r,t) \in \mathscr{R}$, then X is said to be α-I (α-II) type closed w.r.t. $\mathscr{R}$. Note that X is I-I type closed w.r.t. (r,t) if and only if X is I-II type closed w.r.t. (r,t).

Theorem 9.2.1 *Let $X \in \mathscr{F}_L(\mathscr{F}_p)$. If X is α-i type closed w.r.t. $\mathscr{R}$, then for any $\beta \in L$, $Y = \beta \to X$ is α-i type closed w.r.t. $\mathscr{R}$ for $i = \mathrm{I}, \mathrm{II}$.*

Proof. Firstly, we prove Y is α-I type closed w.r.t. $\mathscr{R}$. Let $(r,t) \in \mathscr{R}_n$, $n \in \mathbb{N}$, for any $(p_1, \cdots, p_n) \in D_n(r)$,

$$\begin{aligned} Y(r(p_1, \cdots, p_n)) &= \beta \to X(r(p_1, \cdots, p_n)) \\ &\geqslant \beta \to (\alpha \otimes t(X(p_1), \cdots, X(p_n))) \\ &\geqslant \alpha \otimes (\beta \to t(X(p_1), \cdots, X(p_n))) \\ &\geqslant \alpha \otimes t(\beta \to X(p_1), \cdots, \beta \to X(p_n)) \\ &= \alpha \otimes t(Y(p_1), \cdots, Y(p_n)). \end{aligned}$$

Similarly, we can prove Y is α-II type closed w.r.t. $\mathscr{R}$ if X is α-II type closed w.r.t. $\mathscr{R}$. In fact, let $(r,t) \in \mathscr{R}_n$, $n \in \mathbb{N}$, for any $(p_1, \cdots, p_n) \in D_n(r)$,

$$\begin{aligned} Y(r(p_1, \cdots, p_n)) &= \beta \to X(r(p_1, \cdots, p_n)) \\ &\geqslant \beta \to t(\alpha \otimes X(p_1), \cdots, \alpha \otimes X(p_n)) \\ &\geqslant t(\beta \to (\alpha \otimes X(p_1)), \cdots, \beta \to (\alpha \otimes X(p_n))) \\ &\geqslant t(\alpha \otimes (\beta \to X(p_1)), \cdots, \alpha \otimes (\beta \to X(p_n))) \\ &= t(\alpha \otimes Y(p_1), \cdots, \alpha \otimes Y(p_n)). \quad \square \end{aligned}$$

Theorem 9.2.2 *Let $\alpha \in L$. If*

$$\emptyset \neq \mathscr{U} \subseteq \{X \mid X \text{ is } \alpha\text{-}i \text{ type closed w.r.t. } \mathscr{R}\},$$

then $\bigcap_{X \in \mathscr{U}} X$ is α-i type closed w.r.t. $\mathscr{R}$, $i = \mathrm{I}, \mathrm{II}$.

Proof. For $i = \mathrm{I}$. Let $(r,t) \in \mathscr{R}_n$, $n \in \mathbb{N}$, for any $(p_1, \cdots, p_n) \in D_n(r)$.

$$\begin{aligned}\left(\bigcap_{X\in\mathscr{U}} X\right)(r(p_1,\cdots,p_n)) &= \bigwedge_{X\in\mathscr{U}} (X(r(p_1,\cdots,p_n)))\\ &\geqslant \bigwedge_{X\in\mathscr{U}} (\alpha\otimes t(X(p_1),\cdots,X(p_n)))\\ &\geqslant \alpha\otimes \bigwedge_{X\in\mathscr{U}} t(X(p_1),\cdots,X(p_n))\\ &\geqslant \alpha\otimes t\left(\bigwedge_{X\in\mathscr{U}} X(p_1),\cdots,\bigwedge_{X\in\mathscr{U}} X(p_n)\right)\\ &= \alpha\otimes t\left(\left(\bigcap_{X\in\mathscr{U}} X\right)(p_1),\cdots,\left(\bigcap_{X\in\mathscr{U}} X\right)(p_n)\right).\end{aligned}$$

The result for $i = \mathrm{II}$ can be proved similarly. □

Theorem 9.2.3 *If X is α-i type closed w.r.t. $\mathscr{R}$ for $i = \mathrm{I}, \mathrm{II}$, then the following conclusions hold.*

(1). If X is α-i type closed w.r.t. $\mathscr{R}$ and $\beta \leqslant \alpha$, then X is β-i type closed w.r.t. $\mathscr{R}$.

(2). If X is α_j-i type closed w.r.t. $\mathscr{R}$ for $j = 1, 2$, then X is $(\alpha_1 \vee \alpha_2)$-I type closed w.r.t. $\mathscr{R}$, and X is $(\alpha_1 \wedge \alpha_2)$-$i$ type closed w.r.t. $\mathscr{R}$. Moreover, X is $(\alpha_1 \otimes \alpha_2)$-$i$ type closed w.r.t. $\mathscr{R}$.

(3). If X is $(\alpha \to \beta)$-i type closed w.r.t. $\mathscr{R}$, then X is β-i type closed and α'-i type closed w.r.t. $\mathscr{R}$. □

Example 9.2.1 *Let L be a lattice H implication algebra in L_{vpl} and $\alpha \in L$. If let*

$$\begin{aligned} t_2^0\colon\ & L\times L \longrightarrow L,\\ & (\beta,\gamma) \longmapsto \beta\otimes\gamma,\end{aligned}$$

then for any $T \in \mathscr{T}_H$, T is α-i type closed w.r.t. (r_2^0, t_2^0) for $i = \mathrm{I}, \mathrm{II}$.

Proof. Firstly, t_2^0 is a 2-ary truth-valued operation. In fact, for any $\theta, \beta, \gamma \in L$, because $\theta \vee \theta' = I$ and

$$\begin{aligned} &((\theta\to\beta)\otimes(\theta\to\gamma))\to(\theta\to(\beta\otimes\gamma))\\ \geqslant\ &(\theta\to((\theta\to\beta)\otimes\gamma))\to(\theta\to(\beta\otimes\gamma))\\ \geqslant\ &(\theta\to(\theta\to(\beta\otimes\gamma)))\to(\theta\to(\beta\otimes\gamma))\\ =\ &\theta\vee(\theta\to(\beta\otimes\gamma))\\ \geqslant\ &\theta\vee\theta'\\ =\ &I,\end{aligned}$$

we have

$$\begin{aligned} t_2^0(\theta \to \beta, \theta \to \gamma) &= (\theta \to \beta) \otimes (\theta \to \gamma) \\ &\leqslant \theta \to (\beta \otimes \gamma) \\ &= \theta \to t_2^0(\beta, \gamma). \end{aligned}$$

Obviously, t_2^0 is isotone in each argument.

Secondly, T is α-i type closed w.r.t. (r_2^0, t_2^0), $i = \mathrm{I}, \mathrm{II}$. In fact, for any $p, q \in \mathscr{F}_p$,

$$\begin{aligned} T(r_2^0(p, p \to q)) &= T(q) \\ &\geqslant T(p) \wedge T(q) \\ &= T(p) \otimes (T(p) \to T(q)) \\ &= t_2^0(T(p), T(p \to q)). \end{aligned}$$

For $i = \mathrm{I}$,

$$\begin{aligned} T(r_2^0(p, p \to q)) &\geqslant t_2^0(T(p), T(p \to q)) \\ &\geqslant \alpha \otimes t_2^0(T(p), T(p \to q)). \end{aligned}$$

For $i = \mathrm{II}$,

$$\begin{aligned} T(r_2^0(p, p \to q)) &\geqslant t_2^0(T(p), T(p \to q)) \\ &\geqslant t_2^0(\alpha \otimes T(p), \alpha \otimes T(p \to q)). \ \square \end{aligned}$$

Example 9.2.2 *In L_{vpl}, let $\theta_0 \in L$. Define*

$$\begin{aligned} r_1^{\theta_0}: \ \mathscr{F}_p &\longrightarrow \mathscr{F}_p, \\ p &\longmapsto \theta_0 \to p, \end{aligned}$$

$$\begin{aligned} t_1^{\theta_0}: \ L &\longrightarrow L, \\ \theta &\longmapsto \theta_0 \to \theta, \end{aligned}$$

then for any $T \in \mathscr{T}_H$, T is α-i type closed w.r.t. $(r_1^{\theta_0}, t_1^{\theta_0})$ for $\alpha \in L$ and $i = \mathrm{I}, \mathrm{II}$.

Proof. Firstly, it is easy to know that $t_1^{\theta_0}$ is an unary truth-valued operation. Secondly, for any $T \in \mathscr{T}_H$, T is α-i type closed w.r.t. $(r_1^{\theta_0}, t_1^{\theta_0})$, $i = \mathrm{I}, \mathrm{II}$. In fact, for any $p \in \mathscr{F}_p$,

$$\begin{aligned} T(r_1^{\theta_0}(p)) &= T(\theta_0 \to p) \\ &= \theta_0 \to T(p) \\ &= t_1^{\theta_0}(T(p)). \end{aligned}$$

Hence,

$$\begin{aligned} T(r_1^{\theta_0}(p)) &\geqslant \alpha \otimes t_1^{\theta_0}(T(p)), \\ T(r_1^{\theta_0}(p)) &\geqslant t_1^{\theta_0}(\alpha \otimes T(p)). \ \square \end{aligned}$$

Example 9.2.3 *In L_{vpl}, let $\alpha \in L$ and $\alpha \leqslant \bigwedge_{\theta \in L}(\theta \vee \theta')$. If define (r_2^0, t_2^*) as follows:*

$$\begin{aligned} &r_2^0 \textit{ is the same as Eq. (9.22)},\\ &t_2^*\colon\ L \times L \longrightarrow L,\\ &\qquad (\theta, \beta) \longmapsto \theta \wedge \beta, \end{aligned}$$

then t_2^* is a 2-ary truth-valued operation and for any $T \in \mathscr{T}_H$, T is α-i type closed w.r.t. (r_2^0, t_2^*) for $i = \mathrm{I}, \mathrm{II}$.

Proof. Firstly, it is easy to know that t_2^* is a 2-ary truth-valued operation. In fact, for any $\sigma, \theta, \beta \in L$,

$$\begin{aligned} \sigma \to t_2^*(\theta, \beta) &= \sigma \to (\theta \wedge \beta)\\ &= (\sigma \to \theta) \wedge (\sigma \to \beta)\\ &= t_2^*(\sigma \to \theta, \sigma \to \beta). \end{aligned}$$

Obviously, t_2^* is isotone in each argument.

Secondly, for any $T \in \mathscr{T}_H$, T is α-i type closed w.r.t. (r_2^0, t_2^*), $i = \mathrm{I}, \mathrm{II}$. In fact, for $i = \mathrm{I}$: for any $p, q \in \mathscr{F}_p$,

$$\begin{aligned} \alpha &\leqslant \bigwedge_{\theta \in L}(\theta \vee \theta')\\ &\leqslant T(p) \vee (T(p))' \vee T(q)\\ &\leqslant T(p) \vee (T(p) \to T(q))\\ &\leqslant (T(p) \vee T(q)) \vee (T(p) \to T(q))\\ &= (T(p) \wedge (T(p) \to T(q))) \to T(q)\\ &= t_2^*(T(p), T(p \to q)) \to T(q). \end{aligned}$$

So, $\alpha \otimes t_2^*(T(p), T(p \to q)) \leqslant T(q) = T(r_2^0(p, p \to q))$.

For $i = \mathrm{II}$: note that $\alpha \otimes t_2^*(\theta, \beta) = t_2^*(\alpha \otimes \theta, \alpha \otimes \beta)$. Hence, the property of α-II type closed mentioned here is the same as the property of α-I type closed. □

Example 9.2.4 *In L_{vpl}, if define (r_2^*, t_2^*) as: for any $p, g, q \in \mathscr{F}_p$ and $\alpha \in L$,*

$$\begin{aligned} &r_2^*(p \to g, p \to q) = p \to (g \wedge q),\\ &t_2^* \textit{ is the same as that in Example 9.2.3.} \end{aligned}$$

Then for any $T \in \mathscr{T}_H$, T is α-i type closed w.r.t. (t_2^*, t_2^*), $i = \mathrm{I}, \mathrm{II}$.

Proof. For any $p, g, q \in \mathscr{F}_p$, $\alpha \in L$,

$$\begin{aligned} T(r_2^*(p \to g, p \to q)) &= T(p \to (g \wedge q))\\ &= T(p \to g) \wedge T(p \to q)\\ &= t_2^*(T(p \to g), T(p \to q))\\ &\geqslant \alpha \otimes t_2^*(T(p \to g), T(p \to q)), \end{aligned}$$

and

$$t_2^*(T(p \to g), T(p \to q)) \geqslant t_2^*(\alpha \otimes T(p \to g), \alpha \otimes T(p \to q)). \ \square$$

Example 9.2.5 *In L_{vpl}, if define $(r_2^\triangle, t_2^*)$ as: for any $p, g, q \in \mathscr{F}_p$,*

$$r_2^\triangle(p \to q, q \to g) = p \to g,$$
$$t_2^* \text{ is the same as that in Example 9.2.3,}$$

and if $\alpha \leqslant \bigwedge_{\theta \in L}(\theta \vee \theta')$, then for any $T \in \mathscr{T}_H$, T is α-i type closed w.r.t. $(r_2^\triangle, t_2^*)$, $i = \mathrm{I}, \mathrm{II}$.

Proof. For $i = \mathrm{I}$: for any $T \in \mathscr{T}_H$, we have

$$\begin{aligned}
\alpha &\leqslant \bigwedge_{\theta \in L}(\theta \vee \theta') \\
&\leqslant T(q \vee q') \vee T(p') \vee T(g) \\
&= T(p' \vee q) \vee T(q' \vee g) \vee T(p' \vee g) \\
&\leqslant T(p \to q) \vee T(q \to g) \vee T(p \to g) \\
&= (T(p \to q) \wedge T(q \to g)) \to T(p \to g).
\end{aligned}$$

Hence,

$$\begin{aligned}
\alpha \otimes t_2^*(T(p \to q), T(q \to g)) &= \alpha \otimes (T(p \to q) \wedge T(q \to g)) \\
&\leqslant T(p \to g) \\
&= T(r_2^\triangle(p \to q, q \to g)).
\end{aligned}$$

For $i = \mathrm{II}$: note that

$$t_2^*(\alpha \otimes \theta, \alpha \otimes \beta) = \alpha \otimes t_2^*(\theta, \beta),$$

and by the case of $i = \mathrm{I}$, it is easy to see. $\square$

9.2.3 Semantics

Definition 9.2.4 *Three kinds of maps are defined as follows respectively:*

(1).

$$\begin{aligned}
C_{\mathscr{T}}:\ \mathscr{F}_L(\mathscr{F}_p) &\longrightarrow \mathscr{F}_L(\mathscr{F}_p), \\
X &\longmapsto C_{\mathscr{T}}^X,
\end{aligned}$$

where

$$C_{\mathscr{T}}^X(p) \triangleq \bigwedge_{T \in \mathscr{T}} [\pi(X \subseteq T) \to T(p)]. \tag{9.25}$$

(2).

$$\mathscr{C}_{\mathscr{T}}\colon\ \mathscr{F}_L(\mathscr{F}_p) \longrightarrow \mathscr{F}_L(\mathscr{F}_p),$$
$$X \longmapsto \mathscr{C}_{\mathscr{T}}^X,$$

where

$$\mathscr{C}_{\mathscr{T}}^X(p) \triangleq \bigwedge\{T(p) \mid \pi(X \subseteq T) = I, T \in \mathscr{T}\}. \tag{9.26}$$

In Eqs. (9.25) *and* (9.26),

$$\pi(X \subseteq T) = \bigwedge_{q \in \mathscr{F}_p} (X(q) \to T(q)). \tag{9.27}$$

(3).

$$C^{\beta}_{(C^{\emptyset}_{\mathscr{T}},\mathscr{R}(\alpha-i))}\colon\ \mathscr{F}_L(\mathscr{F}_p) \longrightarrow \mathscr{F}_L(\mathscr{F}_p),$$
$$X \longmapsto C^{\beta,X}_{(C^{\emptyset}_{\mathscr{T}},\mathscr{R}(\alpha-i))},$$

where

$$C^{\beta,X}_{(C^{\emptyset}_{\mathscr{T}},\mathscr{R}(\alpha-i))}(p) \triangleq \bigwedge\{Y(p) \mid Y \supseteq \beta \otimes (C^{\emptyset}_{\mathscr{T}} \cup X),\ Y \text{ is } \alpha\text{-}i \text{ type closed w.r.t.} \mathscr{R}\}, \tag{9.28}$$

where $i = \mathrm{I}, \mathrm{II}$ *and* $\alpha, \beta \in L$.

Theorem 9.2.4 *The mapping* $C_{\mathscr{T}}$ *satisfies the following properties.*

(1). For any $X \in \mathscr{F}_L(\mathscr{F}_p)$ *and for any* $p \in \mathscr{F}_p$,

$$C_{\mathscr{T}}^X(p) = \mathscr{C}_{\mathscr{T}}^X(p) \wedge \bigwedge_{\substack{T \in \mathscr{T} \\ X \nsubseteq T}} (\pi(X \subseteq T) \to T(p)). \tag{9.29}$$

(2). Let $X_1, X_2 \in \mathscr{F}_L(\mathscr{F}_p)$. *Then*

$$C_{\mathscr{T}}^{X_1 \cup X_2} \supseteq C_{\mathscr{T}}^{X_1} \cup C_{\mathscr{T}}^{X_2}, \tag{9.30}$$
$$C_{\mathscr{T}}^{X_1 \cap X_2} \subseteq C_{\mathscr{T}}^{X_1} \cap C_{\mathscr{T}}^{X_2}. \tag{9.31}$$

(3). If $\mathscr{T}_1 \subseteq \mathscr{T}_2 \subseteq \mathscr{F}_L(\mathscr{F}_p)$, *then for any* $X \in \mathscr{F}_L(\mathscr{F}_p)$,

$$C_{\mathscr{T}_1}^X \supseteq C_{\mathscr{T}_2}^X. \tag{9.32}$$

(4). Let $p_1, p_2 \in \mathscr{F}_p$, $X \in \mathscr{F}_L(\mathscr{F}_p)$. *Then*

$$C_{\mathscr{T}_h}^X(p_1 \vee p_2) \geqslant C_{\mathscr{T}_h}^X(p_1) \vee C_{\mathscr{T}_h}^X(p_2), \tag{9.33}$$
$$C_{\mathscr{T}_h}^X(p_1 \wedge p_2) \leqslant C_{\mathscr{T}_h}^X(p_1) \wedge C_{\mathscr{T}_h}^X(p_2). \tag{9.34}$$

(5).

$$C^X_{\mathscr{T}_h}(O) \leqslant \bigvee_{p\in\mathscr{F}_p} X(p), \tag{9.35}$$

$$C^X_{\mathscr{T}_h}(I) = I. \tag{9.36}$$

(6). $C_{\mathscr{T}}$ is a closure operation.
(7). If $T \in \mathscr{T}$, then $C^T_{\mathscr{T}} = T$.

Proof. We only prove Eq. (9.35), (6) and (7), others are easy to see.
For Eq. (9.35):

$$\begin{aligned}
C^X_{\mathscr{T}_h}(O) &= \bigwedge_{T\in\mathscr{T}_h}\left(\bigwedge_{p\in\mathscr{F}_p}(X(p)\to T(p))\to T(O)\right)\\
&= \bigwedge_{T\in\mathscr{T}_h}\left(\bigwedge_{p\in\mathscr{F}_p}(X(p)\to T(p))\right)'\\
&= \bigwedge_{T\in\mathscr{T}_h}\bigvee_{p\in\mathscr{F}_p}(X(p)\to T(p))'\\
&= \bigwedge_{T\in\mathscr{T}_h}\bigvee_{p\in\mathscr{F}_p}(X(p)\otimes(T(p))')\\
&\leqslant \bigwedge_{T\in\mathscr{T}_h}\left(\left(\bigvee_{p\in\mathscr{F}_p}X(p)\right)\otimes\left(\bigvee_{p\in\mathscr{F}_p}T(p')\right)\right)\\
&= \bigwedge_{T\in\mathscr{T}_h}\left(\left(\bigvee_{p\in\mathscr{F}_p}X(p)\right)\otimes I\right)\\
&= \bigvee_{p\in\mathscr{F}_p}X(p). \ \square
\end{aligned}$$

For (6): Firstly, for any $X \in \mathscr{F}_L(\mathscr{F}_p)$, there has $X \subseteq C^X_{\mathscr{T}}$. In fact, for any $p \in \mathscr{F}_p$, we have

$$\begin{aligned}
C^X_{\mathscr{T}}(p) &= \bigwedge_{T\in\mathscr{T}}(\pi(X\subseteq T)\to T(p))\\
&\geqslant \bigwedge_{T\in\mathscr{T}}((X(p)\to T(p))\to T(p))\\
&= \bigwedge_{T\in\mathscr{T}}(X(p)\vee T(p))\\
&\geqslant X(p).
\end{aligned}$$

Secondly, for any $X, Y \in \mathscr{F}_L(\mathscr{F}_p)$, if $X \subseteq Y$, then $C^X_{\mathscr{T}} \subseteq C^Y_{\mathscr{T}}$. In fact, it follows from Eqs. (9.30) and (9.31), and

$$\begin{aligned} C_{\mathscr{T}}^{X} &= C_{\mathscr{T}}^{X\cap Y} \\ &\subseteq C_{\mathscr{T}}^{X} \cap C_{\mathscr{T}}^{Y} \\ &\subseteq C_{\mathscr{T}}^{X} \cup C_{\mathscr{T}}^{Y} \\ &\subseteq C_{\mathscr{T}}^{X\cup Y} \\ &= C_{\mathscr{T}}^{Y}. \end{aligned}$$

Thirdly, for any $X \in \mathscr{F}_L(\mathscr{F}_p)$, then $C_{\mathscr{T}}^{X} = C_{\mathscr{T}}^{C_{\mathscr{T}}^{X}}$.

It suffices to prove $C_{\mathscr{T}}^{X} \supseteq C_{\mathscr{T}}^{C_{\mathscr{T}}^{X}}$. In fact, for any $p \in \mathscr{F}_p$,

$$\begin{aligned} C_{\mathscr{T}}^{C_{\mathscr{T}}^{X}}(p) &= \bigwedge_{T_1\in\mathscr{T}} (\pi(C_{\mathscr{T}}^{X} \subseteq T_1) \to T_1(p)) \\ &= \bigwedge_{T_1\in\mathscr{T}} \left(\left(\bigwedge_{q\in\mathscr{F}_p} (C_{\mathscr{T}}^{X}(q) \to T_1(q))\right) \to T_1(p)\right) \\ &= \bigwedge_{T_1\in\mathscr{T}} \left(\left(\bigwedge_{q\in\mathscr{F}_p} ((\bigwedge_{T_2\in\mathscr{T}} (\pi(X \subseteq T_2) \to T_2(q))) \to T_1(q))\right) \to T_1(p)\right) \\ &\leqslant \bigwedge_{T_1\in\mathscr{T}} \left(\left(\bigwedge_{q\in\mathscr{F}_p} ((\pi(X \subseteq T_1) \to T_1(q)) \to T_1(q))\right) \to T_1(p)\right) \\ &= \bigwedge_{T_1\in\mathscr{T}} \left(\left(\bigwedge_{q\in\mathscr{F}_p} (\pi(X \subseteq T_1) \vee T_1(q))\right) \to T_1(p)\right) \\ &\leqslant \bigwedge_{T_1\in\mathscr{T}} \left(\left(\bigwedge_{q\in\mathscr{F}_p} \pi(X \subseteq T_1)\right) \to T_1(p)\right) \\ &= \bigwedge_{T\in\mathscr{T}} (\pi(X \subseteq T) \to T(p)) \\ &= C_{\mathscr{T}}^{X}(p). \end{aligned}$$

Sum up, (6) holds.

Finally, we prove (7). If $T \in \mathscr{T}$, then for any $p \in \mathscr{F}_p$,

$$\begin{aligned} C_{\mathscr{T}}^{T}(p) &= \bigwedge_{T^*\in\mathscr{T}} (\pi(T \subseteq T^*) \to T^*(p)) \\ &\leqslant \pi(T \subseteq T) \to T(p) \\ &= T(p), \end{aligned}$$

hence, $C_{\mathscr{T}}^{T} \subseteq T$. Moreover, by (6), we know that $T \subseteq C_{\mathscr{T}}^{T}$, *i.e.*, $C_{\mathscr{T}}^{T} = T$. □

Definition 9.2.5 *Let $\mathscr{T} \subseteq \mathscr{F}_L(\mathscr{F}_p)$, $\alpha \in L$, $\mathscr{R}$ is α-i type sound w.r.t. $\mathscr{T}$ if T is α-i type closed w.r.t. $\mathscr{R}$ for any $T \in \mathscr{T}$, $i = \mathrm{I}, \mathrm{II}$.*

Theorem 9.2.5 *Let $\alpha, \alpha_1, \alpha_2, \beta, \beta_1, \beta_2 \in L$, $X, X_1, X_2 \in \mathscr{F}_L(\mathscr{F}_p)$, $\mathscr{T} \subseteq \mathscr{F}_L(\mathscr{F}_p)$. The following conclusions hold for $i = \mathrm{I}, \mathrm{II}$.*

(1). If $\beta_1 \leqslant \beta_2$, $\alpha_1 \leqslant \alpha_2$, and $X_1 \subseteq X_2$, then

$$C^{\beta_1, X_1}_{(C^{\emptyset}_{\mathscr{T}}, \mathscr{R}(\alpha_1 - i))} \subseteq C^{\beta_2, X_2}_{(C^{\emptyset}_{\mathscr{T}}, \mathscr{R}(\alpha_2 - i))}. \tag{9.37}$$

(2). If $\mathscr{R}$ is α-i type sound w.r.t. $\mathscr{T}$, then

$$C^{\beta, X}_{(C^{\emptyset}_{\mathscr{T}}, \mathscr{R}(\alpha - i))} \subseteq C^{\beta \otimes X}_{\mathscr{T}}. \tag{9.38}$$

(3).

$$C^{\beta, C^{\beta, X}_{(C^{\emptyset}_{\mathscr{T}}, \mathscr{R}(\alpha - i))}}_{(C^{\emptyset}_{\mathscr{T}}, \mathscr{R}(\alpha - i))} = C^{\beta, \beta \otimes X}_{(C^{\emptyset}_{\mathscr{T}}, \mathscr{R}(\alpha - i))}. \tag{9.39}$$

Proof. (1). It is easy to see from the following facts:

a) If $\beta_1 \leqslant \beta_2$, then $C^{\beta_1, X}_{(C^{\emptyset}_{\mathscr{T}}, \mathscr{R}(\alpha - i))} \subseteq C^{\beta_2, X}_{(C^{\emptyset}_{\mathscr{T}}, \mathscr{R}(\alpha - i))}$.

b) If $\alpha_1 \leqslant \alpha_2$, then $C^{\beta, X}_{(C^{\emptyset}_{\mathscr{T}}, \mathscr{R}(\alpha_1 - i))} \subseteq C^{\beta, X}_{(C^{\emptyset}_{\mathscr{T}}, \mathscr{R}(\alpha_2 - i))}$.

c) If $X_1 \subseteq X_2$, then $C^{\beta, X_1}_{(C^{\emptyset}_{\mathscr{T}}, \mathscr{R}(\alpha - i))} \subseteq C^{\beta, X_2}_{(C^{\emptyset}_{\mathscr{T}}, \mathscr{R}(\alpha - i))}$.

(2). Note that the following facts:

a) $\beta \otimes X \subseteq C^{\beta \otimes X}_{\mathscr{T}}$.

b) $\beta \otimes C^{\emptyset}_{\mathscr{T}} \subseteq C^{\emptyset}_{\mathscr{T}} \subseteq C^{\beta \otimes X}_{\mathscr{T}}$.

Hence,

$$\beta \otimes (C^{\emptyset}_{\mathscr{T}} \cup X) \subseteq C^{\beta \otimes X}_{\mathscr{T}}.$$

For $i = \mathrm{I}$, we need to prove that $C^{\beta \otimes X}_{\mathscr{T}}$ is α-I type closed w.r.t. $\mathscr{R}$. In fact, for any $(r, t) \in \mathscr{R}_n$, $n \in \mathbb{N}$ and for any $(p_1, \cdots, p_n) \in D_n(r)$,

$$
\begin{aligned}
& C_{\mathscr{T}}^{\beta\otimes X}(r(p_1,\cdots,p_n)) \\
= & \bigwedge_{T\in\mathscr{T}} (\pi(\beta\otimes X\subseteq T)\to T(r(p_1,\cdots,p_n))) \\
\geqslant & \bigwedge_{T\in\mathscr{T}} (\pi(\beta\otimes X\subseteq T)\to (\alpha\otimes t(T(p_1),\cdots,T(p_n)))) \\
\geqslant & \bigwedge_{T\in\mathscr{T}} (\alpha\otimes(\pi(\beta\otimes X\subseteq T)\to t(T(p_1),\cdots,T(p_n)))) \\
\geqslant & \alpha\otimes\bigwedge_{T\in\mathscr{T}} (\pi(\beta\otimes X\subseteq T)\to t(T(p_1),\cdots,T(p_n))) \\
\geqslant & \alpha\otimes\bigwedge_{T\in\mathscr{T}} t((\pi(\beta\otimes X\subseteq T)\to T(p_1)),\cdots, \\
& \qquad (\pi(\beta\otimes X\subseteq T)\to T(p_n))) \\
\geqslant & \alpha\otimes t\Bigg(\bigwedge_{T\in\mathscr{T}} (\pi(\beta\otimes X\subseteq T)\to T(p_1)),\cdots, \\
& \qquad \bigwedge_{T\in\mathscr{T}} (\pi(\beta\otimes X\subseteq T)\to T(p_n))\Bigg) \\
= & \alpha\otimes t(C_{\mathscr{T}}^{\beta\otimes X}(p_1),\cdots,C_{\mathscr{T}}^{\beta\otimes X}(p_n)).
\end{aligned}
$$

Hence, for any $p\in\mathscr{F}_p$,

$$C_{\mathscr{T}}^{\beta\otimes X}(p)\in\{Y(p)\mid Y\supseteq\beta\otimes(C_{\mathscr{T}}^{\emptyset}\cup X), Y \text{ is } \alpha\text{-I closed w.r.t. } \mathscr{R}\}.$$

Consequently,

$$C_{(C_{\mathscr{T}}^{\emptyset},\mathscr{R}(\alpha-\mathrm{I}))}^{\beta,X}(p)\leqslant C_{\mathscr{T}}^{\beta\otimes X}(p).$$

For $i=\mathrm{II}$, we need to prove that $C_{\mathscr{T}}^{\beta\otimes X}$ is α-II type closed w.r.t. $\mathscr{R}$. In fact, for any $(r,t)\in\mathscr{R}_n$, $n\in\mathbb{N}$ and for any $(p_1,\cdots,p_n)\in D_n(r)$,

$$\begin{aligned}
&C_{\mathscr{T}}^{\beta\otimes X}(r(p_1,\cdots,p_n))\\
&=\bigwedge_{T\in\mathscr{T}}(\pi(\beta\otimes X\subseteq T)\to T(r(p_1,\cdots,p_n)))\\
&\geqslant\bigwedge_{T\in\mathscr{T}}(\pi(\beta\otimes X\subseteq T)\to t(\alpha\otimes T(p_1),\cdots,\alpha\otimes T(p_n)))\\
&\geqslant\bigwedge_{T\in\mathscr{T}}t(\pi(\beta\otimes X\subseteq T)\to(\alpha\otimes T(p_1)),\cdots,\\
&\qquad\qquad\pi(\beta\otimes X\subseteq T)\to(\alpha\otimes T(p_n)))\\
&\geqslant\bigwedge_{T\in\mathscr{T}}t(\alpha\otimes(\pi(\beta\otimes X\subseteq T)\to T(p_1)),\cdots,\\
&\qquad\qquad\alpha\otimes(\pi(\beta\otimes X\subseteq T)\to T(p_n)))\\
&\geqslant t(\alpha\otimes\bigwedge_{T\in\mathscr{T}}(\pi(\beta\otimes X\subseteq T)\to T(p_1)),\cdots,\\
&\qquad\qquad\alpha\otimes\bigwedge_{T\in\mathscr{T}}(\pi(\beta\otimes X\subseteq T)\to T(p_n)))\\
&=t(\alpha\otimes C_{\mathscr{T}}^{\beta\otimes X}(p_1),\cdots,\alpha\otimes C_{\mathscr{T}}^{\beta\otimes X}(p_n)).
\end{aligned}$$

Hence, for any $p\in\mathscr{F}_p$,

$$C_{\mathscr{T}}^{\beta\otimes X}(p)\in\{Y(p)\mid Y\supseteq\beta\otimes(C_{\mathscr{T}}^{\emptyset}\cup X),Y\text{ is }\alpha\text{-II type closed w.r.t. }\mathscr{R}\}.$$

Consequently,

$$C_{(C_{\mathscr{T}}^{\emptyset},\mathscr{R}(\alpha-\mathrm{II}))}^{\beta,X}(p)\leqslant C_{\mathscr{T}}^{\beta\otimes X}(p).$$

(3). For any $p\in\mathscr{F}_p$, because

$$\begin{aligned}
C_{(C_{\mathscr{T}}^{\emptyset},\mathscr{R}(\alpha-i))}^{\beta,X}(p)&=\bigwedge\{Y(p)\mid Y\supseteq\beta\otimes(C_{\mathscr{T}}^{\emptyset}\cup X),Y\text{ is }\\
&\qquad\qquad\alpha\text{-}i\text{ type closed w.r.t. }\mathscr{R}\}\\
&\geqslant\beta\otimes(C_{\mathscr{T}}^{\emptyset}(p)\vee X(p))\\
&\geqslant\beta\otimes X(p),
\end{aligned}$$

we know that

$$C_{(C_{\mathscr{T}}^{\emptyset},\mathscr{R}(\alpha-i))}^{\beta,C_{(C_{\mathscr{T}}^{\emptyset},\mathscr{R}(\alpha-i))}^{\beta,X}}\supseteq C_{(C_{\mathscr{T}}^{\emptyset},\mathscr{R}(\alpha-i))}^{\beta,\beta\otimes X}.$$

Moreover, we know that

$$\begin{aligned}
C_{(C_{\mathscr{T}}^{\emptyset},\mathscr{R}(\alpha-i))}^{\beta,C_{(C_{\mathscr{T}}^{\emptyset},\mathscr{R}(\alpha-i))}^{\beta,X}}&=\bigwedge\{Y(p)\mid Y\supseteq\beta\otimes(C_{\mathscr{T}}^{\emptyset}\cup C_{(C_{\mathscr{T}}^{\emptyset},\mathscr{R}(\alpha-i))}^{\beta,X}),\\
&\qquad\qquad Y\text{ is }\alpha\text{-}i\text{ type closed w.r.t. }\mathscr{R}\}.
\end{aligned}$$

So, to prove the including relation $\subseteq$ in Eq. (9.39), it suffices to prove the following conclusions:

a) $C^{\beta,\beta\otimes X}_{(C^{\emptyset}_{\mathscr{T}},\mathscr{R}(\alpha-i))}(p) \geqslant \beta\otimes(C^{\emptyset}_{\mathscr{T}}(p)\vee C^{\beta,X}_{(C^{\emptyset}_{\mathscr{T}},\mathscr{R}(\alpha-i))}(p))$;

b) $C^{\beta,\beta\otimes X}_{(C^{\emptyset}_{\mathscr{T}},\mathscr{R}(\alpha-i))}$ is α-i type closed w.r.t. $\mathscr{R}$.

To prove a), it only needs to prove that

$$C^{\beta,\beta\otimes X}_{(C^{\emptyset}_{\mathscr{T}},\mathscr{R}(\alpha-i))}(p) \geqslant \beta\otimes C^{\beta,X}_{(C^{\emptyset}_{\mathscr{T}},\mathscr{R}(\alpha-i))}(p)$$

for $C^{\beta,\beta\otimes X}_{(C^{\emptyset}_{\mathscr{T}},\mathscr{R}(\alpha-i))}(p) \geqslant \beta\otimes C^{\emptyset}_{\mathscr{T}}(p)$. Furthermore, if we can prove that for any $p\in\mathscr{F}_p$, and for any $Z_0(p)\in\{Z(p)\mid Z\supseteq\beta\otimes(C^{\emptyset}_{\mathscr{T}}\cup(\beta\otimes X)),\ Z$ is α-i type closed w.r.t. $\mathscr{R}\}$, there exists $Y_0(p)\in\{Y(p)\mid Y\supseteq\beta\otimes(C^{\emptyset}_{\mathscr{T}}\cup X),$ Y is α-i type closed w.r.t. $\mathscr{R}\}$, such that

$$Z_0(p)\geqslant\beta\otimes Y_0(p),$$

then

$$\begin{aligned}Z_0(p)&\geqslant\beta\otimes Y_0(p)\\&\geqslant\beta\otimes\bigwedge\{Y(p)\mid Y\supseteq\beta\otimes(C^{\emptyset}_{\mathscr{T}}\cup X), Y\text{ is }\alpha\text{-}i\text{ type closed w.r.t. }\mathscr{R}\}\\&=\beta\otimes C^{\beta,X}_{(C^{\emptyset}_{\mathscr{T}},\mathscr{R}(\alpha-i))}(p).\end{aligned}$$

Since Z_0 is arbitrary, we know that

$$C^{\beta,\beta\otimes X}_{(C^{\emptyset}_{\mathscr{T}},\mathscr{R}(\alpha-i))}\geqslant\beta\otimes C^{\beta,X}_{(C^{\emptyset}_{\mathscr{T}},\mathscr{R}(\alpha-i))}(p).$$

Then, we can take $Y_0=\beta\to Z_0$. Therefore, $Z_0\geqslant\beta\otimes Y_0$. In the following, we are going to prove $Y_0(p)\in\{Y(p)\mid Y\supseteq\beta\otimes(C^{\emptyset}_{\mathscr{T}}\cup X),\ Y$ is α-i type closed w.r.t. $\mathscr{R}\}$. For any $q\in\mathscr{F}_p$, we have

$$\begin{aligned}Y_0(q)&=\beta\to Z_0(q)\\&\geqslant\beta\to(\beta\otimes(C^{\emptyset}_{\mathscr{T}}(q)\vee(\beta\otimes X(q))))\\&=\beta'\vee(C^{\emptyset}_{\mathscr{T}}(q)\vee(\beta\otimes X(q)))\\&\geqslant\beta'\vee((\beta\otimes C^{\emptyset}_{\mathscr{T}}(q))\vee(\beta\otimes X(q)))\\&\geqslant(\beta\otimes C^{\emptyset}_{\mathscr{T}}(q))\vee(\beta\otimes X(q))\\&=\beta\otimes(C^{\emptyset}_{\mathscr{T}}(q)\vee X(q)).\end{aligned}$$

Since Z_0 is α-i type closed w.r.t. $\mathscr{R}$, then by Theorem 9.2.1, we immediately obtain that Y_0 is α-i type closed w.r.t $\mathscr{R}$.

b) is easy to prove from Theorem 9.2.2. □

9.2.4 Syntax

Definition 9.2.6 *Let $X\in\mathscr{F}_L(\mathscr{F}_p)$, $\mathscr{T}\subseteq\mathscr{F}_L(\mathscr{F}_p)$, $p\in\mathscr{F}_p$, $\theta,\alpha,\beta\in L$. $(P^{\mathrm{I}},(n),X,\ (p,\theta)-(\alpha,\beta))$ is said to be an (α,β)-I type proof with the truth* *degree θ from X to p (shortly, $\theta-(\alpha,\beta)$-I type proof from X to p), if the mapping*

$$\begin{aligned} P^{\mathrm{I}}\colon\ (n) &\longrightarrow \mathscr{F}_p\times L \qquad ((n)=\{1,2,\cdots,n\}),\\ i &\longmapsto (p_i,\theta_i), \end{aligned}$$

satisfies

(1). $(p_n,\theta_n)=(p,\theta)$ *and*
(2). $\theta_i=\beta\otimes C^{\emptyset}_{\mathscr{T}}(p_i)$, *or*
(3). $\theta_i=\beta\otimes X(p_i)$, *or*
(4). there exist $i_1,\cdots,i_k\leqslant i$, *and* $(r,t)\in\mathscr{R}_k$ *such that*

$$(p_i,\theta_i)=(r(p_{i_1},\cdots,p_{i_k}),\alpha\otimes t(\theta_{i_1},\cdots,\theta_{i_k})),$$

where n is said to be the length of $\theta-(\alpha,\beta)$-I type proof from X to p under P^{I}, and denoted as $l(P^{\mathrm{I}})$.

Definition 9.2.7 *Let $X\in\mathscr{F}_L(\mathscr{F}_p)$, $\mathscr{T}\subseteq\mathscr{F}_L(\mathscr{F}_p)$, $p\in\mathscr{F}_p$, $\theta,\alpha,\beta\in L$. $(P^{\mathrm{II}},(n),X,\ (p,\theta)-(\alpha,\beta))$ is said to be a (α,β)-II type proof with the truth degree θ from X to p (shortly, $\theta-(\alpha,\beta)$-II type proof from X to p), if the mapping*

$$\begin{aligned} P^{\mathrm{II}}\colon\ (n) &\longrightarrow \mathscr{F}_p\times L \qquad ((n)=\{1,2,\cdots,n\}),\\ i &\longmapsto (p_i,\theta_i), \end{aligned}$$

satisfies

(1). $(p_n,\theta_n)=(p,\theta)$ *and*
(2). $\theta_i=\beta\otimes C^{\emptyset}_{\mathscr{T}}(p_i)$, *or*
(3). $\theta_i=\beta\otimes X(p_i)$, *or*
(4). there exist $i_1,\cdots,i_k\leqslant i$, *and* $(r,t)\in\mathscr{R}_k$ *such that*

$$(p_i,\theta_i)=(r(p_{i_1},\cdots,p_{i_k}),t(\alpha\otimes\theta_{i_1},\cdots,\alpha\otimes\theta_{i_k})),$$

where n is said to be the length of $\theta-(\alpha,\beta)$-II type proof from X to p under P^{II}, and denoted as $l(P^{\mathrm{II}})$.

Theorem 9.2.6 *Let $X\in\mathscr{F}_L(\mathscr{F}_p)$, $\mathscr{T}\subseteq\mathscr{F}_L(\mathscr{F}_p)$, $\alpha,\beta\in L$, and the truth-valued operations in $\mathscr{R}$ satisfy finite semicontinuity, then for any $p\in\mathscr{F}_p$,*

$$C^{\beta,X}_{(C^{\emptyset}_{\mathscr{T}},\mathscr{R}(\alpha-i))}(p)=\bigvee\{\theta\mid\ \text{there exists}\ (P^i,(n),X,(p,\theta)-(\alpha,\beta))\}\qquad(9.40)$$

for $i=\mathrm{I},\mathrm{II}$.

Proof. For $i=\mathrm{I}$, we write

$$\bigvee\{\theta\mid\ \text{there exists}\ (P^{\mathrm{I}},(n),X,(p,\theta)-(\alpha,\beta))\}\triangleq\overline{X}(p).$$

Firstly, we prove that $\overline{X}(p) \leqslant C^{\beta,X}_{(C^{\emptyset}_{\mathscr{T}},\mathscr{R}(\alpha-\mathrm{I}))}(p)$. It suffices to prove that

$$\theta \leqslant C^{\beta,X}_{(C^{\emptyset}_{\mathscr{T}},\mathscr{R}(\alpha-i))}(p)$$

holds for any $(P^{\mathrm{I}}, (n), X, (p, \theta) - (\alpha, \beta))$. Further, it suffices to prove that $\theta \leqslant Y_0(p)$ holds for any $Y_0 \in \{Y \in \mathscr{F}_L(\mathscr{F}_p) \mid Y \supseteq \beta \otimes (C^{\emptyset}_{\mathscr{T}} \cup X), Y$ is α-I type closed w.r.t $\mathscr{R}\}$. They can be obtained from the following.

If $n = 1$, then

$$\begin{aligned}&p_n = p_1 = p,\\&\theta_1 = \theta = \beta \otimes C^{\emptyset}_{\mathscr{T}}(p_1) \qquad \text{or} \qquad \theta_1 = \theta = \beta \otimes X(p_1).\end{aligned}$$

Hence,

$$\begin{aligned}\theta &\leqslant (\beta \otimes C^{\emptyset}_{\mathscr{T}}(p_1)) \vee (\beta \otimes X(p_1))\\&= \beta \otimes (C^{\emptyset}_{\mathscr{T}}(p_1) \vee X(p_1))\\&\leqslant Y_0(p_1).\end{aligned}$$

Suppose it holds for $l(P^{\mathrm{I}}) < m$. When $l(P^{\mathrm{I}}) = m$, if $\theta_m = \theta = \beta \otimes C^{\emptyset}_{\mathscr{T}}(p_m)$ or $\theta_m = \theta = \beta \otimes X(p_m)$, then

$$\begin{aligned}\theta &\leqslant (\beta \otimes C^{\emptyset}_{\mathscr{T}}(p_m)) \vee (\beta \otimes X(p_m))\\&= \beta \otimes (C^{\emptyset}_{\mathscr{T}}(p_m) \vee X(p_m))\\&\leqslant Y_0(p_m).\end{aligned}$$

If there exist $i_1, \cdots, i_k \leqslant m$, and $(r, t) \in \mathscr{R}_k$, such that

$$(p_m, \theta_m) = (r(p_{i_1}, \cdots, p_{i_k}), \alpha \otimes t(\theta_{i_1}, \cdots, \theta_{i_k})),$$

then

$$\begin{aligned}\theta &= \theta_m\\&= \alpha \otimes t(\theta_{i_1}, \cdots, \theta_{i_k})\\&\leqslant \alpha \otimes t(Y_0(p_{i_1}), \cdots, Y_0(p_{i_k}))\\&\leqslant Y_0(r(p_{i_1}, \cdots, p_{i_k}))\\&= Y_0(p_m).\end{aligned}$$

Secondly, for any $q \in \mathscr{F}_p$, we prove $C^{\beta,X}_{(C^{\emptyset}_{\mathscr{T}},\mathscr{R}(\alpha-i))}(p) \leqslant \overline{X}(p)$. Since $(P^{\mathrm{I}}, (1), X, (q, \beta \otimes C^{\emptyset}_{\mathscr{T}}(q)) - (\alpha, \beta))$ and $(P^{\mathrm{I}}_*, (1), X, (q, \beta \otimes X(q)) - (\alpha, \beta))$ are (α, β)-I type proofs from X to q, with the truth degree $\beta \otimes C^{\emptyset}_{\mathscr{T}}(q)$ and $\beta \otimes X(q)$, respectively, hence,

$$\begin{aligned}\beta \otimes (C^{\emptyset}_{\mathscr{T}}(q) \vee X(q)) &= (\beta \otimes C^{\emptyset}_{\mathscr{T}}(q)) \vee (\beta \otimes X(q))\\&\leqslant \overline{X}(q).\end{aligned}$$

Consequently, $\beta \otimes (C^{\emptyset}_{\mathscr{T}} \cup X) \subseteq \overline{X}$.

If we can prove that $\overline{X}$ is α-I type closed w.r.t. $\mathscr{R}$, then $\overline{X} \in \{Y \mid Y \supseteq \beta \otimes (C^{\emptyset}_{\mathscr{T}} \cup X)$, Y is α-I type closed w.r.t. $\mathscr{R}\}$. So we have

$$\begin{aligned}\overline{X}(p) &\geqslant \bigwedge\{Y(p) \mid Y \supseteq \beta \otimes (C^{\emptyset}_{\mathscr{T}} \cup X), Y \text{ is } \alpha\text{-I} \\ &\qquad\qquad \text{type closed w.r.t. } \mathscr{R}\} \\ &= C^{\beta,X}_{(C^{\emptyset}_{\mathscr{T}},\mathscr{R}(\alpha-\mathrm{I}))}(p).\end{aligned}$$

Then we need to prove $\overline{X}$ is α-I type closed w.r.t. $\mathscr{R}$. In fact, for any $(r,t) \in \mathscr{R}$, let $(r,t) \in \mathscr{R}_n$, and for any $(p_1,\cdots,p_n) \in D_n(r)$, it suffices to prove

$$\alpha \otimes t(\overline{X}(p_1),\cdots,\overline{X}(p_n)) \leqslant \overline{X}(r(p_1,\cdots,p_n)).$$

By the definition of $\overline{X}$, we know that for any $k \in \{1,\cdots,n\}$, there exists a family of all (α,β)-I type proofs from X to p_k

$$\{(P^{\mathrm{I}}_{j_k},(n_{j_k}),X,(p_k,\theta_{j_k})-(\alpha,\beta)) \mid j_k \in J_k\},$$

then $\overline{X}(p_k) = \bigvee\limits_{j_k \in J_k} \theta_{j_k}$. Hence,

$$\begin{aligned}&\{(P^{\mathrm{I}}_{(j_1,\cdots,j_n)},\left(\sum_{k=1}^{n} n_{j_k}+1\right),X,(r(p_1,\cdots,p_n),\alpha\otimes t(\theta_{j_1},\cdots,\theta_{j_n})) \\ &\quad -(\alpha,\beta)) \mid (j_1,\cdots,j_n) \in J_1\times\cdots\times J_n\}\end{aligned}$$

is a family of all (α,β)-I type proofs from X to $r(p_1,\cdots,p_n)$, where

$$P^{\mathrm{I}}_{(j_1,\cdots,j_n)}\colon \left(\sum_{k=1}^{n} n_{j_k}+1\right) \longrightarrow \mathscr{F}_p \times L$$

satisfies

$$P^{\mathrm{I}}_{(j_1,\cdots,j_n)}(i) = \begin{cases} P^{\mathrm{I}}_{j_k}\left(i-\sum\limits_{s=1}^{k-1} n_{j_s}\right), & \sum\limits_{s=1}^{k-1} n_{j_s}+1 \leqslant i \leqslant \sum\limits_{s=1}^{k} n_{j_s}, \\ & k=1,2,\cdots,n, \\ (r(p_1,\cdots,p_n),\alpha\otimes t(\theta_{j_1},\cdots,\theta_{j_n})), & i=\sum\limits_{s=1}^{n} n_{j_s}+1, \end{cases}$$

then,

$$\begin{aligned}\alpha \otimes t(\overline{X}(p_1),\cdots,\overline{X}(p_n)) &= \alpha \otimes t(\bigvee_{j_1\in J_1}\theta_{j_1},\cdots,\bigvee_{j_n\in J_n}\theta_{j_n}) \\ &= \alpha \otimes \bigvee_{j_1\in J_1}\cdots\bigvee_{j_n\in J_n} t(\theta_{j_1},\cdots,\theta_{j_n}) \\ &= \bigvee_{j_1\in J_1}\cdots\bigvee_{j_n\in J_n}(\alpha\otimes t(\theta_{j_1},\cdots,\theta_{j_n})) \\ &\leqslant \overline{X}(r(p_1,\cdots,p_n)).\end{aligned}$$

The case for $i = \mathrm{II}$ can be proved in the same way. □

Theorem 9.2.7 *Let $X \in \mathscr{F}_L(\mathscr{F}_p)$, $\mathscr{T} \subseteq \mathscr{F}_L(\mathscr{F}_p)$, $\alpha, \beta \in L$, $(r_2^0, t_2^*) \in \mathscr{R}$. For any $p, q \in \mathscr{F}_p$, let*

$$\overline{X}_i(p) = \bigvee\{\theta \mid \textit{ there exists } (P^i, (n), X, (p, \theta) - (\alpha, \beta))\}.$$

Then

$$\alpha \otimes \overline{X}_i(p \to q) \leqslant \overline{X}_i(p) \to \overline{X}_i(q), \qquad i = \mathrm{I}, \mathrm{II}. \tag{9.41}$$

Proof. Note that

$$\overline{X}_i(p) \to \overline{X}_i(q) = \bigwedge_{\theta_1} \bigvee_{\theta_2} \{\theta_1 \to \theta_2 \mid \text{ there exist } (P_1^i, (n_1), X, (p, \theta_1) - (\alpha, \beta)), (P_2^i, (n_2), X, (q, \theta_2) - (\alpha, \beta))\}.$$

In the following, we prove that for any $(P^i, (n), X, (p \to q, \theta) - (\alpha, \beta))$ and $(P_1^i, (n_1), X, (p, \theta_1) - (\alpha, \beta))$, there exists $(P_2^i, (n_2), X, (q, \theta_2) - (\alpha, \beta))$ such that,

$$\alpha \otimes \theta \leqslant \theta_1 \to \theta_2.$$

In fact, take $(P_2^i, (n_2), X, (q, \theta_2) - (\alpha, \beta))$ as

$$n_2 = n + n_1 + 1,$$

$$\begin{aligned} p_2^i : (n_2) &\longrightarrow \mathscr{F}_p \times L, \\ j &\longmapsto P_2^i(j), \end{aligned}$$

$$P_2^i(j) = \begin{cases} P^i(j), & 1 \leqslant j \leqslant n, \\ P_1^i(j - n), & n + 1 \leqslant j \leqslant n_1 + n, \\ (r_2^0(p, p \to q) = q, \alpha \otimes (\theta \wedge \theta_1)), & j = n_2, \end{cases}$$

$$\theta_2 = \alpha \otimes (\theta \wedge \theta_1),$$

then we have

$$\begin{aligned} \theta_1 \to \theta_2 &= \theta_1 \to \alpha \otimes (\theta \wedge \theta_1) \\ &= (\theta_1 \to (\alpha \otimes \theta)) \wedge (\theta_1 \to (\alpha \otimes \theta_1)) \\ &\geqslant (\alpha \otimes (\theta_1 \to \theta)) \wedge (\alpha \otimes (\theta_1 \to \theta_1)) \\ &= (\alpha \otimes (\theta_1 \to \theta)) \wedge \alpha \\ &= \alpha \otimes (\theta_1 \to \theta) \\ &\geqslant \alpha \otimes \theta. \end{aligned}$$

Hence, for any $(P^i, (n), X, (p \to q, \theta) - (\alpha, \beta))$ and $(P_1^i, (n_1), X, (p, \theta_1) - (\alpha, \beta))$,

$$\alpha \otimes \theta \leqslant \theta_1 \to \bigvee\{\theta_2 \mid \text{ there exist } (P_2^i, (n_2), X, (q, \theta_2) - (\alpha, \beta))\}.$$

So, for any $(P^i,(n),X,(p\to q,\theta)-(\alpha,\beta))$,

$$\alpha\otimes\theta\leqslant\overline{X}_i(p)\to\overline{X}_i(q).$$

Consequently,

$$\alpha\otimes\overline{X}_i(p\to q)\leqslant\overline{X}_i(p)\to\overline{X}_i(q).\ \square$$

Corollary 9.2.1 *Let $X\in\mathscr{F}_L(\mathscr{F}_p)$, $\mathscr{T}\subseteq\mathscr{F}_L(\mathscr{F}_p)$, $\alpha,\beta\in L$, $(r_2^0,t_2^*)\in\mathscr{R}$, and the truth-valued operations in $\mathscr{R}$ satisfy finite semicontinuity, then for any $p,q\in\mathscr{F}_p$,*

$$\alpha\otimes C^{\beta,X}_{(C^{\emptyset}_{\mathscr{T}},\mathscr{R}(\alpha-i))}(p\to q)\leqslant C^{\beta,X}_{(C^{\emptyset}_{\mathscr{T}},\mathscr{R}(\alpha-i))}(p)\to C^{\beta,X}_{(C^{\emptyset}_{\mathscr{T}},\mathscr{R}(\alpha-i))}(q).\ \square\tag{9.42}$$

Theorem 9.2.8 *Let $\mathscr{T}\subseteq\mathscr{F}_L(\mathscr{F}_p)$, $\alpha\in L$. Then $\mathscr{R}$ is α-i type sound w.r.t. $\mathscr{T}$ if and only if*

$$C^{\beta,X}_{(C^{\emptyset}_{\mathscr{T}},\mathscr{R}(\alpha-i))}\subseteq C^{\beta\otimes X}_{\mathscr{T}}\tag{9.43}$$

holds for any $X\in\mathscr{F}_L(\mathscr{F}_p)$ and $\beta\in L$, $i=\mathrm{I},\mathrm{II}$.

Proof. We need only to prove the sufficiency by Theorem 9.2.5. We only prove the case for $i=\mathrm{I}$, while the case for $i=\mathrm{II}$ can be proved similarly.

For any $T\in\mathscr{T}$, $(r,t)\in\mathscr{R}_n$, $(p_1\cdots,p_n)\in D_n(r)$,

$$\begin{aligned}\alpha\otimes t(T(p_1),\cdots,T(p_n))&\leqslant\alpha\otimes t(C^{I,T}_{(C^{\emptyset}_{\mathscr{T}},\mathscr{R}(\alpha-\mathrm{I}))}(p_1),\cdots,C^{I,T}_{(C^{\emptyset}_{\mathscr{T}},\mathscr{R}(\alpha-\mathrm{I}))}(p_n))\\&\leqslant C^{I,T}_{(C^{\emptyset}_{\mathscr{T}},\mathscr{R}(\alpha-\mathrm{I}))}(r(p_1,\cdots,p_n))\\&\leqslant C^{T}_{\mathscr{T}}(r(p_1,\cdots,p_n))\\&=T(r(p_1,\cdots,p_n)).\ \square\end{aligned}$$

Theorem 9.2.9 *Let $\mathscr{T}\subseteq\mathscr{F}_L(\mathscr{F}_p)$. If for any $\alpha,\beta\in L$, and $X\in\mathscr{F}_L(\mathscr{F}_p)$, $\mathscr{R}$ is α-i type sound w.r.t. $\mathscr{T}$, and $C^{\beta,X}_{(C^{\emptyset}_{\mathscr{T}},\mathscr{R}(\alpha-i))}\in\mathscr{T}$, then*

$$C^{\beta,X}_{(C^{\emptyset}_{\mathscr{T}},\mathscr{R}(\alpha-i))}=C^{\beta\otimes X}_{\mathscr{T}}\tag{9.44}$$

for $i=\mathrm{I},\mathrm{II}$.

Proof. By Theorem 9.2.8, it suffices to prove

$$C^{\beta,X}_{(C^{\emptyset}_{\mathscr{T}},\mathscr{R}(\alpha-i))}\supseteq C^{\beta\otimes X}_{\mathscr{T}}.$$

Note that

(1). $C^{\beta,X}_{(C^{\emptyset}_{\mathscr{T}},\mathscr{R}(\alpha-i))}\in\mathscr{T}$;

(2). $C^{\beta,X}_{(C^{\emptyset}_{\mathscr{T}},\mathscr{R}(\alpha-i))}\supseteq\beta\otimes X$.

Hence, for any $p \in \mathscr{F}_p$,

$$\begin{aligned} C_{\mathscr{T}}^{\beta\otimes X}(p) &\leqslant \mathscr{C}_{\mathscr{T}}^{\beta\otimes X}(p) \\ &= \bigwedge\{T(p) \mid T \in \mathscr{T}, \beta \otimes X \subseteq T\} \\ &\leqslant C^{\beta,X}_{(C^{\emptyset}_{\mathscr{T}},\mathscr{R}(\alpha-i))}(p). \ \square \end{aligned}$$

Theorem 9.2.10 ((α,β)-i Type Completeness and Soundness) *Let $\mathscr{T} \subseteq \mathscr{F}_L(\mathscr{F}_p)$, the truth-valued operations in $\mathscr{R}$ satisfy finite semicontinuity. If for any $\alpha, \beta \in L$, and $X \in \mathscr{F}_L(\mathscr{F}_p)$, $\mathscr{R}$ is α-i type sound w.r.t. $\mathscr{T}$ and $C^{\beta,X}_{(C^{\emptyset}_{\mathscr{T}},\mathscr{R}(\alpha-i))} \in \mathscr{T}$, then for any $p \in \mathscr{F}_p$,*

$$C_{\mathscr{T}}^{\beta\otimes X}(p) = \bigvee\{\theta \mid \text{ there exist } (P^i, (n), X, (p,\theta) - (\alpha,\beta))\}, \tag{9.45}$$

where $i = \mathrm{I}, \mathrm{II}$. $\square$

9.2.5 Satisfiability and Consistency

In the following, we always assume that $\alpha \in L$, $\mathscr{T} \subseteq \mathscr{F}_L(\mathscr{F}_p)$, $\mathscr{R}$ is α-i type sound w.r.t. $\mathscr{T}$, and $C^{\beta,X}_{(C^{\emptyset}_{\mathscr{T}},\mathscr{R}(\alpha-i))} \in \mathscr{T}$ for any $X \in \mathscr{F}_L(\mathscr{F}_p)$, $i = \mathrm{I}, \mathrm{II}$.

Definition 9.2.8 *Let $X \in \mathscr{F}_L(\mathscr{F}_p)$, $\delta \in L$, $T_0 \in \mathscr{T}$. If for any $p \in \mathscr{F}_p$,*

(1). $T_0(p') = (T_0(p))'$,
(2). $X(p) \to T_0(p) \geqslant \delta$,

then T_0 is called δ-i type satisfies X. If there exists T_0, T_0 δ-i type satisfies X, then X is said to be δ-i type satisfiable, $i = \mathrm{I}, \mathrm{II}$.

Definition 9.2.9 *Let $X \in \mathscr{F}_L(\mathscr{F}_p)$, $\tau \in L$. If*

$$\bigvee\{C^{\beta,X}_{(C^{\emptyset}_{\mathscr{T}},\mathscr{R}(\alpha-i))}(p) \otimes C^{\beta,X}_{(C^{\emptyset}_{\mathscr{T}},\mathscr{R}(\alpha-i))}(p') \mid p \in \mathscr{F}_p\} \leqslant \tau, \tag{9.46}$$

then X is said to be τ'-i type consistent w.r.t. (α, β).

Theorem 9.2.11 ((α,β)-i Type Consistent) *Let $X \in \mathscr{F}_L(\mathscr{F}_p)$. If X is δ-i type satisfiable, then X is $\delta \otimes \delta$-i type consistent w.r.t. (α, β).*

Proof. Since X is δ-i type satisfiable, there exists $T_0 \in \mathscr{T}$ such that $X \otimes \delta \subseteq T_0$. For any $p \in \mathscr{F}_p$, we notice that $\delta \leqslant \pi(\beta \otimes X \subseteq X \otimes \delta)$, hence

$$\begin{aligned} C_{\mathscr{T}}^{\beta\otimes X}(p) &= \bigwedge_{T\in\mathscr{T}} (\pi(\beta \otimes X \subseteq T) \to T(p)) \\ &\leqslant \pi(\beta \otimes X \subseteq T_0) \to T_0(p) \\ &\leqslant \pi(\beta \otimes X \subseteq X \otimes \delta) \to T_0(p) \\ &\leqslant \delta \to T_0(p). \end{aligned}$$

By Theorem 9.2.9, we know

$$\begin{aligned}C^{\beta,X}_{(C^{\emptyset}_{\mathscr{T}},\mathscr{R}(\alpha-i))}(p)\otimes C^{\beta,X}_{(C^{\emptyset}_{\mathscr{T}},\mathscr{R}(\alpha-i))}(p') &= C^{\beta\otimes X}_{\mathscr{T}}(p)\otimes C^{\beta\otimes X}_{\mathscr{T}}(p')\\ &\leqslant (\delta\to T_0(p))\otimes(\delta\to T_0(p'))\\ &\leqslant \delta\to((\delta\to T_0(p))\otimes T_0(p'))\\ &\leqslant \delta\to(\delta\to(T_0(p')\otimes T_0(p)))\\ &= \delta\to(\delta\to((T_0(p))'\otimes T_0(p)))\\ &= \delta\to(\delta\to O)\\ &= \delta\to\delta'.\end{aligned}$$

Hence,

$$\bigvee\left\{C^{\beta,X}_{(C^{\emptyset}_{\mathscr{T}},\mathscr{R}(\alpha-i))}(p)\otimes C^{\beta,X}_{(C^{\emptyset}_{\mathscr{T}},\mathscr{R}(\alpha-i))}(p')\mid p\in\mathscr{F}_p\right\}\leqslant\delta\to\delta'.$$

Consequently, X is $\delta\otimes\delta$-i type consistent w.r.t. (α,β). □

Definition 9.2.10 *It is called that X has a δ-i model if there exists $T_0\in\mathscr{T}$,*

$$T_0(p')=(T_0(p))'$$

for any $p\in\mathscr{F}_p$, $C^{\delta\otimes X}_{\mathscr{T}}\subseteq T_0$. T_0 is called a δ-i model of X, $i=\mathrm{I},\mathrm{II}$.

Theorem 9.2.12 *If there exists a minimal element T_0 in $(\mathscr{T},\subseteq)$ such that for any $p\in\mathscr{F}_p$,*

$$T_0(p')=(T_0(p))',$$

then X is $\beta\otimes\beta\otimes\tau'$-$i$ type satisfiable when X is τ'-i type consistent w.r.t. (α,β).

Proof. If X is τ'-i type consistent w.r.t. (α,β), then for any $p\in\mathscr{F}_p$,

$$C^{\beta,X}_{(C^{\emptyset}_{\mathscr{T}},\mathscr{R}(\alpha-i))}(p)\otimes C^{\beta,X}_{(C^{\emptyset}_{\mathscr{T}},\mathscr{R}(\alpha-i))}(p')\leqslant\tau.$$

Since, $\beta\otimes X(p)\leqslant C^{\beta,X}_{(C^{\emptyset}_{\mathscr{T}},\mathscr{R}(\alpha-i))}(p)$ and

$$\begin{aligned}\beta\otimes(T_0(p))' &= \beta\otimes T_0(p')\\ &= \beta\otimes\left(\bigcap_{T\in\mathscr{T}}T\right)(p')\\ &= \beta\otimes\bigwedge_{T\in\mathscr{T}}T(p')\\ &= \beta\otimes C^{\emptyset}_{\mathscr{T}}(p').\end{aligned}$$

So that

$$\begin{aligned}&(\beta\otimes X(p))\otimes(\beta\otimes(T_0(p))')\\ &\leqslant C^{\beta,X}_{(C^{\emptyset}_{\mathscr{T}},\mathscr{R}(\alpha-i))}(p)\otimes C^{\beta,X}_{(C^{\emptyset}_{\mathscr{T}},\mathscr{R}(\alpha-i))}(p')\\ &\leqslant\tau.\end{aligned}$$

Hence,

$$\begin{aligned}\beta\otimes\beta\otimes X(p)&\leqslant(T_0(p))'\to\tau\\ &=\tau'\to T_0(p).\end{aligned}$$

Therefore,

$$\beta\otimes\beta\otimes\tau'\leqslant X(p)\to T_0(p).\ \square$$

Theorem 9.2.13 *T_0 δ-i type satisfies X if and only if T_0 is a δ-i model of X.*

Proof. Note that for any $p\in\mathscr{F}_p$,

$$\begin{aligned}C^{\delta\otimes X}_{\mathscr{T}}(p)&=\bigwedge_{T\in\mathscr{T}}(\pi(\delta\otimes X\subseteq T)\to T(p))\\ &\leqslant\pi(\delta\otimes X\subseteq T_0)\to T_0(p)\\ &=T_0(p).\end{aligned}$$

$$\delta\otimes X\subseteq C^{\delta,X}_{(C^{\emptyset}_{\mathscr{T}},\mathscr{R}(\alpha-i))}\subseteq C^{\delta\otimes X}_{\mathscr{T}}\subseteq T_0.\ \square$$

Corollary 9.2.2 *If $T_0\in\mathscr{T}_H$, then $\delta\otimes X\subseteq T_0$ if and only if $C^{\delta\otimes X}_{\mathscr{T}}\subseteq T_0$.* $\square$

Theorem 9.2.14 *(1). Suppose X is τ'-i type consistent w.r.t. (α,β), then X has a $\beta\otimes\beta\otimes\tau'$-$i$ model T_0.*
(2). If $\beta\otimes X=\beta\otimes\tau'\to T_0$, then

$$C^{\beta,X}_{(C^{\emptyset}_{\mathscr{T}},\mathscr{R}(\alpha-i))}=\beta\otimes X.$$

Proof. (1). By Theorem 9.2.12, X is $\beta\otimes\beta\otimes\tau'$-$i$ type satisfiable. By Theorem 9.2.13, X has a $\beta\otimes\beta\otimes\tau'$-$i$ type model T_0.

(2). First, note that

$$\beta\otimes X\subseteq C^{\beta,X}_{(C^{\emptyset}_{\mathscr{T}},\mathscr{R}(\alpha-i))}\subseteq C^{\beta\otimes X}_{\mathscr{T}}.$$

For any $q\in\mathscr{F}_p$, we have

$$\begin{aligned} C_{\mathscr{T}}^{\beta\otimes X}(q) &= \bigwedge_{T\in T} (\pi(\beta\otimes X\subseteq T)\to T(q)) \\ &\leqslant \pi(\beta\otimes X\subseteq T_0)\to T_0(q) \\ &= \bigwedge_{r\in\mathscr{F}_p} (\beta\otimes X(r)\to T_0(r))\to T_0(q) \\ &\leqslant \bigwedge_{r\in\mathscr{F}_p} (\beta\otimes\tau')\to T_0(q) \\ &= (\beta\otimes\tau')\to T_0(q) \\ &= \beta\otimes X(q). \end{aligned}$$

So

$$C^{\beta,X}_{(C^{\emptyset}_{\mathscr{T}},\mathscr{R}(\alpha-i))} = \beta\otimes X. \ \square$$

9.2.6 Deduction Theorem

Let $\mathscr{R}$ is α-i type sound w.r.t. $\mathscr{T}$, and $C^{\beta,X}_{(C^{\emptyset}_{\mathscr{T}},\mathscr{R}(\alpha-i))}\in\mathscr{T}$ for any $X\in\mathscr{F}_L(\mathscr{F}_p)$, $i=\mathrm{I},\mathrm{II}$.

Theorem 9.2.15 ((α,β,θ)-Deduction) *Let $(r_2^0,t_2)\in\mathscr{R}$, $X\in\mathscr{F}_L(\mathscr{F}_p)$, $p,q\in\mathscr{F}_p$, and $\sigma,\theta\in L$. Then*

$$C_{\mathscr{T}}^{\beta\otimes(X\cup\{\theta/p\})}(q)\geqslant\alpha\otimes t_2(\beta\otimes\theta,\sigma) \tag{9.47}$$

if $C_{\mathscr{T}}^{\beta\otimes X}(p\to q)\geqslant\sigma$ for $i=\mathrm{I}$.

$$C_{\mathscr{T}}^{\beta\otimes(X\cup\{\theta/p\})}(q)\geqslant t_2(\alpha\otimes\beta\otimes\theta,\alpha\otimes\sigma) \tag{9.48}$$

if $C_{\mathscr{T}}^{\beta\otimes X}(p\to q)\geqslant\sigma$ for $i=\mathrm{II}$.

$$C_{\mathscr{T}_h}^{\beta\otimes(X\cup\{\theta/p\})}(q)\leqslant C_{\mathscr{T}_h}^{\beta\otimes X}(p\to q) \tag{9.49}$$

if $\mathscr{T}_h\subseteq\mathscr{T}_H$.

Proof. For $i=\mathrm{I}$.

$$C_{\mathscr{T}}^{\beta\otimes(X\cup\{\theta/p\})}(p\to q)\geqslant C_{\mathscr{T}}^{\beta\otimes X}(p\to q)\geqslant\sigma.$$

$$\begin{aligned}
C_{\mathscr{T}}^{\beta\otimes(X\cup\{\theta/p\})}(p) &= \bigwedge_{T\in\mathscr{T}} (\pi(\beta\otimes(X\cup\{\theta/p\})\subseteq T)\to T(p)) \\
&= \bigwedge_{T\in\mathscr{T}} ((\pi(\beta\otimes X\subseteq T)\wedge\pi(\beta\otimes\{\theta/p\}\subseteq T))\to T(p)) \\
&\geqslant \bigwedge_{T\in\mathscr{T}} (\pi(\beta\otimes\{\theta/p\}\subseteq T)\to T(p)) \\
&= \bigwedge_{T\in\mathscr{T}} ((\beta\otimes\theta\to T(p))\to T(p)) \\
&= \bigwedge_{T\in\mathscr{T}} ((\beta\otimes\theta)\vee T(p)) \\
&\geqslant \beta\otimes\theta.
\end{aligned}$$

Notice that for any $Y\in\mathscr{F}_L(\mathscr{F}_p)$, $C_{\mathscr{T}}^{\beta\otimes Y}$ is α-I type closed w.r.t. $\mathscr{R}$, and $(r_2^0,t_2)\in\mathscr{R}$. Hence

$$\begin{aligned}
C_{\mathscr{T}}^{\beta\otimes(X\cup\{\theta/p\})}(q) &= C_{\mathscr{T}}^{\beta\otimes(X\cup\{\theta/p\})}(r_2^0(p,p\to q)) \\
&\geqslant \alpha\otimes t_2(C_{\mathscr{T}}^{\beta\otimes(X\cup\{\theta/p\})}(p),C_{\mathscr{T}}^{\beta\otimes(X\cup\{\theta/p\})}(p\to q)) \\
&\geqslant \alpha\otimes t_2(\beta\otimes\theta,\sigma).
\end{aligned}$$

For $i=\text{II}$, it can be proved in the same way.

Let $\pi(\beta\otimes X\subseteq T)\to T(q)=\sigma_T(q)$, then

$$\begin{aligned}
C_{\mathscr{T}_h}^{\beta\otimes X}(p\to q) &= \bigwedge_{T\in\mathscr{T}_h} (\pi(\beta\otimes X\subseteq T)\to T(p\to q)) \\
&= \bigwedge_{T\in\mathscr{T}_h} (T(p)\to\sigma_T(q)).
\end{aligned}$$

$$
\begin{aligned}
C_{\mathscr{T}_h}^{\beta\otimes(X\cup\{\theta/p\})}(q) &= \bigwedge_{T\in\mathscr{T}_h} (\pi(\beta\otimes(X\cup\{\theta/p\})\subseteq T)\to T(q))\\
&= \bigwedge_{T\in\mathscr{T}_h} ((\pi(\beta\otimes X\subseteq T)\wedge\pi(\beta\otimes\{\theta/p\}\subseteq T))\to T(q))\\
&= \bigwedge_{T\in\mathscr{T}_h} (\sigma_T(q)\vee(\pi(\beta\otimes\{\theta/p\}\subseteq T)\to T(q)))\\
&= \bigwedge_{T\in\mathscr{T}_h} (\sigma_T(q)\vee((\beta\otimes\theta\to T(p))\to T(q)))\\
&= \bigwedge_{T\in\mathscr{T}_h} ((((\beta\otimes\theta\to T(p))\to T(q))\to\sigma_T(q))\to\sigma_T(q))\\
&\leqslant \bigwedge_{T\in\mathscr{T}_h} (((T(p)\to T(q))\to\sigma_T(q))\to\sigma_T(q))\\
&\leqslant \bigwedge_{T\in\mathscr{T}_h} (((T(p)\to T(q))\to T(q))\to\sigma_T(q))\\
&= \bigwedge_{T\in\mathscr{T}_h} ((T(p)\vee T(q))\to\sigma_T(q))\\
&\leqslant \bigwedge_{T\in\mathscr{T}_h} (T(p)\to\sigma_T(q))\\
&= C_{\mathscr{T}_h}^{\beta\otimes X}(p\to q). \ \square
\end{aligned}
$$

9.2.7 Compactness

Definition 9.2.11 *$C_{\mathscr{T}_h}$ is said to be compact on $\mathscr{F}_p$, if for any $X\in\mathscr{F}_L(\mathscr{F}_p)$ and $p\in\mathscr{F}_p$, there exists $Q\in\mathscr{F}_L(\mathscr{F}_p)$ and $|\text{supp } Q|<\aleph_0$, such that*

$$C_{\mathscr{T}_h}^{X}(p)=C_{\mathscr{T}_h}^{X\cap Q}(p).$$

Theorem 9.2.16 *If $C_{\mathscr{T}_h}$ is compact on $\mathscr{F}_p$, then L satisfies the Ascending Chain Condition and the Descending Chain Condition.*

Proof. We prove that if $C_{\mathscr{T}_h}$ is compact on $\mathscr{F}_p$, then L satisfies the Ascending Chain Condition. In fact, if L does not satisfy the Ascending Chain Condition, then there exists an infinite ascending chain in L:

$$\alpha_0<\alpha_1<\cdots<\alpha_n<\cdots.$$

Let $\alpha=\bigvee_{n=0}^{+\infty}\alpha_n$, $X_0\in\mathscr{F}_L(\mathscr{F}_p)$, and

$$X_0(p)=\begin{cases} I, & p=\alpha_n\to\alpha_0, n\in\mathbb{N},\\ O, & \text{otherwise.}\end{cases}$$

For any $m\in\mathbb{N}$,

$$\begin{aligned}
C_{\mathscr{T}_h}^{X_0}(\alpha_0) &= \bigwedge_{T\in\mathscr{T}_h}\left(\left(\bigwedge_{q\in\mathscr{F}_p}(X_0(q)\to T(q))\right)\to T(\alpha_0)\right)\\
&\geqslant \bigwedge_{T\in\mathscr{T}_h}\left(\left(\bigwedge_{n\in\mathbb{N}}T(\alpha_n\to\alpha_0)\right)\to T(\alpha_0)\right)\\
&= \bigwedge_{T\in\mathscr{T}_h}\left(\left(\bigwedge_{n\in\mathbb{N}}(\alpha_n\to\alpha_0)\right)\to \alpha_0\right)\\
&= \left(\bigwedge_{n\in\mathbb{N}}(\alpha_n\to\alpha_0)\right)\to\alpha_0\\
&\geqslant (\alpha_m\to\alpha_0)\to\alpha_0\\
&= \alpha_m\vee\alpha_0\\
&= \alpha_m.
\end{aligned}$$

So, we know that

$$C_{\mathscr{T}_h}^{X_0}(\alpha_0)\geqslant\bigvee_{m\in\mathbb{N}}\alpha_m=\alpha.$$

On the other hand, for any $Q\in\mathscr{F}_L(\mathscr{F}_p)$, if $|\text{supp } Q|<\aleph_0$, then

(1). if there exist some formulae in the form $\alpha_n\to\alpha_0$ in supp Q, then there exist at most finite ones, denoted as $\alpha_{n_1}\to\alpha_0,\cdots,\alpha_{n_{k_0}}\to\alpha_0$. Moreover, let $\alpha_{n_1}<\cdots<\alpha_{n_{k_0}}$. Hence we have

$$\begin{aligned}
C_{\mathscr{T}_h}^{X\cap Q}(\alpha_0) &= \bigwedge_{T\in\mathscr{T}_h}\left(\left(\bigwedge_{q\in\text{supp } Q}(X_0(q)\wedge Q(q)\to T(q))\right)\to T(\alpha_0)\right)\\
&\leqslant \bigwedge_{T\in\mathscr{T}_h}\left(\left(\bigwedge_{q\in\text{supp } Q}(X_0(q)\to T(q))\right)\to T(\alpha_0)\right)\\
&= \bigwedge_{T\in\mathscr{T}_h}\left(\left(\bigwedge_{j=1}^{k_0}(X_0(\alpha_{n_j}\to\alpha_0)\to T(\alpha_{n_j}\to\alpha_0))\right)\to T(\alpha_0)\right)\\
&= \bigwedge_{T\in\mathscr{T}_h}\left(\left(\bigwedge_{j=1}^{k_0}T(\alpha_{n_j}\to\alpha_0)\right)\to T(\alpha_0)\right)\\
&= \left(\bigwedge_{j=1}^{k_0}(\alpha_{n_j}\to\alpha_0)\right)\to\alpha_0\\
&= (\alpha_{n_{k_0}}\to\alpha_0)\to\alpha_0\\
&= \alpha_{n_{k_0}}\vee\alpha_0\\
&< \alpha.
\end{aligned}$$

(2). if there exists no formula in the form $\alpha_n \to \alpha_0$ in supp Q, then for any $p \in \mathscr{F}_p$, $(X_0 \cap Q)(p) = O$. Hence

$$\begin{aligned} C_{\mathscr{T}_h}^{X_0 \cap Q}(\alpha_0) &= \bigwedge_{T \in \mathscr{T}_h} T(\alpha_0) \\ &= \alpha_0 \\ &< \alpha. \end{aligned}$$

Consequently, $C_{\mathscr{T}_h}$ is not compact on $\mathscr{F}_p$.

If L does not satisfy the Descending Chain Condition, then there exists an infinite descending chain in L:

$$\beta_0 > \beta_1 > \cdots > \beta_n > \cdots .$$

Hence

$$\beta_0' < \beta_1' < \cdots < \beta_n' < \cdots .$$

Set $\beta_i' = \alpha_i$, $i \in \mathbb{N}$, we have

$$\alpha_0 < \alpha_1 < \cdots < \alpha_n < \cdots .$$

This is the case of that L does not satisfy the Ascending Chain Condition. □

Definition 9.2.12 *X is said to be δ-i type finite satisfiable, if for any $Q \in \mathscr{F}_L(\mathscr{F}_p)$, when $|\text{supp } Q| < \aleph_0$, $X \cap Q$ is δ-i type satisfiable.*

Definition 9.2.13 *X is said to be δ-i type finite consistent w.r.t. (α, β), if for any $Q \in \mathscr{F}_L(\mathscr{F}_p)$, when $|\text{supp } Q| < \aleph_0$, $X \cap Q$ is δ-i type consistent w.r.t. (α, β).*

Theorem 9.2.17 (Compactness on Consistency) *Let*

$$\bigvee_{q \in \mathscr{F}_p} (C_{\mathscr{T}}^{\beta \otimes X}(q) \otimes C_{\mathscr{T}}^{\beta \otimes X}(q'))$$

and

$$\bigvee \{\theta \mid \text{ there exists } (P^i, (n), X, (p, \theta) - (\alpha, \beta))\}$$

be reachable for any $p \in \mathscr{F}_p$, then X is δ-i type consistent w.r.t.(α, β) if and only if X is δ-i type finite consistent w.r.t. (α, β), $i = \text{I}, \text{II}$.

Proof. (necessity). If there exists $Q_0 \in \mathscr{F}_L(\mathscr{F}_p)$, $|\text{supp } Q_0| < \aleph_0$, $X \cap Q_0$ is not δ-i type consistent w.r.t. (α, β), then there exist $p, p' \in \mathscr{F}_p$ such that

(1). $C_{\mathscr{T}}^{\beta \otimes (X \cap Q_0)}(p) \otimes C_{\mathscr{T}}^{\beta \otimes (X \cap Q_0)}(p') > \delta'$, or
(2). $C_{\mathscr{T}}^{\beta \otimes (X \cap Q_0)}(p) \otimes C_{\mathscr{T}}^{\beta \otimes (X \cap Q_0)}(p') // \delta'$.

For (1), we obtain

$$C_{\mathscr{T}}^{\beta\otimes X}(p)\otimes C_{\mathscr{T}}^{\beta\otimes X}(p')>\delta',$$

it contradicts to that X is δ-i type consistent w.r.t. (α,β).

For (2), we obtain

$$C_{\mathscr{T}}^{\beta\otimes X}(p)\otimes C_{\mathscr{T}}^{\beta\otimes X}(p')//\delta'.$$

[In fact, if $C_{\mathscr{T}}^{\beta\otimes X}(p)\otimes C_{\mathscr{T}}^{\beta\otimes X}(p')\leqslant\delta'$, then

$$C_{\mathscr{T}}^{\beta\otimes(X\cap Q_0)}(p)\otimes C_{\mathscr{T}}^{\beta\otimes(X\cap Q_0)}(p')\leqslant\delta',$$

it contradicts to (2). If $C_{\mathscr{T}}^{\beta\otimes X}(p)\otimes C_{\mathscr{T}}^{\beta\otimes X}(p')>\delta'$, then

$$\bigvee_{q\in\mathscr{F}_p}(C_{\mathscr{T}}^{\beta\otimes X}(q))\otimes C_{\mathscr{T}}^{\beta\otimes X}(q')>\delta',$$

it contradicts to X is δ-i type consistent w.r.t. (α,β).] Then, we get

$$\bigvee_{q\in\mathscr{F}_p}\left(C_{\mathscr{T}}^{\beta\otimes X}(q)\otimes C_{\mathscr{T}}^{\beta\otimes X}(q')\right)//\delta'.$$

[In fact, if $\bigvee_{q\in\mathscr{F}_p}\left(C_{\mathscr{T}}^{\beta\otimes X}(q)\otimes C_{\mathscr{T}}^{\beta\otimes X}(q')\right)\leqslant\delta'$, then

$$C_{\mathscr{T}}^{\beta\otimes X}(p)\otimes C_{\mathscr{T}}^{\beta\otimes X}(p')\leqslant\delta',$$

this contradicts to $\left(C_{\mathscr{T}}^{\beta\otimes X}(p)\otimes C_{\mathscr{T}}^{\beta\otimes X}(p')\right)//\delta'$. If

$$\bigvee_{q\in\mathscr{F}_p}\left(C_{\mathscr{T}}^{\beta\otimes X}(q)\otimes C_{\mathscr{T}}^{\beta\otimes X}(q')\right)>\delta',$$

this contradicts to that X is δ-i type consistent w.r.t. (α,β).]

Sum up, we get that X is not δ-i type consistent w.r.t. (α,β), this is a contradiction, *i.e.*, X is δ-i type finite consistent w.r.t. (α,β).

(sufficiency). For any $Q\in\mathscr{F}_L(\mathscr{F}_p)$, if $|\text{supp }Q|<\aleph_0$, then

$$\bigvee_{q\in\mathscr{F}_p}\left(C_{\mathscr{T}}^{\beta\otimes(X\cap Q)}(q)\otimes C_{\mathscr{T}}^{\beta\otimes(X\cap Q)}(q')\right)\leqslant\delta'. \tag{9.50}$$

To prove

$$\bigvee_{q\in\mathscr{F}_p}\left(C_{\mathscr{T}}^{\beta\otimes X}(q)\otimes C_{\mathscr{T}}^{\beta\otimes X}(q')\right)\leqslant\delta'.$$

If

$$\bigvee_{q\in\mathscr{F}_p}\left(C_{\mathscr{T}}^{\beta\otimes X}(q)\otimes C_{\mathscr{T}}^{\beta\otimes X}(q')\right)>\delta' \tag{9.51}$$

or

$$\bigvee_{q\in\mathscr{F}_p} \left(C_{\mathscr{T}}^{\beta\otimes X}(q) \otimes C_{\mathscr{T}}^{\beta\otimes X}(q')\right) //\delta', \tag{9.52}$$

for Eq. (9.51), according to the condition of reachability, suppose there exists $p \in \mathscr{F}_p$, $C_{\mathscr{T}}^{\beta\otimes X}(p) \otimes C_{\mathscr{T}}^{\beta\otimes X}(p') > \delta'$. By the reachability of Eq. (9.45), let

$$\left(P^i, (n), X, (p, C_{\mathscr{T}}^{\beta\otimes X}(p)) - (\alpha, \beta)\right)$$

be a $(C_{\mathscr{T}}^{\beta\otimes X}(p) - (\alpha, \beta))$-$i$ type proof from X to p, which is

$$(q_1, \beta_1), \cdots, (q_n, \beta_n).$$

$$\left(P^i, (m), X, (p', C_{\mathscr{T}}^{\beta\otimes X}(p')) - (\alpha, \beta)\right)$$

be a $(C_{\mathscr{T}}^{\beta\otimes X}(p') - (\alpha, \beta))$-$i$ type proof from X to p', which is

$$(s_1, \theta_1), \cdots, (s_m, \theta_m).$$

Suppose P is $(p_1, \alpha_1), \cdots, (p_{n+m}, \alpha_{n+m})$, where

$$(p_i, \alpha_i) = \begin{cases} (q_i, \beta_i), & 1 \leqslant i \leqslant n, \\ (s_{i-n}, \theta_{i-n}), & n+1 \leqslant i \leqslant n+m. \end{cases}$$

Set

$$Q = \{p_1, \cdots, p_{n+m}\}.$$

We define

$$Q_0(q) = \begin{cases} X(q), & q \in Q, \\ O, & q \notin Q. \end{cases}$$

Then, $|\text{supp } Q_0| < \aleph_0$, and $Q_0 \subseteq X$,

$$C_{\mathscr{T}}^{\beta\otimes (X\cap Q_0)}(p) \otimes C_{\mathscr{T}}^{\beta\otimes (X\cap Q_0)}(p') = C_{\mathscr{T}}^{\beta\otimes X}(p) \otimes C_{\mathscr{T}}^{\beta\otimes X}(p') > \delta',$$

this contradicts to Eq. (9.50).

For Eq. (9.52), by the condition of reachability, suppose there exists $p \in \mathscr{F}_p$, such that

$$C_{\mathscr{T}}^{\beta\otimes X}(p) \otimes C_{\mathscr{T}}^{\beta\otimes X}(p') // \delta'.$$

Similar to Eq. (9.51), there exists S_0 such that $|\text{supp } S_0| < \aleph_0$, so

$$C_{\mathscr{T}}^{\beta\otimes (X\cap S_0)}(p) \otimes C_{\mathscr{T}}^{\beta\otimes (X\cap S_0)}(p') = C_{\mathscr{T}}^{\beta\otimes X}(p) \otimes C_{\mathscr{T}}^{\beta\otimes X}(p') // \delta',$$

this also contradicts to Eq. (9.50). Hence,

$$\bigvee_{q\in\mathscr{F}_p} (C_{\mathscr{T}}^{\beta\otimes X}(q) \otimes C_{\mathscr{T}}^{\beta\otimes X}(q')) \leqslant \delta'. \ \square$$

Theorem 9.2.18 *(1). If X is δ-i type satisfiable, then X is δ-i type finite satisfiable.*

(2). Suppose there exists a minimal element T_0 in $(\mathscr{T}, \subseteq)$ such that $T_0(p') = (T_0(p))'$,

$$\bigvee_{q \in \mathscr{F}_p} (C_{\mathscr{T}}^{\beta \otimes X}(q) \otimes C_{\mathscr{T}}^{\beta \otimes X}(q'))$$

and

$$\bigvee \{\theta \mid \text{ there exist } (P^i, (n), X, (p, \theta) - (\alpha, \beta))\}$$

are reachable for any $p \in \mathscr{F}_p$. If X is δ-i type finite satisfiable, then X is $(\beta \otimes \beta \otimes \delta \otimes \delta)$-$i$ type satisfiable.

Proof. (1). For any $Q \in \mathscr{F}_L(\mathscr{F}_p)$, if $|\text{supp } Q| < \aleph_0$, for any $p \in \mathscr{F}_p$,

$$(X \cap Q)(p) \to T_0(p) \geqslant X(p) \to T_0(p) \geqslant \delta.$$

So X is δ-i type finite satisfiable.

(2). By Theorems 9.2.11 X is $(\delta \otimes \delta)$-i type finite consistent w.r.t. (α, β). By Theorem 9.2.17, X is $(\delta \otimes \delta)$-i type consistent w.r.t. (α, β). By Theorem 9.2.12, X is $(\beta \otimes \beta \otimes \delta \otimes \delta)$-$i$ type satisfiable. □

9.2.8 Examples

Example 9.2.6 *In L_{vpl}, let $\alpha \leqslant \bigwedge_{\theta \in L}(\theta \vee \theta')$ and*

$$\mathscr{R} = \{(r_1^{\theta_0}, t_1^{\theta_0}) \mid \theta_0 \in L\} \cup \{(r_2^0, t_2^*), (r_2^*, t_2^*), (r_2^{\triangle}, t_2^*)\},$$

$$\mathscr{T}_i = \{T \mid T \in \mathscr{F}_L(\mathscr{F}_p), T \text{ is } \alpha\text{-}i \text{ type closed w.r.t. } \mathscr{R}\},\ i = \text{I}, \text{II}.$$

Then we obtain the following results.

(1). $\mathscr{T}_H \subseteq \mathscr{T}_i$, $i = \text{I}, \text{II}$.

(2). $C^{\beta, X}_{(C^{\emptyset}_{\mathscr{T}_i}, \mathscr{R}(\alpha - i))} \in \mathscr{T}_i$ holds for any $X \in \mathscr{F}_L(\mathscr{F}_p)$ and $\beta \in L$, $i = \text{I}, \text{II}$.

(3). $\mathscr{R}$ is α-i type sound w.r.t. $\mathscr{T}_i$, $i = \text{I}, \text{II}$.

(4). $C^{\beta, X}_{(C^{\emptyset}_{\mathscr{T}_i}, \mathscr{R}(\alpha - i))} = C^{\beta \otimes X}_{\mathscr{T}_i}$ holds for any $X \in \mathscr{F}_L(\mathscr{F}_p)$ and $\beta \in L$, $i = \text{I}, \text{II}$.

(5). $C^{\beta \otimes X}_{\mathscr{T}_i}(p) = \bigvee \{\theta \mid$ there exist $(P^i, (n), X, (p, \theta) - (\alpha, \beta))\}$ holds for any $X \in \mathscr{F}_L(\mathscr{F}_p)$, $\beta \in L$, and $p \in \mathscr{F}_p$, $i = \text{I}, \text{II}$.

(6). For any $X \in \mathscr{F}_L(\mathscr{F}_p)$, and $\delta \in L$, if X is δ-i type satisfiable, then X is $(\delta \otimes \delta)$-i type consistent w.r.t. (α, β), $i = \text{I}, \text{II}$.

(7). a) For any $X \in \mathscr{F}_L(\mathscr{F}_p)$, $p, q \in \mathscr{F}_p$ and $\sigma, \theta \in L$, if $C^{\beta \otimes X}_{\mathscr{T}_p}(p \to q) \geqslant \sigma$, then

$$C^{\beta \otimes (X \cup \{\theta / p\})}_{\mathscr{T}_i}(q) \geqslant \alpha \otimes ((\beta \otimes \theta) \wedge \sigma), i = \text{I}, \text{II}.$$

b) If $\mathscr{T}_h \subseteq \mathscr{T}_H$, then

$$C^{\beta \otimes (X \cup \{\theta / p\})}_{\mathscr{T}_h}(q) \leqslant C^{\beta \otimes X}_{\mathscr{T}_h}(p \to q).$$

Example 9.2.7 (Łukasiewicz Propositional Logic L_u) *In L_{vpl}, if the complete lattice implication algebra $L = ([0,1], \vee, \wedge, ', \rightarrow)$ is the Łukasiewicz implication algebra, then this L_{vpl} is called as Łukasiewicz propositional logic system, denoted as L_u, where*

$$\mathscr{R} = \{(r_1^{\theta_0}, t_1^{\theta_0}) \mid \theta_0 \in L\} \cup \{(r_2^0, t_2^*), (r_2^*, t_2^*), (r_2^\triangle, t_2^*)\},$$

and $\alpha \leqslant \bigwedge_{\theta \in [0,1]} (\theta \vee \theta') = 0.5$,

$$\mathscr{T}_i = \{T \mid T \in \mathscr{F}(\mathscr{F}_p), T \text{ is } \alpha\text{-}i \text{ type closed w.r.t. } \mathscr{R}\}, i = \mathrm{I}, \mathrm{II}.$$

Then we can obtain the similar results as in Example 9.2.6.

Example 9.2.8 (F-Valued Łukasiewicz Propositional Logic FL_u) *If the complete lattice implication algebra L in L_{vpl} is $(\mathscr{F}([0,1]), \vee, \wedge, ', \rightarrow)$ as Example 2.1.7, then this L_{vpl} is called as F-valued Łukasiewicz propositional logic system, denoted as FL_u, which satisfies*

(1). $(\mathscr{F}([0,1]), \vee, \wedge, ', \rightarrow)$ is an infinitely distributive non-chain complete lattice implication algebra.

(2). Let

$$\mathscr{R} = \{(r_1^{\theta_0}, t_1^{\theta_0}) \mid \theta_0 \in L\} \cup \{(r_2^0, t_2^*), (r_2^*, t_2^*), (r_2^\triangle, t_2^*)\},$$

$\alpha \in \mathscr{F}([0,1])$ *and* $\alpha(x) \leqslant 0.5$ *for any* $x \in [0,1]$.

$$\mathscr{T}_i = \{T \mid T \in \mathscr{F}_{\mathscr{F}([0,1])}(\mathscr{F}_p), T \text{ is } \alpha\text{-}i \text{ type closed w.r.t. } \mathscr{R}\}, i = \mathrm{I}, \mathrm{II}.$$

Then we can obtain the similar results as in Example 9.2.7.

Example 9.2.9 *If the lattice implication algebra in L_{vpl} is C_5, then this L_{vpl} is also called $\mathscr{C}_5$, where*

$$\mathscr{R} = \{(r_1^{\theta_0}, t_1^{\theta_0}) \mid \theta_0 \in L\} \cup \{(r_2^0, t_2^*), (r_2^*, t_2^*), (r_2^\triangle, t_2^*)\},$$

$$\mathscr{T}_i = \{T \mid T \in \mathscr{F}_L(\mathscr{F}_p), T \text{ is } \alpha\text{-}i \text{ type closed w.r.t. } \mathscr{R}\}, i = \mathrm{I}, \mathrm{II},$$

and $\alpha \leqslant b$. Then the similar results as in Example 9.2.6 can be obtained.

10 Lattice-Valued First-Order Logics

In Chapter 9, we discussed the lattice-valued propositional logics based on lattice implication algebra and their properties. In this chapter, we discuss the lattice-valued first-order logic based on lattice implication algebra. In Section 10.1, a lattice-valued first-order logic $LF(X)$ is given. In Section 10.2, a gradational lattice-valued first-order logic L_{vfl} is given.

10.1 Lattice-Valued First-Order Logic $LF(X)$

10.1.1 Language

The alphabet of $LF(X)$ is composed of:

(1). individual variables set: $X = \{x_i \mid i \in \mathbb{N}\}$;
(2). individual constants set: $C = \{c_k \mid k \in K\}$, K is an index set;
(3). function symbols set: $\{f_{m_j}^{G_j} \mid j \in J\}$, J is an index set, $f_{m_j}^{G_j}$ is an m_j-ary function symbol, G_j is the super index of $f_{m_j}^{G_j}$;
(4). relation symbols set: $\{p_{m_i}^{(i)} \mid i \in M\}$, M is an index set, $p_{m_i}^{(i)}$ is an m_i-ary relation symbol. In particular, the equality symbol $=$ is considered as a 2-ary relation symbol;
(5). logical connectives: $\vee$, $\wedge$, $'$, $\rightarrow$;
(6). quantifiers: $\forall$, $\exists$;
(7). auxiliary symbols:), (,,(comma).

The set T of terms of $LF(X)$ is defined as the smallest set T^* satisfying the conditions:

(1). $X \cup C \subseteq T^*$ (individual variables and constants are called super-index-free terms);
(2). for any $j \in J$ and $t_1, \cdots, t_{m_j} \in T^*$, if every t_q $(1 \leqslant q \leqslant m_j)$ is super-index-free or has super-index G_j, then $f_{m_i}^{G_j}(t_1, \cdots, t_{m_j}) \in T^*$ and is called a term with super index G_j.

The set $\mathscr{F}$ of well-formed formulas (wff.) is defined as the smallest set $\mathscr{F}^*$ satisfying the conditions:

(1). for any $\alpha \in L$, $\alpha \in \mathscr{F}^*$;

(2). for any $i \in M$ and $t_1, \cdots, t_{m_i} \in T$,

$$p_{m_i}^{(i)}(t_1, \cdots, t_{m_i}) \in \mathscr{F}^*,$$

in the following, $=(t_1, t_2)$ is denoted by $t_1 = t_2$;

(3). for any $\varphi, \psi \in \mathscr{F}^*$, $\varphi \wedge \psi, \varphi \vee \psi, \varphi \rightarrow \psi, \varphi' \in \mathscr{F}^*$;

(4). for any $\varphi \in \mathscr{F}^*$ and $x \in X$, $(\forall x)\varphi, (\exists x)\varphi \in \mathscr{F}^*$.

The notions of free and bound variables are similar to those of classical logic.

10.1.2 Interpretation

By an interpretation of the language of $LF(X)$, we mean a structure:

$$\mathcal{D} = \langle D, f_D, \{p_{iD} \mid i \in M\}, \{f_{jD} \mid j \in J\}\rangle,$$

where

(1). D is a non-empty set and called an interpretation domain;

(2). $f_D\colon\ X \cup C \longrightarrow D$ is a function;

(3). for any $j \in J$, $f_{jD}\colon\ D_p^{m_j} \longrightarrow D_p$ is an m_j-ary function;

(4). for any $i \in M$, $p_{iD}:\ D_p^{m_i} \longrightarrow L$ is an m_i-ary L-fuzzy relation on D_p, where D_p is the set of L-fuzzy points on D, *i.e.*,

$$D_p = \{d_\alpha \mid d \in D, \alpha \in L \setminus \{O\}\}.$$

Let $\mathcal{D}$ be an interpretation. Under $\mathcal{D}$, every term t is corresponded with an L-fuzzy point $\mathcal{D}(t)$ in D, defined by

(1). if $t \in X \cup C$, then $\mathcal{D}(t) = f_D(t)_I$;

(2). if $t = f_{m_j}^{G_j}(t_1, \cdots, t_{m_j})$, then $\mathcal{D}(t) = f_{jD}(\mathcal{D}(t_1), \cdots, \mathcal{D}(t_{m_j}))$.

Every formula $\varphi \in \mathscr{F}$ corresponds to an element $v(\varphi)_{\mathcal{D}}$ of L, defined by:

(1). if φ is α, $\alpha \in L$, then $v(\varphi)_{\mathcal{D}} = \alpha$;

(2). if φ is $p_{m_i}^{(i)}(t_1, \cdots, t_{m_i})$, then $v(\varphi)_{\mathcal{D}} = p_{iD}(\mathcal{D}(t_1), \cdots, \mathcal{D}(t_{m_i}))$;

(3). if φ is $\psi \vee \gamma$, then $v(\varphi)_{\mathcal{D}} = v(\psi)_{\mathcal{D}} \vee v(\gamma)_{\mathcal{D}}$;

(4). if φ is $\psi \wedge \gamma$, then $v(\varphi)_{\mathcal{D}} = v(\psi)_{\mathcal{D}} \wedge v(\gamma)_{\mathcal{D}}$;

(5). if φ is $\psi \rightarrow \gamma$, then $v(\varphi)_{\mathcal{D}} = v(\psi)_{\mathcal{D}} \rightarrow v(\gamma)_{\mathcal{D}}$;

(6). if φ is ψ', then $v(\varphi)_{\mathcal{D}} = (v(\psi)_{\mathcal{D}})'$;

(7). if φ is $(\forall x)\psi$, then $v(\varphi)_{\mathcal{D}} = \bigwedge_{d \in D} v(\psi)_{\mathcal{D}(x/d)}$;

(8). if φ is $(\exists x)\psi$, then $v(\varphi)_{\mathcal{D}} = \bigvee_{d \in D} v(\psi)_{\mathcal{D}(x/d)}$,

where, $\mathcal{D}(x/d)$ is the following interpretation:

$$\mathcal{D}(x/d) = \langle D, f_D^*, \{p_{iD} \mid i \in M\}, \{f_{jD} \mid j \in J\}\rangle$$

with

$$f^*_D(y) = \begin{cases} f_D(y), & y \in X \cup C, y \neq x, \\ d, & y = x, \end{cases}$$

$v(\varphi)_\mathcal{D}$ is called the value of φ under $\mathcal{D}$.

An interpretation $\mathcal{D} = \langle D, f_D, \{p_{iD} \mid i \in M\}, \{f_{jD} \mid j \in J\}\rangle$ is called normal if the equality symbol $=$ is interpreted as the equality relation on D_p. In the following, we consider normal interpretations only.

10.1.3 Semantics

Definition 10.1.1 *Let $\mathcal{D}$ be an interpretation, $A \in \mathscr{F}_L(\mathscr{F})$. $\mathcal{D}$ is said to be a model of A (or satisfy A), if for any $\varphi \in \mathscr{F}$,*

$$A(\varphi) \leqslant v(\varphi)_\mathcal{D}.$$

Definition 10.1.2 *Let $A \in \mathscr{F}_L(\mathscr{F})$, $\varphi \in \mathscr{F}$, $\alpha \in L$.*

(1). φ is said to be α-true in A, if

$$\alpha = \bigwedge\{v(\varphi)_\mathcal{D} \mid \mathcal{D} \textit{ is an interpretation and satisfies } A\}.$$

We write this $A \mid\models_\alpha \varphi$.

(2). If $A \mid\models_\alpha \varphi$, then for any $\beta \leqslant \alpha$, we say that A semantically implies φ by truth value level β and denoted by $A \models_\beta \varphi$.

In the following, $A \mid\models_I \varphi$, $A \models_I \varphi$, $\emptyset \mid\models_\alpha \varphi$ and $\emptyset \models_\beta \varphi$ are simply denoted by $A \mid\models \varphi$, $A \models \varphi$, $\mid\models_\alpha \varphi$ and $\models_\beta \varphi$, respectively. If $\mid\models \varphi$, then φ is called valid.

Corollary 10.1.1 $\models_\alpha \varphi$ *if and only if for any interpretation $\mathcal{D}$, $v(\varphi)_\mathcal{D} \geqslant \alpha$.* □

Corollary 10.1.2 $\mid\models \varphi$ *if and only if* $\models \varphi$. □

By the properties of the lattice implication algebra, it is easy to prove that:

Theorem 10.1.1 *For any $\varphi, \psi \in \mathscr{F}$, $\models \varphi \to \psi$ if and only if for any interpretation $\mathcal{D}$, $v(\varphi)_\mathcal{D} \leqslant v(\psi)_\mathcal{D}$.* □

Theorem 10.1.2 *For any $\varphi, \psi \in \mathscr{F}$, $\alpha \in L$, $A \in \mathscr{F}_L(\mathscr{F})$, $A \models_\alpha \varphi \to \psi$ if and only if*

$$\alpha \otimes v(\varphi)_\mathcal{D} \leqslant v(\psi)_\mathcal{D}$$

holds for any interpretation $\mathcal{D}$ which satisfies A. □

Theorem 10.1.3 *For any $\varphi, \psi, \gamma \in \mathscr{F}$, $x \in X$, $\alpha \in L$,*

(1). $\models \varphi \to (\psi \to \varphi)$;
(2). $\models (\varphi \to (\psi \to \gamma)) \to (\psi \to (\varphi \to \gamma))$;

(3). $\models (\varphi \to \psi) \to ((\psi \to \gamma) \to (\varphi \to \gamma))$;
(4). $\models \varphi \to \varphi$;
(5). $\models \varphi \to I$;
(6). $\models (\varphi \to \psi) \to (\psi' \to \varphi')$;
(7). $\models \varphi \to (\varphi')'$;
(8). $\models (\varphi \wedge \psi) \to \varphi$, $\models (\psi \wedge \varphi) \to \psi$;
(9). $\models (\varphi \to \psi) \to ((\varphi \to \gamma) \to (\varphi \to (\psi \wedge \gamma)))$;
(10). $\models \varphi \to (\varphi \vee \psi)$, $\models \psi \to (\varphi \vee \psi)$;
(11). $\models (\varphi \to \gamma) \to ((\psi \to \gamma) \to ((\varphi \vee \psi) \to \gamma))$;
(12). $\models (\varphi \wedge \psi) \to (\varphi' \vee \psi')'$;
(13). $\models \varphi \to (\exists x)\varphi$;
(14). $\models (\forall x)\varphi \to \varphi$;
(15). $\models_\alpha \alpha$;
(16). $\models \varphi \to (\psi \to \varphi \otimes \psi)$;
(17). $\models (\varphi \to (\psi \to \gamma)) \to (\varphi \otimes \psi \to \gamma)$;
(18). $\models (\varphi^m \to (\psi \to \gamma)) \to ((\varphi^n \to \psi) \to (\varphi^{m+n} \to \gamma))$,

where, $\varphi^2 = \varphi \otimes \varphi$, $\varphi^{m+1} = \varphi^m \otimes \varphi$, $\varphi \otimes \psi = (\varphi \to \psi')'$.

Theorem 10.1.4 *For any* $j \in J$, $x_1, \cdots, x_{m_j}, y_1, \cdots, y_{m_j} \in X$,

$$\models (x_1 = y_1) \to ((x_2 = y_2) \to \cdots \to ((x_{m_j} = y_{m_j}) \to (f_{m_j}^{G_j}(x_1, \cdots, x_{m_j}) = f_{m_j}^{G_j}(y_1, \cdots, y_{m_j}))\cdots).$$

Proof. Without losing generality, let $m_j = 2$. It needs to prove:

$$\models (x_1 = y_1) \to ((x_2 = y_2) \to (f_2^{G_j}(x_1, x_2) = f_2^{G_j}(y_1, y_2))).$$

For any interpretation $\mathcal{D}$,

(1). if $\mathcal{D}(x_1) \neq \mathcal{D}(y_1)$, then

$$v(x_1 = y_1)_\mathcal{D} = O$$

and hence $v((x_1 = y_1) \to ((x_2 = y_2) \to (f_2^{G_j}(x_1, x_2) = f_2^{G_j}(y_1, y_2))))_\mathcal{D} = v(x_1 = y_1)_\mathcal{D} \to v((x_2 = y_2) \to (f_2^{G_j}(x_1, x_2) = f_2^{G_j}(y_1, y_2)))_\mathcal{D} = I$;

(2). if $\mathcal{D}(x_1) = \mathcal{D}(y_1)$, consider $\mathcal{D}(x_2)$ and $\mathcal{D}(y_2)$.

a) If $\mathcal{D}(x_2) \neq \mathcal{D}(y_2)$, then

$$v(x_2 = y_2)_\mathcal{D} = O$$

and similarly, we can prove $v((x_1 = y_1) \to ((x_2 = y_2) \to (f_2^{G_j}(x_1, x_2) = f_2^{G_j}(y_1, y_2)))_\mathcal{D}) = I$.

b) If $\mathcal{D}(x_2) = \mathcal{D}(y_2)$, then

$$\begin{aligned} \mathcal{D}(f_2^{G_j}(x_1, x_2)) &= f_{jD}(\mathcal{D}(x_1), \mathcal{D}(x_2)) \\ &= f_{jD}(\mathcal{D}(y_1), \mathcal{D}(y_2)) \\ &= \mathcal{D}(f_2^{G_j}(y_1, y_2)) \end{aligned}$$

and hence

$$\begin{aligned} & v((x_1 = y_1) \to ((x_2 = y_2) \to (f_2^{G_j}(x_1, x_2) = f_2^{G_j}(y_1, y_2))))_{\mathcal{D}} \\ = & v((x_1 = y_1))_{\mathcal{D}} \to v((x_2 = y_2) \to (f_2^{G_j}(x_1, x_2) = f_2^{G_j}(y_1, y_2)))_{\mathcal{D}} \\ = & I. \ \square \end{aligned}$$

Similarly, we can obtain the following theorem.

Theorem 10.1.5 *For any $i \in M$, $x_1, \cdots, x_{m_i}, y_1, \cdots, y_{m_i} \in X$, then*

$$\begin{aligned} \models (x_1 = y_1) \to ((x_2 = y_2) \to \cdots \to ((x_{m_i} = y_{m_i}) \to \\ (p_{m_i}^{(i)}(x_1, \cdots, x_{m_i}) \leftrightarrow p_{m_i}^{(i)}(y_1, \cdots, y_{m_i}))) \cdots), \end{aligned}$$

where $\varphi \leftrightarrow \psi$ is defined as $(\varphi \to \psi) \wedge (\psi \to \varphi)$. $\square$

Definition 10.1.3 *Let $\varphi \in \mathscr{F}$, the free variables occurring in φ be $y_1, \cdots, y_n$. $\forall y_1 \cdots \forall y_n \varphi$ is called the universal closure of φ and denoted by $ucl(\varphi)$.*

Theorem 10.1.6 *For any $\varphi \in \mathscr{F}$, $\alpha \in L$, $\models_\alpha \varphi$ if and only if $\models_\alpha ucl(\varphi)$.*

Proof. If there is no free variable occurring in φ (such formula φ is called a sentence), then the conclusion obviously holds. Otherwise, suppose there is only one free variable y occurring in φ, without losing generality. If $\models_\alpha ucl(\varphi)$, then $\models_\alpha \varphi$ is trivial. If $\models_\alpha \varphi$, then for any interpretation $\mathcal{D}$ and $d \in D$, $v(\varphi)_{\mathcal{D}(y/d)} \geqslant \alpha$ and hence

$$\begin{aligned} v(ucl(\varphi))_{\mathcal{D}} &= v((\forall y)\varphi)_{\mathcal{D}} \\ &= \bigwedge_{d \in D} v(\varphi)_{\mathcal{D}(y/d)} \\ &\geqslant \alpha. \ \square \end{aligned}$$

In the following, let S be the set of all sentences in $LF(X)$.

Theorem 10.1.7 *Suppose $\varphi, \psi \in \mathscr{F}$, x is not a free variable in φ, then*

(1). $\models (\forall x)(\varphi \to \psi) \leftrightarrow (\varphi \to (\forall x)\psi)$;
(2). $\models (\exists x)(\varphi \to \psi) \leftrightarrow (\varphi \to (\exists x)\psi)$;
(3). $\models (\forall x)(\psi \to \varphi) \leftrightarrow ((\exists x)\psi \to \varphi)$;
(4). $\models (\exists x)(\psi \to \varphi) \leftrightarrow ((\forall x)\psi \to \varphi)$.

Proof. Because x is not a free variable in φ, for every interpretation $\mathcal{D}$ and $d \in D$, $v(\varphi)_{\mathcal{D}(x/d)} = v(\varphi)_{\mathcal{D}}$, it follows that

$$\begin{aligned}
v((\exists x)\psi \to \varphi)_{\mathcal{D}} &= v((\exists x)\psi)_{\mathcal{D}} \to v(\varphi)_{\mathcal{D}} \\
&= \bigvee_{d\in D} v(\psi)_{\mathcal{D}(x/d)} \to v(\varphi)_{\mathcal{D}} \\
&= \bigwedge_{d\in D} (v(\psi)_{\mathcal{D}(x/d)} \to v(\varphi)_{\mathcal{D}}) \\
&= \bigwedge_{d\in D} (v(\psi)_{\mathcal{D}(x/d)} \to v(\varphi)_{\mathcal{D}(x/d)}) \\
&= \bigwedge_{d\in D} (v(\psi \to \varphi)_{\mathcal{D}(x/d)}) \\
&= v((\forall x)(\psi \to \varphi))_{\mathcal{D}}.
\end{aligned}$$

The others can be proved similarly. □

Theorem 10.1.8 *For any* $\varphi, \psi \in \mathscr{F}$, $x, y \in X$,

(1). $\models (\forall x)(\forall y)\varphi \leftrightarrow (\forall y)(\forall x)\varphi$;
(2). $\models (\exists x)(\exists y)\varphi \leftrightarrow (\exists y)(\exists x)\varphi$;
(3). $\models (\forall x)(\varphi \wedge \psi) \leftrightarrow ((\forall x)\varphi) \wedge (\forall x)\psi$;
(4). $\models (\exists x)(\varphi \vee \psi) \leftrightarrow ((\exists x)\varphi) \vee (\exists x)\psi$. □

Theorem 10.1.9 *Suppose* $\varphi, \psi \in \mathscr{F}$, x *is not a free variable in* φ, *then*

(1). $\models (\forall x)(\varphi \vee \psi) \leftrightarrow (\varphi \vee (\forall x)\psi)$;
(2). $\models (\forall x)(\varphi \wedge \psi) \leftrightarrow (\varphi \wedge (\forall x)\psi)$;
(3). $\models (\exists x)(\varphi \vee \psi) \leftrightarrow (\varphi \vee (\exists x)\psi)$;
(4). $\models (\exists x)(\varphi \wedge \psi) \leftrightarrow (\varphi \wedge (\exists x)\psi)$. □

Suppose $A \in \mathscr{F}_L(\mathscr{F})$, $\varphi \in \mathscr{F}$. $C(A) \in \mathscr{F}_L(\mathscr{F})$ is defined as:

$$C(A)(\varphi) \triangleq \bigwedge\{v(\varphi)_{\mathcal{D}} \mid \mathcal{D} \text{ is a model of } A\}.$$

Theorem 10.1.10 C *is a closure operation.* □

Definition 10.1.4 *Suppose* $\varphi, \psi \in \mathscr{F}$, φ *is called equivalent to* ψ *and denoted by* $\varphi \approx \psi$, *if* $v(\varphi)_{\mathcal{D}} = v(\psi)_{\mathcal{D}}$ *for any interpretation* $\mathcal{D}$.

Corollary 10.1.3 *For any* $\varphi, \psi, \gamma \in \mathscr{F}$,

(1). $\varphi \to (\psi \to \gamma) \approx \psi \to (\varphi \to \gamma)$;
(2). $\varphi \to \psi \approx \psi' \to \varphi'$;
(3). $(\varphi \to \psi) \to \psi \approx (\psi \to \varphi) \to \varphi$;
(4). $\varphi \to \varphi \approx O \to \varphi \approx \varphi \to I \approx I$;
(5). $\varphi \to O \approx \varphi'$, $I \to \varphi \approx \varphi$;
(6). $(\varphi \vee \psi) \to \gamma \approx (\varphi \to \gamma) \wedge (\psi \to \gamma)$;
(7). $(\varphi \wedge \psi) \to \gamma \approx (\varphi \to \gamma) \vee (\psi \to \gamma)$;
(8). $\varphi \to (\psi \vee \gamma) \approx (\varphi \to \psi) \vee (\varphi \to \gamma)$;
(9). $\varphi \to (\psi \wedge \gamma) \approx (\varphi \to \psi) \vee (\varphi \to \gamma)$;

(10). $(\varphi \to \psi) \to (\gamma \to \psi) \approx \gamma \to (\varphi \vee \psi) \approx (\psi \to \varphi) \to (\gamma \to \varphi)$;
(11). $(\varphi \to \psi) \to (\varphi \to \gamma) \approx (\varphi \wedge \psi) \to \gamma \approx (\psi \to \varphi) \to (\psi \to \gamma)$;
(12). $(\forall x)(\forall y)\varphi \approx (\forall y)(\forall x)\varphi$;
(13). $(\exists x)(\exists y)\varphi \approx (\exists y)(\exists x)\varphi$;
(14). $(\forall x)(\varphi \wedge \psi) \approx (\forall x)\varphi \wedge (\forall x)\psi$;
(15). $(\exists x)(\varphi \vee \psi) \approx (\exists x)\varphi \vee (\exists x)\psi$. □

It is trivial to verify that $\approx$ is a congruence relation on $\mathscr{F}$. Let

$$\bar{\varphi} = \{\psi \mid \psi \in \mathscr{F}, \varphi \approx \psi\},$$
$$\bar{\mathscr{F}} = \mathscr{F}/_{\approx} = \{\bar{\varphi} \mid \varphi \in \mathscr{F}\}.$$

Define operations on $\bar{\mathscr{F}}$ as follows: for any $\bar{\varphi}, \bar{\psi} \in \bar{\mathscr{F}}$,

$$(\bar{\varphi})' = \overline{(\varphi')},$$
$$\bar{\varphi} \wedge \bar{\psi} = \overline{\varphi \wedge \psi},$$
$$\bar{\varphi} \vee \bar{\psi} = \overline{\varphi \vee \psi},$$
$$\bar{\varphi} \to \bar{\psi} = \overline{\varphi \to \psi}.$$

Theorem 10.1.11 *$(\bar{\mathscr{F}}, \vee, \wedge, ', \to)$ is a lattice implication algebra.* □

10.1.4 Syntax

Theorem 10.1.12 *For any $\varphi, \psi, \gamma \in \mathscr{F}$ and $m, n \in \mathbb{N}^+$, the following formulae are valid:*

(1). $\varphi \to I$;
(2). $\varphi \to \varphi$;
(3). $(\varphi \to \psi) \to ((\psi \to \gamma) \to (\varphi \to \gamma))$;
(4). $(\varphi \to (\psi \to \gamma)) \to (\psi \to (\varphi \to \gamma))$;
(5). $(\varphi \wedge \psi) \to \varphi$;
(6). $(\varphi \wedge \psi) \to \psi$;
(7). $(\varphi \to \gamma) \to ((\varphi \to \psi) \to (\varphi \to (\gamma \wedge \psi)))$;
(8). $\varphi \to (\varphi \vee \psi)$;
(9). $\psi \to (\varphi \vee \psi)$;
(10). $(\varphi \to \gamma) \to ((\psi \to \gamma) \to ((\varphi \vee \psi) \to \gamma))$;
(11). $\varphi \to (\psi \to \varphi \otimes \psi)$;
(12). $\varphi \to (\psi \to (\varphi \wedge \psi))$;
(13). $\varphi \to (\psi \to \varphi)$;
(14). $(\varphi^m \to (\psi \to \gamma)) \to ((\varphi^n \to \psi) \to (\varphi^{m+n} \to \gamma))$;
(15). $(\varphi \to (\psi \to \gamma)) \leftrightarrow (\varphi \otimes \psi \to \gamma)$;
(16). $(\varphi \to \psi) \to (\psi' \to \varphi')$;
(17). $\varphi \to (\exists x)\varphi$;
(18). $(\forall x)\varphi \to \varphi$;
(19). $x = x$, *x is an individual variable;*

(20). $(x_1 = y_1) \to ((x_2 = y_2) \to \cdots \to ((x_{m_j} = y_{m_j}) \to (f^{G_j}_{m_j}(x_1, \cdots, x_{m_j}) = f^{G_j}_{m_j}(y_1, \cdots, y_{m_j}))) \cdots)$, *where* $j \in J$, $x_1, \cdots, x_{m_j}$, $y_1, \cdots, y_{m_j}$ *are individual variables;*

(21). $(x_1 = y_1) \to ((x_2 = y_2) \to \cdots \to ((x_{m_i} = y_{m_i}) \to (p^{(i)}_{m_i}(x_1, \cdots, x_{m_i}) \to p^{(i)}_{m_i}(y_1, \cdots, y_{m_i}))) \cdots)$, *where* $j \in J$, $x_1, \cdots, x_{m_j}$, $y_1, \cdots, y_{m_j}$ *are individual variables.* □

Let $A_L \in \mathscr{F}_L(\mathscr{F})$ satisfy the following condition: for any $\varphi \in \mathscr{F}$,

$$A_L(\varphi) = \begin{cases} I, & \varphi \text{ is a formula of forms (1) – (21) in Theorem 10.1.12,} \\ \alpha, & \varphi = \alpha, \alpha \in L, \\ O, & \text{otherwise.} \end{cases}$$

Then A_L is called the axiom of $LF(X)$.

Definition 10.1.5 *Let* $A \in \mathscr{F}_L(\mathscr{F})$, $\varphi \in \mathscr{F}$. *A formal proof* ω *of* φ *from* A *is a finite sequence of the following form:*

$$\omega\colon\ (\varphi_1, \alpha_1), \cdots, (\varphi_n, \alpha_n),$$

where $\varphi_n = \varphi$, *for any* i, $1 \leqslant i \leqslant n$, $\varphi_i \in \mathscr{F}$, $\alpha_i \in L$ *and*

(1). $A_L(\varphi_i) = \alpha_i$, *or*
(2). $A(\varphi_i) = \alpha_i$, *or*
(3). there exist $j, k < i$, *such that* $\varphi_j = \varphi_k \to \varphi_i$ *and* $\alpha_i = \alpha_j \otimes \alpha_k$, *or*
(4). there exist $j < i$ *and* $\alpha \in L$ *such that* $\varphi_i = \alpha \to \varphi_j$ *and* $\alpha_i = \alpha \to \alpha_j$, *or*
(5). there exists $j < i$ *such that* $\varphi_j = \psi_1 \to \psi_2$, $\varphi_i = \psi_1 \to (\forall x)\psi_2$, $\alpha_i = \alpha_j$, x *is free in* ψ_2 *and bound in* ψ_1, *or*
(6). there exists $j < i$ *such that* $\varphi_j = \psi_1 \to \psi_2$, $\varphi_i = (\exists x)\psi_1 \to \psi_2$, $\alpha_i = \alpha_j$, x *is free in* ψ_1 *and bound in* ψ_2.

In Definition 10.1.5, n is called the length of ω and denoted by $l(\omega)$, α_n is called the value of ω and denoted by $\mathrm{val}(\omega)$.

Definition 10.1.6 *Let* $A \in \mathscr{F}_L(\mathscr{F})$, $\varphi \in \mathscr{F}$, $\alpha \in L$, φ *is called an* α*-theorem of* A, *denoted by* $A \vdash_\alpha \varphi$, *if*

$$\alpha \leqslant \bigvee\{\mathrm{val}(\omega) \mid \omega \textit{ is a formal proof of } \varphi \textit{ from } A\}.$$

If α *satisfies*

$$\alpha = \bigvee\{\mathrm{val}(\omega) \mid \omega \textit{ is a formal proof of } \varphi \textit{ fromA}\},$$

then it is denoted by $A \Vdash_\alpha \varphi$.

Theorem 10.1.13 (Soundness) *Suppose* $A \in \mathscr{F}_L(\mathscr{F})$, $\varphi \in \mathscr{F}$, $\alpha \in L$, $\mathrm{supp}\ A \subseteq S$. *If* $\mathcal{D}$ *is a model of* A *and* $A \vdash_\alpha \varphi$, *then* $v(\varphi)_{\mathcal{D}} \geqslant \alpha$.

Proof. For any formal proof ω of φ from A:

$$\omega:\ (\varphi_1, \alpha_1), \cdots, (\varphi_n, \alpha_n),$$

we prove

$$v(\varphi)_{\mathcal{D}^*} \geqslant \text{val}(\omega)$$

by induction over the length of ω for any model $\mathcal{D}^*$ of A.

(1). If $n = 1$, then $A_L(\varphi_n) = \alpha_n$ or $A(\varphi_n) = \alpha_n$. If $A_L(\varphi_n) = \alpha_n$, then

$$\begin{aligned} v(\varphi)_{\mathcal{D}^*} &= v(\varphi_n)_{\mathcal{D}^*} \\ &\geqslant A_L(\varphi_n) \\ &= \alpha_n \\ &= \text{val}(\omega). \end{aligned}$$

If $A(\varphi_n) = \alpha_n$, then

$$\begin{aligned} v(\varphi)_{\mathcal{D}^*} &= v(\varphi_n)_{\mathcal{D}^*} \\ &\geqslant A(\varphi_n) \\ &= \alpha_n \\ &= \text{val}(\omega). \end{aligned}$$

(2). Suppose that the conclusion holds for any formal proof of φ from A whose length is less than n. Then for the formal proof:

$$\omega:\ (\varphi_1, \alpha_1), \cdots, (\varphi_n, \alpha_n),$$

we have

a) if $A_L(\varphi_n) = \alpha_n$ or $A(\varphi_n) = \alpha_n$, then the conclusion can be proved similarly to that of (1).

b) If there exist $i, j < n$ such that $\varphi_i = \varphi_j \to \varphi_n$ and $\alpha_n = \alpha_i \otimes \alpha_j$, then by inductive hypothesis, $v(\varphi_j)_{\mathcal{D}^*} \geqslant \alpha_j$, $v(\varphi_i)_{\mathcal{D}^*} \geqslant \alpha_i$ and

$$\begin{aligned} v(\varphi_j)_{\mathcal{D}^*} \to v(\varphi_n)_{\mathcal{D}^*} &= v(\varphi_j \to \varphi_n)_{\mathcal{D}^*} \\ &= v(\varphi_i)_{\mathcal{D}^*} \\ &\geqslant \alpha_i, \end{aligned}$$

hence

$$\begin{aligned} v(\varphi)_{\mathcal{D}^*} &= v(\varphi_n)_{\mathcal{D}^*} \geqslant \alpha_i \otimes \alpha_j \\ &= \alpha_n \\ &= \text{val}(\omega). \end{aligned}$$

c) If there exist $i < n$, $\alpha_0 \in L$ such that $\varphi_n = \alpha_0 \to \varphi_i$ and $\alpha_n = \alpha_0 \to \alpha_i$, then by inductive hypothesis, $v(\varphi_i)_{\mathcal{D}^*} \geqslant \alpha_i$ and

$$\begin{aligned} v(\varphi)_{\mathcal{D}^*} &= v(\varphi_n)_{\mathcal{D}^*} \\ &= v(\alpha_0 \to \varphi_i)_{\mathcal{D}^*} \\ &= \alpha_0 \to v(\varphi_i)_{\mathcal{D}^*} \\ &\geqslant \alpha_0 \to \alpha_i \\ &= \alpha_n \\ &= \mathrm{val}(\omega). \end{aligned}$$

d) If there exists $i < n$ such that $\varphi_i = \psi_1 \to \psi_2$, $\varphi_n = \psi_1 \to (\forall x)\psi_2$, $\alpha_n = \alpha_i$, x is free in ψ_2 and bound in ψ_1, then by inductive hypothesis, $v(\psi_1 \to \psi_2)_{\mathcal{D}^*} \geqslant \alpha_i$. For any $d \in D^*$, $\mathcal{D}^*(x/d)$ is also a model of A, so

$$v(\psi_1 \to \psi_2)_{\mathcal{D}^*(x/d)} \geqslant \alpha_i,$$

and hence

$$\begin{aligned} v(\varphi)_{\mathcal{D}^*} &= v(\varphi_n)_{\mathcal{D}^*} \\ &= v(\psi_1 \to (\forall x)\psi_2)_{\mathcal{D}^*} \\ &= v(\psi_1)_{\mathcal{D}^*} \to \bigwedge_{d\in D^*} v(\psi_2)_{\mathcal{D}^*(x/d)} \\ &= \bigwedge_{d\in D^*} (v(\psi_1)_{\mathcal{D}^*} \to v(\psi_2)_{\mathcal{D}^*(x/d)}) \\ &= \bigwedge_{d\in D^*} (v(\psi_1)_{\mathcal{D}^*(x/d)} \to v(\psi_2)_{\mathcal{D}^*(x/d)}) \\ &= \bigwedge_{d\in D^*} v(\psi_1 \to \psi_2)_{\mathcal{D}^*(x/d)} \\ &\geqslant \alpha_i \\ &= \alpha_n \\ &= \mathrm{val}(\omega). \end{aligned}$$

e) If there exists $i < n$ such that $\varphi_i = \psi_1 \to \psi_2$, $\varphi_n = (\exists x)\psi_1 \to \psi_2$, $\alpha_n = \alpha_i$, x is free in ψ_1 and bound in ψ_2, then the conclusion can be proved similarly to that of (2)-d.

So we have proved $v(\varphi)_{\mathcal{D}} \geqslant \mathrm{val}(\omega)$ and hence

$$v(\varphi)_{\mathcal{D}} \geqslant \bigvee\{\mathrm{val}(\omega) \mid \omega \text{ is a formal proof of } \varphi \text{ from } A\} \geqslant \alpha. \ \square$$

Corollary 10.1.4 *Let $A \in \mathscr{F}_L(\mathscr{F})$, $\varphi \in \mathscr{F}$, $\alpha \in L$,* supp *$A \subseteq S$. If $A \vdash_\alpha \varphi$, then $A \models_\alpha \varphi$.* □

Corollary 10.1.5 *For any $\varphi \in \mathscr{F}$, $\alpha \in L$, if $\vdash_\alpha \varphi$, then $\models_\alpha \varphi$.* □

Theorem 10.1.14 (Deduction) *Suppose* $A, A^* \in \mathscr{F}_L(\mathscr{F})$, $\varphi \in S$, $\psi \in \mathscr{F}$, $\alpha \in L$. $A^* = A \cup \{I/\varphi\}$.

(1). If $A \vdash_\alpha \varphi \to \psi$, *then* $A^* \vdash_\alpha \psi$.

(2). If ω_ψ *is a formal proof of* ψ *from* A^*, *then there exist* $n \in \mathbb{N}$ *and a formal proof* $\omega_{\varphi^n \to \psi}$ *of* $\varphi^n \to \psi$ *from* A *such that*

$$\mathrm{val}(\omega_\psi) \leqslant \mathrm{val}(\omega_{\varphi^n \to \psi}).$$

Proof. (1). For any formal proof $\omega_{\varphi \to \psi}$ of $\varphi \to \psi$ from A:

$$\omega_{\varphi \to \psi}\colon\ (\varphi_1, \alpha_1), \cdots, (\varphi_n, \alpha_n),$$

the sequence

$$(\varphi_1, \alpha_1), \cdots, (\varphi_n, \alpha_n), (\varphi, I), (\psi, \alpha_n)$$

is a formal proof of ψ from A^*, and hence

$$\begin{aligned}\alpha &\leqslant \bigvee\{\mathrm{val}(\omega) \mid \omega \text{ is a formal proof of } \varphi \to \psi \text{ from } A\}\\ &\leqslant \bigvee\{\mathrm{val}(\omega^*) \mid \omega^* \text{ is a formal proof of } \psi \text{ from } A^*\},\end{aligned}$$

i.e., $A^* \vdash_\alpha \psi$.

(2). Suppose

$$\omega_\psi\colon\ (\psi_1, \alpha_1), \cdots, (\psi_m, \alpha_m)$$

is a formal proof of ψ from A^*, $\psi_m = \psi$, $\mathrm{val}(\omega_\psi) = \alpha_m$. We prove by induction over $l(\omega_\psi)$ that there exist $n \in \mathbb{N}$ and a formal proof $\omega_{\varphi^n \to \psi}$ of $\varphi^n \to \psi$ from A such that

$$\mathrm{val}(\omega_\psi) \leqslant \mathrm{val}(\omega_{\varphi^n \to \psi}).$$

If $m = 1$, then $A_L(\psi_m) = \alpha_m$ or $A^*(\psi_m) = \alpha_m$. If $A_L(\psi_m) = \alpha_m$, then the sequence

$$\begin{aligned}&(\psi_m, \alpha_m),\\ &(\psi_m \to (\varphi \to \psi_m), I),\\ &(\varphi \to \psi_m, \alpha_m)\end{aligned}$$

is a formal proof as required. If $A^*(\psi_m) = \alpha_m$, $\psi_m \neq \varphi$, then

$$A(\psi_m) = A^*(\psi_m) = \alpha_m,$$

and above sequence is also a formal proof as required. If $A^*(\psi_m) = \alpha_m$, $\psi_m = \varphi$, then the sequence $(\varphi \to \psi_m, I)$ is a formal proof as required. Suppose the conclusion holds for any formal proof whose length is less than m, then for

$$\omega_\psi\colon\ (\psi_1, \alpha_1), \cdots, (\psi_m, \alpha_m),$$

we have:

a) If $A_L(\psi_m) = \alpha_m$ or $A^*(\psi_m) = \alpha_m$, then the conclusion is obvious.
b) If there exist $i, j < m$ such that $\psi_i = \psi_j \to \psi_m$ and $\alpha_m = \alpha_i \otimes \alpha_j$, then there exist $s, t \in \mathbb{N}$ such that $(\varphi_1, a_1), \cdots, (\varphi_{m_1}, a_{m_1})$ is a formal proof of $\varphi^s \to \psi_i$ from A with $\alpha_i \leqslant a_{m_1}$ and $(\psi_1, b_1), \cdots, (\psi_{m_2}, b_{m_2})$ is a formal proof of $\varphi^t \to \psi_j$ from A with $\alpha_j \leqslant b_{m_2}$. It follows that the sequence

$$\begin{aligned}
&(\varphi_1, a_1), \cdots, (\varphi_{m_1}, a_{m_1}), (\psi_1, b_1), \cdots, (\psi_{m_2}, b_{m_2}),\\
&((\varphi^s \to (\psi_j \to \psi_m)) \to ((\varphi^t \to \psi_j) \to (\varphi^{s+t} \to \psi_m)), I),\\
&((\varphi^t \to \psi_j) \to (\varphi^{s+t} \to \psi_m), a_{m_1}),\\
&(\varphi^{s+t} \to \psi_m, a_{m_1} \otimes b_{m_2})
\end{aligned}$$

is a formal proof of $\varphi^{s+t} \to \psi$ from A and

$$\begin{aligned}
a_{m_1} \otimes b_{m_2} &\geqslant \alpha_i \otimes \alpha_j\\
&= \alpha_m\\
&= \mathrm{val}(\omega_\psi).
\end{aligned}$$

c) If there exist $i < m$, $\alpha_0 \in L$ such that $\psi_m = \alpha_0 \to \psi_i$ and $\alpha_m = \alpha_0 \to \alpha_i$, then there exist $s \in \mathbb{N}$ and a formal proof $(\varphi_1, a_1), \cdots, (\varphi_t, a_t)$ of $\varphi^s \to \psi_i$ from A such that $\alpha_i \leqslant a_t$, it follows that

$$\begin{aligned}
&(\varphi_1, a_1), \cdots, (\varphi_t, a_t),\\
&(\alpha_0 \to (\varphi^s \to \psi_i), \alpha_0 \to a_t),\\
&((\alpha_0 \to (\varphi^s \to \psi_i)) \to (\varphi^s \to (\alpha_0 \to \psi_i)), I),\\
&(\varphi^s \to (\alpha_0 \to \psi_i), \alpha_0 \to a_t)
\end{aligned}$$

is a formal proof of $\varphi^s \to \psi_m$ from A and

$$\begin{aligned}
\alpha_0 \to a_t &\geqslant \alpha_0 \to \alpha_i\\
&= \alpha_m\\
&= \mathrm{val}(\omega_\psi).
\end{aligned}$$

d) If there exists $i < m$ such that $\psi_i = p \to q$, $\psi_m = p \to (\forall x)q$, $\alpha_m = \alpha_i$, x is free in q and bound in p, then there exist $s \in \mathbb{N}$ and a formal proof $(p_1, a_1), \cdots, (p_t, a_t)$ of $\varphi^s \to \psi_i$ from A such that $\alpha_i \leqslant a_t$, so the sequence

$$\begin{aligned}
&(p_1, a_1), \cdots, (p_t, a_t),\\
&((\varphi^s \to (p \to q)) \to (\varphi^s \otimes p \to q), I),\\
&(\varphi^s \otimes p \to q, a_t),\\
&(\varphi^s \otimes p \to (\forall x)q, a_t),\\
&((\varphi^s \otimes p \to (\forall x)q) \to (\varphi^s \to (p \to (\forall x)q)), I),\\
&(\varphi^s \to (p \to (\forall x)q), a_t)
\end{aligned}$$

is a formal proof of $\varphi^s \to \psi_m$ from A and

$$a_t \geqslant \alpha_i = \alpha_m = \mathrm{val}(\omega_\psi).$$

e) If there exists $i < m$ such that $\psi_i = p \to q$, $\psi_m = (\exists x)p \to q$, $\alpha_m = \alpha_i$, x is free in p and bound in q, then there exist $s \in \mathbb{N}$ and a formal proof $(p_1, a_1), \cdots, (p_t, a_t)$ of $\varphi^s \to \psi_i$ from A such that $\alpha_i \leqslant a_t$, it follows that

$$\begin{aligned}
&(p_1, a_1), \cdots, (p_t, a_t),\\
&((\varphi^s \to (p \to q)) \to (p \to (\varphi^s \to q)), I),\\
&(p \to (\varphi^s \to q), a_t),\\
&((\exists x)p \to (\varphi^s \to q), a_t),\\
&(((\exists x)p \to (\varphi^s \to q)) \to (\varphi^s \to ((\exists x)p \to q)), I),\\
&(\varphi^s \to ((\exists x)p \to q), a_t)
\end{aligned}$$

is a formal proof of $\varphi^s \to \psi_m$ from A and $a_t \geqslant \alpha_i = \mathrm{val}(\omega_\psi)$. □

Theorem 10.1.15 *Suppose $A \in \mathscr{F}_L(\mathscr{F})$, $\varphi, \psi \in \mathscr{F}$, $\alpha_0, \alpha, \beta \in L$.*

(1). If $A \vdash_\alpha \varphi \to \psi$, $A \vdash_\beta \varphi$, then $A \vdash_{\alpha\otimes\beta} \psi$.
(2). If $A \vdash_\alpha \varphi$, then $A \vdash_{\alpha_0 \to \alpha} \alpha_0 \to \varphi$.
(3). If $A \vdash_\alpha \varphi \to \psi$, x is free in ψ and bound in φ, then $A \vdash_\alpha \varphi \to (\forall x)\psi$.
(4). If $A \vdash_\alpha \varphi \to \psi$, x is free in φ and bound in ψ, then $A \vdash_\alpha (\exists x)\varphi \to \psi$.

Proof. We only prove (1), others can be proved similarly. For any formal proof $\omega_1 :\ (\varphi_1, \alpha_1), \cdots, (\varphi_m, \alpha_m)$ of $\varphi \to \psi$ from A and any formal proof ω_2: $(\psi_1, \beta_1), \cdots, (\psi_n, \beta_n)$ of φ from A, the sequence

$$(\varphi_1, \alpha_1), \cdots, (\varphi_m, \alpha_m), (\psi_1, \beta_1), \cdots, (\psi_n, \beta_n), (\psi, \alpha_m \otimes \beta_n)$$

is a formal proof of ψ from A, so

$$\alpha_m \otimes \beta_n \leqslant \bigvee\{\mathrm{val}(\omega) \mid \omega \text{ is a formal proof of } \psi \text{ from } A\}.$$

By $A \vdash_\alpha \varphi \to \psi$, $A \vdash_\beta \varphi$, it follows that

$$\alpha \leqslant \bigvee\{\mathrm{val}(\omega_1) \mid \omega_1 \text{ is a formal proof of } \varphi \to \psi \text{ from } A\},$$

$$\beta \leqslant \bigvee\{\mathrm{val}(\omega_2) \mid \omega_2 \text{ is a formal proof of } \varphi \text{ from } A\}.$$

And hence

$$\begin{aligned}
\alpha \otimes \beta \leqslant \bigvee\{&\mathrm{val}(\omega_1) \otimes \mathrm{val}(\omega_2) \mid \omega_1, \omega_2 \text{ are formal proof of}\\
&\qquad \varphi \to \psi \text{ and } \varphi \text{ from } A \text{ respectively}\}\\
\leqslant \bigvee\{&\mathrm{val}(\omega) \mid \omega \text{ is a formal proof of } \psi \text{ from } A\},
\end{aligned}$$

i.e., $A \vdash_{\alpha\otimes\beta} \psi$. □

Definition 10.1.7 *Let $A \in \mathscr{F}_L(\mathscr{F})$. A is called λ-consistent, if*

$$\lambda' = \bigvee\{\alpha \otimes \beta \mid A \vdash_\alpha \varphi, A \vdash_\beta \varphi', \varphi \in \mathscr{F}\}.$$

If A is I-consistent, then A is called consistent for short.

Corollary 10.1.6 *If A is λ-consistent, then*

$$\lambda \leqslant A(\varphi') \to (A(\varphi))'$$

for any $\varphi \in \mathscr{F}$. □

Corollary 10.1.7 *If A is λ-consistent, $\varphi \in \mathscr{F}$, ω_φ is a formal proof of φ from A, $\omega_{\varphi'}$ is a formal proof of φ' from A, then $\lambda \leqslant \mathrm{val}(\omega_{\varphi'}) \to (\mathrm{val}(\omega_\varphi))'$.* □

Corollary 10.1.8 *If A is λ-consistent, then $\lambda' \leqslant \bigvee\{\alpha \mid A \models_\alpha \varphi \otimes \varphi', \varphi \in \mathscr{F}\}$.* □

Theorem 10.1.16 *Let $A \in \mathscr{F}_L(\mathscr{F})$,* supp $A \subseteq S$. *If A is satisfiable, then A is consistent.* □

Proof. Let $\mathcal{D}$ is a model of A, for any $\varphi \in \mathscr{F}$, if $A \mid\!-_\alpha \varphi$, $A \mid\!-_\beta \varphi'$, then $v(\varphi)_\mathcal{D} \geqslant \alpha$, $v(\varphi')_\mathcal{D} \geqslant \beta$, it follows that

$$\alpha \otimes \beta \leqslant v(\varphi)_\mathcal{D} \otimes v(\varphi')_\mathcal{D} = O,$$

$$\lambda' = \bigvee\{\alpha \otimes \beta \mid A \vdash_\alpha \varphi, A \vdash_\beta \varphi', \varphi \in \mathscr{F}\} = O. \ \square$$

10.1.5 Properties of Model Theory

In the following, we discuss the model theory of $LF(X)$ and show the theorems of compactness and Löwenheim-Skölem of $LF(X)$.

Let

$$S^* = \{\varphi \mid \varphi \in S, \text{ for any } \alpha \in L, \alpha \text{ not occurring in } \varphi\}.$$

Definition 10.1.8 *$T = \{T_\lambda \mid \lambda \in L\}$ is called an L-dividing of S^*, if*

(1). for any $\lambda \in L$, $T_\lambda \subseteq S^*$;
(2). for any $\lambda_1, \lambda_2 \in L$, if $\lambda_1 \neq \lambda_2$, then $T_{\lambda_1} \cap T_{\lambda_2} = \emptyset$;
(3). $\bigcup_{\lambda \in L} T_\lambda = S^*$.

Remark 12. Here may be $T_\lambda = \emptyset$, for some $\lambda \in L$.

Definition 10.1.9 *Let T, T^* be L-dividings of S^*. T is called a subdividing of T^* if it satisfies: for any $\lambda \in L$,*

$$T_\lambda \subseteq T^*_\lambda \text{ when } \lambda > O.$$

Definition 10.1.10 *Let T be a dividing of S^*. Then $|S^* \setminus T_O|$ is called the potency of T and denoted by $\|T\|$.*

If $\|T\| < \omega$, then T is called a finite dividing.

Definition 10.1.11 *Let T be a dividing of S^*. T is called satisfiable if there exists an interpretation $\mathcal{D}$ of $LF(X)$ such that*

$$\varphi \in T_\lambda \qquad \text{iff} \qquad v(\varphi)_\mathcal{D} = \lambda,$$

for any $\varphi \in S^*$ and $\lambda > O$, where $v(\varphi)_\mathcal{D}$ is the truth-value of φ under the interpretation $\mathcal{D}$.

L-dividing T is called finite satisfiable if every finite subdividing of T is satisfiable.

Definition 10.1.12 *Suppose κ is an infinite cardinal number, S^* is called κ compact if for any dividing T of S^*, then T is satisfiable when T is finite satisfiable such that $\|T\| \leqslant \kappa$.*

S^* is called compact if S^* is κ compact for any infinite cardinal number κ.

Definition 10.1.13 *Let R be an index set, $r \in R$,and L_r a lattice implication algebras, $L = \prod_{r\in R} L_r$ the product algebra of $\{L_r \mid r \in R\}$. For any $r \in R$, suppose*

$$\mathcal{D}_r = \langle D_r, f_{D_r}, \{P_{iD_r} \mid i \in M\}, \{f_{jD_r} \mid j \in J\}\rangle$$

is an interpretation of $LF(X)$ w.r.t. truth-value field L_r, and define

$$\mathcal{D} = \langle D, f_D, \{P_{iD} \mid i \in M\}, \{f_{jD} \mid j \in J\}\rangle \triangleq \prod_{r\in R} \mathcal{D}_r$$

is the interpretation of $LF(X)$ w.r.t. truth-value field L with the form

(1). $D = \prod_{r\in R} D_r$ is the interpretation field of $\mathcal{D}$, where D_r is the interpretation field of $\mathcal{D}_r$;

(2). $f_D\colon\ X \cup C \longrightarrow D$ satisfies

$$f_D(y)(r) = f_{D_r}(y)$$

for any $y \in X \cup C$, $r \in R$;

(3). for any $i \in M$, $P_{iD}\colon\ D_p^{m_i} \longrightarrow L$ satisfies for any $a_1, \cdots, a_{m_i} \in D_p$ and $r \in R$,

$$P_{iD}(a_1, \cdots, a_{m_i})(r) = P_{iD_r}\left(\alpha_1(r)/d_1(r), \cdots, \alpha_{m_i}(r)/d_{m_i}(r)\right),$$

where $a_t = \alpha_t/d_t$, $\alpha_t \in L \setminus \{O\}$, $d_t \in D$, $1 \leqslant t \leqslant m_i$;

(4). for any $j \in J$, $f_{jD}\colon D_p^{m_j} \longrightarrow D_p$ satisfies: for any $a_1, \cdots, a_{m_j} \in D_p$ and $r \in R$,

$$a_t = \alpha_t/d_t,\ \alpha_t \in L \setminus \{O\},\ d_t \in D,\ 1 \leqslant t \leqslant m_j,$$

then

$$f_{jD}(a_1, \cdots, a_{m_j}) = \alpha/d,$$

where α, d satisfy: for any $r \in R$,

$$\alpha(r)/d(r) = f_{jD_r}\left(\alpha_1(r)/d_1(r), \cdots, \alpha_{m_j}(r)/d_{m_j}(r)\right).$$

In the following, if the lattice-valued logic taking L as its truth values field, then it is called L-valued logic for short.

Theorem 10.1.17 *Let $L = \prod_{r \in R} L_r$ be the lattice implication product algebra of $\{L_r \mid r \in R\}$, κ be an infinite cardinal number. If S^* is κ compact in L_r for any $r \in R$, then S^* is κ compact in L.*

Proof. Suppose S^* is κ compact in L_r for any $r \in R$. Let

$$T = \{T_\lambda \mid \lambda \in L\}$$

be a dividing of S^* in L and finite satisfiable, $\|T\| \leqslant \kappa$. For any $r \in R$, $\beta \in L_r$, let

$$T_\beta^{(r)} = \bigcup \{T_u \mid u \in L \text{ and } u(r) = \beta\},$$

hence, $T^{(r)} = \{T_\beta^{(r)} \mid \beta \in L_r\}$ is obviously a dividing of S^* in L_r, and

$$\begin{aligned}
\|T^{(r)}\| &= \left|\bigcup \{T_\beta^{(r)} \mid \beta \in L_r \text{ and } \beta > O_r\}\right| \\
&= \left|\bigcup \{\bigcup \{T_u \mid u \in L \text{ and } u(r) = \beta\}; \beta \in L_r \text{ and } \beta > O_r\}\right| \\
&= \left|\bigcup \{T_u \mid u \in L \text{ and } u(r) > O_r\}\right| \\
&\leqslant \left|\bigcup \{T_u \mid u \in L \text{ and } u > O\}\right| \\
&\leqslant \kappa.
\end{aligned}$$

For any subdividing $T^{(r)^*}$ of $T^{(r)}$,

$$T^{(r)^*} = \{T_\beta^{(r)^*} \mid \beta \in L_r\},$$

let $T^* = \{T_\lambda^* \mid \lambda \in L\}$, where

$$T_\lambda^* = \begin{cases}
\emptyset, & \lambda > O \text{ and } \lambda(r) = O_r, \\
T_\beta^{(r)^*} \cap T_\lambda, & \lambda > O, \lambda(r) = \beta > O_r, \\
S^* \setminus \bigcup \{T_\beta^{(r)^*} \cap T_\xi \mid \xi \in L, \xi(r) = \beta > O_r\}, & \lambda = O.
\end{cases}$$

For any $\lambda \in L$ and $\lambda > O$, we have $T_\lambda^* \subseteq T_\lambda$ and

$$\bigcup\{T_\lambda^* \mid \lambda > O\} = \bigcup\{T_\beta^{(r)^*} \cap T_\lambda \mid \lambda(r) = \beta > O, \lambda > O\}$$
$$\subseteq \bigcup\{T_\beta^{(r)^*} \mid \beta \in L_r, \beta > O_r\}.$$

Hence T^* is a finite subdividing of T. For T is finite satisfiable, then in the L-valued logic, there exists an interpretation $\mathcal{D}$ of $LF(X)$:

$$\mathcal{D} = \langle D, f_D, \{P_{iD} \mid i \in M\}, \{f_{jD} \mid j \in J\}\rangle,$$

such that $\varphi \in T_\lambda^*$ if and only if $v(\varphi)_\mathcal{D} = \lambda$ for any $\lambda > O$ and $\varphi \in S^*$.

In L_r-valued logic, consider the following interpretation:

$$\mathcal{D}_r = \langle D_r, f_{D_r}, \{P_{iD_r} \mid i \in M\}, \{f_{jD_r} \mid j \in J\}\rangle,$$

where,

a). $D_r = D$;
b). $f_{D_r} = f_D$;
c). for any $j \in J$, $f_{jD_r} = f_{jD}$;
d). for any $i \in M$, $P_{iD_r}:\ D_{r_p}^{m_i} \longrightarrow L_r$, satisfies: for any $a_1, \cdots, a_{m_i} \in D_{r_p}$,

$$P_{iD_r}(a_1, \cdots, a_{m_i}) = P_{iD}(a_1, \cdots, a_{m_i})(r),$$

then $v(\varphi)_{\mathcal{D}_r} = v(\varphi)_\mathcal{D}(r)$ for any $\varphi \in S^*$.

In fact, for any term t, $\mathcal{D}(t) = \mathcal{D}_r(t)$ is obvious.

(1). If $\varphi = P_{m_i}^{(i)}(t_1, \cdots, t_{m_i})$, $t_1, \cdots, t_{m_i}$ are terms, then

$$\begin{aligned} v(\varphi)_{\mathcal{D}_r} &= P_{iD_r}(\mathcal{D}_r(t_1), \cdots, \mathcal{D}_r(t_{m_i})) \\ &= P_{iD_r}(\mathcal{D}(t_1), \cdots, \mathcal{D}(t_{m_i})) \\ &= P_{iD}(\mathcal{D}(t_1), \cdots, \mathcal{D}(t_{m_i})) \\ &= v(\varphi)_\mathcal{D}(r). \end{aligned}$$

(2). If $\varphi = \gamma \vee \psi$, then

$$\begin{aligned} v(\varphi)_{\mathcal{D}_r} &= v(\gamma \vee \psi)_{\mathcal{D}_r} \\ &= v(\gamma)_{\mathcal{D}_r} \vee v(\psi)_{\mathcal{D}_r} \\ &= v(\gamma)_\mathcal{D}(r) \vee v(\psi)_\mathcal{D}(r) \\ &= v(\gamma \vee \psi)_\mathcal{D}(r) \\ &= v(\varphi)_\mathcal{D}(r). \end{aligned}$$

(3). If $\varphi = \gamma \wedge \psi$ or $\varphi = \gamma \to \psi$, $\varphi = \gamma'$, it can be proved similarly to (2).
(4). If $\varphi = (\forall x)\psi$, then

$$\begin{aligned}
v(\varphi)_{\mathcal{D}} &= \bigwedge_{d\in D_r} v(\psi)_{\mathcal{D}_r(x/d)} \\
&= \bigwedge_{d\in D_r} v(\psi(x/d))_{\mathcal{D}_r} \\
&= \bigwedge_{d\in D} v(\psi(x/d))_{\mathcal{D}}(r) \\
&= \bigwedge_{d\in D} v(\psi)_{\mathcal{D}(x/d)}(r) \\
&= v(\psi)_{\mathcal{D}}(r).
\end{aligned}$$

(5). If $\varphi = (\exists x)\psi$, it is similar to (4).

For any $\beta \in L_r$, $\beta > O_r$,

$$\begin{aligned}
&\bigcup\{T_u^* \mid u \in L \text{ and } u(r) = \beta\} \\
=&\bigcup\{T_\beta^{(r)^*} \cap T_u \mid u \in L \text{ and } u(r) = \beta\} \\
=&\, T_\beta^{(r)^*} \cap (\bigcup\{T_u \mid u \in L \text{ and } u(r) = \beta\}) \\
=&\, T_\beta^{(r)^*} \cap T_\beta^{(r)} \\
=&\, T_\beta^{(r)^*},
\end{aligned}$$

hence for any $\varphi \in S^*$,

$$\begin{aligned}
\varphi \in T_\beta^{(r)^*} \quad &\text{iff} \quad \text{there exists } u \in L, u(r) = \beta, \text{ such that } \varphi \in T_u^* \\
&\text{iff} \quad \text{there exists } u \in L, u(r) = \beta, \text{ such that } v(\varphi)_{\mathcal{D}} = u \\
&\text{iff} \quad \text{there exists } u \in L, v(\varphi)_{\mathcal{D}_r} = u(r) = \beta,
\end{aligned}$$

i.e., $T^{(r)^*}$ is satisfiable. So $T^{(r)}$ is finite satisfiable. Moreover, $T^{(r)}$ is satisfiable for S^* being κ compact in L_r-valued logic and $\|T^{(r)}\| \leqslant \kappa$.

In L_r-valued logic, let

$$\bar{\mathcal{D}}_r = \langle \bar{D}_r, f_{\bar{D}_r}, \{P_{i\bar{D}_r} \mid i \in M\}, \{f_{j\bar{D}_r} \mid j \in J\}\rangle$$

be an interpretation of $LF(X)$ such that

$$v(\varphi)_{\bar{\mathcal{D}}_r} = \beta \text{ iff } \varphi \in T_\beta^{(r)}$$

for any $\varphi \in S^*$, $\beta \in L_r$, and $\beta > O_r$.

In L_r-valued logic, consider the interpretation $\bar{\mathcal{D}} = \prod_{r\in R} \bar{\mathcal{D}}_r$ in a sense of Definition 10.1.13.

For any $\lambda \in L$, $\lambda > O$ and $r \in R$, it is obvious that

$$T_\lambda \subseteq \bigcup\{T_u \mid u \in L, u(r) = \lambda(r)\} = T_{\lambda(r)}^{(r)}.$$

For any $\lambda \in L$, $\lambda > O$, if $\varphi \in T_\lambda$, then $\varphi \in T^{(r)}_{\lambda(r)}$ for any $r \in R$. Hence

$$v(\varphi)_{\bar{\mathcal{D}}}(r) = v(\varphi)_{\bar{\mathcal{D}}_r} = \lambda(r),$$

i.e., $v(\varphi)_{\bar{\mathcal{D}}} = \lambda$.

On the contrary, if $v(\varphi)_{\bar{\mathcal{D}}} = \lambda$, then for arbitrary $r \in R$, there holds

$$v(\varphi)_{\bar{\mathcal{D}}_r} = v(\varphi)_{\bar{\mathcal{D}}}(r) = \lambda(r),$$

and hence $\varphi \in T^{(r)}_{\lambda(r)}$, *i.e.*, $\varphi \in \bigcap T^{(r)}_{\lambda(r)}$.

For $r \in R$, let

$$B_r = \{u \mid u \in L, u(r) = \lambda(r)\}.$$

It follows that

$$\bigcap_{r\in R} B_r = \{u \mid u \in L, u(r) = \lambda(r) \text{ for any } r \in R\} = \{\lambda\}$$

and hence

$$\begin{aligned}\bigcap_{r\in R} T^{(r)}_{\lambda(r)} &= \bigcap_{r\in R} \left(\bigcup\{T_u \mid u \in L, u(r) = \lambda(r)\}\right)\\ &= \bigcap_{r\in R}\bigcup_{u\in B_r} T_u\\ &= T_\lambda.\end{aligned}$$

So, we proved that for any $\varphi \in S^*$, $\lambda \in L$, $\lambda > O$,

$$\varphi \in T_r \text{ iff } v(\varphi)_{\bar{\mathcal{D}}} = \lambda,$$

then T is satisfiable. □

Corollary 10.1.9 *Given the conditions in Theorem. 10.1.17, if S^* is compact in L_r-valued logic for any $r \in R$, then S^* is compact in L-valued logic.* □

Definition 10.1.14 *Let κ be an infinite cardinal number. S^* has the property of κ-Löwenheim-Skölem in L-valued logic, if for any $\varphi \in S^*$, $\lambda \in L$, and $\lambda > O$, there exists an interpretation $\mathcal{D}^*$ such that $v(\varphi)_{\mathcal{D}^*} = \lambda$ and $|\mathcal{D}^*| \leqslant \kappa$ when there exists an interpretation $\mathcal{D}$ such that $v(\varphi)_{\mathcal{D}} = \lambda$.*

Corollary 10.1.10 *Suppose $L = \prod_{r\in R} L_r$ is a lattice implication product algebra of $\{L_r \mid r \in R\}$, κ is an infinite cardinal number. If S^* has the property of κ-Löwenheim-Skölem in L_r-valued logic for any $r \in R$, then S^* holds the property of $\kappa^{|R|}$-Löwenheim-Skölem in L-valued logic. Moreover, if $|R| \leqslant \omega$, then S^* holds property of κ-Löwenheim-Skölem.* □

10.2 Gradational Lattice-Valued First-Order Logic L_{vfl}

In this section, we give the gradational lattice-valued first-order logic L_{vfl} based on lattice implication algebra, which is corresponding to L_{vpl} discussed in Section 9.2.

10.2.1 Language

The symbols in L_{vfl} is composed of:

(1). variable set: $\mathbf{V} = \{x_i \mid i \in \mathbb{N}\}$;
(2). constant set: $\mathbf{C} = \{c_k \mid k \in K\}$, where K is an index set;
(3). function symbols set: $\{f_{m_j} \mid j \in J\}$, where J is an index set and $m_j \in \mathbb{N}^+$, f_{m_j} is an m_j-ary function symbol;
(4). relation symbols set: $\{p_{m_s} \mid s \in M\}$, where M is an index set and $m_s \in \mathbb{N}^+$, p_{m_s} is an m_s-ary relation symbol. Specially, $=$ is the 2-ary identity relation symbols;
(5). logical connectives: $\vee$, $\wedge$, $'$, $\rightarrow$;
(6). quantifiers: $\forall$, $\exists$, $\mathbf{Q}_u$, $u \in U$ ($\mathbf{Q}_u$ is called a generalized quantifier and U is an index set);
(7). auxiliary symbols:), (,,(comma).

The set $\mathbf{T}$ of terms in L_{vfl} is the minimum set $\mathbf{T}^*$ which satisfies the following conditions:

(1). $\mathbf{V} \cup \mathbf{C} \subseteq \mathbf{T}^*$;
(2). for any $j \in J$, if $t_1, \cdots, t_{m_j} \in \mathbf{T}^*$, then $f_{m_j}(t_1, \cdots, t_{m_j}) \in \mathbf{T}^*$.

The set $\mathcal{F}_f$ of well-formed formulas (wffs) in L_{vfl} is the minimum set $\mathcal{F}_f^*$ which satisfies the following conditions:

(1). $L \subseteq \mathcal{F}_f^*$;
(2). for any $s \in M$, if $t_1, \cdots, t_{m_s} \in \mathbf{T}$, then $p_{m_s}(t_1, \cdots, t_{m_s}) \in \mathcal{F}_f^*$;
(3). if $\varphi, \psi \in \mathcal{F}_f^*$, then $\varphi \vee \psi$, $\varphi \wedge \psi$, φ', $\varphi \rightarrow \psi \in \mathcal{F}_f^*$;
(4). if $\varphi \in \mathcal{F}_f^*$, then $(\forall x)\varphi$, $(\exists x)\varphi \in \mathcal{F}_f^*$, and for any $u \in U$, $(\mathbf{Q}_u x)\varphi \in \mathcal{F}_f^*$.

A well-formed formula is called a formula for short.

Let $\varphi, \psi \in \mathcal{F}_f$, we also denote

$$\varphi \leftrightarrow \psi = (\varphi \rightarrow \psi) \wedge (\psi \rightarrow \varphi),$$
$$\varphi \otimes \psi = (\varphi \rightarrow \psi')'.$$

10.2.2 Interpretation

An interpretation of symbols of L_{vfl} is a quaternion:

$$\mathcal{D} = (D, f_D, \{p_{sD} \mid s \in M\}, \{f_{jD} \mid j \in J\}),$$

where,

(1). $D(\neq \emptyset)$ is the domain of interpretation;
(2). $f_D\colon\ \mathbf{V} \cup \mathbf{C} \longrightarrow D$ is a function;
(3). for any $j \in J$, $f_{jD}\colon\ D^{m_j} \longrightarrow D$ is a mapping;
(4). for any $s \in M$, $p_{sD}\colon\ D^{m_s} \longrightarrow L$ is a mapping, also called the L-valued m_s-ary relation. Especially $=$ is interpreted as an identical relation.

The interpretation of terms under $\mathcal{D}$ in L_{vfl} is a mapping $\mathcal{D}_{\mathbf{T}}\colon\ \mathbf{T} \longrightarrow D$ such that

$$\mathcal{D}_{\mathbf{T}}(t) = \begin{cases} f_D(t), & t \in \mathbf{V} \cup \mathbf{C}, \\ f_{jD}(\mathcal{D}_{\mathbf{T}}(t_1), \cdots, \mathcal{D}_{\mathbf{T}}(t_{m_j})), & t = f_{m_j}(t_1, \cdots, t_{m_j}). \end{cases} \tag{10.1}$$

The interpretation of wffs under $\mathcal{D}$ in L_{vfl} is a mapping $\mathcal{D}_{\mathcal{F}_f}\colon\ \mathcal{F}_f \longrightarrow L$ such that

(1). $\mathcal{D}_{\mathcal{F}_f}(\alpha) = \alpha$, $\alpha \in L$.
(2). $\mathcal{D}_{\mathcal{F}_f}(p_{m_s}(t_1, \cdots, t_{m_s})) = p_{sD}(\mathcal{D}_T(t_1), \cdots, \mathcal{D}_T(t_{m_s}))$, where, $t_1, \cdots, t_{m_s} \in \mathbf{T}$. Specially, if p_2 is $=$, then

$$\mathcal{D}_{\mathcal{F}_f}(t_1 = t_2) = \begin{cases} I, & \mathcal{D}_T(t_1) = \mathcal{D}_T(t_2), \\ O, & \text{otherwise.} \end{cases}$$

(3). $\mathcal{D}_{\mathcal{F}_f}(\varphi \vee \psi) = \mathcal{D}_{\mathcal{F}_f}(\varphi) \vee \mathcal{D}_{\mathcal{F}_f}(\psi)$.
(4). $\mathcal{D}_{\mathcal{F}_f}(\psi \wedge \psi) = \mathcal{D}_{\mathcal{F}_f}(\varphi) \wedge \mathcal{D}_{\mathcal{F}_f}(\psi)$.
(5). $\mathcal{D}_{\mathcal{F}_f}(\varphi \rightarrow \psi) = \mathcal{D}_{\mathcal{F}_f}(\varphi) \rightarrow \mathcal{D}_{\mathcal{F}_f}(\psi)$.
(6). $\mathcal{D}_{\mathcal{F}_f}(\varphi') = (\mathcal{D}_{\mathcal{F}_f}(\varphi))'$.
(7). $\mathcal{D}_{\mathcal{F}_f}((\forall x)(\varphi) = \bigwedge\limits_{d \in D} \mathcal{D}_{(x/d)\mathcal{F}_f}(\varphi)$.
(8). $\mathcal{D}_{\mathcal{F}_f}((\exists x)\varphi) = \bigvee\limits_{d \in D} \mathcal{D}_{(x/d)\mathcal{F}_f}(\varphi)$.
(9). $\mathcal{D}_{\mathcal{F}_f}((\mathbf{Q}_u x)\varphi) = \underset{d \in D}{Q_u}\, \mathcal{D}_{(x/d)\mathcal{F}_f}(\varphi)$.

$\mathcal{D}_{(x/d)}$ denotes the following interpretation

$$\mathcal{D}_{(x/d)} = (D, f_{D(x/d)}, \{p_{sD} \mid s \in M\}, \{f_{jD} \mid j \in J\}).$$

Here,

$$f_{D(x/d)}\colon\ \mathbf{V} \cup \mathbf{C} \longrightarrow D,$$

$$f_{D(x/d)}(y) = \begin{cases} f_D(y), & y \neq x, \\ d, & y = x. \end{cases}$$

Obviously, $f_{D(x/f_D(x))} = f_D$. Q_u $(u \in U)$ is a generalized operation, *i.e.*, a mapping

$$Q_u\colon\ P(L) \longrightarrow L$$

such that

$$Q_u(\{\alpha\}) = \alpha, \qquad (\alpha \in L),$$
$$\underset{\alpha\in\Gamma}{Q_u}(\alpha \otimes \beta) \leqslant \underset{\alpha\in\Gamma}{Q_u}\alpha \otimes \beta,$$
$$\left(\underset{\alpha\in\Gamma}{Q_u}(\alpha \otimes \beta)'\right)' \leqslant \left(\underset{\alpha\in\Gamma}{Q_u}\alpha'\right)' \otimes \beta,$$

and

$$\bigwedge_{\alpha\in\Gamma} \alpha \leqslant \underset{\alpha\in\Gamma}{Q_u}\alpha \leqslant \bigvee_{\alpha\in\Gamma} \alpha,$$

where $\Gamma \in P(L)$.

In general, for $\{x_1, \cdots, x_n\} \subseteq \mathbf{V}$, the interpretation $\mathcal{D}_{(x_1/d_1,\cdots,x_n/d_n)}$ can be defined as:

$$\mathcal{D}_{(x_1/d_1,\cdots,x_n/d_n)} = (D, f_{D(x_1/d_1,\cdots,x_n/d_n)}, \{p_{sD} \mid s \in M\}, \{f_{jD} \mid j \in J\}),$$

where

$$f_{D(x_1/d_1,\cdots,x_n/d_n)}\colon\ \mathbf{V} \cup \mathbf{C} \longrightarrow D,$$
$$f_{D(x_1/d_1,\cdots,x_n/d_n)}(y) = \begin{cases} f_D(y), & y \notin \{x_1, \cdots, x_n\}, \\ d_i, & y = x_i, i \in \{1, 2, \cdots, n\}. \end{cases}$$

We write

$$\mathcal{I}_h \subseteq \mathcal{I}_H \triangleq \{\mathcal{D}_{\mathcal{F}_f} \mid \mathcal{D}_{\mathcal{F}_f} \text{ is an interpretation of wffs under } \mathcal{D}, \\ \mathcal{D} \text{ is an interpretation of symbols in } L_{vfl}\}.$$

In the following, we assume that $\mathcal{I} \subseteq \mathscr{F}_L(\mathcal{F}_f)$.

10.2.3 Semantics

Definition 10.2.1 *Let*

(1).

$$C_{\mathcal{I}}\colon\ \mathscr{F}_L(\mathcal{F}_f) \longrightarrow \mathscr{F}_L(\mathcal{F}_f),$$
$$X \longmapsto C_{\mathcal{I}}^X,$$

$$C_{\mathcal{I}}^X(\varphi) \triangleq \bigwedge_{T\in\mathcal{I}} (\pi(X \subseteq T) \to T(\varphi)). \tag{10.2}$$

(2).

$$C_{\mathcal{I}}^*\colon\ \mathscr{F}_L(\mathcal{F}_f) \longrightarrow \mathscr{F}_L(\mathcal{F}_f),$$
$$X \longmapsto C_{\mathcal{I}}^{*X},$$

$$C_{\mathcal{I}}^{*X}(\varphi) \triangleq \bigwedge\{T(\varphi) \mid \pi(X \subseteq T) = I, T \in \mathcal{I}\}. \tag{10.3}$$

In Eqs. (10.2) *and* (10.3),

$$\pi(X \subseteq T) = \bigwedge_{\varphi \in \mathcal{F}_f} (X(\varphi) \to T(\varphi)). \tag{10.4}$$

Obviously, Eq. (10.3) is a special case of Eq. (10.2).

Definition 10.2.2 *Let $X \in \mathscr{F}_L(\mathcal{F}_f)$, $\varphi \in \mathcal{F}_f$, $\alpha \in L$.*

(1). A formula φ is α-true in X (w.r.t. $\mathcal{I}$), denoted by $X \Vdash_\alpha^{\mathcal{I}} \varphi$, if

$$C_{\mathcal{I}}^X(\varphi) = \alpha. \tag{10.5}$$

(2). A formula φ is said to be semantically inferred from X (w.r.t. $\mathcal{I}$) with the degree α, denoted by $X \models_\alpha^{\mathcal{I}} \varphi$, if

$$\alpha \leqslant C_{\mathcal{I}}^X(\varphi). \tag{10.6}$$

It is obvious that

(1). If $X \Vdash_\alpha^{\mathcal{I}} \varphi$, then $X \models_\alpha^{\mathcal{I}} \varphi$.
(2). Let $\alpha_1 \leqslant \alpha_2$, if $X \models_{\alpha_2}^{\mathcal{I}} \varphi$, then $X \models_{\alpha_1}^{\mathcal{I}} \varphi$.

Specially, we write $X \Vdash_I^{\mathcal{I}} \varphi$, $X \models_I^{\mathcal{I}} \varphi$, $\emptyset \Vdash_\alpha^{\mathcal{I}} \varphi$, $\emptyset \models_\alpha^{\mathcal{I}}$ as $X \Vdash^{\mathcal{I}} \varphi$, $X \models^{\mathcal{I}} \varphi$ $\Vdash_\alpha^{\mathcal{I}} \varphi$, $\models_\alpha^{\mathcal{I}} \varphi$, respectively.

Theorem 10.2.1 *In L_{vfl}, we have*

(1). Let $X_1, X_2 \in \mathscr{F}_L(\mathcal{F}_f)$, $\alpha \in L$.
 a) If $X_1 \models_{\alpha_1}^{\mathcal{I}} \varphi$, $X_2 \models_{\alpha_2}^{\mathcal{I}} \varphi$, then $X_1 \cup X_2 \models_{\alpha_1 \vee \alpha_2}^{\mathcal{I}} \varphi$.
 b) If $X_1 \cap X_2 \models_\alpha^{\mathcal{I}} \varphi$, then $X_1 \models_\alpha^{\mathcal{I}} \varphi$, $X_2 \models_\alpha^{\mathcal{I}} \varphi$.
(2). Let $\mathcal{I}_1 \subseteq \mathcal{I}_2 \subseteq \mathscr{F}_L(\mathcal{F}_f)$, $\varphi \in \mathcal{F}_f$, $X \in \mathscr{F}_L(\mathcal{F}_f)$, $\alpha \in L$. If $X \models_\alpha^{\mathcal{I}_2} \varphi$, then $X \models_\alpha^{\mathcal{I}_1} \varphi$.
(3). Let $\varphi_1, \varphi_2 \in \mathcal{F}_f$, $X \in \mathscr{F}_L(\mathcal{F}_f)$, $\alpha_1, \alpha_2, \alpha \in L$.
 a) If $X \models_{\alpha_1}^{\mathcal{I}_h} \varphi_1$, $X \models_{\alpha_2}^{\mathcal{I}_h} \varphi_2$, then $X \models_{\alpha_1 \vee \alpha_2}^{\mathcal{I}_h} \varphi_1 \vee \varphi_2$.
 b) If $X \models_\alpha^{\mathcal{I}_h} \varphi_1 \wedge \varphi_2$, then $X \models_\alpha^{\mathcal{I}_h} \varphi_1$, $X \models_\alpha^{\mathcal{I}_h} \varphi_2$.
(4). Let $\varphi \in \mathcal{F}_f$, $X \in \mathscr{F}_L(\mathcal{F}_f)$, $\alpha \in L$.
 a) If $X \models_\alpha^{\mathcal{I}_h} O$, then $\alpha \leqslant \bigvee_{\varphi \in \mathcal{F}_f} X(\varphi)$.
 b) $X \Vdash^{\mathcal{I}_h} I$.
(5). Let $\varphi \in \mathcal{F}_f$, $X, X_1, X_2 \in \mathscr{F}_L(\mathcal{F}_f)$, $\alpha \in L$.
 a) $X \models_{X(\varphi)}^{\mathcal{I}} \varphi$.
 b) If $X_1 \subseteq X_2$ and $X_1 \models_\alpha^{\mathcal{I}} \varphi$, then $X_2 \models_\alpha^{\mathcal{I}} \varphi$.
 c) $X \Vdash_\alpha^{\mathcal{I}} \varphi$ if and only if $C_{\mathcal{I}}^X \Vdash_\alpha^{\mathcal{I}} \varphi$.
(6). Let $\alpha \in L$, $T \in \mathcal{I}$. Then for any $\varphi \in \mathcal{F}_f$, $T \Vdash_\alpha^{\mathcal{I}} \varphi$ if and only if $\alpha = T(\varphi)$.
□

Remark 13. $C_{\mathcal{I}}$ is a closure operation.

Proof. We only prove (4)-a, (5) and (6), others can be proved in the same way. Since

$$\begin{aligned}
C^X_{\mathcal{I}_h}(O) &= \bigwedge_{T\in\mathcal{I}_h}\left(\bigwedge_{\varphi\in\mathcal{F}_f}(X(\varphi)\to T(\varphi))\to T(O)\right)\\
&= \bigwedge_{T\in\mathcal{I}_h}\left(\bigwedge_{\varphi\in\mathcal{F}_f}(X(\varphi)\to T(\varphi))\right)'\\
&= \bigwedge_{T\in\mathcal{I}_h}\bigvee_{\varphi\in\mathcal{F}_f}(X(\varphi)\to T(\varphi))'\\
&= \bigwedge_{T\in\mathcal{I}_h}\bigvee_{\varphi\in\mathcal{F}_f}(X(\varphi)\otimes(T(\varphi))')\\
&\leqslant \bigwedge_{T\in\mathcal{I}_h}\left(\left(\bigvee_{\varphi\in\mathcal{F}_f}X(\varphi)\right)\otimes\left(\bigvee_{\varphi\in\mathcal{F}_f}T(\varphi')\right)\right)\\
&= \bigwedge_{T\in\mathcal{I}_h}\left(\left(\bigvee_{\varphi\in\mathcal{F}_f}X(\varphi)\right)\otimes I\right)\\
&= \bigvee_{\varphi\in\mathcal{F}_f}X(\varphi).
\end{aligned}$$

So (4)-a is verified.

In order to prove (5), we consider:

(1). for any $X\in\mathscr{F}_L(\mathcal{F}_f)$, $X\subseteq C^X_{\mathcal{I}}$. In fact, for any $\varphi\in\mathcal{F}_f$,

$$\begin{aligned}
C^X_{\mathcal{I}}(\varphi) &= \bigwedge_{T\in\mathcal{I}}(\pi(X\subseteq T)\to T(\varphi))\\
&\geqslant \bigwedge_{T\in\mathcal{I}}((X(\varphi)\to T(\varphi))\to T(\varphi))\\
&= \bigwedge_{T\in\mathcal{I}}(X(\varphi)\vee T(\varphi))\\
&\geqslant X(\varphi).
\end{aligned}$$

(2). For any $X_1,X_2\in\mathscr{F}_L(\mathcal{F}_f)$, if $X_1\subseteq X_2$, then $C^{X_1}_{\mathcal{I}}\subseteq C^{X_2}_{\mathcal{I}}$. It follows from (1), that

$$\begin{aligned}
C^{X_1}_{\mathcal{I}} &= C^{X_1\cap X_2}_{\mathcal{I}}\\
&\subseteq C^{X_1}_{\mathcal{I}}\cap C^{X_2}_{\mathcal{I}}\\
&\subseteq C^{X_1}_{\mathcal{I}}\cup C^{X_2}_{\mathcal{I}}\\
&\subseteq C^{X_1\cup X_2}_{\mathcal{I}}\\
&= C^{X_2}_{\mathcal{I}}.
\end{aligned}$$

(3). For any $X \in \mathscr{F}_L(\mathcal{F}_f)$, $C_{\mathcal{I}}^X = C_{\mathcal{I}}^{C_{\mathcal{I}}^X}$. It follows that $C_{\mathcal{I}}^X \subseteq C_{\mathcal{I}}^{C_{\mathcal{I}}^X}$. Now we prove $C_{\mathcal{I}}^X \supseteq C_{\mathcal{I}}^{C_{\mathcal{I}}^X}$. In fact, for any $\varphi \in \mathcal{F}_f$,

$$\begin{aligned}
C_{\mathcal{I}}^{C_{\mathcal{I}}^X}(\varphi) &= \bigwedge_{T_1 \in \mathcal{I}} (\pi(C_{\mathcal{I}}^X \subseteq T_1) \to T_1(\varphi)) \\
&= \bigwedge_{T_1 \in \mathcal{I}} \left(\left(\bigwedge_{\psi \in \mathcal{F}_f} (C_{\mathcal{I}}^X(\psi) \to T_1(\psi)) \right) \to T_1(\varphi) \right) \\
&\leqslant \bigwedge_{T_1 \in \mathcal{I}} \left(\left(\bigwedge_{\psi \in \mathcal{F}_f} ((\pi(X \subseteq T_1) \to T_1(\psi)) \to T_1(\psi)) \right) \to T_1(\varphi) \right) \\
&= \bigwedge_{T_1 \in \mathcal{I}} \left(\left(\bigwedge_{\psi \in \mathcal{F}_f} (\pi(X \subseteq T_1) \vee T_1(\psi)) \right) \to T_1(\varphi) \right) \\
&\leqslant \bigwedge_{T_1 \in \mathcal{I}} \left(\left(\bigwedge_{\psi \in \mathcal{F}_f} \pi(X \subseteq T_1) \right) \to T_1(\varphi) \right) \\
&= \bigwedge_{T \in \mathcal{I}} (\pi(X \subseteq T) \to T(\varphi)) \\
&= C_{\mathcal{I}}^X(\varphi).
\end{aligned}$$

In order to prove (6), we only need to prove that $C_{\mathcal{I}}^T(\varphi) = T(\varphi)$ holds for any $\varphi \in \mathcal{F}_f$. Firstly,

$$\begin{aligned}
C_{\mathcal{I}}^T(\varphi) &= \bigwedge_{T^* \in \mathcal{I}} (\pi(T \subseteq T^*) \to T^*(\varphi)) \\
&\leqslant \pi(T \subseteq T) \to T(\varphi) \\
&= T(\varphi),
\end{aligned}$$

and it follows from (5)-a that $T(\varphi) \leqslant C_{\mathcal{I}}^T(\varphi)$. □

Theorem 10.2.2 *Let $\varphi, \psi \in \mathcal{F}_f$, $\alpha \in L$, $X \in \mathscr{F}_L(\mathcal{F}_f)$.*

(1). $X \models_\alpha^{\mathcal{I}} \varphi$ if and only if $T(\varphi) \geqslant \alpha \otimes \pi(X \subseteq T)$ holds for any $T \in \mathcal{I}$.
(2). $X \Vert\!\models^{\mathcal{I}} \varphi$ if and only if $X \models^{\mathcal{I}} \varphi$.
(3). $X \models^{\mathcal{I}_h} \varphi \to \psi$ if and only if for any $T \in \mathcal{I}_h$, $T(\varphi) \otimes \pi(X \subseteq T) \leqslant T(\psi)$.
(4). $X \models^{\mathcal{I}_h} \varphi \leftrightarrow \psi$, if and only if for any $T \in \mathcal{I}_h$, $T(\varphi) \otimes \pi(X \subseteq T) \leqslant T(\psi)$ and $T(\psi) \otimes \pi(X \subseteq T) \leqslant T(\varphi)$.

Proof. (1). Note that

$$\begin{aligned}
X \models_\alpha^{\mathcal{I}} \varphi \quad &\text{iff} \quad \text{for any } T \in \mathcal{I}, \pi(X \subseteq T) \to T(\varphi) \geqslant \alpha \\
&\text{iff} \quad \text{for any } T \in \mathcal{I}, \alpha \otimes \pi(X \subseteq T) \leqslant T(\varphi).
\end{aligned}$$

(2). Note that

$$X \models\!\!\models^{\mathcal{I}} \varphi \quad \text{iff} \quad C_{\mathcal{I}}^{X}(\varphi) = I$$
$$\text{iff} \quad C_{\mathcal{I}}^{X}(\varphi) \geqslant I,$$

i.e. $X \models^{\mathcal{I}} \varphi$.

(3). Note that

$$X \models^{\mathcal{I}_h} \varphi \to \psi \quad \text{iff} \quad \text{for any } T \in \mathcal{I}_h, \pi(X \subseteq T) \leqslant T(\varphi \to \psi)$$
$$\text{iff} \quad \text{for any } T \in \mathcal{I}_h, T(\varphi) \otimes \pi(X \subseteq T) \leqslant T(\psi).$$

(4). It follows from (3). □

Definition 10.2.3 *Two generalized operations Q_{u_1} and Q_{u_2} are said to be dual, if for any $\Gamma \in P(L)$,*

$$\underset{\alpha\in\Gamma}{Q_{u_1}} \alpha = \left(\underset{\alpha\in\Gamma}{Q_{u_2}} \alpha'\right)'. \tag{10.7}$$

Two generalized quantifiers $\mathbf{Q}_{u_1}$ and $\mathbf{Q}_{u_2}$ are said to be dual, if for any interpretation of L_{vfl}, the corresponding generalized operations Q_{u_1} and Q_{u_2} are dual.

Theorem 10.2.3 *Let $\varphi, \psi, \gamma \in \mathcal{F}_f$, $X \in \mathscr{F}_L(\mathcal{F}_f)$, $\alpha \in L$. We have*

(1). $X \models\!\!\models^{\mathcal{I}_h} \varphi \to (\psi \to \varphi)$.
(2). $X \models\!\!\models^{\mathcal{I}_h} (\varphi \to \psi) \to ((\psi \to \gamma) \to (\varphi \to \gamma))$.
(3). $X \models\!\!\models^{\mathcal{I}_h} (\varphi \to (\psi \to \gamma)) \to (\psi \to (\varphi \to \gamma))$.
(4). $X \models\!\!\models^{\mathcal{I}_h} \varphi \to \varphi$.
(5). $X \models\!\!\models^{\mathcal{I}_h} \varphi \to I$.
(6). $X \models\!\!\models^{\mathcal{I}_h} (\varphi \to \psi) \leftrightarrow (\psi' \to \varphi')$.
(7). $X \models\!\!\models^{\mathcal{I}_h} \varphi \leftrightarrow (\varphi')'$.
(8). $X \models\!\!\models^{\mathcal{I}_h} (\varphi \wedge \psi) \to \varphi$.
(9). $X \models\!\!\models^{\mathcal{I}_h} (\varphi \wedge \psi) \to \psi$.
(10). $X \models\!\!\models^{\mathcal{I}_h} \varphi \to (\varphi \vee \psi)$.
(11). $X \models\!\!\models^{\mathcal{I}_h} \psi \to (\varphi \vee \psi)$.
(12). $X \models\!\!\models^{\mathcal{I}_h} (\varphi \wedge \psi) \leftrightarrow (\psi \wedge \varphi)$.
(13). $X \models\!\!\models^{\mathcal{I}_h} (\varphi \vee \psi) \leftrightarrow (\psi \vee \varphi)$.
(14). $X \models\!\!\models^{\mathcal{I}_h} (\varphi \to \psi) \to ((\varphi \to \gamma) \to (\varphi \to (\psi \wedge \gamma)))$.
(15). $X \models\!\!\models^{\mathcal{I}_h} (\varphi \to \gamma) \to ((\psi \to \gamma) \to ((\varphi \vee \psi) \to \gamma))$.
(16). $X \models\!\!\models^{\mathcal{I}_h} (\varphi \wedge \psi) \leftrightarrow (\varphi' \vee \psi')'$.
(17). $X \models\!\!\models^{\mathcal{I}_h} (\varphi' \to \psi) \leftrightarrow (\psi' \to \varphi)$.
(18). $X \models\!\!\models^{\mathcal{I}_h} \varphi \to (\exists x)\varphi$.
(19). $X \models\!\!\models^{\mathcal{I}_h} (\forall x)\varphi \to \varphi$.
(20). $X \models\!\!\models^{\mathcal{I}_h} (\mathbf{Q}x)\varphi \leftrightarrow ((\bar{\mathbf{Q}}x)\varphi')'$, *where $\mathbf{Q}$ and $\bar{\mathbf{Q}}$ are dual.*
(21). $X \models\!\!\models^{\mathcal{I}_h} (\forall x)(\varphi \leftrightarrow \psi) \to \mathbf{Q}x(\varphi \leftrightarrow \psi)$.

(22). $X \Vdash^{\mathcal{I}_h} \varphi \to (\psi \to (\varphi \otimes \psi))$.
(23). $X \Vdash^{\mathcal{I}_h} (\varphi \to (\psi \to \gamma)) \to ((\varphi \otimes \psi) \to \gamma)$.
(24). $X \Vdash^{\mathcal{I}_h} (\varphi^m \to (\psi \to \gamma)) \to ((\varphi^n \to \psi) \to (\varphi^{m+n} \to \gamma))$.
(25). $X \Vdash^{\mathcal{I}_h} (\varphi \otimes \psi) \to \varphi$.
(26). $X \Vdash^{\mathcal{I}_h}_{\alpha} \alpha$.
(27). $X \Vdash^{\mathcal{I}_h} (x_1 = y_1) \to ((x_2 = y_2) \to \cdots \to ((x_{m_j} = y_{m_j}) \to (f_{m_j}(x_1, \cdots, x_{m_j}) = f_{m_j}(y_1, \cdots, y_{m_j})))\cdots)$.
(28). $X \Vdash^{\mathcal{I}_h} (x_1 = y_1) \to ((x_2 = y_2) \to \cdots \to ((x_{m_i} = y_{m_i}) \to (p_{m_i}(x_1, \cdots, x_{m_i}) \leftrightarrow p_{m_i}(y_1, \cdots, y_{m_i})))\cdots)$.

In (27), (28), $x_1, \cdots, x_{m_i}, y_1, \cdots, y_{m_i}, x_1, \cdots, x_{m_j}, y_1, \cdots, y_{m_j} \in \mathbf{V}$.

Proof. We only prove (18), (19), (21), (22), (26) , (27) and (28), others can be proved in the same way. From Theorem 10.2.1 (2) and (5)-b, we need only to prove the case of $X = \emptyset$ and $\mathcal{I}_h = \mathcal{I}_H$.

For (18).

$$\begin{aligned} C^{\emptyset}_{\mathcal{I}_H}(\varphi \to (\exists x)\varphi) &= \bigwedge_{\mathcal{D}_{\mathcal{F}_f} \in \mathcal{I}_H} \mathcal{D}_{\mathcal{F}_f}(\varphi \to (\exists x)\varphi) \\ &= \bigwedge_{\mathcal{D}_{\mathcal{F}_f} \in \mathcal{I}_H} \left(\mathcal{D}_{\mathcal{F}_f}(\varphi) \to \mathcal{D}_{\mathcal{F}_f}((\exists x)\varphi)\right). \end{aligned}$$

Note that $\mathcal{D}_{\mathcal{F}_f}(\varphi) = \mathcal{D}_{(x/f_D(x))\mathcal{F}_f}(\varphi)$, so

$$\mathcal{D}_{\mathcal{F}_f}(\varphi) \leqslant \bigvee_{d \in D} \mathcal{D}_{(x/d)\mathcal{F}_f}(\varphi) = \mathcal{D}_{\mathcal{F}_f}((\exists x)\varphi).$$

Hence, $C^{\emptyset}_{\mathcal{I}_H}(\varphi \to (\exists x)\varphi) = I$.

For (19).

$$\begin{aligned} \mathcal{D}_{\mathcal{F}_f}(\varphi) &= \mathcal{D}_{(x/f_D(x))\mathcal{F}_f}(\varphi) \\ &\geqslant \bigwedge_{d \in D} \mathcal{D}_{(x/d)\mathcal{F}_f}(\varphi) \\ &= \mathcal{D}_{\mathcal{F}_f}((\forall x)\varphi). \end{aligned}$$

Hence,

$$\begin{aligned} C^{\emptyset}_{\mathcal{I}_H}((\forall x)\varphi \to \varphi) &= \bigwedge_{\mathcal{D}_{\mathcal{F}_f} \in \mathcal{I}_H} \mathcal{D}_{\mathcal{F}_f}((\forall x)\varphi \to \varphi) \\ &= \bigwedge_{\mathcal{D}_{\mathcal{F}_f} \in \mathcal{I}_H} (\mathcal{D}_{\mathcal{F}_f}((\forall x)\varphi) \to \mathcal{D}_{\mathcal{F}_f}(\varphi)) \\ &= I. \end{aligned}$$

(21) follows from the fact that

$$\bigwedge_{\alpha \in \Gamma} \alpha \leqslant Q_{u \atop \alpha \in \Gamma} \alpha.$$

(22) follows from the fact that

$$\begin{aligned}\mathcal{D}_{\mathcal{F}_f}(\varphi \to (\psi \to (\varphi \otimes \psi))) &\geqslant \mathcal{D}_{\mathcal{F}_f}(\varphi \to (\varphi \otimes (\psi \to \psi))) \\ &= v(\varphi \to \varphi) \\ &= I.\end{aligned}$$

For (26).

$$C^{\emptyset}_{\mathcal{I}_H}(\alpha) = \bigwedge_{\mathcal{D}_{\mathcal{F}_f} \in \mathcal{I}_H} \mathcal{D}_{\mathcal{F}_f}(\alpha) = \alpha.$$

For (27). It only needs to prove the case of $m_j = 2$. For any interpretation $\mathcal{D}$: if $\mathcal{D}_{\mathbf{T}}(x_1) \neq \mathcal{D}_{\mathbf{T}}(y_1)$, then $\mathcal{D}_{\mathcal{F}_f}(x_1 = y_1) = O$. So,

$$\begin{aligned}&\mathcal{D}_{\mathcal{F}_f}((x_1 = y_1) \to ((x_2 = y_2) \to (f_2(x_1, x_2) = f_2(y_1, y_2)))) \\ = &\mathcal{D}_{\mathcal{F}_f}((x_1 = y_1)) \to \mathcal{D}_{\mathcal{F}_f}((x_2 = y_2) \to (f_2(x_1, x_2) = f_2(y_1, y_2))) \\ = &I.\end{aligned}$$

If $\mathcal{D}_{\mathbf{T}}(x_1) = \mathcal{D}_{\mathbf{T}}(y_1)$, and

(i) if $\mathcal{D}_{\mathbf{T}}(x_2) \neq \mathcal{D}_{\mathbf{T}}(y_2)$, then $\mathcal{D}_{\mathcal{F}_f}(x_2 = y_2) = O$. So,

$$\begin{aligned}&\mathcal{D}_{\mathcal{F}_f}((x_1 = y_1) \to ((x_2 = y_2) \to (f_2(x_1, x_2) = f_2(y_1, y_2)))) \\ = &I \to (O \to \mathcal{D}_{\mathcal{F}_f}(f_2(x_1, x_2) = f_2(y_1, y_2))) \\ = &I;\end{aligned}$$

(ii) if $\mathcal{D}_{\mathbf{T}}(x_2) = \mathcal{D}_{\mathbf{T}}(y_2)$, then

$$\begin{aligned}&\mathcal{D}_{\mathcal{F}_f}(f_2(x_1, x_2) = f_2(y_1, y_2)) \\ = &p_{2D}(\mathcal{D}_{\mathbf{T}}(f_2(x_1, x_2)), \mathcal{D}_{\mathbf{T}}(f_2(y_1, y_2))) \\ = &p_{2D}(f_{2D}(\mathcal{D}_{\mathbf{T}}(x_1), \mathcal{D}_{\mathbf{T}}(x_2)), f_{2D}(\mathcal{D}_{\mathbf{T}}(y_1), \mathcal{D}_{\mathbf{T}}(y_2))) \\ = &p_{2D}(f_{2D}(\mathcal{D}_{\mathbf{T}}(x_1), \mathcal{D}_{\mathbf{T}}(x_2)), f_{2D}(\mathcal{D}_{\mathbf{T}}(x_1), \mathcal{D}_{\mathbf{T}}(x_2))) \\ = &I.\end{aligned}$$

Hence,

$$\mathcal{D}_{\mathcal{F}_f}((x_1 = y_1) \to ((x_2 = y_2) \to (f_2(x_1, x_2) = f_2(y_1, y_2)))) = I.$$

For (28). It also only needs to prove the case of $m_i = 2$. For any interpretation $\mathcal{D}$:

(i) If $\mathcal{D}_{\mathbf{T}}(x_1) \neq \mathcal{D}_{\mathbf{T}}(y_1)$ or $\mathcal{D}_{\mathbf{T}}(x_2) \neq \mathcal{D}_{\mathbf{T}}(y_2)$, then

$$\mathcal{D}_{\mathcal{F}_f}(x_1 = y_1) = O \quad \text{or} \quad \mathcal{D}_{\mathcal{F}_f}(x_2 = y_2) = O.$$

(ii) If $\mathcal{D}_{\mathbf{T}}(x_1) = \mathcal{D}_{\mathbf{T}}(y_1)$ and $\mathcal{D}_{\mathbf{T}}(x_2) = \mathcal{D}_{\mathbf{T}}(y_2)$, then

$$\begin{aligned}\mathcal{D}_{\mathcal{F}_f}(p_2(x_1,x_2)) &= p_{2D}(\mathcal{D}_{\mathbf{T}}(x_1), \mathcal{D}_{\mathbf{T}}(x_2)) \\ &= p_{2D}(\mathcal{D}_{\mathbf{T}}(y_1), \mathcal{D}_{\mathbf{T}}(y_2)) \\ &= \mathcal{D}_{\mathcal{F}_f}(p_2(y_1,y_2)).\end{aligned}$$

And it holds for both of them that

$$\mathcal{D}_{\mathcal{F}_f}(x_1 = y_1) \to (\mathcal{D}_{\mathcal{F}_f}(x_2 = y_2) \to (\mathcal{D}_{\mathcal{F}_f}(p_2(x_1,x_2)) \leftrightarrow \mathcal{D}_{\mathcal{F}_f}(p_2(y_1,y_2)))) = I. \square$$

Definition 10.2.4 *Let $\varphi \in \mathcal{F}_f$, the free variables occurring in φ are $y_1, \cdots, y_n$. The well-formed formula*

$$(\forall y_1) \cdots (\forall y_n)\varphi$$

is called a universal closure of φ, denoted by $ucl(\varphi)$.

Theorem 10.2.4 *Let $\varphi \in \mathcal{F}_f$, $\alpha \in L$. Then*

$$\models_{\alpha}^{\mathcal{I}_H} \varphi \qquad \text{iff} \qquad \models_{\alpha}^{\mathcal{I}_H} ucl(\varphi).$$

Proof. If there exists no free variable in φ, then the assertion is obvious. We can suppose that there exists one free variable x in φ, then $ucl(\varphi) = (\forall x)\varphi$. If $\models_{\alpha}^{\mathcal{I}_H} ucl(\varphi)$, then

$$\begin{aligned}\alpha &\leqslant C_{\mathcal{I}_H}^{\emptyset}(ucl(\varphi)) \\ &= \bigwedge_{\mathcal{D}_{\mathcal{F}_f} \in \mathcal{I}_H} \mathcal{D}_{\mathcal{F}_f}(ucl(\varphi)) \\ &= \bigwedge_{\mathcal{D}_{\mathcal{F}_f} \in \mathcal{I}_H} \bigwedge_{d \in D} \mathcal{D}_{(x/d)\mathcal{F}_f}(\varphi) \\ &\leqslant \bigwedge_{\mathcal{D}_{\mathcal{F}_f} \in \mathcal{I}_H} \mathcal{D}_{(x/f_D(x))\mathcal{F}_f}(\varphi) \\ &= \bigwedge_{\mathcal{D}_{\mathcal{F}_f} \in \mathcal{I}_H} \mathcal{D}_{\mathcal{F}_f}(\varphi).\end{aligned}$$

If $\models_{\alpha}^{\mathcal{I}_H} \varphi$, then for any $\mathcal{D}_{\mathcal{F}_f} \in \mathcal{I}_H$, and for any $d \in D$,

$$\mathcal{D}_{(x/d)\mathcal{F}_f}(\varphi) \geqslant \alpha.$$

Hence,

$$\bigwedge_{\mathcal{D}_{\mathcal{F}_f} \in \mathcal{I}_H} \bigwedge_{d \in D} \mathcal{D}_{(x/d)\mathcal{F}_f}(\varphi) \geqslant \alpha. \square$$

Theorem 10.2.5 *Let $\varphi, \psi \in \mathcal{F}_f$, x do not occur freely in φ, $X \in \mathscr{F}_L(\mathcal{F}_f)$. Then*

(1). $X \models^{\mathcal{I}_h} (\forall x)(\varphi \to \psi) \leftrightarrow (\varphi \to (\forall x)\psi)$.
(2). $X \models^{\mathcal{I}_h} (\exists x)(\varphi \to \psi) \leftrightarrow (\varphi \to (\exists x)\psi)$.
(3). $X \models^{\mathcal{I}_h} (\forall x)(\varphi \to \psi) \leftrightarrow ((\exists x)\psi \to \varphi)$.
(4). $X \models^{\mathcal{I}_h} (\exists x)(\psi \to \varphi) \leftrightarrow ((\forall x)\psi \to \varphi)$.
(5). $X \models^{\mathcal{I}_h} (\varphi \to (\mathbf{Q}x)\varphi) \to (\mathbf{Q}x)(\varphi \to \psi)$.
(6). $X \models^{\mathcal{I}_h} ((\mathbf{Q}_2 x)\psi \to \varphi) \to (\mathbf{Q}_1 x)(\psi \to \varphi)$.

Where $\mathbf{Q}_1$ *and* $\mathbf{Q}_2$ *are dual.*

Proof. In view of Theorem 10.2.1, it only needs to prove the cases of $X = \emptyset$ and $\mathcal{I}_h = \mathcal{I}_H$. We only prove (1), (5) and (6), others can be proved in the same way.

For any interpretation $\mathcal{D}$, since x is not a free variable in φ, so

$$\mathcal{D}_{(x/d)\mathcal{F}_f}(\varphi) = \mathcal{D}_{\mathcal{F}_f}(\varphi)$$

holds for any $d \in D$.

For (1).

$$\begin{aligned}
\mathcal{D}_{\mathcal{F}_f}((\forall x)(\varphi \to \psi)) &= \bigwedge_{d\in D} \mathcal{D}_{(x/d)\mathcal{F}_f}(\varphi \to \psi) \\
&= \bigwedge_{d\in D} (\mathcal{D}_{\mathcal{F}_f}(\varphi) \to \mathcal{D}_{(x/d)\mathcal{F}_f}(\psi)) \\
&= \mathcal{D}_{\mathcal{F}_f}(\varphi) \to \bigwedge_{d\in D} \mathcal{D}_{(x/d)\mathcal{F}_f}(\psi) \\
&= \mathcal{D}_{\mathcal{F}_f}(\varphi \to (\forall x)\psi).
\end{aligned}$$

For (5).

$$\begin{aligned}
\mathcal{D}_{\mathcal{F}_f}((\mathbf{Q}x)(\varphi \to \psi)) &= \underset{d\in D}{Q} (\mathcal{D}_{(x/d)\mathcal{F}_f}(\varphi \to \psi)) \\
&= \underset{d\in D}{Q} (\mathcal{D}_{\mathcal{F}_f}(\varphi) \to \mathcal{D}_{(x/d)\mathcal{F}_f}(\psi)) \\
&= \underset{d\in D}{Q} (\mathcal{D}_{\mathcal{F}_f}(\varphi) \otimes (\mathcal{D}_{(x/d)\mathcal{F}_f}(\psi))')' \\
&\geqslant \left(\mathcal{D}_{\mathcal{F}_f}(\varphi) \otimes \left(\underset{d\in D}{Q} \mathcal{D}_{(x/d)\mathcal{F}_f}(\psi)\right)'\right)' \\
&= \mathcal{D}_{\mathcal{F}_f}(\varphi) \to \underset{d\in D}{Q} (\mathcal{D}_{(x/d)\mathcal{F}_f}(\psi)) \\
&= \mathcal{D}_{\mathcal{F}_f}(\varphi \to (\mathbf{Q}x)\psi).
\end{aligned}$$

For (6).

$$\begin{aligned}
\mathcal{D}_{\mathcal{F}_f}((\mathbf{Q}_1 x)(\psi \to \varphi)) &= \underset{d\in D}{Q_1} (\mathcal{D}_{(x/d)\mathcal{F}_f}(\psi \to \varphi)) \\
&= \underset{d\in D}{Q_1} (\mathcal{D}_{(x/d)\mathcal{F}_f}(\psi) \to \mathcal{D}_{\mathcal{F}_f}(\varphi)) \\
&= \underset{d\in D}{Q_1} (\mathcal{D}_{(x/d)\mathcal{F}_f}(\psi) \otimes (\mathcal{D}_{\mathcal{F}_f}(\varphi))')' \\
&\geqslant \left(\left(\underset{d\in D}{Q_1} (\mathcal{D}_{(x/d)\mathcal{F}_f}(\psi))' \right)' \otimes (\mathcal{D}_{\mathcal{F}_f}(\varphi))' \right)' \\
&= \left(\underset{d\in D}{Q_2}\, \mathcal{D}_{(x/d)\mathcal{F}_f}(\psi) \otimes (\mathcal{D}_{\mathcal{F}_f}(\varphi))' \right)' \\
&= \underset{d\in D}{Q_2}\, \mathcal{D}_{(x/d)\mathcal{F}_f}(\psi) \to \mathcal{D}_{\mathcal{F}_f}(\varphi) \\
&= \mathcal{D}_{\mathcal{F}_f}((\mathbf{Q}_2 x)\psi \to \varphi). \ \square
\end{aligned}$$

Analogous to Theorem 10.2.5, the following theorems can be proved.

Theorem 10.2.6 *Let* $\varphi \in \mathcal{F}_f$, $x, y \in \mathbf{V}$, $X \in \mathscr{F}_L(\mathcal{F}_f)$. *Then*

(1). $X \models^{\mathcal{I}_h} (\forall x)(\forall y)\varphi \leftrightarrow (\forall y)(\forall x)\varphi$.
(2). $X \models^{\mathcal{I}_h} (\exists x)(\exists y)\varphi \leftrightarrow (\exists y)(\exists x)\varphi$. $\square$

Theorem 10.2.7 *Let* $\varphi, \psi \in \mathcal{F}_f$, $x, y \in \mathbf{V}$, $X \in \mathscr{F}_L(\mathcal{F}_f)$. *Then*

(1). $X \models^{\mathcal{I}_h} (\forall x)(\varphi \wedge \psi) \leftrightarrow (\forall x)\varphi \wedge (\forall x)\psi$.
(2). $X \models^{\mathcal{I}_h} (\forall x)(\varphi \vee \psi) \leftrightarrow (\exists x)\varphi \vee (\exists x)\psi$.
(3). $X \models^{\mathcal{I}_h} (\mathbf{Q}_1 x)\varphi \leftrightarrow (\mathbf{Q}_2 x)\varphi$, *where* $\mathbf{Q}_1$ *and* $\mathbf{Q}_2$ *are dual.* $\square$

Theorem 10.2.8 *Let* $\varphi, \psi \in \mathcal{F}_f$, $x \in \mathbf{V}$, *and* x *do not occur freely in* φ, $X \in \mathscr{F}_L(\mathcal{F}_f)$. *Then*

(1). $X \models^{\mathcal{I}_h} (\forall x)(\varphi \vee \psi) \leftrightarrow (\varphi \vee (\forall x)\psi)$.
(2). $X \models^{\mathcal{I}_h} (\forall x)(\varphi \wedge \psi) \leftrightarrow (\varphi \wedge (\forall x)\psi)$.
(3). $X \models^{\mathcal{I}_h} (\exists x)(\varphi \vee \psi) \leftrightarrow (\varphi \vee (\exists x)\psi)$.
(4). $X \models^{\mathcal{I}_h} (\exists x)(\varphi \wedge \psi) \leftrightarrow (\varphi \wedge (\exists x)\psi)$.
(5). $X \models^{\mathcal{I}_h} (\forall x)(\varphi \otimes \psi) \leftrightarrow (\varphi \otimes (\forall x)\psi)$.
(6). $X \models^{\mathcal{I}_h} (\exists x)(\varphi \otimes \psi) \leftrightarrow (\varphi \otimes (\exists x)\psi)$. $\square$

Theorem 10.2.9 *Let* $\varphi, \psi \in \mathcal{F}_f$, x *the unique free variable in* φ, y *the unique free variable in* ψ, $X \in \mathscr{F}_L(\mathcal{F}_f)$, *then*

(1). $X \models^{\mathcal{I}_h} (\forall x)(\forall y)(\varphi \vee \psi) \leftrightarrow ((\forall x)\varphi) \vee ((\forall y)\psi)$,
(2). $X \models^{\mathcal{I}_h} (\exists x)(\exists y)(\varphi \wedge \psi) \leftrightarrow ((\exists x)\varphi) \wedge ((\exists y)\psi)$.

Proof. (1). In view of Theorems 10.2.1 and 10.2.2, it only needs to prove the conclusion under the hypothesis that $X = \emptyset$ and $\mathcal{I}_h = \mathcal{I}_H$. For any interpretation $\mathcal{D}$, because y is not free in φ and x is not free in ψ, it follows that

$$\mathcal{D}_{\mathcal{F}_f}((\forall x)(\forall y)(\varphi \vee \psi)) = \bigwedge_{d_1 \in D} \bigwedge_{d_2 \in D} \mathcal{D}_{(x/d_1, y/d_2)}(\varphi \vee \psi)$$
$$= \mathcal{D}_{\mathcal{F}_f}(((\forall x)\varphi) \vee ((\forall y)\psi)).$$

(2). It can be proved similarly. □

Definition 10.2.5 *Let $\varphi, \psi \in \mathcal{F}_f$. φ and ψ are said to be identically equal, denoted by $\varphi = \psi$, if for any interpretation $\mathcal{D}$,*

$$\mathcal{D}_{\mathcal{F}_f}(\varphi) = \mathcal{D}_{\mathcal{F}_f}(\psi).$$

φ is less than or equal to ψ, denoted by $\varphi \leqslant \psi$, if for any interpretation $\mathcal{D}$,

$$\mathcal{D}_{\mathcal{F}_f}(\varphi) \leqslant \mathcal{D}_{\mathcal{F}_f}(\psi).$$

By Theorem 10.2.2, we have

Corollary 10.2.1 *Let $\varphi, \psi \in \mathcal{F}_f$. Then*

(1). $\models^{\mathcal{I}_H} \varphi \rightarrow \psi$ if and only if $\varphi \leqslant \psi$.
(2). $\models^{\mathcal{I}_H} \varphi \leftrightarrow \psi$ if and only if $\varphi = \psi$. □

Hence, from Theorems 10.2.1–10.2.8, we can obtain the relations between the wffs expressed by $=$ or $\leqslant$. It is easy to prove that $=$ is a congruence relation on $(\mathcal{F}_f, \vee, \wedge, ', \rightarrow)$. And since $L \subseteq \mathcal{F}_f$, we have the following theorem.

Theorem 10.2.10 *$(\mathcal{F}_f/_=, \vee, \wedge, ', \rightarrow)$ is a lattice implication algebra, and $(L, \vee, \wedge, ', \rightarrow, O, I)$ can be embedded into $(\mathcal{F}_f/_=, \vee, \wedge, ', \rightarrow)$. Here, for any $\bar{\varphi}, \bar{\psi} \in \mathcal{F}_f/_=$,*

$$\bar{\varphi} \vee \bar{\psi} = \overline{\varphi \vee \psi},$$
$$\bar{\varphi} \wedge \bar{\psi} = \overline{\varphi \wedge \psi},$$
$$\bar{\varphi} \rightarrow \bar{\psi} = \overline{\varphi \rightarrow \psi},$$
$$(\bar{\varphi})' = \overline{\varphi'}.$$ □

10.2.4 Standardization of Formulae

Definition 10.2.6 *A formula $\varphi \in \mathcal{F}_f$ is called a g-literal if its congruence class $\bar{\varphi} \in \mathcal{F}_f/_=$ contains no connectives $\vee$ or $\wedge$. A conjunction of finite g-literals is said to be a g-phrase and a disjunction of finite g-literals is said to be a g-clause; A conjunction of finite g-clauses is called a g-conjunctive normal form; A disjunction of finite g-phrases is called a g-disjunctive normal form.*

Definition 10.2.7 *Suppose G is a formula in the form*

$$\mathbf{Q}_1 x_1 \cdots \mathbf{Q}_n x_n G^*,$$

where $\mathbf{Q}_1, \cdots, \mathbf{Q}_n$ are quantifiers and G^* is a formula containing no quantifiers. Then

(1). G is said to be a g-prenex normal form.
(2). If G^* is a g-disjunctive normal form, then G is said to be a g-prenex disjunctive normal form.
(3). If G^* is a g-conjunctive normal form, then G is said to be a g-prenex conjunctive normal form.

Definition 10.2.8 *Let a formula G be a g-prenex normal form*

$$(\mathbf{Q}_1 x_1)\cdots(\mathbf{Q}_n x_n)G^*,$$

where each $\mathbf{Q}_i x_i$, $i = 1, 2, \cdots, n$, is either $\forall x_i$ or $\exists x_i$. Then

(1). G is said to be a basic g-prenex normal form.
(2). If G^* is a g-disjunctive normal form, then G is said to be a basic g-prenex disjunctive normal form.
(3). If G^* is a g-conjunctive normal form, then G is said to be a basic g-prenex conjunctive normal form.

Definition 10.2.9 *If a formula contains no quantifiers other than $\forall$ and $\exists$, then it is said to be a basic formula.*

Theorem 10.2.11 *Suppose $X \in \mathscr{F}_L(\mathcal{F}_f)$. If G is a basic formula, then there exists a basic g-prenex normal form $(\mathbf{Q}_1 x_1)\cdots(\mathbf{Q}_n x_n)G^*$ such that*

$$X \mid\models^{\mathcal{T}_H} G \leftrightarrow (\mathbf{Q}_1 x_1)\cdots(\mathbf{Q}_n x_n)G^*. \tag{10.8}$$

Proof. G can be transformed into a basic g-prenex normal form by the following two steps:

Step 1: Rename the bound variables if necessary.

Step 2: Repeatedly using Theorems 10.2.5–10.2.9 to move the quantifier $\forall$ or $\exists$ to the left of the whole formula, thus we can transform a given formula into a basic g-prenex normal form. □

By mutual distributivity of $\vee$ and $\wedge$, each formula can be transformed into a g-disjunctive normal form as well as a g-conjunctive normal form. To study the α-resolution principle based on L_{vfl}, we introduce the following necessary conceptions.

Definition 10.2.10 *Suppose*

$$G = (\mathbf{Q}_1 x_1)\cdots(\mathbf{Q}_n x_n)M$$

is a basic g-prenex conjunctive normal form, the formula G^* obtained by the following steps is called a g-Skölem standard form of G:

(1). If $\mathbf{Q}_r$ is an existential quantifier and there is not any universal quantifier on its left, we choose a new constant c different from other constants occurring in M, replace all x_r appearing in M by c, and then delete $(\mathbf{Q}_r x_r)$ from the prefix.

(2). If $\mathbf{Q}_{k_1}, \cdots, \mathbf{Q}_{k_m}$ *are all the universal quantifiers appearing before* $\mathbf{Q}_r$ *(* $m \geqslant 1$, $1 \leqslant k_1 < \cdots < k_m < r$*), we choose a new m-ary function symbol* f *different from all other functional symbols appearing in* M*, replace all* x_r *in* M *by* $f(x_{k_1}, \cdots, x_{k_m})$ *and then delete* $\mathbf{Q}_r x_r$ *from the prefix.*
(3). Do not repeat (2) until there is no existential quantifier in G*.*

Definition 10.2.11 *Let* $X \in \mathscr{F}_L(\mathcal{F}_f)$*,* $\varphi \in \mathcal{F}_f$*,* $\alpha \in L$*.* φ *is said to be strongly semantically inferred from* X *(w.r.t.* $\mathcal{I}$*) in the degree* α*, denoted by* $X \models^{\mathcal{I}}_{\underset{\cdot}{\alpha}} \varphi$*, if* $C^X_{\mathcal{I}}(\varphi) > \alpha$*.*

In the resolution principle based on classical logic, the unsatisfiability of a formula G is equivalent to that of the Skölem standard form of G. For studying the α-resolution principle of L_{vfl}, we give the following theorem.

Theorem 10.2.12 *Let* $\alpha \in L$*,* $G, G^* \in \mathcal{F}_f$*,* G^* *is a g-Skölem standard form of* G*, then*

$$\models^{\mathcal{I}_H}_{\underset{\cdot}{\alpha}} G' \quad \textit{iff} \quad \models^{\mathcal{I}_H}_{\underset{\cdot}{\alpha}} G^{*\prime}.$$

Proof. We can suppose that G is a basic g-prenex normal form

$$G = (\mathbf{Q}_1 x_1) \cdots (\mathbf{Q}_n x_n) M(x_1, \cdots, x_n).$$

Suppose $\mathbf{Q}_r$ is the first existential quantifier in $\mathbf{Q}_1, \cdots, \mathbf{Q}_n$ (from left to right) and

$$\begin{aligned} G_1 = {} & (\forall x_1) \cdots (\forall x_{r-1})(\mathbf{Q}_{r+1} x_{r+1}) \cdots (\mathbf{Q}_n x_n) \\ & M(x_1, \cdots, x_{r-1}, f(x_1, \cdots, x_{r-1}), x_{r+1}, \cdots, x_n). \end{aligned}$$

In what follows we prove that

$$\models^{\mathcal{I}_H}_{\underset{\cdot}{\alpha}} G' \quad \text{iff} \quad \models^{\mathcal{I}_H}_{\underset{\cdot}{\alpha}} G_1',$$

i.e.,

$$C^{\emptyset}_{\mathcal{I}_H}(G') > \alpha \quad \text{iff} \quad C^{\emptyset}_{\mathcal{I}_H}(G_1') > \alpha,$$

or

$$\bigvee_{\mathcal{D}_{\mathcal{F}_f} \in \mathcal{I}_H} \mathcal{D}_{\mathcal{F}_f}(G) < \alpha' \quad \text{iff} \quad \bigvee_{\mathcal{D}_{\mathcal{F}_f} \in \mathcal{I}_H} \mathcal{D}_{\mathcal{F}_f}(G_1) < \alpha'.$$

(1). If

$$\bigvee_{\mathcal{D}_{\mathcal{F}_f} \in \mathcal{I}_H} \mathcal{D}_{\mathcal{F}_f}(G) < \alpha',$$

and there exists an interpretation $\mathcal{D}$ of G_1 such that $\mathcal{D}_{\mathcal{F}_f} \in \mathcal{I}_H$ and $\mathcal{D}_{\mathcal{F}_f}(G_1) \nless \alpha'$.

a) If $\mathcal{D}_{\mathcal{F}_f}(G_1) \geqslant \alpha'$, then for any $(x_1^0, \cdots, x_{r-1}^0) \in D^{r-1}$, it follows that $f(x_1^0, \cdots, x_{r-1}^0) \in D$ such that

$$\mathcal{D}_{\mathcal{F}_f}((\mathbf{Q}_{r+1}x_{r+1}) \cdots (\mathbf{Q}_n x_n) \\ M(x_1^0, \cdots, x_{r-1}^0, f(x_1^0, \cdots, x_{r-1}^0), x_{r+1}, \cdots, x_n)) \geqslant \alpha'.$$

Because $\mathcal{D}$ is also an interpretation of G, it follows that $\mathcal{D}_{\mathcal{F}_f}(G) \geqslant \alpha'$, which is a contradiction.

b) If $\mathcal{D}_{\mathcal{F}_f}(G_1)//\alpha'$, then it can be proved similarly that $\mathcal{D}_{\mathcal{F}_f}(G_1)//\alpha'$, which is a contradiction.

(2). If $\bigvee_{\mathcal{D}_{\mathcal{F}_f} \in \mathcal{I}_H} \mathcal{D}_{\mathcal{F}_f}(G_1) < \alpha'$ and there exists an interpretation $\mathcal{D}$ of G such that

$$\mathcal{D}_{\mathcal{F}_f} \in \mathcal{I}_H, \mathcal{D}_{\mathcal{F}_f}(G) \nless \alpha'.$$

a) If $\mathcal{D}_{\mathcal{F}_f}(G) \geqslant \alpha'$, then for any $(x_1^0, \cdots, x_{r-1}^0) \in D^{r-1}$, there exists $x_r^0 \in D$ such that

$$\mathcal{D}_{\mathcal{F}_f}(\mathbf{Q}_{r+1}x_{r+1} \cdots \mathbf{Q}_n x_n M(x_1^0, \cdots, x_{r-1}^0, x_r^0, x_{r+1}, \cdots, x_n)) \geqslant \alpha'$$

Extending $\mathcal{D}$ to $\mathcal{D}^*$ such that

$$f(x_1^0, \cdots, x_{r-1}^0) = x_r^0$$

for any $(x_1^0, \cdots, x_{r-1}^0) \in D^{r-1}$, it follows that $\mathcal{D}^*_{\mathcal{F}_f}(G_1) \geqslant \alpha'$, which is a contradiction.

b) If $\mathcal{D}_{\mathcal{F}_f}(G)//\alpha'$, then it can be proved in the same way as (2)-a that $\mathcal{D}^*_{\mathcal{F}_f}(G_1)//\alpha'$.

It follows that

$$\models_{\underset{\cdot}{\alpha}}^{\mathcal{I}_H} G' \quad \text{iff} \quad \models_{\underset{\cdot}{\alpha}}^{\mathcal{I}_H} G_1'.$$

Suppose there are m existential quantifiers in G. Let $G_0 = G$ and G_k be the g-Skölem standard form of G_{k-1} w.r.t. its first existential quantifiers (from left to right), $(k = 1, 2, \cdots, m)$, it follows similarly that

$$\begin{aligned} \models_{\underset{\cdot}{\alpha}}^{\mathcal{I}_H} G_1' \ &\text{iff} \models_{\underset{\cdot}{\alpha}}^{\mathcal{I}_H} G_2', \\ &\text{iff} \models_{\underset{\cdot}{\alpha}}^{\mathcal{I}_H} G_3', \\ &\cdots \\ &\text{iff} \models_{\underset{\cdot}{\alpha}}^{\mathcal{I}_H} G_{m-1}', \\ &\text{iff} \models_{\underset{\cdot}{\alpha}}^{\mathcal{I}_H} G_m'. \end{aligned}$$

It is clear $G_m = G^*$, and hence

$$\models_{\underset{\cdot}{\alpha}}^{\mathcal{I}_H} G' \text{ iff } \models_{\underset{\cdot}{\alpha}}^{\mathcal{I}_H} G^{*\prime}. \ \square$$

By this theorem, the α'-unsatisfiability of G is equivalent to that of G^*, the g-Skölem standard form of G. This provides a foundation for establishing the α-resolution principle in L_{vfl}.

10.2.5 Syntax

Definition 10.2.12 *Let $D_n \subseteq \mathcal{F}_f^n$. A mapping*

$$r_n\colon\ D_n \longrightarrow \mathcal{F}_f$$

is called an n-ary partial operation of $\mathcal{F}_f$, where D_n is the domain of r_n, also denoted by $D_n(r_n)$.

We denote

$$R_n \subseteq \{r_n \mid r_n \text{ is an } n\text{-ary partial operation of } \mathcal{F}_f\},$$

$$\begin{aligned} &T_n \subseteq \{t_n \mid t_n \text{ is an } n\text{-ary truth-valued operation on } L\},\\ &\mathscr{R}_n \subseteq R_n \times T_n,\\ &\mathscr{R} \subseteq \bigcup_{n=0}^{+\infty} \mathscr{R}_n. \end{aligned}$$

If $(r,t) \in \mathscr{R}_n$, then (r,t) is called an n-ary rule of inference in L_{vfl}.

In the following, we give three special rules and five classes of rules in L_{vfl}. Let $\varphi, \psi \in \mathscr{F}_L(\mathcal{F}_f)$, $\sigma, \delta, \theta_0 \in L$, $x \in \mathbf{V}$, $\mathbf{Q}_u (u \in U)$ be generalized quantifier.

Let

$$\begin{aligned} \mathscr{R}^* &= \{(r_2^0, t_2^*), (r_2^*, t_2^*), (r_2^{\triangle}, t_2^*)\} \cup \{(r_1^{\theta_0}, t_1^{\theta_0}) \mid \theta_0 \in L\}\\ &\quad \cup\{(r_1^u, t_1) \mid u \in U\} \cup \{(r_2^u, t_1) \mid u \in U\}\\ &\quad \cup\{(r_3^u, t_1) \mid u \in U\} \cup \{(r_4^u, t_1) \mid u \in U\}\\ &\subseteq \mathscr{R}, \end{aligned}$$

where, the forms of the rules (r_2^0, t_2^*), (r_2^*, t_2^*), $(r_2^{\triangle}, t_2^*)$, $(r_1^{\theta_0}, t_1^{\theta_0})$ are the same as those in Section 9.2, the rule class (r_1^u, t_1) is

$$\begin{cases} r_1^u(\varphi) = (\mathbf{Q}_u x)\varphi,\\ t_1(\sigma) = \sigma. \end{cases} \tag{10.9}$$

The rule class (r_2^u, t_1) is

$$\begin{cases} r_2^u(\varphi \to \psi) = \varphi \to (\mathbf{Q}_u x)\psi, x \text{ is not free in } \varphi,\\ t_1(\sigma) = \sigma. \end{cases} \tag{10.10}$$

The rule class (r_3^u, t_1) is

$$\begin{cases} r_3^u(\varphi \to \psi) = (\mathbf{Q}_u x)\varphi \to \psi, x \text{ is not free in } \psi,\\ t_1(\sigma) = \sigma. \end{cases} \tag{10.11}$$

The rule class (r_4^u, t_1) is

$$\begin{cases} r_4^u((\mathbf{Q}_u x)(\varphi \otimes \psi)) = ((\mathbf{Q}_u x)\varphi) \otimes \psi, x \text{ is not free in } \psi,\\ t_1(\sigma) = \sigma. \end{cases} \tag{10.12}$$

Definition 10.2.13 *Let $\mathbf{Q}_u (u \in U)$ be a generalized quantifier.*

(1). *If for any $\varphi \in D(r_1^u)$, $\models_\alpha^{\mathcal{T}_h} \varphi \to (\mathbf{Q}_u x)\varphi$, then $\mathbf{Q}_u$ is said to be $r_1^u - h$ regular at the level α.*
(2). *If for any $\varphi, \psi \in \mathcal{F}_f$, when $\varphi \to \psi \in D(r_2^u)$, $\models_\alpha^{\mathcal{T}_h} (\varphi \wedge \psi) \to (\mathbf{Q}_u x)\psi$, then $\mathbf{Q}_u$ is said to be $r_2^u - h$ regular at the level α.*
(3). *If for any $\varphi, \psi \in \mathcal{F}_f$, when $\varphi \to \psi \in D(r_3^u)$, $\models_\alpha^{\mathcal{T}_h} (\mathbf{Q}_u x)\varphi \to (\varphi \vee \psi)$, then $\mathbf{Q}_u$ is said to be $r_3^u - h$ regular at the level α.*

Definition 10.2.14 *Let $X \in \mathscr{F}_L(\mathcal{F}_f)$, $(r,t) \in \mathscr{R}_n$, $\alpha \in L$. If*

$$X \circ r \supseteq \alpha \otimes \left(t \circ \prod^n X \right) \tag{10.13}$$

holds in $D_n(r)$, then X is said to be α-I type closed w.r.t. (r,t). If

$$X \circ r \supseteq t \circ \prod^n (\alpha \otimes X) \tag{10.14}$$

holds in $D_n(r)$, then X is said to be α-II type closed w.r.t. (r,t).

If for any $(r,t) \in \mathscr{R}$, X is α-I (α-II) type closed w.r.t. (r,t), then X is said to be α-I (α-II) type closed w.r.t. $\mathscr{R}$.

It can be proved by Definition 10.2.14 that X is I-I type closed w.r.t. (r,t) if and only if X is I-II type closed w.r.t. (r,t).

Similar to Theorem 9.2.1, we can have the following theorem.

Theorem 10.2.13 *Let $X \in \mathscr{F}_L(\mathcal{F}_f)$. If X is α-I (α-II) type closed w.r.t. $\mathscr{R}$, then for any $\beta \in L$, $Y = \beta \to X$ is α-I (α-II) type closed w.r.t. $\mathscr{R}$.* □

Similar to Theorem 9.2.2, we can have the following theorem.

Theorem 10.2.14 *Let $\alpha \in L$. If*

$$\emptyset \neq \mathscr{U} \subseteq \{X \mid X \text{ is } \alpha\text{-}i \text{ type closed w.r.t. } \mathscr{R}\},$$

then $\bigcap_{X \in \mathscr{U}} X$ is α-i type closed w.r.t. $\mathscr{R}$, $i = \mathrm{I}, \mathrm{II}$. □

Similar to Theorem 9.2.3, we can have the following theorem.

Theorem 10.2.15 *If X is α-i type closed w.r.t. $\mathscr{R}$, $i = \mathrm{I}, \mathrm{II}$, then the following properties hold:*

(1). *If X is α-i type closed w.r.t. $\mathscr{R}$ and $\beta \leqslant \alpha$, then X is β-i type closed w.r.t. $\mathscr{R}$.*
(2). *If X is α_j-i ($j = 1,2$) type closed w.r.t. $\mathscr{R}$, then X is $(\alpha_1 \vee \alpha_2)$-I type closed w.r.t. $\mathscr{R}$, and X is $(\alpha_1 \wedge \alpha_2)$-$i$ type closed w.r.t. $\mathscr{R}$. Moreover, X is $(\alpha_1 \otimes \alpha_2)$-$i$ type closed w.r.t. $\mathscr{R}$.*
(3). *If X is $(\alpha \to \beta)$-i type closed w.r.t. $\mathscr{R}$, then X is β-i type closed w.r.t. $\mathscr{R}$ and α'-i type closed w.r.t. $\mathscr{R}$.* □

Theorem 10.2.16 *Let $x \in \mathbf{V}$, $\alpha, \theta \in L$, $\mathbf{Q}_u (u \in U)$ be a generalized quantifier, $i = \mathrm{I}, \mathrm{II}$.*

(1). $\mathbf{Q}_u$ is $r_1^u - h$ regular at the level α, if and only if for any interpretation $\mathcal{D}$, $\mathcal{D}_{\mathcal{F}_f}$ is α-i type closed w.r.t. (r_1^u, t_1).
(2). $\mathbf{Q}_u$ is $r_2^u - h$ regular at the level α, if and only if for any interpretation $\mathcal{D}$, $\mathcal{D}_{\mathcal{F}_f}$ is α-i type closed w.r.t. (r_2^u, t_1).
(3). $\mathbf{Q}_u$ is $r_3^u - h$ regular at the level α, if and only if for any interpretation $\mathcal{D}$, $\mathcal{D}_{\mathcal{F}_f}$ is α-i type closed w.r.t. (r_3^u, t_1).
(4). For any interpretation $\mathcal{D}$, $\mathcal{D}_{\mathcal{F}_f}$ is α-i type closed w.r.t. (r_4^u, t_1).
(5). Let $\alpha \leqslant \bigwedge_{\theta \in L} (\theta \vee \theta')$. Then for any interpretation $\mathcal{D}$, $\mathcal{D}_{\mathcal{F}_f}$ is α-i type closed w.r.t. both (r_2^0, t_2^*) and $(r_2^{\triangle}, t_2^*)$.
(6). For any interpretation $\mathcal{D}$, $\mathcal{D}_{\mathcal{F}_f}$ is α-i type closed w.r.t. both (r_2^*, t_2^*) and $(r_1^{\theta_0}, t_1^{\theta_0})$ $(\theta_0 \in L)$.

Proof. It is obvious that t_1 is an unary truth-valued operation.

(1). For any $\varphi \in D(r_1^u)$, $\models_{\alpha}^{\mathcal{I}_h} \varphi \to (\mathbf{Q}_u x)\varphi$ if and only if for any $\varphi \in D(r_1^u)$,

$$\begin{aligned} \alpha &\leqslant C_{\mathcal{I}_h}^{\emptyset}(\varphi \to (\mathbf{Q}_u x)\varphi) \\ &= \bigwedge_{\mathcal{D}_{\mathcal{F}_f} \in \mathcal{I}_h} \mathcal{D}_{\mathcal{F}_f}(\varphi \to (\mathbf{Q}_u x)\varphi), \end{aligned}$$

if and only if for any $\varphi \in D(r_1^u)$, for any interpretation $\mathcal{D}$, if $\mathcal{D}_{\mathcal{F}_f} \in \mathcal{I}_h$, then

$$\alpha \leqslant \mathcal{D}_{\mathcal{F}_f}(\varphi) \to \mathcal{D}_{\mathcal{F}_f}((\mathbf{Q}_u x)\varphi),$$

if and only if for any $\varphi \in D(r_1^u)$, for any interpretation $\mathcal{D}$, if $\mathcal{D}_{\mathcal{F}_f} \in \mathcal{I}_h$, then

$$\begin{aligned} \alpha \otimes t_1(\mathcal{D}_{\mathcal{F}_f}(\varphi)) &= \alpha \otimes \mathcal{D}_{\mathcal{F}_f}(\varphi) \\ &\leqslant \mathcal{D}_{\mathcal{F}_f}((\mathbf{Q}_u x)\varphi) \\ &= \mathcal{D}_{\mathcal{F}_f}(r_1^u(\varphi)). \end{aligned}$$

Note that $t_1(\alpha \otimes \sigma) = \alpha \otimes t_1(\sigma)$. Hence, it also holds for the case i=II.

(2). For any $\varphi, \psi \in \mathcal{F}_f$, if $\varphi \to \psi \in D(r_2^u)$, then

$$\models_{\alpha}^{\mathcal{I}_h} (\varphi \wedge \psi) \to (\mathbf{Q}_u x)\varphi,$$

if and only if for any $\varphi, \psi \in \mathcal{F}_f$, if $\varphi \to \psi \in D(r_2^u)$, then

$$\begin{aligned} \alpha &\leqslant C_{\mathcal{I}_h}^{\emptyset}((\varphi \wedge \psi) \to (\mathbf{Q}_u x)\varphi) \\ &= \bigwedge_{\mathcal{D}_{\mathcal{F}_f} \in \mathcal{I}_h} \mathcal{D}_{\mathcal{F}_f}((\varphi \wedge \psi) \to (\mathbf{Q}_u x)\varphi), \end{aligned}$$

if and only if for any $\varphi, \psi \in \mathcal{F}_f$ and any interpretation $\mathcal{D}$, if $\varphi \to \psi \in D(r_2^u)$ and $\mathcal{D}_{\mathcal{F}_f} \in \mathcal{I}_h$, then

$$\alpha \leqslant \mathcal{D}_{\mathcal{F}_f}((\varphi \wedge \psi) \to (\mathbf{Q}_u x)\varphi)$$
$$= (\mathcal{D}_{\mathcal{F}_f}(\varphi) \to \mathcal{D}_{\mathcal{F}_f}(\psi)) \to (\mathcal{D}_{\mathcal{F}_f}(\varphi) \to \mathcal{D}_{\mathcal{F}_f}((\mathbf{Q}_u x)\varphi)),$$

if and only if for any $\varphi, \psi \in \mathcal{F}_f$ and any interpretation $\mathcal{D}$, if $\varphi \to \psi \in D(r_2^u)$ and $\mathcal{D}_{\mathcal{F}_f} \in \mathcal{I}_h$, then

$$\begin{aligned}\alpha \otimes t_1(\mathcal{D}_{\mathcal{F}_f}(\varphi \to \psi)) &= \alpha \otimes t_1(\mathcal{D}_{\mathcal{F}_f}(\varphi) \to \mathcal{D}_{\mathcal{F}_f}(\psi))\\ &= \alpha \otimes (\mathcal{D}_{\mathcal{F}_f}(\varphi) \to \mathcal{D}_{\mathcal{F}_f}(\psi))\\ &\leqslant \mathcal{D}_{\mathcal{F}_f}(\varphi) \to \mathcal{D}_{\mathcal{F}_f}((\mathbf{Q}_u x)\varphi)\\ &= \mathcal{D}_{\mathcal{F}_f}(r_2^u(\varphi \to \psi)).\end{aligned}$$

It can be proved in the same way for the case i =II.

(3). Similar to (2).

(4). For any $\varphi, \psi \in \mathcal{F}_f$, if $(\mathbf{Q}_u x)(\varphi \otimes \psi) \in D(r_4^u)$, then

$$\begin{aligned}\mathcal{D}_{\mathcal{F}_f}(r_4^u((\mathbf{Q}_u x)(\varphi \otimes \psi))) &= \mathcal{D}_{\mathcal{F}_f}((\mathbf{Q}_u x)\varphi \otimes \psi)\\ &= \mathcal{D}_{\mathcal{F}_f}((\mathbf{Q}_u x)\varphi) \otimes \mathcal{D}_{\mathcal{F}_f}(\psi)\\ &= \left(\underset{d \in D}{Q_u}\right) \mathcal{D}_{(x/d)\mathcal{F}_f}(\varphi) \otimes \mathcal{D}_{\mathcal{F}_f}(\psi)\\ &\geqslant \left(\underset{d \in D}{Q_u}\right) \mathcal{D}_{(x/d)\mathcal{F}_f}(\varphi \otimes \psi)\\ &= \mathcal{D}_{\mathcal{F}_f}((\mathbf{Q}_u x)(\varphi \otimes \psi))\\ &\geqslant \alpha \otimes t_1(\mathcal{D}_{\mathcal{F}_f}((\mathbf{Q}_u x)(\varphi \otimes \psi)))\\ &= t_1(\alpha \otimes \mathcal{D}_{\mathcal{F}_f}((\mathbf{Q}_u x)(\varphi \otimes \psi))).\end{aligned}$$

(5). For the case w.r.t. (r_2^0, t_2^*). It is easy to prove that t_2^* is a 2-ary truth-valued operation. Then for any interpretation $\mathcal{D}$, $\mathcal{D}_{\mathcal{F}_f}$ is α-i type closed w.r.t. (r_2^0, t_2^*), $i = \mathrm{I}, \mathrm{II}$. In fact, if i =I: for any $\varphi, \psi \in L$,

$$\begin{aligned}\alpha &\leqslant \bigwedge_{\theta \in L} (\theta \vee \theta')\\ &\leqslant \mathcal{D}_{\mathcal{F}_f}(\varphi) \vee (\mathcal{D}_{\mathcal{F}_f}(\varphi))' \vee \mathcal{D}_{\mathcal{F}_f}(\psi)\\ &\leqslant \mathcal{D}_{\mathcal{F}_f}(\varphi) \vee (\mathcal{D}_{\mathcal{F}_f}(\varphi) \to \mathcal{D}_{\mathcal{F}_f}(\psi))\\ &\leqslant (\mathcal{D}_{\mathcal{F}_f}(\varphi) \vee \mathcal{D}_{\mathcal{F}_f}(\psi)) \vee (\mathcal{D}_{\mathcal{F}_f}(\varphi) \to \mathcal{D}_{\mathcal{F}_f}(\psi))\\ &= (\mathcal{D}_{\mathcal{F}_f}(\varphi) \wedge (\mathcal{D}_{\mathcal{F}_f}(\varphi) \to \mathcal{D}_{\mathcal{F}_f}(\psi))) \to \mathcal{D}_{\mathcal{F}_f}(\psi)\\ &= t_2^*(\mathcal{D}_{\mathcal{F}_f}(\varphi), \mathcal{D}_{\mathcal{F}_f}(\varphi \to \psi)) \to \mathcal{D}_{\mathcal{F}_f}(\psi).\end{aligned}$$

Hence,

$$\alpha \otimes t_2^*(\mathcal{D}_{\mathcal{F}_f}(\varphi), \mathcal{D}_{\mathcal{F}_f}(\varphi \to \psi)) \leqslant \mathcal{D}_{\mathcal{F}_f}(\psi) = \mathcal{D}_{\mathcal{F}_f}(r_2^0(\varphi, \varphi \to \psi)).$$

If i=II: note that

$$\alpha \otimes t_2^*(\theta, \beta) = t_2^*(\alpha \otimes \theta, \alpha \otimes \beta),$$

so the case for $i =$ II is the same as $i =$ I.

For the case w.r.t. $(r_2^{\triangle}, t_2^*)$. For $i =$I: for any interpretation $\mathcal{D}$,

$$\begin{aligned} \alpha &\leqslant \bigwedge_{\theta\in L} (\theta \vee \theta') \\ &\leqslant \mathcal{D}_{\mathcal{F}_f}(\psi \vee \psi') \vee \mathcal{D}_{\mathcal{F}_f}(\varphi') \vee \mathcal{D}_{\mathcal{F}_f}(\gamma) \\ &= \mathcal{D}_{\mathcal{F}_f}(\varphi' \vee \psi) \vee \mathcal{D}_{\mathcal{F}_f}(\psi' \vee \gamma) \vee \mathcal{D}_{\mathcal{F}_f}(\varphi' \vee \gamma) \\ &\leqslant \mathcal{D}_{\mathcal{F}_f}(\varphi \to \psi) \vee \mathcal{D}_{\mathcal{F}_f}(\psi \to \gamma) \vee \mathcal{D}_{\mathcal{F}_f}(\varphi \to \gamma) \\ &= (\mathcal{D}_{\mathcal{F}_f}(\varphi \to \psi) \wedge \mathcal{D}_{\mathcal{F}_f}(\psi \to \gamma)) \to \mathcal{D}_{\mathcal{F}_f}(\varphi \to \gamma). \end{aligned}$$

So

$$\begin{aligned} &\alpha \otimes t_2^*(\mathcal{D}_{\mathcal{F}_f}(\varphi \to \psi), \mathcal{D}_{\mathcal{F}_f}(\psi \to \gamma)) \\ &= \alpha \otimes (\mathcal{D}_{\mathcal{F}_f}(\varphi \to \psi) \wedge \mathcal{D}_{\mathcal{F}_f}(\psi \to \gamma)) \\ &\leqslant \mathcal{D}_{\mathcal{F}_f}(\varphi \to \gamma) \\ &= \mathcal{D}_{\mathcal{F}_f}(r_2^{\triangle}(\varphi \to \psi, \psi \to \gamma)). \end{aligned}$$

For $i =$II: note that

$$t_2^*(\alpha \otimes \theta, \alpha \otimes \beta) = \alpha \otimes t_2^*(\theta, \beta),$$

so it follows from the case for $i =$ I.

(6). For the case w.r.t. (r_2^*, t_2^*): for any interpretation $\mathcal{D}$, $\varphi, \psi, \gamma \in \mathcal{F}_f$, and $\alpha \in L$,

$$\begin{aligned} &\mathcal{D}_{\mathcal{F}_f}(r_2^*(\varphi \to \gamma, \varphi \to \psi)) \\ &= \mathcal{D}_{\mathcal{F}_f}(\varphi \to (\gamma \wedge \psi)) \\ &= \mathcal{D}_{\mathcal{F}_f}(\varphi \to \gamma) \wedge \mathcal{D}_{\mathcal{F}_f}(\varphi \to \psi) \\ &= t_2^*(\mathcal{D}_{\mathcal{F}_f}(\varphi \to \gamma), \mathcal{D}_{\mathcal{F}_f}(\varphi \to \psi)) \\ &\geqslant \alpha \otimes t_2^*(\mathcal{D}_{\mathcal{F}_f}(\varphi \to \gamma), \mathcal{D}_{\mathcal{F}_f}(\varphi \to \psi)) \\ &= t_2^*(\alpha \otimes \mathcal{D}_{\mathcal{F}_f}(\varphi \to \gamma), \alpha \otimes \mathcal{D}_{\mathcal{F}_f}(\varphi \to \psi)). \end{aligned}$$

For the case w.r.t. $(r_1^{\theta_0}, t_1^{\theta_0})$ ($\theta_0 \in L$): it is easy to prove that $t_1^{\theta_0}$ is an unary truth-valued operation.

For any interpretation $\mathcal{D}$, and $\varphi \in \mathcal{F}_f$, since

$$\begin{aligned} \mathcal{D}_{\mathcal{F}_f}(r_1^{\theta_0}(\varphi)) &= \mathcal{D}_{\mathcal{F}_f}(\theta_0 \to \varphi) \\ &= \theta_0 \to \mathcal{D}_{\mathcal{F}_f}(\varphi) \\ &= t_1^{\theta_0}(\mathcal{D}_{\mathcal{F}_f}(\varphi)), \end{aligned}$$

hence

$$\begin{aligned} &\mathcal{D}_{\mathcal{F}_f}(r_1^{\theta_0}(\varphi)) \geqslant \alpha \otimes t_1^{\theta_0}(\mathcal{D}_{\mathcal{F}_f}(\varphi)), \\ &\mathcal{D}_{\mathcal{F}_f}(r_1^{\theta_0}(\varphi)) \geqslant t_1^{\theta_0}(\alpha \otimes \mathcal{D}_{\mathcal{F}_f}(\varphi)). \ \square \end{aligned}$$

Definition 10.2.15 *Let $\alpha, \beta \in L$. We define a mapping:*

$$\begin{aligned} C^{\beta}_{(C^{\emptyset}_{\mathcal{I}}, \mathscr{R}(\alpha-i))} \colon \mathscr{F}_L(\mathcal{F}_f) &\longrightarrow \mathscr{F}_L(\mathcal{F}_f), \\ X &\longmapsto C^{\beta,X}_{(C^{\emptyset}_{\mathcal{I}}, \mathscr{R}(\alpha-i))}, \end{aligned} \tag{10.15}$$

where for any $\varphi \in \mathcal{F}_f$,

$$\begin{aligned} C^{\beta,X}_{(C^{\emptyset}_{\mathcal{I}}, \mathscr{R}(\alpha-i))}(\varphi) \triangleq \bigwedge\{Y(\varphi) \mid Y \supseteq \beta \otimes (C^{\emptyset}_{\mathcal{I}} \cup X), Y \\ \text{is } \alpha\text{-}i \text{ type closed w.r.t. } \mathscr{R}\}, i = \mathrm{I}, \mathrm{II}. \end{aligned}$$

Definition 10.2.16 *Let $\alpha \in L$, $\mathscr{R}$ is said to be α-i type sound w.r.t. $\mathcal{I}$, if T is α-i type closed w.r.t. $\mathscr{R}$ holds for any $T \in \mathcal{I}$, $i = \mathrm{I}, \mathrm{II}$.*

Similar to Theorem 9.2.5, we have the following theorem.

Theorem 10.2.17 *Let $\beta, \beta_1, \beta_2, \alpha, \alpha_1, \alpha_2 \in L$, $i = \mathrm{I}, \mathrm{II}$, $X, X_1, X_2 \in \mathscr{F}_L(\mathcal{F}_f)$.*

(1). If $\beta_1 \leqslant \beta_2$, $\alpha_1 \leqslant \alpha_2$ and $X_1 \subseteq X_2$, then

$$C^{\beta_1,X_1}_{(C^{\emptyset}_{\mathcal{I}}, \mathscr{R}(\alpha_1-i))} \subseteq C^{\beta_2,X_2}_{(C^{\emptyset}_{\mathcal{I}}, \mathscr{R}(\alpha_2-i))}. \tag{10.16}$$

(2). If $\mathscr{R}$ is α-i type closed w.r.t. $\mathcal{I}$, then

$$C^{\beta,X}_{(C^{\emptyset}_{\mathcal{I}}, \mathscr{R}(\alpha-i))} \subseteq C^{\beta\otimes X}_{\mathcal{I}}. \tag{10.17}$$

(3).

$$C^{\beta, C^{\beta,X}_{(C^{\emptyset}_{\mathcal{I}}, \mathscr{R}(\alpha-i))}}_{(C^{\emptyset}_{\mathcal{I}}, \mathscr{R}(\alpha-i))} = C^{\beta,\beta\otimes X}_{(C^{\emptyset}_{\mathcal{I}}, \mathscr{R}(\alpha-i))}. \ \square \tag{10.18}$$

Definition 10.2.17 *Let $X \in \mathscr{F}_L(\mathcal{F}_f)$, $\varphi \in \mathcal{F}_f$, $\theta, \alpha, \beta \in L$. $(P^{\mathrm{I}}, (n), X, (\varphi, \theta) -(\alpha, \beta))$ is said to be an (α, β)-I type proof of φ from X with the truth-valued degree θ (shortly, $\theta - (\alpha, \beta)$-I type proof of φ from X), if the mapping*

$$\begin{aligned} P^{\mathrm{I}} \colon (n) &\longrightarrow \mathcal{F}_f \times L, \qquad (n) = \{1, 2, \cdots, n\}, \\ i &\longmapsto (\varphi_i, \theta_i), \end{aligned}$$

satisfies:

(1). $(\varphi_n, \theta_n) = (\varphi, \theta)$ and
(2). $\theta_i = \beta \otimes C^{\emptyset}_{\mathcal{I}}(\varphi_i)$, or
(3). $\theta_i = \beta \otimes X(\varphi_i)$, or
(4). there exist $i_1, \cdots, i_k < i$, and $(r, t) \in \mathscr{R}_k$, such that

$$(\varphi_i, \theta_i) = (r(\varphi_{i_1}, \cdots, \varphi_{i_k}), \alpha \otimes t(\theta_{i_1}, \cdots, \theta_{i_k})),$$

where n is said to be the length of $\theta - (\alpha, \beta)$-I type proof of φ from X under P^{I}, and is denoted by $l(P^{\mathrm{I}})$.

Definition 10.2.18 *Let* $X \in \mathscr{F}_L(\mathcal{F}_f)$, $\varphi \in \mathcal{F}_f$, $\theta, \alpha, \beta \in L$. $(P^{\mathrm{II}}, (n), X, (\varphi, \theta) -(\alpha, \beta))$ *is said to be an* (α, β)-II *type proof of* φ *from* X *with the truth-valued degree* θ *(shortly,* $\theta - (\alpha, \beta)$-II *type proof of* φ *from* X*), if the mapping*

$$P^{\mathrm{II}}\colon\ (n) \longrightarrow \mathcal{F}_f \times L, \qquad (n) = \{1, 2, \cdots, n\},$$
$$i \longmapsto (\varphi_i, \theta_i),$$

satisfies:

(1). $(\varphi_n, \theta_n) = (\varphi, \theta)$ *and*
(2). $\theta_i = \beta \otimes C_{\mathcal{I}}^{\emptyset}(\varphi_i)$, *or*
(3). $\theta_i = \beta \otimes X(\varphi_i)$, *or*
(4). there exist $i_1, \cdots, i_k < i$, *and* $(r, t) \in \mathscr{R}_k$, *such that*

$$(\varphi_i, \theta_i) = (r(\varphi_{i_1}, \cdots, \varphi_{i_k}), t(\alpha \otimes \theta_{i_1}, \cdots, \alpha \otimes \theta_{i_k})),$$

where n *is said to be the length of* $\theta - (\alpha, \beta)$-II *type proof of* φ *from* X *under* P^{II}, *and is denoted by* $l(P^{\mathrm{II}})$.

Similar to Theorem 9.2.1, we have the following theorem.

Theorem 10.2.18 *Let* $X \in \mathscr{F}_L(\mathcal{F}_f)$, $\alpha, \beta \in L$, *and the truth-valued operations in* $\mathscr{R}$ *satisfy the finite semicontinuity. Then for any* $\varphi \in \mathcal{F}_f$,

$$C^{\beta, X}_{(C^{\emptyset}_{\mathcal{I}}, \mathscr{R}(\alpha - i))}(\varphi) = \bigvee \{\theta \mid \text{ there exists } (P^i, (n), X, (\varphi, \theta) - (\alpha, \beta))\}, i = \mathrm{I}, \mathrm{II}. \ \square \tag{10.19}$$

Similar to Theorem 9.2.7, we have the following theorem.

Theorem 10.2.19 *Let* $X \in \mathscr{F}_L(\mathcal{F}_f)$, $\alpha, \beta \in L$. *For any* $\varphi \in \mathcal{F}_f$, *let*

$$\bar{X}_i(\varphi) = \bigvee \{\theta \mid \text{ there exists } (P^i, (n), X, (\varphi, \theta) - (\alpha, \beta))\}.$$

Then for any $\varphi, \psi \in \mathcal{F}_f$,

$$\alpha \otimes \bar{X}_i(\varphi \to \psi) \leqslant \bar{X}_i(\varphi) \to \bar{X}_i(\psi), i = \mathrm{I}, \mathrm{II}. \ \square \tag{10.20}$$

Corollary 10.2.2 *Let* $X \in \mathscr{F}_L(\mathcal{F}_f)$, $\alpha, \beta \in L$, *and the truth-valued operations in* $\mathscr{R}$ *satisfy the finite semicontinuity. Then for arbitrary* $\varphi, \psi \in \mathcal{F}_f$,

$$\alpha \otimes C^{\beta, X}_{(C^{\emptyset}_{\mathcal{I}}, \mathscr{R}(\alpha - i))}(\varphi \to \psi) \leqslant C^{\beta, X}_{(C^{\emptyset}_{\mathcal{I}}, \mathscr{R}(\alpha - i))}(\varphi) \to C^{\beta, X}_{(C^{\emptyset}_{\mathcal{I}}, \mathscr{R}(\alpha - i))}(\psi). \ \square \tag{10.21}$$

10.2.6 Soundness and Completeness

Similar to Theorem 9.2.2, we have the following theorem.

Theorem 10.2.20 *Let* $\alpha \in L$. *Then* $\mathscr{R}$ *is* α-i *type sound w.r.t.* $\mathcal{I}$ *if and only if*

$$C^{\beta, X}_{(C^{\emptyset}_{\mathcal{I}}, \mathscr{R}(\alpha - i))} \subseteq C^{\beta \otimes X}_{\mathcal{I}}, i = \mathrm{I}, \mathrm{II}.$$

holds for any $X \in \mathscr{F}_L(\mathcal{F}_f)$, $\beta \in L$. $\square$

Similar to Theorem 9.2.9, we have the following theorem.

Theorem 10.2.21 *Let $\alpha, \beta \in L$, and for any $X \in \mathscr{F}_L(\mathcal{F}_f)$, $\mathscr{R}$ is α-i type sound w.r.t. $\mathcal{I}$, and $C^{\beta,X}_{(C^{\emptyset}_{\mathcal{I}},\mathscr{R}(\alpha-\mathrm{i}))} \in \mathcal{I}$. Then*

$$C^{\beta,X}_{(C^{\emptyset}_{\mathcal{I}},\mathscr{R}(\alpha-i))} = C^{\beta\otimes X}_{\mathcal{I}}, \qquad i = \mathrm{I}, \mathrm{II}. \ \square \tag{10.22}$$

By Theorems 10.2.20 and 10.2.21, we immediately have the following theorem.

Theorem 10.2.22 ((α, β)-*i* Type Completeness and Soundness) *Let the truth-valued operations in $\mathscr{R}$ satisfy the finite semicontinuity. If for any $\alpha, \beta \in L$, $X \in \mathscr{F}_L(\mathcal{F}_f)$, $\mathscr{R}$ is α-i type sound w.r.t. $\mathcal{I}$, and $C^{\beta,X}_{(C^{\emptyset}_{\mathcal{I}},\mathscr{R}(\alpha-i))} \in \mathcal{I}$, then for any $\varphi \in \mathcal{F}_f$,*

$$C^{\beta\otimes X}_{\mathcal{I}}(\varphi) = \bigvee\{\theta \mid \textit{there exists } (P^i, (n), X, (\varphi, \theta) - (\alpha, \beta))\}, i = \mathrm{I}, \mathrm{II}. \ \square \tag{10.23}$$

10.2.7 Satisfiability and Consistency

Definition 10.2.19 *Let $X \in \mathscr{F}_L(\mathcal{F}_f)$, $\delta \in L$. If there exists $T_0 \in \mathcal{I}$, such that for any $\varphi \in \mathcal{F}_f$,*

(1). $T_0(\varphi') = (T_0(\varphi))'$,
(2). $X(\varphi) \to T_0(\varphi) \geqslant \delta$,
(3). T_0 *is α-i type closed w.r.t. $\mathscr{R}$,*

then X is said to be δ-i type satisfiable or T_0 is a δ-i model of X, $i = \mathrm{I}, \mathrm{II}$.

Definition 10.2.20 *Let $X \in \mathscr{F}_L(\mathcal{F}_f)$, $\tau \in L$. If*

$$\bigvee\{C^{\beta,X}_{(C^{\emptyset}_{\mathcal{I}},\mathscr{R}(\alpha-i))}(\varphi) \otimes C^{\beta,X}_{(C^{\emptyset}_{\mathcal{I}},\mathscr{R}(\alpha-i))}(\varphi') \mid \varphi \in \mathcal{F}_f\} \leqslant \tau,$$

then X is said to be τ'-i type consistent w.r.t. (α, β), $i = \mathrm{I}, \mathrm{II}$.

Theorem 10.2.23 ((α, β)-*i* Type Consistent) *Let $\mathscr{R}$ be α-i type sound w.r.t. $\mathcal{I}$, $X \in \mathscr{F}_L(\mathcal{F}_f)$, $\delta, \alpha, \beta \in L$, $C^{\beta,X}_{(C^{\emptyset}_{\mathcal{I}},\mathscr{R}(\alpha-i))} \in \mathcal{I}$.*

(1). If X is δ-i type satisfiable, then X is $(\delta\otimes\delta)$-i type consistent w.r.t. (α, β).
(2). If there exists $T^* \in \mathcal{I}_H \cap \mathcal{I}$,

$$\bigwedge_{T\in\mathcal{I}} (\pi(\beta \otimes X \subseteq T))' = (\pi(\beta \otimes X \subseteq T^*))' \triangleq \bar{\beta}$$

and if X is τ'-i type consistent w.r.t. (α, β), then $\beta \otimes X$ is $(\bar{\beta} \otimes \tau')$-i type satisfiable.

Proof. (1). It is similar to the proof of Theorem 9.2.11.

(2). For any $\varphi \in \mathcal{F}_f$,

$$(\pi(\beta \otimes X \subseteq T^*))' \leqslant \bar{\beta} \to \tau,$$

i.e.,

$$\begin{aligned}\pi(\beta \otimes X \subseteq T^*) &\geqslant (\bar{\beta} \to \tau)' \\ &= \bar{\beta} \otimes \tau'. \ \square\end{aligned}$$

Theorem 10.2.24 *Let $X \in \mathscr{F}_L(\mathcal{F}_f)$, $\varphi \in \mathcal{F}_f$, $\delta, \theta \in L$. If $C_{\mathcal{I}}^X(\varphi) \geqslant \theta$, then for any δ-i model T of X, $\delta \otimes \theta \leqslant T(\varphi)$.*

Proof. Since T is a δ-i model of X, we know that $\delta \leqslant \pi(X \subseteq T)$. So,

$$\delta \to T(\varphi) \geqslant \pi(X \subseteq T) \to T(\varphi) \geqslant \theta,$$

i.e., $\delta \otimes \theta \leqslant T(\varphi)$. $\square$

10.2.8 Deduction Theorem

Similar to Theorem 9.2.15, we have the following theorem.

Theorem 10.2.25 (**(α, β, θ)-deduction**) *Let $\mathscr{R}$ be α-i type sound w.r.t. $\mathcal{I}$, $\varphi, \psi \in \mathcal{F}_f$, $\sigma, \theta \in L$, and for any $X \in \mathscr{F}_L(\mathcal{F}_f)$, $C^{\beta,X}_{(C_{\mathcal{I}}^{\emptyset}, \mathscr{R}(\alpha-i))} \in \mathcal{I}$, $i = \mathrm{I}, \mathrm{II}$, $(r_2^0, t_2) \in \mathscr{R}_2$.*

(1). For $i = \mathrm{I}$, if $C_{\mathcal{I}}^{\beta \otimes X}(\varphi \to \psi) \geqslant \sigma$, then

$$C_{\mathcal{I}}^{\beta \otimes (X \cup \{\theta/\varphi\})}(\psi) \geqslant \alpha \otimes t_2(\beta \otimes \theta, \sigma). \tag{10.24}$$

For $i = \mathrm{II}$, if $C_{\mathcal{I}}^{\beta \otimes X}(\varphi \to \psi) \geqslant \sigma$, then

$$C_{\mathcal{I}}^{\beta \otimes (X \cup \{\theta/\varphi\})}(\psi) \geqslant t_2(\alpha \otimes \beta \otimes \theta, \alpha \otimes \sigma). \tag{10.25}$$

(2).

$$C_{\mathcal{I}_h}^{\beta \otimes (X \cup \{\theta/\varphi\})}(\psi) \leqslant C_{\mathcal{I}_h}^{\beta \otimes X}(\varphi \to \psi). \ \square \tag{10.26}$$

10.2.9 Compactness

Definition 10.2.21 *$C_{\mathcal{I}_h}$ is said to be compact on $\mathcal{F}_f$, if for any $X \in \mathscr{F}_L(\mathcal{F}_f)$, there exists $Q \in \mathscr{F}_L(\mathcal{F}_f)$ and $|\mathrm{supp}\ Q| < \aleph_0$, such that*

$$C_{\mathcal{I}_h}^X = C_{\mathcal{I}_h}^{X \cap Q}.$$

Similar to Theorem 9.2.16, we have the following theorem.

Theorem 10.2.26 *If $C_{\mathcal{I}_h}$ is compact on $\mathcal{F}_f$, then L satisfies the Ascending Chain Condition and the Descending Chain Condition.* $\square$

10.2.10 Examples

Example 10.2.1 *In* L_{vfl}, *let* $\alpha \in L$ *and* $\alpha \leqslant \bigwedge_{\theta \in L}(\theta \vee \theta')$, *for any* $u \in U$, $\mathbf{Q}_u$ *is* $r_1^u - h$ *regular,* $r_2^u - h$ *regular and* $r_3^u - h$ *regular at the level* α.
Let $\mathscr{R} = \mathscr{R}^*$,

$$\mathcal{I}_i = \{T \mid T \in \mathscr{F}_L(\mathcal{F}_f), T \text{ is } \alpha\text{-}i \text{ type closed w.r.t. } \mathscr{R}\}, (i = \text{I}, \text{II}),$$

then we have the following results.

(1). $\mathcal{I}_H \subseteq \mathcal{I}_i$, $i = \text{I}, \text{II}$.
(2). For any $X \in \mathscr{F}_L(\mathcal{F}_f)$ *and* $\beta \in L$,

$$C^{\beta,X}_{(C^{\emptyset}_{\mathcal{I}_i}, \mathscr{R}(\alpha-i))} \in \mathcal{I}_i, i = \text{I}, \text{II}.$$

(3). $\mathscr{R}$ *is* α-i *type sound w.r.t.* $\mathcal{I}_i$, $i = \text{I}, \text{II}$.
(4). For any $X \in \mathscr{F}_L(\mathcal{F}_f)$ *and* $\beta \in L$,

$$C^{\beta,X}_{(C^{\emptyset}_{\mathcal{I}_i}, \mathscr{R}(\alpha-i))} = C^{\beta \otimes X}_{\mathcal{I}_i}, i = \text{I}, \text{II}.$$

(5). For any $X \in \mathscr{F}_L(\mathcal{F}_f)$, $\beta \in L$, *and* $\varphi \in \mathcal{F}_f$,

$$C^{\beta \otimes X}_{\mathcal{I}_i}(\varphi) = \bigvee \{\theta \mid \text{ there exists } (\varphi^i, (n), X, (\varphi, \theta) - (\alpha, \beta))\}, i = \text{I}, \text{II}.$$

(6). For any $X \in \mathscr{F}_L(\mathcal{F}_f)$, $\delta \in L$, *if* X *is* δ-i *type satisfiable, then* X *is* $(\delta \otimes \delta)$-i *type consistent w.r.t.* (α, β), $i = \text{I}, \text{II}$.
(7). For any $X \in \mathscr{F}_L(\mathcal{F}_f)$, $\varphi, \psi \in \mathcal{F}_f$, $\sigma, \theta \in L$, *we have*
a) If $C^{\beta \otimes X}_{\mathcal{I}_i}(\varphi \to \psi) \geqslant \sigma$, *then*

$$C^{\beta \otimes (X \cup \{\theta/\varphi\})}_{\mathcal{I}_i}(\psi) \geqslant \alpha \otimes ((\beta \otimes \theta) \wedge \sigma), i = \text{I}, \text{II}.$$

b) $C^{\beta \otimes (X \cup \{\theta/\varphi\})}_{\mathcal{I}_h}(\psi) \leqslant C^{\beta \otimes X}_{\mathcal{I}_h}(\varphi \to \psi)$.

Example 10.2.2 (Łukasiewicz First-Order Logic L_{uf}) *In* L_{vfl}, *if the complete lattice implication algebra* $L = ([0,1], \vee, \wedge, ', \to)$ *is Łukasiewicz implication algebra, then this* L_{vfl} *is called a Łukasiewicz first-order logic system, denoted by* L_{uf}. *In this case, we have*

$$\mathscr{R} = \mathscr{R}^*,$$
$$\alpha \leqslant \bigwedge_{\theta \in [0,1]} (\theta \vee \theta') = 0.5,$$
$$\mathcal{I}_i = \{T \mid T \in \mathscr{F}(\mathcal{F}_f), T \text{ is } \alpha\text{-}i \text{ type closed w.r.t. } \mathscr{R}\}, i = \text{I}, \text{II}.$$

Then we can obtain the similar results as those in Example 10.2.1.

Example 10.2.3 (**F-Valued Łukasiewicz First-Order Logic FL_{uf}**) *In L_{vfl}, if the complete lattice implication algebra L is $(\mathscr{F}([0,1]), \vee, \wedge, ', \rightarrow)$ defined as Example 2.1.7, then this L_{vfl} is called F-valued Łukasiewicz first-order logic, denoted by FL_{uf}. Moreover, take $\mathscr{R} = \mathscr{R}^*$ and for any $x \in [0,1]$, $\alpha \in \mathscr{F}([0,1])$ such that:*

$$\alpha(x) \leqslant 0.5,$$

and

$$\mathcal{I}_i = \{T \mid T \in \mathscr{F}_{\mathscr{F}([0,1])}(\mathcal{F}_f), T \textit{ is } \alpha\text{-}i \textit{ type closed w.r.t. } \mathscr{R}\}, i = \mathrm{I}, \mathrm{II}.$$

Then the similar results as those in Example 10.2.1 can be obtained.

Example 10.2.4 *In L_{vfl}, let the lattice implication algebra is defined as C_5, then this L_{vfl} is denoted as $\mathscr{C}_{5f}$, where*

$$\mathscr{R} = \mathscr{R}^*,$$
$$\alpha \leqslant b,$$
$$\mathcal{I}_i = \{T \mid T \in \mathscr{F}_{C_5}(\mathcal{F}_f), T \textit{ is } \alpha\text{-}i \textit{ type closed w.r.t. } \mathscr{R}\}, i = \mathrm{I}, \mathrm{II}.$$

Then the similar results as those in Example 10.2.1 can be obtained.

11 Uncertainty and Automated Reasoning

In this chapter, based on lattice-valued logic, we discuss some basis of applied theory for fuzziness and incomparability processing.

In Section 11.1, we briefly discuss uncertainty reasoning with four reasoning models based on $LP(X)$. In Section 11.2, we discuss uncertainty reasoning theory and method of L_{vpl}. In Sections 11.3 and 11.4, we establish resolution-based automated reasoning theory and method of $LP(X)$ and $LF(X)$, respectively.

11.1 Uncertainty Reasoning Based on $LP(X)$

In $LP(X)$, we have presented the inference rule **MP** as follows:

$$\begin{array}{l} (p \to q, \alpha_1) \\ (p, \alpha_2) \\ \hline (q, \alpha_1 \otimes \alpha_2) \end{array} \tag{11.1}$$

and it means that if we have two assumptions $p \to q$ and p with the truth degree α_1, and the truth degree α_2, respectively, then we can get the conclusion q with the truth degree $\alpha_1 \otimes \alpha_2$ by the inference rule **MP**.

In the following, let L be the lattice implication algebra defined in Example 2.1.7, RM a reasoning method and $A \overset{RM}{\Longrightarrow} B$ have the meaning that we can get the conclusion B from the assumption A by the reasoning method RM.

Definition 11.1.1 *A reasoning method RM is called increasing, if for any $A_1, A_2, B_1, B_2 \in \mathscr{F}([0,1])$, $A_1 \overset{RM}{\Longrightarrow} B_1$, $A_2 \overset{RM}{\Longrightarrow} B_2$, and $A_1 \subseteq A_2$ imply*

$$B_1 \subseteq B_2.$$

RM is called decreasing, if for any $A_1, A_2, B_1, B_2 \in \mathscr{F}([0,1])$, $A_1 \overset{RM}{\Longrightarrow} B_1$, $A_2 \overset{RM}{\Longrightarrow} B_2$, and $A_1 \subseteq A_2$ imply

$$B_2 \subseteq B_1.$$

RM is called monotone, if it is increasing or decreasing.

Definition 11.1.2 *Suppose a reasoning method RM with the following form:*

$$\begin{array}{l} A \longrightarrow B \\ \underline{A^* \qquad\qquad} \\ \qquad\quad B^*. \end{array}$$

Then RM is called quasi-increasing, if $A^* \subseteq A$ implies $B^* \subseteq B$ and $A \subseteq A^*$ implies $B \subseteq B^*$. RM is called quasi-decreasing, if $A^* \subseteq A$ implies $B \subseteq B^*$ and $A \subseteq A^*$ implies $B^* \subseteq B$.

In the following, we investigate four inference models. Let u, u_i, v, v_j are names of objects, $A, A^*, A_1, \cdots, A_n \in \mathscr{F}(U)$, $B, B^*, B_1, \cdots, B_n \in \mathscr{F}(V)$, $i, j, n \in \mathbb{N}^+$, and U, V are universe of discourse.

Model One ($M1$):

$$\begin{array}{rl} \textit{Rule}: & \text{if "}u \text{ is } A,\text{" then "}v \text{ is } B\text{"} \\ \textit{Fact}: & \text{"}u \text{ is } A^*\text{"} \\ \hline \textit{Conclusion}: & \text{"}v \text{ is } B^*.\text{"} \end{array}$$

This model can be rewritten as:

$$\begin{array}{l} A \longrightarrow B \\ \underline{A^* \qquad\qquad} \\ \qquad\quad B^*. \end{array}$$

For the model $M1$, we consider the language truth value $\tau(A^*|A)$ of A^* restricted by A

$$\tau(A^*|A)(x) = \sup_{A(u)=x} A^*(u)$$

for any $x \in [0, 1]$. Intuitively, if taking A as the standard of "true," then $\tau(A^*|A)$ means the truth degree of A^* measured with A. In this sense, the assumptions of model $M1$ can be looked on as follows:

$$\begin{aligned} &(A \rightarrow B, I), \\ A^* &= (A, \tau(A^*|A)), \end{aligned}$$

where $I \in \mathscr{F}([0, 1])$ and $I(x) = 1$ for any $x \in [0, 1]$. By Eq. (11.1), we know

$$I \otimes \tau(A^*|A) = \tau(A^*|A),$$

hence, we can get

$$B^* = (B, \tau(A^*|A)).$$

That means the truth degree of B^* measured with B is $\tau(A^*|A)$ if taking B as the standard of "true," *i.e.*,

$$B(B^*) \triangleq \tau(B^*|B) = \tau(A^*|A).$$

By Zadeh's extension principle of fuzzy sets, we can take

$$B^* = B^{-1}(\tau(A^*|A)).$$

Hence, we have the following reasoning method for the model $M1$:

(1). Calculate the language truth value $\tau(A^*|A)$ of A^* restricted by A

$$\tau(A^*|A)(x) = \sup_{A(u)=x} A^*(u)$$

for any $x \in [0,1]$.

(2). Take inverse operation of restricted truth value

$$B^*(v) = B^{-1}(\tau(A^*|A))(v) = \tau(A^*|A)(B(v))$$

for any $v \in V$.

It is easy to prove that this reasoning method for model $M1$ is increasing and compatibles with **MP**, *i.e.*, if $A^* = A$, then $B^* = B$.

Model Two ($M2$):

$$\begin{array}{rl} \textit{Rule } 1: & \text{if "}u \text{ is } A_1\text{," then "}v \text{ is } B_1\text{,"} \\ & \vdots \\ \textit{Rule } n: & \text{if "}u \text{ is } A_n\text{," then "}v \text{ is } B_n\text{,"} \\ \textit{Fact}: & \text{"}u \text{ is } A\text{,"} \\ \hline \textit{Conclusion}: & \text{"}v \text{ is } B\text{."} \end{array}$$

This model can be rewritten as:

$$\begin{array}{l} A_1 \longrightarrow B_1, \\ \qquad \vdots \\ A_n \longrightarrow B_n, \\ A \\ \hline \qquad\quad B. \end{array} \tag{11.2}$$

For the model $M2$, we get the following reasoning method like the model $M1$:

(1). For any rule in Eq. (11.2), by combining the fact A, we get $B_1^*, \cdots, B_n^*$ by the reasoning method for $M1$, *i.e.*,

$$B_i^*(v) = \tau(A_i|A)(B_i(v)), v \in V, 1 \leqslant i \leqslant n.$$

(2). Compute B by $B_1^*, \cdots, B_n^*$. We can take many methods, such as: for any $v \in V$,

$$B(v) = \sqrt[n]{B_1^*(v) B_2^*(v) \cdots B_n^*(v)},$$

or

$$B(v) = \frac{1}{n} \sum_{i=1}^{n} B_i^*(v),$$

or

$$B(v) = \sup_{1 \leqslant i \leqslant n} B_i^*(v),$$

or

$$B(v) = \min_{1 \leqslant i \leqslant n} B_i^*(v),$$

or

$$B(v) = \sum_{i=1}^{n} \alpha_i B_i^*(v),$$

where, α_i is the weight of "Rule i" and $\sum_{i=1}^{n} \alpha_i = 1$.

Model Three ($M3$):

$$\begin{array}{rl} \mathit{Rule}: & \text{if "}u_1 \text{ is } A_1, u_2 \text{ is } A_2, \cdots, u_n \text{ is } A_n,\text{" then "}v \text{ is } B,\text{"} \\ \mathit{Fact}: & \text{"}u_1 \text{ is } A_1^*, u_2 \text{ is } A_2^*, \cdots, u_n \text{ is } A_n^*,\text{"} \\ \hline \mathit{Conclusion}: & \text{"}v \text{ is } B^*.\text{"} \end{array}$$

For this model we have two reasoning methods as follows:

Method 1: Converting the *Rule* and the *Fact* of $M3$ to

$$\begin{array}{rl} \mathit{Rule}: & \text{if "}(u_1, u_2, \cdots, u_n) \text{ is } A,\text{" then "}v \text{ is } B,\text{"} \\ \mathit{Fact}: & \text{"}(u_1^*, u_2^*, \cdots, u_n^*) \text{ is } A^*,\text{"} \end{array}$$

where $A = A_1 \times A_2 \times \cdots \times A_n$ and $A^* = A_1^* \times A_2^* \times \cdots \times A_n^*$. Then we can compute B^* according to the model $M1$: for any $v \in V$,

$$B^*(v) = \tau(A^*|A)(B(v)).$$

Method 2: Dividing the model $M3$ into n submodels $M_1^3, M_2^3, \cdots, M_n^3$ and for any i, M_i^3 has the form:

$$\begin{array}{ll} A_i & \longrightarrow B \\ A_i^* & \\ \hline & B_i^*. \end{array}$$

Hence we can get B_i^* for any submodel M_i^3, respectively. Then B^* can be obtained from $B_1^*, \cdots, B_n^*$ like this:

$$B^*(v) = \bigwedge_{1 \leqslant i \leqslant n} B_i^*.$$

Model Four ($M4$):

$$\begin{array}{rl} \mathit{Rule}: & \text{"if } u \text{ is } A, \text{ then } v \text{ is } B\text{" is } \tau_1, \\ \mathit{Fact}: & \text{"}u \text{ is } A^*\text{" is } \tau_2, \\ \hline \mathit{Conclusion}: & \text{"}v \text{ is } B^*,\text{"} \end{array}$$

where, $\tau_1, \tau_2 \in \mathscr{F}([0,1])$, τ_1 and τ_2 are the truth degrees of the rule and the fact, respectively. This model can be rewritten as:

$$\begin{array}{l} \tau_1(A \longrightarrow B), \\ \underline{\tau_2(A^*),\qquad\qquad} \\ B^*. \end{array}$$

Then we have two reasoning methods as follows:

Method 1: Note that this model is more like the model $M1$, hence, we can take B_0 as a conclusion in the model M_1 and compute it, *i.e.*,

$$B_0(v) = \tau(A^*|A)(B(v))$$

for any $v \in V$. Taking $B^* = (B_0, \tau_1 \otimes \tau_2)$ in this model, hence B^* can be computed as follows:

$$\begin{aligned} B^*(v) &= B_0^{-1}(\tau_1 \otimes \tau_2)(v) \\ &= (\tau_1 \otimes \tau_2)(B_0(v)) \\ &= 0 \vee (\tau_1(B_0(v)) + \tau_2(B_0(v)) - 1). \end{aligned}$$

Method 2: Because τ_2 means truth degree of A^*, we can take truth degree of $\tau_2(A^*)$ w.r.t. A as $\tau_2 \otimes \tau(A^*|A)$, *i.e.*,

$$d(A, \tau_2(A^*)) = \tau_2 \otimes \tau(A^*|A).$$

So *Rule* and *Fact* of this model can be rewritten as:

$$\begin{array}{l} (A \rightarrow B, \tau_1), \\ (A, d(A, \tau_2(A^*))), \end{array}$$

respectively. Hence the conclusion can be computed as:

$$(B, \tau_1 \otimes d(A, \tau_2(A^*)))$$

by the inference rule **MP** and B^* can be obtained by the extension principle. In details, this method can be described as follows:

(1). Compute $d(A, \tau_2(A^*))$:

$$d(A, \tau_2(A^*)) = \tau_2 \otimes \tau(A^*|A).$$

For any $x \in [0, 1]$,

$$\begin{aligned} d(A, \tau_2(A^*))(x) &= \tau_2(x) \otimes \tau(A^*|A)(x) \\ &= 0 \vee (\tau_2(x) + \tau(A^*|A)(x) - 1) \\ &= 0 \vee (\tau_2(x) + \sup_{A(u)=x} A^*(u) - 1). \end{aligned}$$

(2). Compute B^*: for any $v \in V$,

$$\begin{aligned} B^*(v) &= B^{-1}(\tau_1 \otimes d(A, \tau_2(A^*)))(v) \\ &= (\tau_1 \otimes d(A, \tau(A^*)))(B(v)) \\ &= 0 \vee (\tau_1(B(v)) + d(A, \tau_2(A^*))(B(v)) - 1). \end{aligned}$$

11.2 Uncertainty Reasoning Based on L_{vpl}

11.2.1 Another Kind of Interpretation of $X \models\!\!\models Y$

In classical logic, we write $\mathscr{F}_2$ as the set of all well-formed formulae (shortly wffs), $\mathscr{T}^*$ as the set of all valuations. Let $X, Y \subseteq \mathscr{F}_2$. A set Y of wffs is said to be logically inferred from the set X of wffs, written by $X \models\!\!\models Y$, if for any $q \in Y$, $T(X) \subseteq \{I\}$ for any $T \in \mathscr{T}^*$ implies $T(q) = I$. Hence, it is easy to prove that

$$\begin{aligned} X \models\!\!\models Y \ \text{ iff } \ & X \models\!\!\models q \text{ for any } q \in Y \\ \text{iff } \ & \bigwedge_{T\in\mathscr{T}^*} ((\bigwedge_{p\in X} T(p)) \to T(q)) \geqslant Y(q) \\ \text{iff } \ & \bigwedge_{T\in\mathscr{T}^*} \{(\bigwedge_{p\in\mathscr{F}_2} (X(p) \to T(p))) \to T(q)\} \geqslant Y(q), \end{aligned} \tag{11.3}$$

where $\to$ is the Kleene's implication operator and

$$X(p) = \begin{cases} I, & p \in X, \\ O, & p \notin X. \end{cases}$$

$$Y(p) = \begin{cases} I, & p \in Y, \\ O, & p \notin Y. \end{cases}$$

According to the notation in L_{vpl}, Eq. (11.3) means that

$$C^X_{\mathscr{T}^*} \supseteq Y. \tag{11.4}$$

As for $X \models\!\!\models Y$, X and Y should be consistent. Otherwise, the correct inference consequence may not be obtained even using a correct inference. From Eq. (11.4), we know that:

(1). In classical logic, whether $X \models\!\!\models Y$ holds or not completely depends on whether Eq. (11.3), which is constructed by $\mathscr{T}^*$ and $\to$, holds or not. Therefore, it is one reason for implication operators to be paid considerable attention.

(2). On the other hand, a propositional symbol has no meaning without being given a truth value. Hence, the reasoning in classical logic can be regarded as a transformation among truth-values. Moreover, the reasoning among uncertainty can also be looked on as a transformation of truth-value. Hence, it is rational and feasible to take a logical system and law of truth-value transformation as a foundation for the modelling of uncertainty reasoning.

(3). Note that Eq. (11.3) is only a semantic form. It is easy to see from the soundness and completeness in classical logic that there exists a corresponding syntax form for Eq. (11.3). Moreover, note that Eq. (11.3) can be extended expediently to uncertainty reasoning, it is feasible to extend the corresponding syntax system to uncertainty reasoning. So, for the uncertainty reasoning, we can establish a kind of formal logical system in which semantics is compatible with syntax. Therefore, uncertainty reasoning may be implemented according to semantics in this logical system and the uncertainty reasoning consequence can be strictly proved by using the syntax in this logical system. Consequently, this kind of uncertainty reasoning theory and method can be established on the strict and solid logical foundation.

In L_{vpl}, we established a logical system with a semantic system and the corresponding syntactical system compatible with each other constructed by more general semantic interpretation and more general implication operator $\to$. Based on L_{vpl} and Eq.(11.3), it is certainly reasonable and feasible to construct two kinds of uncertainty reasoning theories and methods.

11.2.2 Basic Theory

Since a single-input-single-output uncertainty reasoning theory and method can be extended to multiple, multi-dimension as well as multi-dimension and multiple cases by using the similar method in [558], in this chapter we only discuss single-input-single-output uncertainty reasoning models.

In the following, let $i = \text{I}, \text{II}$, $\alpha, \beta, \tau \in L$, $X, \tilde{X}, Y, \tilde{Y} \in \mathscr{F}_L(\mathscr{F}_p)$, and $\mathscr{T} \subseteq \mathscr{F}_L(\mathscr{F}_p)$.

Single-input-single-output (SISO) uncertainty reasoning model:

$$\begin{array}{rl} \textit{Rule}: & \text{If } X, \text{ then } Y, \\ \textit{Fact}: & \tilde{X}, \\ \hline \textit{Conclusion}: & \tilde{Y}. \end{array} \tag{11.5}$$

Since truth-valued field L is fixed in L_{vpl}, the consistency of semantics and syntax is mainly adjusted by $\mathscr{T}$. Therefore, we focus on $\mathscr{T}$ and rewrite that X is τ'-i type consistent w.r.t. (α, β) as X is τ'-i type consistent w.r.t. $(\alpha, \beta, \mathscr{T})$.

Definition 11.2.1 *The rule in Eq.* (11.5) *is said to be* $(\alpha, \beta, \tau, \mathscr{T})$*-$i$ type representable in* L_{vpl}*, if there exist* $\alpha, \beta, \tau \in L$, $\mathscr{T} \subseteq \mathscr{F}_L(\mathscr{F}_p)$ *such that* X, Y *are* τ'*-i type consistent w.r.t.* $(\alpha, \beta, \mathscr{T})$ *and*

$$C^X_{\mathscr{T}} \supseteq \tau' \otimes Y. \tag{11.6}$$

Theorem 11.2.1 *Let* $X, Y \in \mathscr{F}_L(\mathscr{F}_p)$ *and* $\mathscr{T} \subseteq \mathscr{F}_L(\mathscr{F}_p)$*. If* Y *is* τ'*-i type consistent w.r.t.* $(\alpha, \beta, \mathscr{T})$ *and* $X \subseteq Y$*, then so is* X*.* □

Corollary 11.2.1 *Let* $X, Y \in \mathscr{F}_L(\mathscr{F}_p)$ *and* $\mathscr{T} \subseteq \mathscr{F}_L(\mathscr{F}_p)$*. If* X, Y *are* τ'*-i type consistent w.r.t.* $(\alpha, \beta, \mathscr{T})$*, then* $X \cap Y$*,* $X \otimes Y$*,* $\theta \otimes X$ $(\theta \in L)$ *are all* τ'*-i type consistent w.r.t.* $(\alpha, \beta, \mathscr{T})$*.* □

It follows from Theorems 11.2.1 and 9.2.4

Theorem 11.2.2 *In Eq. (11.5), if $Y \subseteq X$ and X is τ'-i type consistent w.r.t. $(\alpha, \beta, \mathscr{T})$, then the rule in Eq. (11.5) is $(\alpha, \beta, \tau, \mathscr{T})$-i type representable in L_{vpl}.* □

Theorem 11.2.3 *In Eq. (11.5), if $X \subseteq Y$, and there exist $T_0 \in \mathscr{T}_H$, $Y \subseteq T_0$, then for any $\alpha, \beta, \tau \in L$, the rule in Eq. (11.5) is $(\alpha, \beta, \tau, \mathscr{T})$-i type representable in L_{vpl}, where $\mathscr{T} = \{T \mid T \supseteq Y\}$.*

Proof. For any $\alpha, \beta, \tau \in L$. Because

$$\begin{aligned} C_{\mathscr{T}}^{\beta \otimes Y} &= C_{\mathscr{T}}^{\beta \otimes \bigcap_{T \in \mathscr{T}} T} \\ &\subseteq C_{\mathscr{T}}^{\bigcap_{T \in \mathscr{T}} (\beta \otimes T)} \\ &\subseteq \bigcap_{T \in \mathscr{T}} C_{\mathscr{T}}^{\beta \otimes T} \\ &\subseteq \bigcap_{T \in \mathscr{T}} C_{\mathscr{T}}^{T} \\ &= \bigcap_{T \in \mathscr{T}} T \\ &\subseteq T_0, \end{aligned}$$

hence, for any $p \in \mathscr{F}_p$, there have

$$\begin{aligned} C_{\mathscr{T}}^{\beta \otimes Y}(p) \otimes C_{\mathscr{T}}^{\beta \otimes Y}(p') &\leqslant T_0(p) \otimes T_0(p') \\ &= T_0(p) \otimes (T_0(p))' \\ &= O \\ &\leqslant \tau. \end{aligned}$$

Consequently, X and Y are all τ'-i type consistent w.r.t. $(\alpha, \beta, \mathscr{T})$.

Note that $X \subseteq Y$ and $\mathscr{T} = \{T \mid T \supseteq Y\}$, then

$$C_{\mathscr{T}}^{X} = C_{\mathscr{T}}^{\emptyset} = Y \supseteq \tau' \otimes Y. \ \square$$

Corollary 11.2.2 *In Eq. (11.5), if $Y \in \mathscr{T}_H$, $X \subseteq Y$, then for any $\alpha, \beta, \tau \in L$, the rule in Eq. (11.5) is $(\alpha, \beta, \tau, \mathscr{T})$-i type representable in L_{vpl}, where $\mathscr{T} = \{T \mid T \supseteq Y\}$.* □

Corollary 11.2.3 *In Eq. (11.5), if $X \subseteq Y$ and Y is τ'-i type consistent w.r.t. $(\alpha, \beta, \mathscr{T})$, then the rule in Eq. (11.5) is $(\alpha, \beta, \tau, \mathscr{T})$-i type representable in L_{vpl}, where $\mathscr{T} = \{T | T \supseteq Y\}$.* □

Theorem 11.2.4 *In Eq. (11.5), let*

$$\mathscr{T} = \{T \mid T \supseteq X \rightarrow (\tau' \otimes Y \otimes T)\}.$$

Let X and Y be τ'-i type consistent w.r.t. $(\alpha, \beta, \mathscr{T})$, then the rule in Eq. (11.5) is $(\alpha, \beta, \tau, \mathscr{T})$-i type representable in L_{vpl}.

Proof. It only needs to prove that $C_{\mathscr{T}}^X \supseteq \tau' \otimes Y$. For any $q \in \mathscr{F}_p$,

$$\begin{aligned}
C_{\mathscr{T}}^X(q) &= \bigwedge_{T\in\mathscr{T}} (\pi(X \subseteq T) \to T(q)) \\
&\geqslant \bigwedge_{T\in\mathscr{T}} ((X(q) \to T(q)) \to T(q)) \\
&\geqslant (X(q) \to T(q)) \to (X(q) \to (\tau' \otimes Y(q) \otimes T(q))) \\
&\geqslant (X(q) \to T(q)) \to ((\tau' \otimes Y(q)) \otimes (X(q) \to T(q))) \\
&\geqslant (\tau' \otimes Y(q)) \otimes ((X(q) \to T(q)) \to (X(q) \to T(q))) \\
&= \tau' \otimes Y(q). \ \square
\end{aligned}$$

Definition 11.2.2 *Let $\mathscr{T} \subseteq \mathscr{F}_L(\mathscr{F}_p)$, $X \in \mathscr{F}_L(\mathscr{F}_p)$, $\beta \in L$. We define recursively $(f_{\mathscr{T}}^\beta)^{(n)}(X)$ as follows: for $n \in \mathbb{N}$,*

$$\begin{aligned}
&(f_{\mathscr{T}}^\beta)^{(0)}(X) = X, \\
&(f_{\mathscr{T}}^\beta)^{(1)}(X) = C_{\mathscr{T}}^{\beta\otimes X}, \\
&(f_{\mathscr{T}}^\beta)^{(n)}(X) = (f_{\mathscr{T}}^\beta)^{(1)}((f_{\mathscr{T}}^\beta)^{(n-1)}(X)), n = 2, 3, \cdots .
\end{aligned}$$

Theorem 11.2.5 *Let $\mathscr{T} \subseteq \mathscr{F}_L(\mathscr{F}_p)$, $X \in \mathscr{F}_L(\mathscr{F}_p)$, $\beta \in L$, then*

(1). $(f_{\mathscr{T}}^\beta)^{(n)}(X) \subseteq (f_{\mathscr{T}}^\beta)^{(n-1)}(X)$ for $n = 2, 3, \cdots$.

(2). If X is τ'-i type consistent w.r.t. $(\alpha, \beta, \mathscr{T})$, then so are $(f_{\mathscr{T}}^\beta)^{(n)}(X)$ for $n \in \mathbb{N}^+$.

(3). If $\mathscr{T} = \{T | T \supseteq X\}$, then $(f_{\mathscr{T}}^\beta)^{(n)}(X) = X$ for $n \in \mathbb{N}$.

Proof. We only prove (1) and (3).

For (1). If $n = 2$, then

$$\begin{aligned}
(f_{\mathscr{T}}^\beta)^{(2)}(X) &= C_{\mathscr{T}}^{\beta\otimes C_{\mathscr{T}}^{\beta\otimes X}} \\
&\subseteq C_{\mathscr{T}}^{C_{\mathscr{T}}^{\beta\otimes X}} \\
&= C_{\mathscr{T}}^{\beta\otimes X} \\
&= (f_{\mathscr{T}}^\beta)^{(1)}(X).
\end{aligned}$$

Suppose that when $n \leqslant k$, $k > 2$,

$$(f_{\mathscr{T}}^\beta)^{(k)}(X) \subseteq (f_{\mathscr{T}}^\beta)^{(k-1)}(X).$$

When $n = k + 1$,

$$\begin{aligned}
(f_{\mathscr{T}}^\beta)^{(k+1)}(X) &= C_{\mathscr{T}}^{\beta\otimes (f_{\mathscr{T}}^\beta)^{(k)}(X)} \\
&\subseteq C_{\mathscr{T}}^{\beta\otimes (f_{\mathscr{T}}^\beta)^{(k-1)}(X)} \\
&= (f_{\mathscr{T}}^\beta)^{(k)}(X).
\end{aligned}$$

For (3). If $n = 1$, there has

$$(f_{\mathscr{T}}^{\beta})^{(1)}(X) = C_{\mathscr{T}}^{\beta\otimes X} \subseteq C_{\mathscr{T}}^{X} = X.$$

For any $q \in \mathscr{F}_p$,

$$\begin{aligned} C_{\mathscr{T}}^{\beta\otimes X}(q) &= \bigwedge_{T\in\mathscr{T}} [\pi(\beta\otimes X \subseteq T) \to T(q)] \\ &\geqslant \bigwedge_{T\in\mathscr{T}} [(\beta\otimes X(q) \to T(q)) \to T(q)] \\ &= \bigwedge_{T\in\mathscr{T}} [(\beta\otimes X(q)) \vee T(q)] \\ &\geqslant \bigwedge_{T\in\mathscr{T}} T(q) \\ &= X(q). \end{aligned}$$

Hence, $(f_{\mathscr{T}}^{\beta})^{(1)}(X) = X$. Let $(f_{\mathscr{T}}^{\beta})^{(n)}(X) = X$, then

$$(f_{\mathscr{T}}^{\beta})^{(n+1)}(X) = (f_{\mathscr{T}}^{\beta})^{(1)}(X) = X. \ \square$$

If $\mathscr{T}$ is given, then for any $X \in \mathscr{F}_L(\mathscr{F}_p)$ and any $\alpha, \beta, \tau \in L$, we can obtain the following rule chain:

$$\begin{aligned} &\textit{Rule } 1: \text{ if } (f_{\mathscr{T}}^{\beta})^{(0)}(X), \text{then } (f_{\mathscr{T}}^{\beta})^{(1)}(X); \\ &\textit{Rule } n: \text{ if } (f_{\mathscr{T}}^{\beta})^{(n-1)}(X), \text{then } (f_{\mathscr{T}}^{\beta})^{(n)}(X); \quad (n = 2, 3, \ldots). \end{aligned}$$

which is called a rule chain generated by X. The following Theorem 11.2.6 shows every rule in the rule chain generated by X is $(\alpha, \beta, \tau, \mathscr{T})$-*i* type representable in L_{vpl}.

Theorem 11.2.6 *In the rule chain above, if X is τ'-i type consistent w.r.t. $(\alpha, \beta, \mathscr{T})$, then every rule is $(\alpha, \beta, \tau, \mathscr{T})$-i type representable in L_{vpl}.*

Proof. Since X is τ'-*i* type consistent w.r.t. $(\alpha, \beta, \mathscr{T})$, then it follows from Theorems 11.2.5 and 11.2.1 that $(f_{\mathscr{T}}^{\beta})^{(n)}(X)$ is τ'-*i* type consistent w.r.t. $(\alpha, \beta, \mathscr{T})$, $n \in \mathbb{N}^+$. Also

$$\begin{aligned} C_{\mathscr{T}}^{(f_{\mathscr{T}}^{\beta})^{(n-1)}(X)} &\supseteq C_{\mathscr{T}}^{\beta\otimes(f_{\mathscr{T}}^{\beta})^{(n-1)}(X)} \\ &= (f_{\mathscr{T}}^{\beta})^{(n)}(X) \\ &\supseteq \tau' \otimes (f_{\mathscr{T}}^{\beta})^{(n)}(X) \end{aligned}$$

for $n \in \mathbb{N}^+$. Hence, every rule in the above rule chain is $(\alpha, \beta, \tau, \mathscr{T})$-*i* type representable in L_{vpl}. $\square$

Remark 14. If $\beta = I$, then there exists only one rule in the above rule chain, *i.e.*, If X, then $C_{\mathscr{T}}^{X}$.

Remark 15. If $\beta = I$, and $X \in \mathscr{T}$, then there exists unique rule, so the above rule chains are changed to If X, then X.

Theorem 11.2.7 *Let $\mathscr{T}^* \subseteq \mathscr{F}_L(\mathscr{F}_p)$. If $\mathscr{T} \subseteq \{T | T$ is τ'-i type consistent w.r.t. $(\alpha, \beta, \mathscr{T}^*)\}$, then*

$$Rule : if \bigcap_{T \in \mathscr{T}} T, \ then \bigcap_{T \in \mathscr{T}} (\theta \otimes T), \ for\ any\ \theta \in L,$$

$$Rule : if \bigcap_{T \in \mathscr{T}} T, \ then\ \theta \otimes \bigcap_{T \in \mathscr{T}} T, \ for\ any\ \theta \in L,$$

are $(\alpha, \beta, \tau, \mathscr{T}^*)$-$i$ type representable in L_{vpl}. □

Just like the reasoning in classical logic, if uncertainty reasoning is based on the rule in Eq. (11.5), then generally it is required that not only the rule should be representable to certain extent, but also new input information should be consistent to certain extent. Otherwise, the reasoning consequence cannot be reasonable. Therefore, we propose the following definition.

Definition 11.2.3 *The uncertainty reasoning model Eq.* (11.5) *is called an $(\alpha, \beta, \tau, \mathscr{T})$-$i$ type regular uncertainty reasoning model, if*

(1). The rule in Eq. (11.5) *is $(\alpha, \beta, \tau, \mathscr{T})$-$i$ type representable.*
(2). New input information $\tilde{X}$ is τ'-i type consistent w.r.t. $(\alpha, \beta, \mathscr{T})$.

Uncertainty reasoning consequence can be given by the following equation:

$$\tilde{Y} = C_{\mathscr{T}}^{\beta \otimes \tilde{X}}, \tag{11.7}$$

where β and $\mathscr{T}$ are obtained from the representation of the rule in Eq. (11.5).

Unless otherwise stated, we always suppose in the following that uncertainty reasoning model Eq. (11.5) is an $(\alpha, \beta, \tau, \mathscr{T})$-$i$ type regular uncertainty reasoning model.

Note that generally α, β, τ and $\mathscr{T}$ are not obtained uniquely by the representation of uncertainty rule. The reasoning consequence depends on the selection of α, β, τ and $\mathscr{T}$ despite uncertainty rule is representable and uncertain reasoning model is regular. This is just because L_{vpl} is a class of lattice-valued logical systems which are different due to the different α, β, τ and $\mathscr{T}$. Hence, the different selection of α, β, τ and $\mathscr{T}$ in uncertainty reasoning means the different uncertainty reasonings in different logical systems. Consequently, the reasoning consequence depends on the selection of α, β, τ and $\mathscr{T}$ (*i.e.*, the corresponding logical system). It follows from L_{vpl} that if there exist α, β, τ and $\mathscr{R}$ such that

$$C_{(C_{\mathscr{T}}^{\emptyset}, \mathscr{R}(\alpha - i))}^{\beta, \tilde{X}} \in \mathscr{T},$$

then the reasoning consequence obtained from Eq. (11.7) can also be obtained by a strict formal deduction in system L_{vpl}.

From the definition of $C^{\beta,\tilde{X}}_{(C^{\emptyset}_{\mathscr{T}},\mathscr{R}(\alpha-i))} \in \mathscr{T}$ and Theorem 9.2.3, we have the following theorem.

Theorem 11.2.8 *Given* $\mathscr{R}$ *and* α. *If*

$$\mathscr{T} = \{T|T \text{ is } \alpha\text{-}i \text{ type closed w.r.t. } \mathscr{R}\},$$

then $C^{\beta,\tilde{X}}_{(C^{\emptyset}_{\mathscr{T}},\mathscr{R}(\alpha-i))} \in \mathscr{T}$. □

Theorem 11.2.9 *Given* $\mathscr{R}$ *and* α. *If* $\mathscr{T} = \{T|T \supseteq X\}$, *then when*

(1). $\beta \otimes \tilde{X} \supseteq X$, *or*
(2). $\beta \otimes \tau' \otimes Y \supseteq X$,

we have

$$C^{\beta,\tilde{X}}_{(C^{\emptyset}_{\mathscr{T}},\mathscr{R}(\alpha-i))} \in \mathscr{T}.$$

Proof. Note that for any

$$Y^{\circ} \in \{Y \mid Y \supseteq \beta \otimes (C^{\emptyset}_{\mathscr{T}} \cup \tilde{X}), Y \text{ is } \alpha\text{-}i \text{ type closed w.r.t. } \mathscr{R}\},$$

we have

$$\begin{aligned} Y^{\circ} &\supseteq \beta \otimes (C^{\emptyset}_{\mathscr{T}} \cup \tilde{X}) \\ &\supseteq \beta \otimes \tilde{X} \\ &\supseteq X, \end{aligned}$$

and

$$\begin{aligned} Y^{\circ} &\supseteq \beta \otimes (C^{\emptyset}_{\mathscr{T}} \cup \tilde{X}) \\ &\supseteq \beta \otimes C^{\emptyset}_{\mathscr{T}} \\ &\supseteq \beta \otimes \tau' \otimes Y \\ &\supseteq X. \end{aligned}$$

Hence,

$$\bigcap\{Y \mid Y \supseteq \beta \otimes (C^{\emptyset}_{\mathscr{T}} \cup \tilde{X}), Y \text{ is } \alpha - i \text{ type closed w.r.t. } \mathscr{R}\} \in \mathscr{T},$$

i.e., $C^{\beta,\tilde{X}}_{(C^{\emptyset}_{\mathscr{T}},\mathscr{R}(\alpha-i))} \in \mathscr{T}$. □

Theorem 11.2.10 *Given* $\mathscr{R}$ *and* α. *In Eq.* (11.5), *if* $\mathscr{T} = \{T|T \supseteq \beta \otimes \tilde{X}\}$, *then*

$$C^{\beta,\tilde{X}}_{(C^{\emptyset}_{\mathscr{T}},\mathscr{R}(\alpha-i))} \in \mathscr{T}. \ \square$$

It follows from Theorem 9.2.3 that

Theorem 11.2.11 *The following properties hold in Eq.* (11.7)*:*

(1). $\tilde{Y} = C_{\mathscr{T}}^{\beta\otimes\tilde{X}}$ *is* τ'*-i type consistent w.r.t.* $(\alpha, \beta, \mathscr{T})$.
(2). If $\tilde{X} = X$*, then* $\tilde{Y} = C_{\mathscr{T}}^{\beta\otimes X}$.
(3). Let $X_1, X_2 \in \mathscr{F}_L(\mathscr{F}_p)$*, then*

$$C_{\mathscr{T}}^{\beta\otimes X_1} \cup C_{\mathscr{T}}^{\beta\otimes X_2} \subseteq C_{\mathscr{T}}^{\beta\otimes(X_1\cup X_2)},$$
$$C_{\mathscr{T}}^{\beta\otimes X_1} \cap C_{\mathscr{T}}^{\beta\otimes X_2} \supseteq C_{\mathscr{T}}^{\beta\otimes(X_1\cap X_2)}.$$

If L *is a completely distributive lattice, then* $=$ *holds in the above two expressions respectively.*

(4). If $\mathscr{T}_2 \subseteq \mathscr{T}_1 \subseteq \mathscr{F}_L(\mathscr{F}_p)$, $\alpha_1 \leqslant \alpha_2$, $\beta_1 \leqslant \beta_2$*, and the rule in Eq.* (11.5) *are not only* $(\alpha_1, \beta_1, \tau, \mathscr{T}_1)$*-i type representable in* L_{vpl}*, but also* $(\alpha_2, \beta_2, \tau, \mathscr{T}_2)$*-i type representable in* L_{vpl}*, then for any* $\tilde{X} \in \mathscr{F}_L(\mathscr{F}_p)$,

$$C_{\mathscr{T}_1}^{\beta_1\otimes\tilde{X}} \subseteq C_{\mathscr{T}_2}^{\beta_2\otimes\tilde{X}}.$$

(5). Let $p_1, p_2 \in \mathscr{F}_p$, $\tilde{X} \in \mathscr{F}_L(\mathscr{F}_p)$*. If* $\mathscr{T} = \mathscr{T}_h$*, then*

$$C_{\mathscr{T}_h}^{\beta\otimes\tilde{X}}(p_1) \vee C_{\mathscr{T}_h}^{\beta\otimes\tilde{X}}(p_2) \leqslant C_{\mathscr{T}_h}^{\beta\otimes\tilde{X}}(p_1 \vee p_2),$$
$$C_{\mathscr{T}_h}^{\beta\otimes\tilde{X}}(p_1) \wedge C_{\mathscr{T}_h}^{\beta\otimes\tilde{X}}(p_2) \geqslant C_{\mathscr{T}_h}^{\beta\otimes\tilde{X}}(p_1 \vee p_2),$$

and

$$C_{\mathscr{T}_h}^{\beta\otimes\tilde{X}}(O) \leqslant \bigvee_{p\in\mathscr{F}_p} (\beta \otimes \tilde{X}(p)) \leqslant \beta \otimes \bigvee_{p\in\mathscr{F}_p} \tilde{X}(p),$$

$$C_{\mathscr{T}_h}^{\beta\otimes\tilde{X}}(I) = I.$$

(6). For any $\tilde{X}, X_1, X_2 \in \mathscr{F}_L(\mathscr{F}_p)$*, there have* $\beta \otimes \tilde{X} \subseteq C_{\mathscr{T}}^{\beta\otimes\tilde{X}}$, $X_1 \subseteq X_2$ *implies* $C_{\mathscr{T}}^{\beta\otimes X_1} \subseteq C_{\mathscr{T}}^{\beta\otimes X_2}$*, and* $C_{\mathscr{T}}^{\beta\otimes\tilde{X}} = C_{\mathscr{T}}^{C_{\mathscr{T}}^{\beta\otimes\tilde{X}}}$.
(7). If $T \in \mathscr{T}$*, then* $\beta \otimes T \subseteq C_{\mathscr{T}}^{\beta\otimes T} \subseteq T$*. If* $\beta \otimes T \in \mathscr{T}$*, then* $\beta \otimes T = C_{\mathscr{T}}^{\beta\otimes T}$.

From Theorem 11.2.6, we have the following theorems.

Theorem 11.2.12 *Every rule in the rule chain generated by* X *satisfies Modus Ponens.* □

Theorem 11.2.13 *Given* $\mathscr{T}$*. Suppose* $\mathscr{R}$ *is* α*-i type sound w.r.t.* $\mathscr{T}$*, the following conclusions hold.*

(1). $C_{\mathscr{T}}^{\beta\otimes\tilde{X}}$ *is* α*-i type closed w.r.t.* $\mathscr{R}$.
(2). If $\mathscr{T} = \{T | T$ *is* α*-i type closed w.r.t.* $\mathscr{R}\}$*, then* $C_{\mathscr{T}}^{\beta\otimes\tilde{X}}$ *is provable in* L_{vpl}.

Proof. (1). Since $\mathscr{R}$ is α-i type sound w.r.t. $\mathscr{T}$, then for any $T \in \mathscr{T}$, T is α-i type closed w.r.t. $\mathscr{R}$. It follows from Theorem 9.2.1 that $\pi(\beta \otimes \tilde{X} \subseteq T) \to T$ is α-i type closed w.r.t. $\mathscr{R}$ for any $T \in \mathscr{T}$. Moreover, from Theorem 9.2.2, we have

$$C_{\mathscr{T}}^{\beta\otimes\tilde{X}} = \bigcap_{T\in\mathscr{T}} (\pi(\beta \otimes \tilde{X} \subseteq T) \to T) \text{ is } \alpha\text{-}i \text{ type closed w.r.t. } \mathscr{R}.$$

(2). It follows from Theorems 9.2.3 and 11.2.8 that

$$C_{\mathscr{T}}^{\beta\otimes\tilde{X}} = C_{(C_{\mathscr{T}}^{\emptyset},\mathscr{R}(\alpha-i))}^{\beta,\tilde{X}} \in \mathscr{T}.$$

Hence, $C_{\mathscr{T}}^{\beta\otimes\tilde{X}}$ is provable in L_{vpl}. □

Corollary 11.2.4 *Given $\mathscr{T}$. Suppose that $\mathscr{R}$ is α-i type sound w.r.t. $\mathscr{T}$.*

(1). If $(r_2^0, t_2^0) \in \mathscr{R}$, $p, q \in \mathscr{F}_p$, then for $i = \mathrm{I}$,

$$C_{\mathscr{T}}^{\beta\otimes\tilde{X}}(q) \geqslant \alpha \otimes C_{\mathscr{T}}^{\beta\otimes\tilde{X}}(p) \otimes C_{\mathscr{T}}^{\beta\otimes\tilde{X}}(p \to q);$$

for $i = \mathrm{II}$,

$$C_{\mathscr{T}}^{\beta\otimes\tilde{X}}(q) \geqslant \alpha \otimes \alpha \otimes C_{\mathscr{T}}^{\beta\otimes\tilde{X}}(p) \otimes C_{\mathscr{T}}^{\beta\otimes\tilde{X}}(p \to q).$$

(2). If $(r_2^0, t_2^*) \in \mathscr{R}$, then for any $p, q \in \mathscr{F}_p$,

$$C_{\mathscr{T}}^{\beta\otimes\tilde{X}}(q) \geqslant \alpha \otimes (C_{\mathscr{T}}^{\beta\otimes\tilde{X}}(p) \wedge C_{\mathscr{T}}^{\beta\otimes\tilde{X}}(p \to q)).$$

(3). If $(r_2^*, t_2^*) \in \mathscr{R}$, then for any $p, q, g \in \mathscr{F}_p$,

$$C_{\mathscr{T}}^{\beta\otimes\tilde{X}}(p \to (q \wedge g)) \geqslant \alpha \otimes (C_{\mathscr{T}}^{\beta\otimes\tilde{X}}(p \to g) \wedge C_{\mathscr{T}}^{\beta\otimes\tilde{X}}(p \to q)).$$

(4). If $(r_2^{\triangle}, t_2^*) \in \mathscr{R}$, then for any $g, p, q \in \mathscr{F}_p$,

$$C_{\mathscr{T}}^{\beta\otimes\tilde{X}}(p \to g) \geqslant \alpha \otimes (C_{\mathscr{T}}^{\beta\otimes\tilde{X}}(p \to q) \wedge C_{\mathscr{T}}^{\beta\otimes\tilde{X}}(q \to g)).$$

(5). If $(r_1^{\theta_0}, t_1^{\theta_0}) \in \mathscr{R}$, $p \in \mathscr{F}_p$, then for $i = \mathrm{I}$,

$$C_{\mathscr{T}}^{\beta\otimes\tilde{X}}(\theta_0 \to p) \geqslant \alpha \otimes (\theta_0 \to C_{\mathscr{T}}^{\beta\otimes\tilde{X}}(p));$$

for $i = \mathrm{II}$,

$$C_{\mathscr{T}}^{\beta\otimes\tilde{X}}(\theta_0 \to p) \geqslant \theta_0 \to (\alpha \otimes C_{\mathscr{T}}^{\beta\otimes\tilde{X}}(p)).$$

(6). In general, if $(r_n, t_n) \in \mathscr{R}$, $p_1, \cdots, p_n \in \mathscr{F}_p$, then for $i = \mathrm{I}$,

$$C_{\mathscr{T}}^{\beta\otimes\tilde{X}}(r_n(p_1, \cdots, p_n)) \geqslant \alpha \otimes t_n(C_{\mathscr{T}}^{\beta\otimes\tilde{X}}(p_1), \cdots, C_{\mathscr{T}}^{\beta\otimes\tilde{X}}(p_n));$$

for $i = \mathrm{II}$,

$$C_{\mathscr{T}}^{\beta\otimes\tilde{X}}(r_n(p_1, \cdots, p_n)) \geqslant t_n(\alpha \otimes C_{\mathscr{T}}^{\beta\otimes\tilde{X}}(p_1), \cdots, \alpha \otimes C_{\mathscr{T}}^{\beta\otimes\tilde{X}}(p_n)).$$ □

11.2.3 Examples

In this section, we take

$$\mathscr{R} = \{(r_1^{\theta_0}, t_1^{\theta_0}) | \theta_0 \in L\} \cup \{(r_2^0, t_2^*), (r_2^*, t_2^*), (r_2^{\triangle}, t_2^*)\}, \quad (11.8)$$
$$\mathscr{T}_i = \{T \mid T \in \mathscr{F}_L(\mathscr{F}_p), T \text{ is } \alpha\text{-}i \text{ type closed w.r.t.} \mathscr{R}\}, \quad (11.9)$$

where $i = \text{I, II}$.

It is obvious that if $\alpha \leqslant \bigwedge_{\theta \in L}(\theta \vee \theta')$, then for $i = \text{I, II}$,

(1). $\mathscr{T}_H \subseteq \mathscr{T}_i$;
(2). $C^{\beta,X}_{(C^{\emptyset}_{\mathscr{T}_i}, \mathscr{R}(\alpha-i))} \in \mathscr{T}_i$ for any $X \in \mathscr{F}_L(\mathscr{F}_p)$, $\beta \in L$;
(3). $\mathscr{R}$ is α-i type sound w.r.t. $\mathscr{T}_i$;
(4). $C^{\beta,X}_{(C^{\emptyset}_{\mathscr{T}_i}, \mathscr{R}(\alpha-i))} = C^{\beta \otimes X}_{\mathscr{T}_i}$ for any $X \in \mathscr{F}_L(\mathscr{F}_p)$, $\beta \in L$;
(5). $C^{\beta \otimes X}_{\mathscr{T}_i}(p) = \bigvee\{\theta \mid \text{there exists } (P^i, (n), X, (p, \theta) - (\alpha, \beta))\}$ holds for any $p \in \mathscr{F}_p$, $X \in \mathscr{F}_L(\mathscr{F}_p)$, $\beta \in L$.

Example 11.2.1 *Take L_{vpl} as Łukasiewicz propositional logic L_u defined in Example 9.2.7. Let $X, \tilde{X}, Y \in \mathscr{F}(\mathscr{F}_p)$ be given respectively as follows:*

$$X(p) = \begin{cases} a, & p = a \in [0,1], \\ 0.2, & otherwise. \end{cases}$$
$$Y(p) = \begin{cases} a^2, & p = a \in [0,1], \\ 0.2, & otherwise. \end{cases}$$
$$\tilde{X}(p) = 0.5 \otimes X.$$

Take $\alpha = 0.4$, $\beta = 0.8$. Note that:

(1). there exists $T_0 \in \mathscr{T}_H$, such that

$$\begin{aligned} C^{0.8 \otimes X}_{\mathscr{T}_i}(q) &= \bigwedge_{T \in \mathscr{T}_i} [(\bigwedge_{p \in \mathscr{F}_p} (0.8 \otimes X(p) \to T(p))) \to T(q)] \\ &= \bigwedge_{T \in \mathscr{T}_i} [(\bigwedge_{\theta \geqslant 0.2} (0.8 \otimes \theta \to T(\theta))) \to T(q)] \\ &\leqslant T_0(q). \end{aligned}$$

Hence, X is 1-i type consistent w.r.t. $(0.4, 0.8, \mathscr{T}_i)$, $i = \text{I, II}$.
(2). Since $\tilde{X}, Y \subseteq X$, then $\tilde{X}, Y$ are 1-i type consistent w.r.t. $(0.4, 0.8, \mathscr{T}_i)$, $i = \text{I, II}$.
(3). $C^X_{\mathscr{T}} \supseteq X \supseteq Y = 1 \otimes Y$.

Consequently, we obtain the following $(0.4, 0.8, 0, \mathscr{T})$-$i$ type regular uncertainty reasoning model:

$$\begin{aligned} &Rule: \ if\ X,\ then\ Y, \\ &Fact: \ \tilde{X}. \end{aligned}$$

So, its consequence can be obtained as follows:

$$\tilde{Y}_i(q) = C_{\mathscr{T}_i}^{0.8\otimes\tilde{X}}(q) = \bigwedge_{T\in\mathscr{T}_i}[(\bigwedge_{\theta\geqslant 0.7}(0.3\otimes\theta\rightarrow T(\theta)))\rightarrow T(q)],$$

and it is 1*-i type consistent w.r.t.* $(0.4, 0.8, \mathscr{T})$ *in* L_u, $i = \mathrm{I}, \mathrm{II}$.

Example 11.2.2 *Take* L_{vpl} *as* F*-valued Łukasiewicz propositional logic* FL_u *defined in Example 9.2.8. The minimal element and the maximal element of* $\mathscr{F}([0,1])$ *are respectively:*

$$\underline{0}\colon\ [0,1]\longrightarrow[0,1], \underline{0}(a)=0, \textit{for any } a\in[0,1],$$

$$\underline{1}\colon\ [0,1]\longrightarrow[0,1], \underline{1}(a)=1, \textit{for any } a\in[0,1].$$

Let $X, \tilde{X}, Y \in \mathscr{F}_{\mathscr{F}([0,1])}(\mathscr{F}_p)$ *be defined respectively as follows:*

$$X(p) = \begin{cases} f, & p = f\in\mathscr{F}([0,1]), \\ x\wedge 0.4, & \textit{otherwise}, \end{cases}$$
$$Y = x^2\otimes X,$$
$$\tilde{X} = x^3\wedge X.$$

Take $\alpha = 0.4\wedge x$, $\beta = 0.6\wedge x$, *where,* $x\colon\ [0,1]\longrightarrow[0,1]$, $x(a)=a$. *Note that:*

(1). There exists $T_0\in\mathscr{T}_H$ *such that*

$$\begin{aligned} C_{\mathscr{T}_i}^{(0.6\wedge x)\otimes X}(q) &= \bigwedge_{T\in\mathscr{T}_i}((\bigwedge_{p\in\mathscr{F}_p}((0.6\wedge x)\otimes X(p)\rightarrow T(p)))\rightarrow T(q)) \\ &= \bigwedge_{T\in\mathscr{T}_i}((\bigwedge_{f\in\mathscr{F}([0,1])}((0.6\wedge x)\otimes f\rightarrow T(f)))\rightarrow T(q)) \\ &\leqslant T_0(q). \end{aligned}$$

Hence, X *is* $\underline{1}$*-i type consistent w.r.t.* $(0.4\wedge x, 0.6\wedge x, \mathscr{T}_i)$, $i=\mathrm{I},\mathrm{II}$.

(2). Since $\tilde{X}, Y\subseteq X$, *then* $\tilde{X}, Y$ *are* $\underline{1}$*-i type consistent w.r.t.* $(0.4\wedge x, 0.6\wedge x)$, $i=\mathrm{I},\mathrm{II}$.

(3). $C_{\mathscr{T}}^X\supseteq X\supseteq Y=\underline{1}\otimes Y$.

Consequently, a $(0.4\wedge x, 0.6\wedge x, \underline{0}, \mathscr{T}_i)$*-i type regular uncertainty reasoning model can be obtained:*

$$\begin{aligned} &Rule:\ \textit{If } X,\ \textit{then } Y, \\ &Fact:\ \tilde{X}. \end{aligned}$$

So its consequence can be obtained as follows:

$$
\begin{aligned}
\tilde{Y}_i(q) &= C_{\mathscr{T}_i}^{(0.6\wedge x)\otimes \tilde{X}}(q) \\
&= \bigwedge_{T\in\mathscr{T}_i} [(\bigwedge_{p\in\mathscr{F}_p} ((0.6\wedge x)\otimes (X(p)\wedge x^2)) \to T(p))) \to T(q)] \\
&= \bigwedge_{T\in\mathscr{T}_i} [(\bigwedge_{f\in\mathscr{F}([0,1])} ((0.6\wedge x)\otimes (f\wedge x^2)) \to T(f))) \to T(q)]
\end{aligned}
$$

and it is $\underline{1}$*-i type consistent w.r.t.* $(0.4\wedge x, 0.6\wedge x, \mathscr{T}_i)$, $i = \mathrm{I}, \mathrm{II}$.

Example 11.2.3 *Take* L_{vpl} *as* $\mathscr{C}_5$ *defined in Example 9.2.9. Let* $X, \tilde{X}, Y \in \mathscr{F}_{C_5}(\mathscr{F}_p)$ *be defined respectively as follows:*

$$
\begin{aligned}
X(p) &= \begin{cases} p, & p\in\{O,a,b,c,I\}, \\ O, & otherwise; \end{cases} \\
Y &= C_{\mathscr{T}_i}^X; \\
\tilde{X}(O) &= O; \\
\tilde{X}(a) &= a; \\
\tilde{X}(b) &= a; \\
\tilde{X}(c) &= b; \\
\tilde{X}(I) &= I; \\
\tilde{X}(p) &= O, p\in\mathscr{F}_p\setminus\{O,a,b,c,I\}.
\end{aligned}
$$

Take $\alpha = b$, $\beta = a$. *Note that:*

(1). There exists $T_0 \in \mathscr{T}_H$ *such that*

$$
\begin{aligned}
C_{\mathscr{T}_i}^{\beta\otimes X}(q) &= \bigwedge_{T\in\mathscr{T}_i} [(\bigwedge_{p\in\mathscr{F}_p} (\beta\otimes X(p) \to T(p))) \to T(q)] \\
&= \bigwedge_{T\in\mathscr{T}_i} [(a\to T(I)) \to T(q)] \\
&\leqslant T_0(q).
\end{aligned}
$$

Hence, X *is* I*-i type consistent w.r.t.* $(b, a, \mathscr{T})$, $i = \mathrm{I}, \mathrm{II}$.

(2). Note that

$$
C_{\mathscr{T}_i}^{\beta\otimes Y} = C_{\mathscr{T}_i}^{\beta\otimes C_{\mathscr{T}_i}^X} \subseteq C_{\mathscr{T}_i}^{C_{\mathscr{T}}^X} = C_{\mathscr{T}_i}^X,
$$

and there exists $T_1 \in \mathscr{T}_H$ *such that*

$$
\begin{aligned}
C_{\mathscr{T}_i}^X(q) &= \bigwedge_{T\in\mathscr{T}_i} [(\bigwedge_{p\in\mathscr{F}_p} (X(p)\to T(p))) \to T(q)] \\
&= \bigwedge_{T\in\mathscr{T}_i} [((a\to T(a))\wedge(b\to T(b))\wedge(c\to T(c))\wedge T(I)) \to T(q)] \\
&\leqslant T_1(q).
\end{aligned}
$$

Hence, $C^{\beta\otimes Y}_{\mathscr{T}_i}(q) \leqslant T_1(q)$. *Consequently,* Y *is* I*-*i *type consistent w.r.t.* $(b, a, \mathscr{T}_i)$, $i = \text{I}, \text{II}$.

(3). Since $\tilde{X} \subseteq X$, *so* $\tilde{X}$ *is* I*-*i *type consistent w.r.t.* $(b, a, \mathscr{T}_i)$, $i = \text{I}, \text{II}$.

(4). $C^{X}_{\mathscr{T}_i} = Y = I \otimes Y$.

Consequently, a $(b, a, O, \mathscr{T}_i)$*-*i *type regular uncertainty reasoning model as follows can be obtained:*

$$\begin{aligned} &Rule: \ if\ X,\ then\ Y, \\ &Fact: \ \tilde{X}. \end{aligned}$$

So, its consequence can be obtained as follows:

$$\begin{aligned} \tilde{Y}_i(q) &= C^{a\otimes\tilde{X}}_{\mathscr{T}_i}(q) \\ &= \bigwedge_{T\in\mathscr{T}_i}((\bigwedge_{p\in\mathscr{F}_p}(a\otimes\tilde{X}(p)\to T(p)))\to T(q)) \\ &= \bigwedge_{T\in\mathscr{T}_i}((a\to T(I))\to T(q)), \end{aligned}$$

and it is I*-*i *type consistent w.r.t.* $(b, a, \mathscr{T})$, $i = \text{I}, \text{II}$, *and can also be proved formally in* $\mathscr{C}_5$.

11.2.4 Multi-Dimensional and Multiple Uncertainty Reasoning

Models and Methods The multi-dimensional and multiple uncertainty reasoning model is

$$\begin{aligned} Rules:\ & \text{If } (u_1 \text{ is } A_{k1}, \beta_{k1}) \text{ and } \cdots \text{ and } (u_j \text{ is } A_{kj}, \beta_{kj}) \\ & \text{and } \cdots \text{ and } (u_n \text{ is } A_{kn}, \beta_{kn}), \text{Then } (y \text{ is } B_k, \theta_k), \\ & \quad k = 1, \ldots, m, \\ Fact:\ & (u_1 \text{ is } A_1, \beta_1) \text{ and } \cdots \text{ and } (u_j \text{ is } A_j, \beta_j) \\ & \text{and } \cdots \text{ and } (u_n \text{ is } A_n, \beta_n), \\ \hline Conclusion:\ & (y \text{ is } B, \theta). \end{aligned} \tag{11.10}$$

Where $\beta_{kj}, \beta_j, \theta_k, \theta \in L$, $(u_j$ is $A_{kj})$, $(u_j$ is $A_j)$, $(y$ is $B_k)$, $(y$ is $B) \in \mathscr{F}_p$, $(u_j$ is A_{kj}, $\beta_{kj})$ shows the truth value of the lattice-valued proposition $(u_j$ is $A_{kj})$ (for short, p_{kj}) is β_{kj}, $(u_j$ is A_j, β_j) shows the truth value of the lattice-valued proposition $(u_j$ is $A_j)$ (for short, p_j) is β_j, $(y$ is B_k, $\theta_k)$ shows the truth value of the lattice-valued proposition $(y$ is $B_k)$ (for short, q_k) is θ_k, $k = 1, \cdots, m$, $j = 1, \cdots, n$, $(y$ is B, $\theta)$ shows the truth value of the lattice-valued proposition $(y$ is $B)$ (for short, q) is θ.

For this uncertainty reasoning model, we have the following method.

Let X_k, $X^{\sim}$, Y_k, $Y^{\sim} \in \mathscr{F}_L(\mathscr{F}_p)$, such that

$$X_k(p_{kj}) = \beta_{kj},$$
$$X^{\sim}(p_j) = \beta_j,$$
$$Y_k(q_k) = \theta_k,$$

where $k = 1, \cdots, m$, $j = 1, \cdots, n$, and

$$Y^{\sim}(q) = \theta.$$

The uncertainty reasoning model Eq. (11.10) can be rewritten as follows

$$\begin{array}{rl} Rules: & \text{If } X_k, \text{ then } Y_k, k = 1, \cdots, m, \\ Fact: & X^{\sim}, \\ \hline Conclusion: & Y^{\sim}. \end{array} \tag{11.11}$$

For $k = 1, \cdots, m$, we call the following model as the k-th submodel and write as M_k:

$$\begin{array}{rl} Rules\ k: & \text{If } X_k, \text{ then } Y_k, \\ Fact: & X^{\sim}, \\ \hline Conclusion: & Y_k^{\sim}, \end{array}$$

where $Y_k^{\sim} = C_{\mathscr{T}(k)}^{\delta_k \otimes X^{\sim}}$.

Definition 11.2.4 *The model M_k is said to be $(\alpha_k, \delta_k, \tau_k, \mathscr{T}(k))$-i type regular, if*

(1). The k-th rule is $(\alpha_k, \delta_k, \tau_k, \mathscr{T}(k))$-i type representable;
(2). $X^{\sim}$ is τ_k'-I type consistent w.r.t. $(\alpha_k, \delta_k, \mathscr{T}(k))$.

The uncertainty reasoning consequence for $(\alpha_k, \delta_k, \tau_k, \mathscr{T}(k))$-$i$ type regular uncertainty reasoning model M_k can be given by the following equation:

$$Y_k^{\sim} = C_{\mathscr{T}(k)}^{\delta_k \otimes X^{\sim}}. \tag{11.12}$$

So in the model Eq. (11.11), $Y^{\sim}$ has two forms:

(1). If we explain the relation among the rules of Eq. (11.11) as disjunction, then

$$Y^{\sim} = \bigcup_{k=1}^{m} C_{\mathscr{T}(k)}^{\delta_k \otimes X^{\sim}}. \tag{11.13}$$

(2). If we explain the relation among the rules of Eq. (11.11) as conjunction, then

$$Y^{\sim} = \bigcap_{k=1}^{m} C_{\mathscr{T}(k)}^{\delta_k \otimes X^{\sim}}. \tag{11.14}$$

Semantical Interpretation and Syntactical Proof

Theorem 11.2.14 *Let $\mathscr{R}$, $\mathscr{R}_k$ are rule sets, $k = 1, 2$, $X \in \mathscr{F}_L(\mathscr{F}_p)$. If denote X is τ'-i type consistent w.r.t. $(\alpha, \beta, \mathscr{T})$ in the rule set $\mathscr{R}$ as X is τ'-i type consistent w.r.t. $(\alpha, \beta, \mathscr{T}, \mathscr{R})$, then*

(1). If $\alpha_1 \leqslant \alpha_2$, and X is τ'-i type consistent w.r.t. $(\alpha_2, \beta, \mathscr{T}, \mathscr{R})$, then X is τ'-i type consistent w.r.t. $(\alpha_1, \beta, \mathscr{T}, \mathscr{R})$.
(2). If $\beta_1 \leqslant \beta_2$, and X is τ'-i type consistent w.r.t. $(\alpha, \beta_2, \mathscr{T}, \mathscr{R})$, then X is τ'-i type consistent w.r.t. $(\alpha, \beta_1, \mathscr{T}, \mathscr{R})$.
(3). If $\tau_1 \leqslant \tau_2$, and X is τ_2'-i type consistent w.r.t. $(\alpha, \beta, \mathscr{T}, \mathscr{R})$, then X is τ_1'-i type consistent w.r.t. $(\alpha, \beta, \mathscr{T}, \mathscr{R})$.
(4). If $\mathscr{T}_1 \subseteq \mathscr{T}_2$, and X is τ'-i type consistent w.r.t. $(\alpha, \beta, \mathscr{T}_1, \mathscr{R})$, then X is τ'-i type consistent w.r.t. $(\alpha, \beta, \mathscr{T}_2, \mathscr{R})$.
(5). If $\mathscr{R}_1 \subseteq \mathscr{R}_2$, and X is τ'-i type consistent w.r.t. $(\alpha, \beta, \mathscr{T}, \mathscr{R}_2)$, then X is τ'-i type consistent w.r.t. $(\alpha, \beta, \mathscr{T}, \mathscr{R}_1)$. □

Theorem 11.2.15 *For Eqs.* (11.13) *and* (11.14), *we have*

$$C^{\left(\bigwedge\limits_{k=1}^{m}\delta_k\right)\otimes X^{\sim}}_{\bigcup\limits_{k=1}^{m}\mathscr{T}(k)} \subseteq Y^{\sim} \subseteq C^{\left(\bigvee\limits_{k=1}^{m}\delta_k\right)\otimes X^{\sim}}_{\bigcap\limits_{k=1}^{m}\mathscr{T}(k)}. \quad \square \tag{11.15}$$

Lemma 11.2.1 *Assume $\mathscr{T}(k) = \{T \mid T$ is α_k-i type closed w.r.t. $\mathscr{R}\}$, then*

(1).

$$\bigcap_{k=1}^{m} \mathscr{T}(k) = \{T \mid T \text{ is } (\bigvee_{k=1}^{m} \alpha_k)\text{-}i \text{ type closed w.r.t. } \mathscr{R}\}. \tag{11.16}$$

(2). If $\bigwedge_{k=1}^{m} \alpha_k = \alpha_{k_0}$, $k_0 \in \{1, 2, \cdots, m\}$, then

$$\bigcup_{k=1}^{m} \mathscr{T}(k) = \{T \mid T \text{ is } (\bigwedge_{k=1}^{m} \alpha_k)\text{-}i \text{ type closed w.r.t. } \mathscr{R}\}. \tag{11.17}$$

Proof. If $T \in \bigcap_{k=1}^{m} \mathscr{T}(k)$, then for any $k \in \{1, \cdots, m\}$, T is α_k-i type closed w.r.t. $\mathscr{R}$. By Theorem 9.2.3, T is $(\bigvee_{k=1}^{m} \alpha_k)$-$i$ type closed w.r.t. $\mathscr{R}$. Conversely, according to Theorem 9.2.3, for any $k \in \{1, \cdots, m\}$, T is α_k-i type closed w.r.t. $\mathscr{R}$. Therefore, $T \in \bigcap_{k=1}^{m} \mathscr{T}(k)$.

Similarly, we can prove (2). □

Theorem 11.2.16 *Assume*

$$\mathscr{T}(k) = \{T \mid T \text{ is } \alpha_k\text{-}i \text{ type closed w.r.t. } \mathscr{R}\}, k \in \{1, \cdots, m\},$$

and

$$\bigvee_{k=1}^{m} \delta_k = \beta, \bigvee_{k=1}^{m} \alpha_k = \alpha, \bigcap_{k=1}^{m} \mathscr{T}(k) = \mathscr{T},$$

then $C_{\mathscr{T}}^{\beta \otimes X}$ is provable in L_{vpl}.

Proof. If T is α-i type closed w.r.t. $\mathscr{R}$, then by Theorem 9.2.3, for any $k \in \{1, \cdots, m\}$, T is α_k-i type closed w.r.t. $\mathscr{R}$, therefore, $C^{\beta,X}_{(C^{\emptyset}_{\mathscr{T}},\mathscr{R}(\alpha-i))}$ is α_k-i type closed w.r.t. $\mathscr{R}$ by Definition 9.2.4 and Theorem 9.2.2. Hence, according to Lemma 11.2.1, we have

$$C^{\beta,X}_{(C^{\emptyset}_{\mathscr{T}},\mathscr{R}(\alpha-i))} \in \bigcap_{k=1}^{m} \mathscr{T}(k) = \mathscr{T}.$$

And then $C^{\beta\otimes X}_{\mathscr{T}}$ is provable in L_{vpl} by Theorems 9.2.9 and 9.2.10. □

Theorem 11.2.16 shows that if $Y^{\sim}$ in Eq. (11.13) is

$$Y^{\sim} = C^{\left(\bigvee\limits_{k=1}^{m} \delta_k\right)\otimes X^{\sim}}_{\bigcap\limits_{k=1}^{m} \mathscr{T}(k)}, \tag{11.18}$$

then the results of uncertainty reasoning obtained from semantics can be proved by a strict formal deduction in L_{vpl}.

Theorem 11.2.17 *Assume*

$$\mathscr{T}(k) = \{T \mid T \text{ is } \alpha_k - i \text{ type closed w.r.t. } \mathscr{R}\}, k \in \{1, \cdots, m\},$$

and

$$\bigcup_{k=1}^{m} \mathscr{T}(k) = \mathscr{T}, \bigwedge_{k=1}^{m} \delta_k = \beta, \bigwedge_{k=1}^{m} \alpha_k = \alpha_{k_0}$$

for $k_0 \in \{1, \cdots, m\}$, *then* $C^{\beta\otimes X}_{\mathscr{T}}$ *is provable in* L_{vpl}.

Proof. By Definition 9.2.4 and $\bigwedge_{k=1}^{m} \alpha_k = \alpha_{k_0}$, $k_0 \in \{1, \cdots, m\}$, we can get

$$C^{\beta,X}_{(C^{\emptyset}_{\mathscr{T}},\mathscr{R}(\bigwedge\limits_{k=1}^{m} \alpha_k - i))} = C^{\beta,X}_{(C^{\emptyset}_{\mathscr{T}},\mathscr{R}(\alpha_{k_0}-i))} \in \mathscr{T}(k_0) \subseteq \mathscr{T}.$$

Therefore, $C^{\beta\otimes X}_{\mathscr{T}}$ is provable in L_{vpl} by Theorems 9.2.9 and 9.2.10. □

Theorem 11.2.17 shows that if $\bigwedge_{k=1}^{m} \alpha_k$ is reachable and $Y^{\sim}$ in Eq. (11.14) is

$$Y^{\sim} = C^{\left(\bigwedge\limits_{k=1}^{m} \delta_k\right)\otimes X^{\sim}}_{\bigcup\limits_{k=1}^{m} \mathscr{T}(k)}, \tag{11.19}$$

then the results of uncertainty reasoning obtained from semantics can be proved from syntax in L_{vpl}.

Lemma 11.2.2 *Let* $\mathscr{T}(k) = \{T \mid T \supseteq \delta_k \otimes X_k\}$, $k = 1, \cdots, m$, *then*

(1).

$$\bigcap_{k=1}^{m} \mathscr{T}(k) = \{T \mid T \supseteq \bigcup_{k=1}^{m} (\delta_k \otimes X_k)\}, \tag{11.20}$$

(2).

$$\bigcup_{k=1}^{m} \mathscr{T}(k) \subseteq \{T \mid T \supseteq \bigcap_{k=1}^{m} (\delta_k \otimes X_k)\}. \ \square \tag{11.21}$$

Proof. (1).

$$\begin{aligned} T \in \bigcap_{k=1}^{m} \mathscr{T}(k) \ &\text{iff} \quad \text{for any } k \in \{1, \cdots, m\}, T \in \mathscr{T}(k), \\ &\text{iff} \quad \text{for any } k \in \{1, \cdots, m\}, T \supseteq \delta_k \otimes X_k, \\ &\text{iff} \ \ T \supseteq \bigcup_{k=1}^{m} (\delta_k \otimes X_k). \end{aligned}$$

(2). Obviously. $\square$

Theorem 11.2.18 *Let* $\mathscr{T}(k) = \{T \mid T \supseteq \delta_k \otimes X_k\}$, $k = 1, \cdots, m$, $\bigvee_{k=1}^{m} \delta_k = \beta$, $\bigvee_{k=1}^{m} \alpha_k = \alpha$, $\bigcap_{k=1}^{m} \mathscr{T}(k) = \mathscr{T}$. *If* $X \supseteq \bigcup_{k=1}^{m} X_k$, *then* $C_{\mathscr{T}}^{\beta \otimes X}$ *is provable in* L_{vpl}.

Proof. From Definition 9.2.4, for any $k \in \{1, \cdots, m\}$,

$$\begin{aligned} C^{\bigvee_{k=1}^{m} \delta_k, X}_{(C^{\emptyset}_{\bigcap_{k=1}^{m} \mathscr{T}(k)}, \mathscr{R}(\bigvee_{k=1}^{m} \alpha_k - i))} &\supseteq \bigvee_{k=1}^{m} \delta_k \otimes X \\ &\supseteq \delta_k \otimes \bigcup_{k=1}^{m} X_k \\ &\supseteq \delta_k \otimes X_k. \end{aligned}$$

Hence, for any $k \in \{1, \cdots, m\}$,

$$C^{\bigvee_{k=1}^{m} \delta_k, X}_{(C^{\emptyset}_{\bigcap_{k=1}^{m} \mathscr{T}(k)}, \mathscr{R}(\bigvee_{k=1}^{m} \alpha_k - i))} \in \mathscr{T}(k).$$

Therefore $C^{\beta, X}_{(C^{\emptyset}_{\mathscr{T}}, \mathscr{R}(\alpha - i))} \in \bigcap_{k=1}^{m} \mathscr{T}(k) = \mathscr{T}$. By Theorems 9.2.9 and 9.2.10, $C_{\mathscr{T}}^{\beta \otimes X}$ is provable in L_{vpl}. $\square$

Theorem 11.2.19 *Let* $\bigwedge_{k=1}^{m} \delta_k = \beta$, $\bigwedge_{k=1}^{m} \alpha_k = \alpha$, $\mathscr{T} = \{T \mid T \supseteq \bigcap_{k=1}^{m} (\delta_k \otimes X_k)\}$. *If* $X \supseteq \bigcup_{k=1}^{m} X_k$, *then* $C_{\mathscr{T}}^{\beta \otimes X}$ *is provable in* L_{vpl}.

Proof.

$$
\begin{aligned}
C^{\bigwedge\limits_{k=1}^{m}\delta_k, X}_{(C^{\emptyset}_{\mathscr{T}}, \mathscr{R}(\alpha-i))} &\supseteq \left(\bigwedge_{k=1}^{m}\delta_k\right)\otimes X \\
&= \bigcap_{k=1}^{m}(\delta_k \otimes X) \\
&\supseteq \bigcap_{k=1}^{m}(\delta_k \otimes X_k).
\end{aligned}
$$

Hence, $C^{\bigwedge_{k=1}^{m}\delta_k, X}_{(C^{\emptyset}_{\mathscr{T}}, \mathscr{R}(\alpha-i))} \in \mathscr{T}$. From Theorems 9.2.9 and 9.2.10, $C^{\beta\otimes X}_{\mathscr{T}}$ is provable in L_{vpl}. □

Theorem 11.2.20 *In the model Eq.* (11.11)*, let* $\mathscr{T}(k) = \{T \mid T \supseteq Y_k\}$*,* $k = 1, \cdots, m$*,* $\bigvee_{k=1}^{m}\delta_k = \beta$*,* $\bigvee_{k=1}^{m}\alpha_k = \alpha$*. If* $Y_k = C^{\delta_k, X_k}_{(C^{\emptyset}_{\mathscr{T}(k)}, \mathscr{R}(\alpha_k-i))} \supseteq X_k$*,* $k = 1, \ldots, m$*, then*

(1). $C^{\beta\otimes\bigcup_{k=1}^{m}X_k}_{\bigcap_{k=1}^{m}\mathscr{T}(k)}$ *is provable in* L_{vpl}*.*

(2).

$$
\bigcup_{k=1}^{m} Y_k = C^{\beta\otimes\bigcup\limits_{k=1}^{m}X_k}_{\bigcap\limits_{k=1}^{m}\mathscr{T}(k)}. \tag{11.22}
$$

Proof. (1) It is easy to know that

$$
\bigcap_{k=1}^{m}\mathscr{T}(k) = \{T \mid T \supseteq \bigcup_{k=1}^{m} Y_k\}.
$$

Denote $\mathscr{T} = \bigcap_{k=1}^{m}\mathscr{T}(k)$,by Theorem 9.2.5, for $k \in \{1, \cdots, m\}$,

$$
C^{\beta, \bigcup\limits_{k=1}^{m}X_k}_{(C^{\emptyset}_{\mathscr{T}}, \mathscr{R}(\alpha-i))} \supseteq C^{\delta_k, X_k}_{(C^{\emptyset}_{\mathscr{T}(k)}, \mathscr{R}(\alpha_k-i))} = Y_k.
$$

Therefore, $C^{\beta, \bigcup_{k=1}^{m}X_k}_{(C^{\emptyset}_{\mathscr{T}}, \mathscr{R}(\alpha-i))} \supseteq \bigcup_{k=1}^{m} Y_k$. Consequently,

$$
C^{\beta, \bigcup\limits_{k=1}^{m}X_k}_{(C^{\emptyset}_{\mathscr{T}}, \mathscr{R}(\alpha-i))} \in \mathscr{T}.
$$

By Theorems 9.2.9 and 9.2.10, $C^{\beta, \bigcup_{k=1}^{m}X_k}_{(C^{\emptyset}_{\mathscr{T}}, \mathscr{R}(\alpha-i))} = C^{\beta\otimes\bigcup_{k=1}^{m}X_k}_{\mathscr{T}}$ and $C^{\beta\otimes\bigcup_{k=1}^{m}X_k}_{\mathscr{T}}$ is provable in L_{vpl}.

(2). For $Y_k \in \mathscr{T}(k)$, from Theorem 9.2.9, $Y_k = C^{\delta_k\otimes X_k}_{\mathscr{T}(k)}$,

$$\begin{aligned}\bigcup_{k=1}^{m} Y_k &= \bigcup_{k=1}^{m} C_{\mathscr{T}(k)}^{\delta_k \otimes X_k} \\ &\subseteq C_{\bigcap\limits_{k=1}^{m} \mathscr{T}(k)}^{\bigcup\limits_{k=1}^{m} (\delta_k \otimes X_k)} \\ &\subseteq C_{\mathscr{T}}^{\bigcup\limits_{k=1}^{m} (\beta \otimes X_k)} \\ &= C_{\mathscr{T}}^{\beta \otimes \bigcup\limits_{k=1}^{m} X_k} \\ &\subseteq C_{\mathscr{T}}^{\bigcup\limits_{k=1}^{m} X_k} \\ &\subseteq C_{\mathscr{T}}^{\bigcup\limits_{k=1}^{m} Y_k} \\ &= \bigcup_{k=1}^{m} Y_k. \ \square\end{aligned}$$

Theorem 11.2.20 shows that in the uncertainty reasoning model Eq. (11.11), for every rule, if we properly select valuation set $\mathscr{T}(k)$ which can make it satisfy Modus Ponens, then the union of conclusion of m rules in Eq. (11.11) can be inferred on semantic and syntax in L_{vpl} by the union of fact of these m rules.

11.3 α-Resolution Principle Based on $LP(X)$

Since Robinson presented the resolution principle in 1965, automated reasoning based on the resolution principle has been extensively studied in areas such as artificial intelligence, logic programming, database theory, problem solving and question answering systems, *etc.*

In this section, we discuss the α-resolution principles based on the lattice-valued propositional logic $LP(X)$.

11.3.1 α-Resolution Principle

Definition 11.3.1 *A lattice-valued propositional logical formula F is called an extremely simple form (ESF), if a lattice-valued propositional logical formula F^* obtained by deleting any constant or literal or implication term occurring in F is not equivalent to F.*

Definition 11.3.2 *A lattice-valued propositional logical formula F is called an indecomposable extremely simple form ($IESF$), if*

(1). F is an ESF containing connective $\rightarrow$ and $'$ at most;

(2). for any $G \in \mathcal{F}$, if $G \in \bar{\mathcal{F}}$ in $\overline{LP(X)}$, then G is an ESF containing connectives $\rightarrow$ and $'$ at most.

Definition 11.3.3 *An IESF F is called an n-IESF if there exist exactly n implication connectives occurring in F.*

Remark 16. For any $p \in \mathcal{F}$, we always consider $p \rightarrow O$ as p'. If $p \in \mathcal{F}$, p has only constants, then p is regarded as one constant and a 0-IESF.

Definition 11.3.4 *All IESFs are called generalized literals.*

Definition 11.3.5 *A lattice-valued propositional logical formula G is called a generalized clause (phrase) if G is a formula of the form*

$$G = g_1 \vee \cdots \vee g_i \vee \cdots \vee g_n, \qquad (G = g_1 \wedge \cdots \wedge g_i \wedge \cdots \wedge g_n) \tag{11.23}$$

where g_i are generalized literals, $i = 1, \cdots, n$.

A conjunction (disjunction) of finite generalized clauses (phrases) is called a generalized conjunctive (disjunctive) normal form.

In the following, that whether α is an element of L or a constant formula in $LP(X)$ can be seen from context.

Definition 11.3.6 (α-resolution) *Let $\alpha \in L$, G_1 and G_2 be two generalized clauses of the form*

$$G_1 = g_1 \vee \cdots \vee g_i \vee \cdots \vee g_m, \tag{11.24}$$

$$G_2 = h_1 \vee \cdots \vee h_j \vee \cdots \vee h_n. \tag{11.25}$$

If $g_i \wedge h_j \leqslant \alpha$, then

$$G = g_1 \vee \cdots \vee g_{i-1} \vee g_{i+1} \vee \cdots \vee g_m \vee h_1 \vee \cdots \vee h_{j-1} \vee h_{j+1} \vee \cdots \vee h_n \tag{11.26}$$

is called an α-resolvent of G_1 and G_2, denoted by $G = R_\alpha(G_1, G_2)$, and (g_i, h_j) called an α-resolution pair, denoted by (g_i, h_j)-α (It can be regarded as the complemented pair in the sense of α-false).

Lemma 11.3.1 *Let x, y, z be literals in $LP(X)$, $\alpha \in L$, $\alpha < I$, and suppose that there exists $\beta \in L$ such that $\beta \wedge (\beta \rightarrow \beta') \nleqslant \alpha$, then $x \wedge (y \rightarrow z) \nleqslant \alpha$.*

Proof. According to Table 11.1, there are nine cases to be considered.

(1). Since there exists a valuation v such that $v(x) = I$, then

$$\begin{aligned} x \wedge (y \rightarrow z) &= x \wedge (x \rightarrow x) \\ &= x \\ &\nleqslant \alpha. \end{aligned}$$

Table 11.1. Nine Cases for x, y, z.

y	z		
	x	x'	$\neq x$ and $\neq x'$
x	case (1)	case (2)	case (3)
x'	case (4)	case (5)	case (6)
$\neq x$ and $\neq x'$	case (7)	case (8)	case (9)

(2). In this case, we have $x \wedge (y \to z) = x \wedge (x \to x')$. Then there exists a valuation v such that $v(x) = \beta$. Hence $v(x \wedge (x \to x')) = \beta \wedge (\beta \to \beta') \nleq \alpha$.
(3). There exists a valuation v such that $v(x) = I$ and $v(z) = I$, so $v(x \wedge (x \to z)) = I \nleq \alpha$.
(4). There exists a valuation v such that $v(x) = I$, so $v(x \wedge (x' \to x)) = I \nleq \alpha$.
(5). There exists a valuation v such that $v(y) = x'$, $v(z) = x'$, so

$$\begin{aligned} x \wedge (y \to z) &= x \wedge (x' \to x') \\ &= x \\ &\nleq \alpha. \end{aligned}$$

(6). There exists a valuation v such that $v(x) = I$, so $v(x \wedge (x' \to z)) = I \nleq \alpha$.
(7). There exists a valuation v such that $v(x) = I$, so

$$\begin{aligned} v(x \wedge (y \to z)) &= v(x \wedge (y \to x)) \\ &= I \nleq \alpha. \end{aligned}$$

(8). There exists a valuation v such that $v(x) = I$ and $v(y) = O$, so

$$\begin{aligned} v(x \wedge (y \to z)) &= v(x \wedge (y \to x')) \\ &= I \nleq \alpha. \end{aligned}$$

(9). There exists a valuation v such that $v(x) = I$ and $v(y) = O$, so

$$v(x \wedge (y \to z)) = I \nleq \alpha. \ \square$$

Corollary 11.3.1 *Let x, y, z be literals in $LP(X)$, and there exist $\beta \in L$, $\alpha < I$ such that $\beta \wedge (\beta \to \beta') \nleq \alpha \in L$, then $(x \to y) \wedge (t \to z) \nleq \alpha$.*

Proof. Since $(x \to y) \wedge (t \to z) \geqslant y \wedge (t \to z)$. It follows from Lemma 11.3.1 that $y \wedge (t \to z) \nleq \alpha$. Hence $(x \to y) \wedge (t \to z) \nleq \alpha$. □

Lemma 11.3.2 *Let $F_m(x_1, \cdots, x_n)$ be an m-IESF ($m \geqslant 1$) with propositional variables $x_1, \cdots, x_n$ and constants $a_1, \cdots, a_t$. Then there exist m_1-IESF $G_1(x_1, \cdots, x_n)$ and m_2-IESF $G_2(x_1, \cdots, x_n)$ such that*

$$F_m(x_1, \cdots, x_n) = G_1(x_1, \cdots, x_n) \to G_2(x_1, \cdots, x_n), \tag{11.27}$$

$$m_1 + m_2 + 1 = m, \tag{11.28}$$

where the propositional variable sets of G_1 *and* G_2 *are subsets of* $\{x_1, \cdots, x_n\}$, *and the constant sets of* G_1 *and* G_2 *are subsets of* $\{a_1, \cdots, a_t\}$.

In the following, $F_m(x_1, \cdots, x_n)$ is denoted as F_m.

Proof. If $m = 1$, then F_1 is a lattice-valued propositional logical formula with exactly one implication connective, *i.e.*, F_1 is of the form

$$F_1 = x_1 \to x_2 \quad \text{or} \quad F_1 = a_1 \to x_1 \quad \text{or} \quad F_1 = x_1 \to a_1,$$

where x_1, x_2, a_1, a_2 are 0-IESF, and Eq. (11.28) holds.

Suppose $F_k = G_1 \to G_2$ and $m_{1k} + m_{2k} + 1 = k$ hold for $m = k$, where G_1, G_2 are m_{1k}-IESF and m_{2k}-IESF, respectively. We prove Eqs. (11.27) and (11.28) hold for $m = k + 1$.

Firstly, note that F_{k+1} has one more implication connective $\to$ than F_k. Hence, we can consider the following cases.

(1). If $F_{k+1} = x \to F_k$, then Eqs. (11.27) and (11.28) obviously hold, and x may be a literal or a constant.

(2). If $F_{k+1} = F_k \to x$, then Eqs. (11.27) and (11.28) obviously hold, and x may be a literal or a contant.

(3). If it is not the cases of (1) and (2), then there exists a k-IESF

$$F_k = G_1 \to G_2,$$

which satisfies Eqs. (11.27) and (11.28). In fact:

a) If we add "$x \to$" or "$\to x$" to G_1 (denoted by G_1^*) such that F_k becomes F_{k+1}, where x may be a literal or a constant, then we have

$$F_{k+1} = G_1^* \to G_2.$$

As G_1 is an m_{1k}-IESF, so G_1^* is an $(m_{1k} + 1)$-IESF. Hence,

$$(m_{1k} + 1) + m_{2k} + 1 = k + 1.$$

Consequently, Eqs. (11.27) and (11.28) hold.

b) If we add "$x \to$" or "$\to x$" to G_2 (denoted by G_2^*) such that F_k becomes F_{k+1}, where x may be a literal or a constant. It can be proved in the same way as G_1. □

Corollary 11.3.2 *Let* $F_m(x_1, \cdots, x_n)$ *be an m-IESF* ($m \geqslant 1$) *with propositional variables* $x_1, \cdots, x_n$ *and constants* $a_1, \cdots, a_t$. *Then there exists a literal* x *or a constant* $a \neq O$ *such that:*

$$F_m(x_1, \cdots, x_n) \geqslant x \quad \text{or} \quad F_m(x_1, \cdots, x_n) \geqslant a. \ \square$$

In the following, we always write $\theta = \bigvee_{a \in L}(a \wedge a')$.

Let α be a dual numerator, there exists $\beta \in L$ such that $\beta \wedge (\beta \rightarrow \beta') \nleqslant \alpha$, $F_m(x_1, \cdots, x_n)$ an m-IESF ($m \geqslant 1$) with propositional variables $x_1, \cdots, x_n$ and constants $a_1, \cdots, a_t$, y a generalized literal, and γ a constant.

If $F_m(x_1, \cdots, x_n) \geqslant x$, then

(1). if $y = x$, then $y \wedge F_m(x_1, \cdots, x_n) \nleqslant \alpha$;
(2). if y is a literal, $y \neq x$, and $y \neq x'$, then $y \wedge F_m(x_1, \cdots, x_n) \nleqslant \alpha$; if $y = x'$, $\theta \nleqslant \alpha$, then $y \wedge F_m(x_1, \cdots, x_n) \nleqslant \alpha$;
(3). if $y = \gamma \nleqslant \alpha$, then $y \wedge F_m(x_1, \cdots, x_n) \nleqslant \alpha$;
(4). if $y = p \rightarrow q$, $y \wedge F_m(x_1, \cdots, x_n) \nleqslant \alpha$.

If $F_m(x_1, \cdots, x_n) \geqslant a$ and $a \nleqslant \alpha$, then

(1). if y is a literal, then $y \wedge F_m(x_1, \cdots, x_n) \nleqslant \alpha$;
(2). if $y = \gamma \nleqslant \alpha$, then $y \wedge F_m(x_1, \cdots, x_n) \nleqslant \alpha$;
(3). if $y = p \rightarrow q$, then $y \wedge F_m(x_1, \cdots, x_n) \nleqslant \alpha$.

Note that $F_m(x_1, \cdots, x_n)$ is an m-IESF ($m \geqslant 1$) with propositional variables $x_1, \cdots, x_n$ and constants $a_1, \cdots, a_t$. Then there exists $x_i \in \{x_1, \cdots, x_n\}$ or $a_j \in \{a_1, \cdots, a_t\}$ such that $F_m(x_1, \cdots, x_n) \geqslant x_i$ or $F_m(x_1, \cdots, x_n) \geqslant x_i'$ or $F_m(x_1, \cdots, x_n) \geqslant a_j$ or $F_m(x_1, \cdots, x_n) \geqslant a_j'$.

Definition 11.3.7 *An IESF F is regular if there exist propositional variables x, y or $a \neq I$ such that $F \geqslant x \rightarrow y$ or $F \geqslant a \rightarrow x$.*

Definition 11.3.8 *An IESF F is called subregular if there exists a propositional variable x such that $F \geqslant x$.*

Obviously, an IESF F is subregular if F is regular.

Definition 11.3.9 *A generalized clause (phrase) G is regular if all non-constant IESFs in G are regular. A generalized conjunctive (disjunctive) normal form S is regular if all non-constant IESFs of S are regular.*

If there is no specification, the IESF mentioned in the following are subregular IESF. Let g be a generalized literal in $LP(X)$, we write

$$D_\alpha(g) = \{h \mid (g, h) - \alpha\},$$

where $D_\alpha(g)$ is called the α-resolution field of g.

Lemma 11.3.3 *Let g be a regular generalized literal of $LP(X)$, $\alpha \in L$, $\alpha < I$, and there exists $\beta \in L$ such that $\beta \wedge (\beta \rightarrow \beta') \nleqslant \alpha$.*

(1). If g is a constant, then
a) if $g \leqslant \alpha$ then $D_\alpha(g) = \{h \mid h$ is a generalized literal of $LP(X)\}$ (g is called an α-resolvent for any generalized literal);
b) if α is a dual numerator, $g \nleqslant \alpha$, then

$$D_\alpha(g) = \{h \mid h \text{ is a constant, } h \leqslant \alpha\}.$$

(2). If g is a literal, then $D_\alpha(g) \supseteq \{h \mid h$ is a constant, $h \leqslant \alpha\}$.
(3). If g is a regular IESF, then $D_\alpha(g) = \{h \mid h$ is a constant, $h \leqslant \alpha\}$.

Proof. (1). a) Since $g \leqslant \alpha$, (g, h) is an α-resolution pair for any generalized literal h.
b) Since α is a dual numerator and $g \nleqslant \alpha$, $g \wedge h \leqslant \alpha$ if and only if $h \leqslant \alpha$ (obviously h is a constant).

(2). Obviously.

(3). a) If h is a constant and $h \leqslant \alpha$, then $h \in D_\alpha(g)$.
b) If h is a constant but $h \nleqslant \alpha$, then we have:
i. if $h > \alpha$, then $g \wedge h \nleqslant \alpha$;
ii. if $h//\alpha$, then there exist x, y such that $g \geqslant x \to y$ and there exists a valuation v such that $v(g) \geqslant v(x \to y) = I$. Hence $v(g \wedge h) = h//\alpha$, and $g \wedge h \nleqslant \alpha$.
Consequently, if h is a constant satisfying $h \nleqslant \alpha$, then $h \notin D_\alpha(g)$.
c) If h is not a constant, then we have
i. if h is a propositional variable, it follows from Lemma 11.3.1 that $g \wedge h \nleqslant \alpha$;
ii. if h is a regular IESF, then h is also a subregular IESF. Hence there exists x such that $h \geqslant x$. It follows from Lemma 11.3.1 that $g \wedge h \nleqslant \alpha$.
Hence, if h is not a constant, then $h \notin D_\alpha(g)$. □

11.3.2 Soundness and Completeness

Definition 11.3.10 *Suppose S is the set of generalized clauses $C_1, \cdots, C_n$ and written as $S = C_1 \wedge \cdots \wedge C_n$, $\alpha \in L$. $\omega = \{D_1, \cdots, D_m\}$ is called an α-resolution deduction from S to generalized clause D_m, if*

(1). $D_i \in \{C_1, \cdots, C_n\}$, or
(2). there exist $j, k < i$, such that $D_i = R_\alpha(D_j, D_k)$.

In the following, if a generalized clause is α-false, then it is called an α-empty clause (for short, denoted by α-◎). If there exists an α-resolution deduction ω from S to an α-◎, then ω is called an α-refutation.

Theorem 11.3.1 (Soundness of α-Resolution Principle) *Suppose a generalized conjunctive normal form $S = C_1 \wedge \cdots \wedge C_n$, $\alpha \in L$, $\{D_1, \cdots, D_m\}$ is an α-resolution deduction from S to a generalized clause D_m. If D_m is α-◎, then $S \leqslant \alpha$, i.e., if $D_m \leqslant \alpha$, then $S \leqslant \alpha$.*

Proof. Suppose S has been resolved as the following form by using of t times α-resolution,

$$S_t = G_1^{(t)} \wedge \cdots \wedge G_{n_t}^{(t)} \wedge S.$$

It needs to prove

(1). if $D_m \leqslant \alpha$, then $S_m \leqslant \alpha$;
(2). if $S_{t+1} \leqslant \alpha$, then $S_t \leqslant \alpha$.

For case (1).

a). If $D_m = C_{i_0}$, then $S_m \leqslant S \leqslant C_{i_0}$. Hence, if $D_m \leqslant \alpha$, then $S_m \leqslant \alpha$.
b). If $D_m = R_\alpha(C_i, C_j)$. Let

$$\begin{aligned} C_i &= p_i \vee C_i^*, \\ C_j &= p_j \vee C_j^*, \end{aligned}$$

where (p_i, p_j) is an α-resolution pair. Then $D_m = C_i^* \vee C_j^*$. Moreover, if $D_m = C_i^* \vee C_j^* \leqslant \alpha$, then

$$\begin{aligned} S_m &\leqslant S \\ &\leqslant C_i \wedge C_j \\ &= (p_i \vee C_i^*) \wedge (p_j \vee C_j^*) \\ &= (p_i \wedge p_j) \vee (p_i \wedge C_j^*) \vee (p_j \wedge C_i^*) \vee (C_i^* \wedge C_j^*) \\ &\leqslant (p_i \wedge p_j) \vee (C_i^* \vee C_j^*) \\ &\leqslant \alpha \vee \alpha \\ &= \alpha. \end{aligned}$$

c). If $D_m = R_\alpha(G_i^{(t)}, C_j)$. Let

$$\begin{aligned} G_i^{(t)} &= p_i \vee G_i^{(t)^*}, \\ C_j &= p_j \vee C_j^*, \end{aligned}$$

where (p_i, p_j) is an α-resolution pair. Then $D_m = G_i^{(t)^*} \vee C_j^*$. Moreover, if $D_m = G_i^{(t)^*} \vee C_j^* \leqslant \alpha$, then

$$\begin{aligned} S_m &\leqslant G_i^{(t)} \wedge C_j \\ &= (p_j \vee G_i^{(t)^*}) \wedge (p_j \vee C_j^*) \\ &\leqslant \alpha. \end{aligned}$$

d). If $D_m = R_\alpha(G_i^{(t)}, G_j^{(t)})$. Let

$$G_i^{(t)} = p_i \vee G_i^{(t)^*}, \qquad G_j^{(t)} = p_j \vee G_j^{(t)^*},$$

where (p_i, p_j) is an α-resolution pair. Then $D_m = G_i^{(t)^*} \vee G_j^{(t)^*}$. Moreover, if $D_m = G_i^{(t)^*} \vee G_j^{(t)^*} \leqslant \alpha$, then $S_m \leqslant G_i^{(t)} \wedge G_j^{(t)} \leqslant \alpha$. Consequently, if $D_m \leqslant \alpha$, then $S_m \leqslant \alpha$.

For case (2).

a). If $D_{t+1} = C_{i_0}$, then $S_{t+1} = S_t$. Hence, if $S_{t+1} \leqslant \alpha$, then $S_t \leqslant \alpha$.
b). If $D_{t+1} = R_\alpha(C_i, C_j)$. Let

$$C_i = p_i \vee C_i^*, \qquad C_j = p_j \vee C_j^*,$$

where (p_i, p_j) is an α-resolution pair, then $S_{t+1} = S_t \wedge (C_i^* \vee C_j^*)$. If $S_{t+1} \leqslant \alpha$, then

$$\begin{aligned} S_t &= G_1^{(t)} \wedge \cdots \wedge G_{n_t}^{(t)} \wedge S \\ &= G_1^{(t)} \wedge \cdots \wedge G_{n_t}^{(t)} \wedge S \wedge C_i \wedge C_j \\ &= S_t \wedge C_i \wedge C_j \\ &\leqslant S_t \wedge ((p_i \wedge p_j) \vee (C_i^* \vee C_j^*)) \\ &\leqslant (p_i \wedge p_j) \vee (S_t \wedge (C_i^* \vee C_j^*)) \\ &\leqslant \alpha \vee \alpha \\ &= \alpha. \end{aligned}$$

c). If $D_{t+1} = R_\alpha(G_i^{(t)}, C_j)$. Let

$$G_i^{(t)} = p_i \vee G_i^{(t)^*}, \qquad C_j = p_j \vee C_j^*,$$

where (p_i, p_j) is an α-resolution pair. Then $S_{t+1} = S_t \wedge (G_i^{(t)^*} \vee C_j^*)$. If $S_{t+1} \leqslant \alpha$, then

$$\begin{aligned} S_t &= S_t \wedge G_i^{(t)} \wedge C_j \\ &\leqslant S_t \wedge ((p_i \wedge p_j) \vee (G_i^{(t)^*} \vee C_j^*)) \\ &\leqslant (p_i \wedge p_j) \vee (S_t \wedge (G_i^{(t)^*} \vee C_j^*)) \\ &\leqslant \alpha \vee \alpha \\ &= \alpha. \end{aligned}$$

d). If $D_{t+1} = R_\alpha(G_i^{(t)}, G_j^{(t)})$. Let

$$G_i^{(t)} = p_i \vee G_i^{(t)^*}, \qquad G_j^{(t)} = p_j \vee G_j^{(t)^*},$$

where (p_i, p_j) is an α-resolution pair. Then $S_{t+1} = S_t \wedge (G_i^{(t)^*} \vee G_j^{(t)^*})$. If $S_{t+1} \leqslant \alpha$, then

$$\begin{aligned} S_t &= S_t \wedge G_i^{(t)} \wedge G_j^{(t)} \\ &\leqslant S_t \wedge ((p_i \wedge p_j) \vee (G_i^{(t)^*} \vee G_j^{(t)^*})) \\ &\leqslant (p_i \wedge p_j) \vee (S_t \wedge (G_i^{(t)^*} \vee G_j^{(t)^*})) \\ &\leqslant \alpha \vee \alpha \\ &= \alpha. \end{aligned}$$

Consequently, if $S_{t+1} \leqslant \alpha$, then $S_t \leqslant \alpha$.

Finally, we have the following conclusions:

$$\begin{aligned}
&\text{if } D_m \leqslant \alpha, \text{ then } S_m \leqslant \alpha;\\
&\text{if } S_m \leqslant \alpha, \text{ then } S_{m-1} \leqslant \alpha;\\
&\qquad\vdots\\
&\text{if } S_{t+1} \leqslant \alpha, \text{ then } S_t \leqslant \alpha;\\
&\qquad\vdots\\
&\text{if } S_1 \leqslant \alpha, \text{ then } S_0 (= S) \leqslant \alpha. \ \square
\end{aligned}$$

Theorem 11.3.2 (Weak Completeness) *Let S be a regular generalized conjunctive normal form, $\alpha \in L$ a dual numerator and $\bigvee_{a\in L}(a \wedge a') \leqslant \alpha < I$. Suppose that there exists $\beta \in L$ such that $\beta \wedge (\beta \to \beta') \nleqslant \alpha$. If $S \leqslant \alpha$, then there exists an α-resolution deduction from S to α-⊚.*

Proof. Firstly, we notice that:

1. S is a regular generalized conjunctive normal form. If S contains some IESFs, then there exist certain 1-generalized conjunctive normal form $S^\star$ (*i.e.*, $S^\star$ contains 1-IESF) such that $S \geqslant S^\star$. It follows from Lemma 11.3.3 that the α-resolution fields of all IESFs of S and $S^\star$ are

$$\{h \mid h \text{ is a constant}, \ h \leqslant \alpha\}.$$

Furthermore, if $F \in \mathcal{F}$ is regular IESF, let $F \geqslant x \to y$ or $F \geqslant \delta \to y$, then for any $b \in L$,

$$\begin{aligned}
b \wedge (x \to y) \leqslant \alpha \ \text{ iff } \ & b \wedge F \leqslant \alpha\\
\text{iff } \ & b \wedge (\delta \to y) \leqslant \alpha\\
\text{iff } \ & b \leqslant \alpha.
\end{aligned}$$

Hence, if we can prove that "*If $S^\star \leqslant \alpha$, then there exists an α-resolution deduction from $S^\star$ to α-⊚*," then we can conclude that "*if $S \leqslant \alpha$, then there exists an α-resolution deduction from S to α-⊚*." Hence, without losing generality, we can suppose that S is a regular 1-generalized conjunctive normal form (in short, generalized conjunctive normal form) without more than one implication connective in every IESF.

2. We can always assume that there exist $\omega \in L$ and C_{i_0} in all the conjunctive terms $C_1, \cdots, C_k$ of S such that ω is a disjunctive term of C_{i_0}. [In fact, if S does not satisfy the assumption, then let $S_0 = \beta \wedge S$, we have:

(1). S_0 satisfies the assumption;
(2). $S \leqslant \alpha$ if and only if $S_0 \leqslant \alpha$;
(3). S can be α-resolved into α-⊚ if and only if S_0 can be α-resolved into α-⊚. (It follows from $\beta \nleqslant \alpha$ that β and a generalized literal G in S can be α-resolved into α-⊚ if and only if $G \leqslant \alpha$.)]

Hence, the following discussion is always based on this assumption: *Suppose that S^* is the generalized conjunctive normal form of S with locks (for a 1-IESF, it only needs to lock the whole implication term and not to lock the generalized literal in it)*. Notice that the value of the same propositional variables (although with different locks) in S^* under a valuation should be equal. Hence

$$S \leqslant \alpha \quad \text{iff} \quad S^* \leqslant \alpha.$$

Let

M_1 – the set of locked literals in S^*,
M_2 – the set of locked constants in S^*,
M_3 – the set of locked 1-IESF in S^*,

and $M = M_1 \cup M_2 \cup M_3$.

In the following, we divide the proof into two parts. In part one, we want to prove that if every generalized conjunctive term is constructed by single generalized literals, then

$$\text{if } S \leqslant \alpha, \text{ then } S \text{ can be } \alpha\text{-resolved into } \alpha\text{-} \circledcirc\,. \tag{11.29}$$

Let

$$S = \alpha_1 \wedge \cdots \wedge \alpha_m \wedge p_1 \wedge \cdots \wedge p_l \wedge (x_1 \to y_1) \wedge \cdots \wedge (x_k \to y_k).$$

Since S is a generalized conjunctive normal form, we suppose that:

- there is no same constant in $\alpha_1, \cdots, \alpha_m$;
- there is no same literal in $p_1, \cdots, p_l$;
- there is no equivalent 1-IESF in $x_1 \to y_1, \cdots, x_k \to y_k$.

Hence, the locked generalized literals in S are not different from the unlocked generalized literals in S (since there exist no equal literals). So S^* is the same as S. We need only to prove the case of S in the first part.

Here let

$$M_1 = \{p_1, \cdots, p_l\},$$
$$M_2 = \{\alpha_1, \cdots, \alpha_m\},$$
$$M_3 = \{x_1 \to y_1, \cdots, x_k \to y_k\}.$$

Now we prove conclusion (11.29) by induction over $|M|$.

(1). $|M| = 1$.

a) $|M_1| = 1$. If $p \in M_1$, then $S = p \nleqslant \alpha$, this is a contradiction to $S \leqslant \alpha$. This case cannot appear.

b) $|M_2| = 1$. If $\alpha_0 \in M_2$, then for $\alpha_0 = S \leqslant \alpha$, α_0 is an α-self-resolvent. Hence S can be α-resolved into α-$\circledcirc$.

c) $|M_3| = 1$. If $x \to y \in M_3$, then $S = x \to y \nleqslant \alpha$, this is a contradiction to $S \leqslant \alpha$. This case cannot appear.

(2). $|M| = 2$. We consider the following six cases.

a) $|M_1| = 2$. If $p, q \in M$, then for $p \wedge q = S \leqslant \alpha$, S can be α-resolved into α-◎ by taking (p, q) as α-resolution pair.

b) $|M_2| = 2$. If $\delta, \gamma \in M_2$, then for $\delta \wedge \gamma = S \leqslant \alpha$, S can be α-resolved into α-◎ by taking (δ, γ) as α-resolution pair.

c) $|M_3| = 2$. If $x \to y, z \to t \in M_3$, then for $(x \to y) \wedge (z \to t) = S \leqslant \alpha$, S can be α-resolved into α-◎ by taking $(x \to y, z \to t)$ as α-resolution pair (Note: this case cannot appear in fact).

d) $|M_1| = |M_2| = 1$. If $p \in M_1$, $\delta \in M_2$, then for $\delta \wedge p = S \leqslant \alpha$, S can be α-resolved into α-◎ by taking (p, δ) as α-resolution pair.

e) $|M_2| = |M_3| = 1$. If $\gamma \in M_2$, $x \to y \in M_3$, then for $\gamma \wedge (x \to y) = S \leqslant \alpha$, S can be α-resolved into α-◎ by taking $(\gamma, x \to y)$ as α-resolution pair.

f) $|M_1| = |M_3| = 1$. If $p \in M_1$, $x \to y \in M_3$, then for $p \wedge (x \to y) = S \leqslant \alpha$, S can be α-resolved into α-◎ by taking $(p, x \to y)$ as α-resolution pair (Note: this case cannot appear in fact).

It is clear that if $S \leqslant \alpha$, then S can be α-resolved into α-◎.

(3). Suppose the result holds for $|M| < n$ ($n \geqslant 3$). Now we need to prove the result for $|M| = n$, and there is no harm in supposing $n = m + l + k$.

a) $M_2 \neq \emptyset$. Let $M_2 = \{\alpha_1, \cdots, \alpha_m\}$.

i. Firstly, we want to prove that if $\alpha_1 \wedge \cdots \wedge \alpha_m \leqslant \alpha$, then there exists an $\alpha_i \in M_2$ such that $\alpha_i \leqslant \alpha$, *i.e.*, α_i is an α-self-resolvent. [In fact, it follows from $\alpha_1 \wedge (\alpha_2 \wedge \cdots \wedge \alpha_m) \leqslant \alpha$ and α is a dual numerator that

$$\alpha_1 \leqslant \alpha \qquad \text{or} \qquad \alpha_2 \wedge \cdots \wedge \alpha_m \leqslant \alpha.$$

So,

A. if $\alpha_1 \leqslant \alpha$, then α_1 is an α-self-resolvent;

B. if $\alpha_1 \nleqslant \alpha$, then $\alpha_2 \wedge \cdots \wedge \alpha_m \leqslant \alpha$.

Repeating the above procedure on $\alpha_2 \wedge \cdots \wedge \alpha_{m-1}$ $m-1$ times, we can conclude that $\alpha_1 \nleqslant \alpha, \cdots, \alpha_{m-1} \nleqslant \alpha$, hence $\alpha_m \leqslant \alpha$, *i.e.*, α_m is an α-self-resolvent.]

ii. Secondly, we want to prove that If $\alpha_1 \wedge \cdots \wedge \alpha_m \nleqslant \alpha$, then $\alpha_1 \nleqslant \alpha$. Notice that

$$S = \alpha_1 \wedge \cdots \wedge \alpha_m \wedge p_1 \wedge \cdots \wedge p_l \wedge (x_1 \to y_1) \wedge \cdots \wedge (x_k \to y_k) \leqslant \alpha.$$

So,

A. if

$$\alpha_2 \wedge \cdots \wedge \alpha_m \wedge p_1 \wedge \cdots \wedge p_l \wedge (x_1 \to y_1) \wedge \cdots \wedge (x_k \to y_k) \leqslant \alpha,$$

then by induction, we have

$$\alpha_2 \wedge \cdots \wedge \alpha_m \wedge p_1 \wedge \cdots \wedge p_l \wedge (x_1 \to y_1) \wedge \cdots \wedge (x_k \to y_k)$$

can be α-resolved into α-⊚. It means that S can be α-resolved into α-⊚;

B. if

$$\alpha_2 \wedge \cdots \wedge \alpha_m \wedge p_1 \wedge \cdots \wedge p_l \wedge (x_1 \to y_1) \wedge \cdots \wedge (x_k \to y_k) \nleqslant \alpha,$$

then there exists a valuation v such that

$$v(\alpha_2 \wedge \cdots \wedge \alpha_m \wedge p_1 \wedge \cdots \wedge p_l \wedge (x_1 \to y_1) \wedge \cdots \wedge (x_k \to y_k)) \nleqslant \alpha.$$

However, it follows from $S \leqslant \alpha$ that

$$\alpha_1 \wedge v(\alpha_2 \wedge \cdots \wedge \alpha_m \wedge p_1 \wedge \cdots \wedge p_l \wedge (x_1 \to y_1) \wedge \cdots \wedge (x_k \to y_k)) \leqslant \alpha.$$

Moreover, since α is a dual numerator, we conclude that $\alpha_1 \leqslant \alpha$, or

$$v(\alpha_2 \wedge \cdots \wedge \alpha_m \wedge p_1 \wedge \cdots \wedge p_l \wedge (x_1 \to y_1) \wedge \cdots \wedge (x_k \to y_k)) \leqslant \alpha.$$

But this contradicts to the fact that $\alpha_1 \nleqslant \alpha$ and

$$v(\alpha_2 \wedge \cdots \wedge \alpha_m \wedge p_1 \wedge \cdots \wedge p_l \wedge (x_1 \to y_1) \wedge \cdots \wedge (x_k \to y_k)) \nleqslant \alpha.$$

Hence,

$$\alpha_2 \wedge \cdots \wedge \alpha_m \wedge p_1 \wedge \cdots \wedge p_l \wedge (x_1 \to y_1) \wedge \cdots \wedge (x_k \to y_k) \leqslant \alpha$$

does not hold.

b) $M_2 = \emptyset$. Select $\delta \nleqslant \alpha$ (specially, we can take $\delta = I$) and let $S_0 = \delta \wedge S$. Notice that

$$S \leqslant \alpha \quad \text{iff} \quad S_0 \leqslant \alpha.$$

We can apply above discussion to S_0 and get *if* $S_0 \leqslant \alpha$, *then* S_0 *can be* α*-resolved into* α-⊚. Hence if $S \leqslant \alpha$, then S_0 can be α-resolved into α-⊚. Since δ is neither α-self-resolvent literal nor literal which can be α-resolved with generalized literal in M, that S_0 can be α-resolved into α-⊚ means S can be α-resolved into α-⊚.

This completed the first part of the proof.

In the second part, we need to prove that *if there exist some generalized conjunctive terms in* S *constructed by at least two generalized literals, and* $S \leqslant \alpha$, *then* S *can be* α*-resolved into* α-⊚.

We prove the conclusion by induction over $|M|$.

When $|M| = 2$, we need to investigate six cases, *i.e.*,

$$|M_1| = 2, \text{ or } |M_2| = 2, \text{ or } |M_3| = 2, \text{ or}$$

$$|M_1| = |M_2| = 1, \text{ or } |M_1| = |M_3| = 1, \text{ or} |M_2| = |M_3| = 1.$$

(1). $|M_1| = 2$. If $p, q \in M_1$, then S only has the form $p \vee q$. But this contradicts to $S \leqslant \alpha$. Hence this case cannot appear.

(2). $|M_2| = 2$. If $\delta, \gamma \in M_2$, then S only has the form $\delta \vee \gamma$. Since $\delta \vee \gamma = S \leqslant \alpha$, $\delta \leqslant \alpha$ and $\gamma \leqslant \alpha$. Hence, both δ and γ are α-self-resolvent, and S can be α-resolved into α-$\circledcirc$ by two α-self-resolutions .

(3). $|M_3| = 2$. If $x \to y, z \to t \in M_3$, then S only has the form $(x \to y) \vee (z \to t)$. But it is a contradiction to $S \leqslant \alpha$. Hence this case cannot appear.

(4). $|M_1| = |M_2| = 1$. If $p \in M_1$, $\gamma \in M_2$, then $S = \gamma \vee p$. But it is a contradiction to $S \leqslant \alpha$. Hence this case cannot appear.

(5). $|M_1| = |M_2| = 1$. Suppose $p \in M_1$, $\gamma \in M_2$. Since we have $S = \delta \vee p$, it is impossible that $S \leqslant \alpha$.

(6). $|M_2| = |M_3| = 1$. Suppose $\gamma \in M_2$ and $x \to y \in M_3$. Since $S = \gamma \vee (x \to y)$, it is impossible that $S \leqslant \alpha$.

Suppose the result holds for $|M| < n$ ($n \geqslant 3$). We need to prove that the result holds for $|M| = n$.

(1). Suppose $M_3 \neq \emptyset$. Let $x \to y \in M_3$ and $S = C_1 \wedge \cdots \wedge C_K$. Without losing generality, we can suppose $C_i = C_i^* \vee (x \to y)$ for $i = t+1, \cdots, K$, and C_i contains no $x \to y$ for $i = 1, \cdots, t$. Hence

$$\begin{aligned} S &= C_1 \wedge \cdots \wedge C_t \wedge ((x \to y) \vee (C_{t+1}^* \wedge \cdots \wedge C_K^*)) \\ &= (C_1 \wedge \cdots \wedge C_t \wedge C_{t+1}^* \wedge \cdots \wedge C_K^*) \vee (C_1 \wedge \cdots \wedge C_t \wedge (x \to y)) \\ &\triangleq S_1 \vee S_2, \end{aligned}$$

where

$$\begin{aligned} S_1 &= C_1 \wedge \cdots \wedge C_t \wedge C_{t+1}^* \wedge \cdots \wedge C_K^*, \\ S_2 &= C_1 \wedge \cdots \wedge C_t \wedge (x \to y). \end{aligned}$$

If we let

$$C_k = \bigvee_{l_k=1}^{L_k} G_{kl_k}, \qquad k = 1, \cdots, t,$$

then

$$S_2 = \bigvee_{f \in \prod\limits_{k \in \{1,\cdots,t\}} \{1,\cdots,L_k\}} \left(\bigwedge_{k \in \{1,\cdots,t\}} G_{kf(k)} \wedge (x \to y) \right).$$

It follows that $S_2 \leqslant \alpha$ from $S \leqslant \alpha$. Hence, we have

$$\bigwedge_{k \in \{1,\cdots,t\}} G_{kf(k)} \wedge (x \to y) \leqslant \alpha,$$

for any $f \in \prod_{k \in \{1,\cdots,t\}} \{1, \cdots, L_k\}$.

a) If we denote the locked generalize literal set of S_1 as M_{S_1} and $M_{S_1} \neq \emptyset$, then for any $f \in \prod_{k\in\{1,\cdots,t\}}\{1,\cdots,L_k\}$, the number of locked generalized literal in $\{G_{1f(1)},\cdots,G_{tf(t)},x \to y\}$ is $t+1<n$.

b) If $M_{S_1}=\emptyset$, *i.e.*, there exists no generalized conjunctive normal form S_1, then $S=S_2$. Hence there exists a C_i in $C_1,\cdots,C_t$ containing at least two generalized literals. Therefore, M contains a locked generalized literal which does not occur in $\{G_{1f(1)},\cdots,G_{tf(t)},x \to y\}$ for any $f \in \prod_{k\in\{1,\cdots,t\}}\{1,\cdots,L_k\}$. Consequently, the number of locked generalized literals in $G_{1f(1)},\cdots,G_{tf(t)},x \to y$ is also $t+1<n$.

From a) and b), and the hypothesis of induction, we can conclude that for any $f \in \prod_{k\in\{1,\cdots,t\}}\{1,\cdots,L_k\}$, $G_{1f(1)}\wedge\cdots\wedge G_{tf(t)}\wedge x \to y\}$ can be α-resolved into α-$\circledcirc$.

Nextly, we only need to prove that

$$S_2 \leqslant \alpha \quad \text{iff} \quad C_1\wedge\cdots\wedge C_t \leqslant \alpha.$$

a) If there exists a constant γ in $G_{1f(1)},\cdots,G_{tf(t)}$ such that $\gamma \leqslant \alpha$, then

$$G_{1f(1)}\wedge\cdots\wedge G_{tf(t)} \leqslant \alpha.$$

b) If there exists no constant in $G_{1f(1)},\cdots,G_{tf(t)}$, or there exists a constant γ such that $\gamma \nleqslant \alpha$, then there is no such generalized literal in $G_{1f(1)},\cdots,G_{tf(t)}$ that itself and $x \to y$ can be α-resolved into α-$\circledcirc$ by Lemmas 11.3.1 and 11.3.2. Hence, it is the only possible case that there exist two generalized literals in $G_{1f(1)},\cdots,G_{tf(t)}$ which can be α-resolved into α-$\circledcirc$.

Suppose $(G_{if(i)},G_{jf(j)})$ is the α-resolution pair, then

$$G_{1f(1)}\wedge\cdots\wedge G_{tf(t)} \leqslant G_{if(i)}\wedge G_{jf(j)} \leqslant \alpha.$$

Sum up a) and b), for any $f \in \prod_{k\in\{1,\cdots,t\}}\{1,\cdots,L_k\}$. we have

$$G_{1f(1)}\wedge\cdots\wedge G_{tf(t)} \leqslant \alpha.$$

Consequently,

$$C_1\wedge\cdots\wedge C_t = \bigvee_{f\in\prod_{k\in\{1,\cdots,t\}}\{1,\cdots,L_k\}} \bigwedge_{k\in\{1,\cdots,t\}} G_{kf(k)} \leqslant \alpha,$$

i.e., we can conclude that $C_1\wedge\cdots\wedge C_t \leqslant \alpha$ from $S_2 \leqslant \alpha$.

Conversely, it is obvious that $S_2 \leqslant \alpha$ can be derived from $C_1\wedge\cdots\wedge C_t \leqslant \alpha$. Hence

$$S_2 \leqslant \alpha \quad \text{iff} \quad C_1\wedge\cdots\wedge C_t \leqslant \alpha.$$

Notice that there exists no $x \to y$ in the locked generalized literal set constructed by $C_1,\cdots,C_t$. Hence, if we write the locked generalized literal set constructed by $C_1,\cdots,C_t$ as M_0, then $|M_0|<n$. Consequently, it follows from the inductive hypothesis that $C_1\wedge\cdots\wedge C_t$ can be α-resolved into α-$\circledcirc$.

(2). Suppose $M_1 \neq \emptyset$. If $p \in M_1$, then except for the occurring of p or p' in implicative terms, S can be rewritten as

$$S = C_1 \wedge \cdots \wedge C_{t_1} \wedge C_{t_1+1} \wedge \cdots \wedge C_{t_2} \wedge C_{t_2+1} \wedge \cdots \wedge C_{t_3} \wedge C_{t3+1} \wedge \cdots \wedge C_K.$$

where

- Every generalized clause in $C_1, \cdots, C_{t_1}$ contains p, but contains no p';
- Every generalized clause in $C_{t_1+1}, \cdots, C_{t_2}$ contains p', but contains no p;
- Every generalized clause in $C_{t_2+1}, \cdots, C_{t_3}$ contains both p and p';
- Every generalized clause in $C_{t_3+1}, \cdots, C_K$ contains neither p nor p'.

Hence

$$\begin{aligned} S &= (p \vee (C_1^* \wedge \cdots \wedge C_{t_1}^*)) \wedge (p' \vee (C_{t_1+1}^* \wedge \cdots \wedge C_{t_2}^*)) \\ &\quad \wedge((p \vee p') \vee (C_{t_2+1}^* \wedge \cdots \wedge C_{t_3}^*)) \wedge (C_{t_3+1} \wedge \cdots \wedge C_K) \\ &\triangleq S_1 \wedge S_2 \wedge S_3 \wedge S_4, \end{aligned}$$

where

$$\begin{aligned} S_1 &= p \vee (C_1^* \wedge \cdots \wedge C_{t_1}^*), \\ S_2 &= p' \vee (C_{t_1+1}^* \wedge \cdots \wedge C_{t_2}^*), \\ S_3 &= (p \vee p') \vee (C_{t_2+1}^* \wedge \cdots \wedge C_{t_3}^*), \\ S_4 &= C_{t_3+1} \wedge \cdots \wedge C_K. \end{aligned}$$

Let

$$\begin{aligned} S_1^* &= C_1^* \wedge \cdots \wedge C_{t_1}^*, \\ S_2^* &= C_{t_1+1}^* \wedge \cdots \wedge C_{t_2}^*, \\ S_3^* &= C_{t_2+1}^* \wedge \cdots \wedge C_{t_3}^*, \\ S_4^* &= S_4, \end{aligned}$$

and write the locked generalized literal sets of S_1, S_2, S_3 and S_4 as M_{S_1}, M_{S_2}, M_{S_3} and M_{S_4}, respectively.

Since $M_1 \neq \emptyset$, without losing generality, we can suppose $M_{S_1} \neq \emptyset$. We investigate the following cases:

a) $M_{S_2} = \emptyset$, $M_{S_3} = \emptyset$, $M_{S_4} = \emptyset$. In this case, $S = S_1 = p \vee S_1^*$ and $S \leqslant \alpha$ cannot hold simultaneously.

b) $M_{S_2} = \emptyset$, $M_{S_3} = \emptyset$, $M_{S_4} \neq \emptyset$. In this case,

$$\begin{aligned} S &= S_1 \wedge S_4 \\ &= (p \vee S_1^*) \wedge S_4 \\ &= (p \wedge S_4) \vee (S_1^* \wedge S_4). \end{aligned}$$

Obviously, if $S \leqslant \alpha$, then $p \wedge S_4 \leqslant \alpha$. So, we only need to prove that

$$S \leqslant \alpha \quad \text{iff} \quad S_4 \leqslant \alpha.$$

i. If $M_{S_1^*} = \emptyset$, then $S = p \wedge S_4$. Therefore, S_4 contains a generalized clause with at least two generalized literals. Set

$$\begin{aligned} S_4 &= C_{t_3+1} \wedge \cdots \wedge C_{t_K} \\ &= \bigvee_{f \in \prod_{k\in\{1,\cdots,K-t_3\}} \{1,\cdots,L_k\}} \bigwedge_{k\in\{1,\cdots,K-t_3\}} G_{kf(k)}. \end{aligned}$$

Hence

$$\begin{aligned} S &= p \wedge S_4 \\ &= \bigvee_{f \in \prod_{k\in\{1,\cdots,K-t_3\}} \{1,\cdots,L_k\}} \bigwedge_{k\in\{1,\cdots,K-t_3\}} (G_{kf(k)} \wedge p) \\ &\leqslant \alpha. \end{aligned}$$

It follows that for any $f \in \prod_{k\in\{1,\cdots,K-t_3\}}\{1,\cdots,L_k\}$,

$$\bigwedge_{k\in\{1,\cdots,K-t_3\}} G_{kf(k)} \wedge p \leqslant \alpha.$$

Note that S_4 contains a generalized clause with at least two generalized literals. Hence, $\bigwedge_{k\in\{1,\cdots,K-t_3\}}(G_{kf(k)} \wedge p)$ can be α-resolved into α-⊚ by inductive hypothesis. Moreover, since each generalized clause in S_4 contains no p' except for the implication term in it, $G_{1f(1)}, \cdots, G_{(K-t_3)f(K-t_3)}$ contains no p' except for the implication term for any $f \in \prod_{k\in\{1,\cdots,K-t_3\}}\{1,\cdots,L_k\}$. Hence, $\bigwedge_{k\in\{1,\cdots,K-t_3\}}(G_{kf(k)} \wedge p)$ can be α-resolved into α-⊚ if and only if $\bigwedge_{k\in\{1,\cdots,K-t_3\}} G_{kf(k)}$ can be α-resolved into α-⊚. Notice that for any $k \in \{1,\cdots,K-t_3\}$, $G_{kf(k)}$ is a generalized literal, therefore there exists an α-resolution pair $(G_{if(i)}, G_{jf(j)})$ such that

$$\bigwedge_{k\in\{1,\cdots,K-t_3\}} G_{kf(k)} \leqslant G_{if(i)} \wedge G_{jf(j)} \leqslant \alpha.$$

Hence, we have $S_4 \leqslant \alpha$.

ii. If $M_{S_1^*} \neq \emptyset$, then

$$S = (p \wedge S_4) \vee (S_1^* \wedge S_4).$$

Hence $p \wedge S_4 \leqslant \alpha$ and the number of the locked generalized literals in $p \wedge S_4$ is less than n. So $p \wedge S_4$ can be α-resolved into α-⊚. Analogous to above, we can prove that $S_4 \leqslant \alpha$. (Notice that for any $f \in \prod_{k\in\{1,\cdots,K-t_3\}}\{1,\cdots,L_k\}$, the number of the locked generalized literals in $\bigwedge_{k\in\{1,\cdots,K-t_3\}}(G_{kf(k)} \wedge p)$ is less than n).

Sum up, we can conclude that $S \leqslant \alpha$ implies $S_4 \leqslant \alpha$. Conversely, $S_4 \leqslant \alpha$ implies $S \leqslant \alpha$, obviously.
Since $p \notin M_{S_4}$, $|M_{S_4}| < n$. It follows from the inductive hypothesis that S_4 can be α-resolved into α-◎.

c) $M_{S_2} = \emptyset$, $M_{S_3} \neq \emptyset$, $M_{S_4} = \emptyset$. In this case,

$$\begin{aligned} S &= S_1 \wedge S_3 \\ &= p \vee (S_1^* \wedge (S_3^* \vee p')). \end{aligned}$$

Hence, $S \leqslant \alpha$ does not hold for there exists a valuation v such that $v(p) = I$ and $S \nleqslant \alpha$.

d) $M_{S_2} = \emptyset$, $M_{S_3} \neq \emptyset$, $M_{S_4} \neq \emptyset$. In this case,

$$\begin{aligned} S &= S_1 \wedge S_3 \wedge S_4 \\ &= (p \vee (S_1^* \wedge (S_3^* \vee p'))) \wedge S_4. \end{aligned}$$

Since $S_1^* \wedge (S_3^* \vee p')$ contains a generalized literal p', $M_{S_1^*(S_3^* \vee p')} \neq \emptyset$. Let $S_1^* \wedge (S_3^* \vee p') = S_{13}^*$, then $S = (p \vee S_{13}^*) \wedge S_4$ and $M_{S_{13}^*} \neq \emptyset$. We can conclude that $S \leqslant \alpha$ if and only if $S_4 \leqslant \alpha$ and S_4 can be α-resolved into α-◎.

e) $M_{S_2} \neq \emptyset$, $M_{S_3} = \emptyset$, $M_{S_4} = \emptyset$. In this case,

$$\begin{aligned} S &= S_1 \wedge S_2 \\ &= (p \vee S_1^*) \wedge (p' \vee S_2^*). \end{aligned}$$

We consider the following cases.

i. If $M_{S_2^*} = \emptyset$, then $S = (p \wedge p') \vee (S_1^* \wedge p') \leqslant \alpha$. Hence, $p \wedge p' \leqslant \alpha$, *i.e.*, (p, p') is an α-resolution pair. Since $S_2 = p'$, so p and p' can be α-resolved into α-◎, *i.e.*, S can be α-resolved into α-◎.
ii. If $M_{S_1^*} = \emptyset$, we can prove that S can be α-resolved into α-◎.
iii. If $M_{S_1^*} \neq \emptyset$ and $M_{S_2^*} \neq \emptyset$. In this case,

$$\begin{aligned} S &= (p \vee S_1^*) \wedge (p' \vee S_2^*) \\ &= (p \wedge p') \vee (p \wedge S_2^*) \vee (p' \wedge S_1^*) \vee (S_1^* \wedge S_2^*) \\ &\leqslant \alpha. \end{aligned}$$

Hence, we can conclude that $p \wedge S_2^* \leqslant \alpha$, $p' \wedge S_1^* \leqslant \alpha$, $p \wedge p' \leqslant \alpha$.
Now we prove the following cases.
case (1) $p \wedge S_2^* \leqslant \alpha$ if and only if $S_2^* \leqslant \alpha$;
case (2) $p' \wedge S_1^* \leqslant \alpha$ if and only if $S_1^* \leqslant \alpha$.
Case (1). Set

$$S_2^* = \bigvee_{f \in \prod_{k \in \{1, \cdots, t_2 - t_1\}} \{1, \cdots, L_k\}} \bigwedge_{k \in \{1, \cdots, t_2 - t_1\}} G_{kf(k)}.$$

Therefore

$$p \wedge S_2^* = \bigvee_{f \in \prod_{k\in\{1,\cdots,t_2-t_1\}}\{1,\cdots,L_k\}} \bigwedge_{k\in\{1,\cdots,t_2-t_1\}} G_{kf(k)} \wedge p \leqslant \alpha.$$

For any $f \in \prod_{k\in\{1,\cdots,t_2-t_1\}}\{1,\cdots,L_k\}$, we have

$$\bigwedge_{k\in\{1,\cdots,t_2-t_1\}} G_{kf(k)} \wedge p \leqslant \alpha,$$

and $p' \notin \{G_{1f(1)},\cdots,G_{(t_2-t_1)f(t_2-t_1)}\}$. Hence, the number of generalized literals in $\bigwedge_{k\in\{1,\cdots,t_2-t_1\}} G_{kf(k)} \wedge p$ is less than n. It follows that $\bigwedge_{k\in\{1,\cdots,t_2-t_1\}} G_{kf(k)} \wedge p$ can be α-resolved into α-$\circledcirc$ by inductive hypothesis. Hence, $\{G_{1f(1)},\cdots,G_{(t_2-t_1)f(t_2-t_1)}\}$ contains a constant $\gamma \leqslant \alpha$ or there exists an α-resolution pair $(G_{if(i)}, G_{jf(j)})$. So,

$$\bigwedge_{k\in\{1,\cdots,t_2-t_1\}} G_{kf(k)} \leqslant \alpha$$

always holds. It follows that $S_2^* \leqslant \alpha$ ($p \wedge S_2^* \leqslant \alpha$ if and only if $S_2^* \leqslant \alpha$). Since the number of generalized literals of $S_2^* < n$, so S_2^* can be α-resolved into α-$\circledcirc$.

Hence, $S_2 = p' \vee S_2^*$ can be α-resolved into α-$\circledcirc$ or p'. Similarly, $S_1 = p \vee S_1^*$ can be α-resolved into α-$\circledcirc$ or p. Moreover, from $p \wedge p' \leqslant \alpha$, S can be α-resolved into α-$\circledcirc$.

f) $M_{S_2} \neq \emptyset$, $M_{S_3} = \emptyset$, $M_{S_4} \neq \emptyset$. In this case, $S = S_1 \wedge S_2 \wedge S_3$.

i. If $M_{S_2^*} = \emptyset$, then $S = (p \wedge S_1^*) \wedge p' \wedge S_4 \leqslant \alpha$. Since $p \in S_1$, $S_2 = p'$, (p, p') is an α-resolution pair and $S_1 \wedge S_2$ can be α-resolved to α-$\circledcirc$. Therefore, S can be α-resolved into α-$\circledcirc$.
If $M_{S_1^*} = \emptyset$, then $S = p \wedge p' \wedge S_4 \leqslant \alpha$. If $p \wedge p' \leqslant \alpha$, then S can be α-resolved into α-$\circledcirc$. If $p \wedge p' \nleqslant \alpha$, set

$$\begin{aligned} S_4 &= C_{t_3+1} \wedge \cdots \wedge C_K \\ &= \bigvee_{f \in \prod_{k\in\{1,\cdots,K-t_3\}}\{1,\cdots,L_k\}} \bigwedge_{k\in\{1,\cdots,K-t_3\}} G_{kf(k)}. \end{aligned}$$

Hence,

$$\begin{aligned} S &= p \wedge p' \wedge S_4 \\ &= \bigvee_{f \in \prod_{k\in\{1,\cdots,K-t_3\}}\{1,\cdots,L_k\}} \bigwedge_{k\in\{1,\cdots,K-t_3\}} (G_{kf(k)} \wedge p \wedge p') \\ &\leqslant \alpha. \end{aligned}$$

So, for any $f \in \prod_{k\in\{1,\cdots,K-t_3\}}\{1,\cdots,L_k\}$, we have

$$\bigwedge_{k\in\{1,\cdots,K-t_3\}} (G_{kf(k)} \wedge p \wedge p') \leqslant \alpha.$$

Note that S_4 contains a clause with at least two generalized literals. It follows that

$$\bigwedge_{k\in\{1,\cdots,K-t_3\}} (G_{kf(k)} \wedge p \wedge p')$$

can be α-resolved into α-$\circledcirc$ by inductive hypothesis. Moreover, note that S_4 contains no p and p' except for the implication term. Therefore, for any $f \in \prod_{k\in\{1,\cdots,K-t_3\}}\{1,\cdots,L_k\}$, $G_{1f(1)}, \cdots, G_{(K-t_3)f(K-t_3)}$ contains no p and p' except for the implication term. Hence, $\bigwedge_{k\in\{1,\cdots,K-t_3\}}(G_{kf(k)} \wedge p \wedge p')$ can be α-resolved into α-$\circledcirc$ if and only if $\bigwedge_{k\in\{1,\cdots,K-t_3\}} G_{kf(k)}$ can be α-resolved into α-$\circledcirc$. Note that for any $k \in \{1,\cdots,K-t_3\}$, $G_{kf(k)}$ is a generalized literal, so there exists an α-resolution pair $(G_{if(i)}, G_{jf(j)})$, *i.e.*,

$$\bigwedge_{k\in\{1,\cdots,K-t_3\}} G_{kf(k)} \leqslant G_{if(i)} \wedge G_{jf(j)} \leqslant \alpha.$$

So we have $S_4 \leqslant \alpha$. And from the hypothesis of induction, we can conclude that S_4 can be α-resolved into α-$\circledcirc$. Consequently, S can be α-resolved into α-$\circledcirc$. If $M_{S_1^*} \neq \emptyset$, then

$$\begin{aligned} S &= (p \vee S_1^*) \wedge p' \wedge S_4 \\ &= (p \wedge p' \wedge S_4) \vee (S_1^* \wedge p' \wedge S_4) \\ &\leqslant \alpha. \end{aligned}$$

Hence, $p \wedge p' \wedge S_4 \leqslant \alpha$. Since $M_{S_1^*} \neq \emptyset$, the number of locked generalized literals in $p \wedge p' \wedge S_4$ is less than n. Hence, $p \wedge p' \wedge S_4$ can be α-resolved into α-$\circledcirc$. Moreover, we can conclude that S_4 can be α-resolved into α-$\circledcirc$. Consequently, S can be α-resolved into α-$\circledcirc$.

ii. If $M_{S_1^*} = \emptyset$, it can be proved analogously to (i).

iii. If $M_{S_1^*} \neq \emptyset$ and $M_{S_2^*} \neq \emptyset$, then

$$\begin{aligned} S &= (p \vee S_1^*) \wedge (p' \vee S_2^*) \wedge S_4 \\ &= (p \wedge p' \wedge S_4) \vee (p \wedge S_2^* \wedge S_4) \vee (p' \wedge S_1^* \wedge S_4) \vee (S_1^* \wedge S_2^* \wedge S_4^*) \\ &\leqslant \alpha. \end{aligned}$$

So,

$$p \wedge p' \wedge S_4 \leqslant \alpha,$$
$$p \wedge S_2^* \wedge S_4 \leqslant \alpha,$$
$$p' \wedge S_1^* \wedge S_4 \leqslant \alpha.$$

We can prove that

$$p \wedge S_2^* \wedge S_4 \leqslant \alpha \text{ iff } S_2^* \wedge S_4 \leqslant \alpha,$$
$$p' \wedge S_1^* \wedge S_4 \leqslant \alpha \text{ iff } S_1^* \wedge S_4 \leqslant \alpha.$$

Since the number of locked generalized literals in $S_2^* \wedge S_4$ is less than n and $S_2^* \wedge S_4 \leqslant \alpha$, $S_2^* \wedge S_4$ can be α-resolved into α-◎ by induction. Hence, $S_2 \wedge S_4$ can be α-resolved into α-◎ or p'. Similarly, $S_1 \wedge S_4$ can be α-resolved into α-◎ or p. If $S_2 \wedge S_4$ can be α-resolved into α-◎ or $S_1 \wedge S_4$ can be α-resolved into α-◎, then S can be α-resolved into α-◎. If $S_2 \wedge S_4$ can be α-resolved into p' and $S_1 \wedge S_4$ can be α-resolved into p, then S can be α-resolved into $p \wedge p'$. Since $p \wedge p' \wedge S_4 \leqslant \alpha$, and by induction, we can conclude that $p \wedge p' \wedge S_4$ can be α-resolved into α-◎, *i.e.*, S can be α-resolved into α-◎.

g) $M_{S_2} \neq \emptyset$, $M_{S_3} \neq \emptyset$, $M_{S_4} = \emptyset$. Here

$$\begin{aligned} S &= (p \vee S_1^*) \wedge (p' \vee S_2^*) \wedge (p \vee p' \vee S_3^*) \\ &= (p \wedge p') \vee (p \wedge S_2^*) \vee (p' \wedge S_1^*) \vee (S_1^* \wedge S_2^* \wedge S_3^*) \\ &\leqslant \alpha. \end{aligned}$$

Hence $p \wedge p' \leqslant \alpha$, $p \wedge S_2^* \leqslant \alpha$, $p' \wedge S_1^* \leqslant \alpha$. The conclusion can be proved similar to e).

h) $M_{S_2} \neq \emptyset$, $M_{S_3} \neq \emptyset$, $M_{S_4} \neq \emptyset$. Here

$$\begin{aligned} S &= (p \wedge p' \wedge S_4) \vee (p \wedge S_2^* \wedge S_4) \vee (p' \wedge S_1^* \wedge S_4) \\ &\quad \vee (S_1^* \wedge S_2^* \wedge ((p \vee p') \vee S_3^*)) \wedge S_4) \\ &\leqslant \alpha. \end{aligned}$$

Hence $(p \wedge p' \wedge S_4) \vee (p \wedge S_2^* \wedge S_4) \leqslant \alpha$ can be proved, so we can conclude that this can be α-resolved into α-◎. The conclusion can be proved similar to f).

(3). If $M_1 = \emptyset$. There are three cases.

a) $M_2 \neq \emptyset$ and $M_3 = \emptyset$. Set $S = C_1 \wedge \cdots \wedge C_K$, where all the generalized literals in $C_1, \cdots, C_K$ are constants and there exists at least one generalized literal with two or more constants.Without losing generality, we suppose that C_K contains more than one constants, *i.e.*,

$$C_K = \beta_1 \vee \cdots \vee \beta_m. \quad (m > 1)$$

In this case,

$$S = (C_1 \wedge \cdots \wedge C_{K-1} \wedge \beta_1) \vee (C_1 \wedge \cdots \wedge C_{K-1}) \wedge (\beta_2 \vee \cdots \vee \beta_m).$$

Since $S \leqslant \alpha$, we have

i. $C_1 \wedge \cdots \wedge C_{K-1} \wedge \beta_1 \leqslant \alpha$ and

ii. $C_1 \wedge \cdots \wedge C_{K-1} \wedge (\beta_2 \vee \cdots \vee \beta_m) \leqslant \alpha$.

For case (i). It follows that $C_1 \wedge \cdots \wedge C_{K-1} \wedge \beta_1$ can be α-resolved into α-◎ by inductive hypothesis. Hence, we replace β_1 by C_K, then S can be α-resolved into α-◎ or $\beta_2 \vee \cdots \vee \beta_m$. If S can be α-resolved into $\beta_2 \vee \cdots \vee \beta_m$, then we turn to the proof of case (2).

For case (ii). $C_1 \wedge \cdots \wedge C_{K-1} \wedge (\beta_2 \vee \cdots \vee \beta_m)$ can be obtained by α-resolution in S. Notice that

$$C_1 \wedge \cdots \wedge C_{K-1} \wedge (\beta_2 \vee \cdots \vee \beta_m) \leqslant \alpha,$$

and the number of generalized literals in $C_1, \cdots, C_{K-1}, (\beta_2 \vee \cdots \vee \beta_m)$ less than n. Hence, by inductive hypothesis $C_1 \wedge \cdots \wedge C_{K-1} \wedge (\beta_2 \vee \cdots \vee \beta_m)$ can be α-resolved into α-◎, *i.e.*, S can be α-resolved into α-◎.

b) $M_2 \neq= \emptyset$ and $M_3 \neq= \emptyset$. In this case,

$$S = C_1 \wedge \cdots \wedge C_{t_1} \wedge C_{t_1+1} \wedge \cdots \wedge C_{t_2} \wedge C_{t_2+1} \wedge \cdots \wedge C_K,$$

where,

- $C_1, \cdots, C_{t_1}$ only have constants;
- $C_{t_1+1}, \cdots, C_{t_2}$ have both constants and 1-implicative terms;
- $C_{t_2+1}, \cdots, C_K$ only have 1-implicative terms.

Let $\delta = C_1 \wedge \cdots \wedge C_{t_1}$, $C_{t_1+i} = \delta_i \vee C^*_{t_1+i}$, $i = 1, \cdots, t_2 = t_1$. Therefore, $S = \delta \wedge ((\delta_i \vee C^*_{t_1+1}) \wedge \cdots \wedge (\delta_{t_2-t_1} \vee C^*_{t_2}) \wedge C_{t_2+1} \wedge \cdots \wedge C_K$ implies $\delta \wedge \delta_1 \wedge \cdots \wedge \delta_{t_2-t_1} \wedge (x_{t_2+1} \rightarrow y_{t_2+1}) \wedge \cdots \wedge (x_K \rightarrow y_K) \leqslant \alpha$. By inductive hypothesis, $\delta \wedge \delta_1 \wedge \cdots \wedge \delta_{t_2-t_1} \wedge (x_{t_2+1} \rightarrow y_{t_2+1}) \wedge \cdots \wedge (x_K \rightarrow y_K) \leqslant \alpha$ can be α-resolved into α-◎. So, there must be β in $\{\delta, \delta_1, \cdots, \delta_{t_2-t_1}\}$ such that $\beta \leqslant \alpha$.

If $\beta = \delta$, then β and any selected generalized literal in S can be α-resolved into α-◎. Therefore, S can be α-resolved into α-◎.

If $\beta \neq \delta$, without losing generality, we can set $\beta = \delta_1$, then (δ, δ_1) is an α-resolution pair and S can be α-resolved into α-◎.

c) $M_2 = \emptyset$ and $M_3 \neq \emptyset$. In this case,

$$\begin{aligned} S &= C_1 \wedge \cdots C_K \\ &= \bigvee_{f \in \prod_{k \in \{1, \cdots, K\}} \{1, \cdots, L_k\}} \bigwedge_{k \in \{1, \cdots, K\}} G_{kf(k)}, \end{aligned}$$

where $G_{kf(k)}$ is 1-IESF, $i = 1, \cdots, K$.

Since $S \leqslant \alpha$, for any $f \in \prod_{k \in \{1, \cdots, K\}} \{1, \cdots, L_k\}$,

$$\bigwedge_{k \in \{1, \cdots, K\}} G_{kf(k)} \leqslant \alpha.$$

Notice that at least one generalized clause in $C_1, \cdots, C_K$, which has at least two generalized literals, the number of generalized literals in $\bigwedge_{k\in\{1,\cdots,K\}} G_{kf(k)}$ less than n. By inductive hypothesis, $\bigwedge_{k\in\{1,\cdots,K\}} G_{kf(k)}$ can be α-resolved into α-⊚. Therefore, there exist $i, j \in \{1, \cdots, K\}$ such that $(G_{if(i)}, G_{jf(j)})$ is an α-resolution pair. Hence,

$$(G_{if(i)}, G_{jf(j)}) \leqslant \alpha. \tag{11.30}$$

By Corollary 11.3.1, Eq. 11.30 cannot be true. Hence $S \nleqslant \alpha$. □

11.4 α-Resolution Principle Based on $LF(X)$

In this section, we discuss the α-resolution principle based on the lattice-valued first-order logic $LF(X)$.

Definition 11.4.1 *A formula G is a generalized-literal, if*

(1). G is a literal, or

(2). G is constructed only by some literals and some implication connectives with the condition that G cannot be represented by connectives $\vee$ or $\wedge$ and G cannot be decomposed into a simple form (G is called an indecomposable implication form).

The conjunction of a finite number of generalized-literals is a generalized-phrase, the disjunction of a finite number of generalized-literals is a generalized-clause. The disjunction of a finite number of generalized-phrase is a generalized-disjunctive normal form, and the conjunction of a finite number of generalized-clauses is a generalized-conjunctive normal form.

Definition 11.4.2 *An indecomposable implication form G is regular, if there exist literals p_1, p_2 (at least either of both is not a constant element) such that $G \geqslant p_1 \to p_2$. The conjunction of a finite number of generalized-literals is a regular generalized-phrase, if all of its indecomposable implication forms are regular. The disjunction of a finite number of generalized-literals is a regular generalized-clause, if all of its indecomposable implication forms are regular.*

Definition 11.4.3 *Suppose G is a formula of the form*

$$Q_1 x_1 \cdots Q_n x_n G^*,$$

where $Q_1, \cdots, Q_n$ are the quantifiers and G^* is a formula without any quantifier. Then

(1). G is said to be a generalized-prenex normal form, if G^* contains indecomposable implication forms;

(2). G is said to be a generalized-prenex disjunctive (or conjunctive) normal form, if G^* is a generalized-disjunctive normal form (or generalized-conjunctive normal form).

Definition 11.4.4 *Suppose a formula $G = Q_1x_1 \cdots Q_nx_nM$ is a generalized-prenex conjunctive normal form. The formula G^* obtained by the following steps is called a generalized-Skölem standard form of G:*

(1). If Q_r is an existential quantifier and without any universal quantifier occurring ahead it in the prefix $Q_1, \cdots, Q_n$ (from left to right), we choose a new constant c different from other constants occurring in M, replace all x_r occurring in M by c, and then delete Q_r from the prefix $Q_1, \cdots, Q_n$.
(2). If Q_r is an existential quantifier and $Q_{k_1}, \cdots, Q_{k_m}$ are all the universal quantifiers occurring ahead Q_r ($m \geqslant 1$, $1 \leqslant k_1 < \cdots < k_m < r$), we choose a new m-ary function symbol $f^{(m)}$ different from all other function symbols occurring in M, replace all x_r in M by $f^{(m)}(x_{k_1}, \cdots, x_{k_m})$ and then delete Q_r from the prefix $Q_1, \cdots, Q_n$.
(3). Repeating (1) and (2) until there is no existential quantifier occurring in the prefix.

Definition 11.4.5 *Suppose G^* is a generalized-Skölem standard form of a formula G. G^* is a regular generalized-Skölem standard form of G, if all generalized-clauses occurring in M are regular generalized-clauses.*

Definition 11.4.6 *$\varphi \in \mathscr{F}$ is said to be closed if there is no free variable in φ.*

In the following, we only consider the closed formulae.

11.4.1 Interpretation of Formulae

In Section 10.1, we gave the interpretation $\mathcal{D}$ of $LF(X)$. In this section, $\mathcal{D}$ is confined to concrete formulae and D_p is restricted to the interpretation domain D. Elements of $V \cup C$ are regarded as 0-ary functions and elements of L occurring in formulae are regarded as 0-ary relations on L.

Suppose $G \in \mathscr{F}$, F_G is the set of all functional symbols occurring in G, P_G is the set of predicate symbols occurring in G, and $D(\neq \emptyset)$ is the domain of interpretation. An interpretation $\mathscr{I}_D$ of G over D is a triple

$$\mathscr{I}_D = \langle D, u_D, v_D \rangle,$$

where

$$\begin{aligned} u_D:\ & F_G \longrightarrow U_D = \{f_D^{(n)}:\ D^n \longrightarrow D \mid n \in \mathbb{N}\}, \\ & f^{(0)} \longmapsto f_D^{(0)}, f_D^{(0)}(D^0) = \{f_D^{(0)}\} \subseteq D, D^0 \text{ is a non-empty set}, \\ & f^{(n)} \longmapsto f_D^{(n)} (n \in \mathbb{N}^+), \\ v_D:\ & P_G \longrightarrow V_D = \{p_D^{(n)}:\ D^n \longrightarrow L | n \in \mathbb{N}\}, \\ & p^{(0)} \longmapsto p_D^{(0)}, p_D^{(0)}(D^0) = \{p_D^{(0)}\} \subseteq L, \\ & p^{(n)} \longmapsto p_D^{(n)} (n \in \mathbb{N}^+). \end{aligned}$$

Definition 11.4.7 *Let $G \in \mathscr{F}$, $\alpha \in L$. G is said to be α-satisfiable, if there exists an interpretation $\mathscr{I}_D = \langle D, u_D, v_D\rangle$ such that $v_D(G) \geqslant \alpha$. G is said to be α-true, if $v_D(G) \geqslant \alpha$ for any interpretation $\mathscr{I}_D$ of G. G is said to be α-false, if $v_D \ngeqslant \alpha$ for any interpretation $\mathscr{I}_D$ of G.*

Theorem 11.4.1 *Let $\alpha \in L$, $G, G^* \in \mathscr{F}$, and G^* be a generalized-Skölem standard form of G. Then G is α-false if and only if G^* is α-false.* □

Other concepts in $LF(X)$, such as ground term, ground term set, ground atom, ground atom set, ground literal, ground clause, ground clause set, ground instance, instance, H-field of a clause set, substitution, unifier, and most general unifier, *etc.*, are formally the same as the corresponding concepts in classical logic.

Definition 11.4.8 (H-interpretation) *Suppose G^* is a generalized-Skölem standard form of a formula G, H is the H-field of G^*, $\mathscr{I}_H = \langle H, u_H, v_H\rangle$ is an interpretation of G^* over H. $\mathscr{I}_H$ is said to be an H-interpretation of G^*, if*

$$\begin{aligned}
u_H:\ & F_{G^*} \longrightarrow U_H = \{f_H^{(n)}:\ H^n \longrightarrow H \mid n \in \mathbb{N}\},\\
& f^{(0)} \longmapsto f_H^{(0)}, (f_H^{(0)}(H^0) = \{f^{(0)}\} \subseteq H_0, H^0 = H_0),\\
& f^{(n)} \longmapsto f_H^{(n)}, f_H^{(n)}(h_1, \cdots, h_n) = f^{(n)}(h_1, \cdots, h_n)(n \in \mathbb{N}^+),\\
v_H:\ & P_{G^*} \longrightarrow V_H = \{p_H^{(n)}:\ H^n \longrightarrow L \mid n \in \mathbb{N}\},\\
& p^{(0)} \longmapsto p_H^{(0)}, (p_H^{(0)}(H^0) = \{p^{(0)}\} \subseteq L, H^0 = H_0),\\
& p^{(n)} \longmapsto p_H^{(n)}, (n \in \mathbb{N}^+).
\end{aligned}$$

Definition 11.4.9 *Suppose G^* is a generalized-Skölem standard form of a formula G, $\mathscr{I}_D = \langle D, u_D, v_D\rangle$ is an interpretation of G^* over D, H is the H-field of G^*. A mapping*

$$g:\ H \longrightarrow D$$

is defined recursively as follows:

(1). g: $H_0 \longrightarrow D$. Let C_{G^*} be the set of all constants occurring in G^*.
  a) If $C_{G^*} = \emptyset$, then let $H_0 = \{a\}$, where $a \in C$ is an arbitrary specified constant symbol. So $g(a) \in D$ can be arbitrarily defined.
  b) If $C_{G^*} \neq \emptyset$, then let $H_0 = C_{G^*}$. We define

$$g(c) = u_D(c),$$

  for any $c \in H_0$.

(2). Suppose that g : $H_{i-1} \longrightarrow D$ ($i \geqslant 1$) has been defined. Since $H_i = H_{i-1} \cup \{f^{(n)}(h_1, \cdots, h_n) \mid f^{(n)} \in F_{G^*}, n \in \mathbb{N}, h_1, \cdots, h_n \in H_{i-1}\}$, we only need to define:

$$g: \{f^{(n)}(h_1,\cdots,h_n) \mid f^{(n)} \in F_{G^*}, n \in \mathbb{N}, h_1,\cdots,h_n \in H_{i-1}\} \longrightarrow D,$$
$$f^{(n)}(h_1,\cdots,h_n) \longmapsto g(f^{(n)}(h_1,\cdots,h_n)),$$

where

$$\begin{aligned} g(f^{(n)}(h_1,\cdots,h_n)) &\triangleq u_D(f^{(n)})(g(h_1),\cdots,g(h_n)) \\ &= f_D^{(n)}(g(h_1),\cdots,g(h_n)). \end{aligned}$$

Definition 11.4.10 *Suppose G^* is a generalized-Skölem standard form of a formula G, $\mathscr{I}_D = \langle D, u_D, v_D\rangle$ is an interpretation of G^* over D, and $g\colon H \longrightarrow D$ is the mapping defined in Definition 11.4.9, C_{G^*} is the set of all constants occurring in G^*. Then an H-interpretation $\mathscr{I}_H = \langle H, u_H, v_H\rangle$ corresponding to $\mathscr{I}_D$ can be defined as follows:*

(1). If $C_{G^*} \neq \emptyset$, then let $C_{G^*} = H_0$ and define

$$\begin{aligned} u_H:\ & C_{G^*} \longrightarrow U_H = \{f_H^{(n)}\colon\ H^n \longrightarrow H \mid n \in \mathbb{N}\}, \\ & f^{(0)} \longmapsto f_H^{(0)}, \\ & (f_H^{(0)}(H^0) = \{f^{(0)}\}, H^0 = H_0) \\ & (g \circ u_H)(f^{(0)}) = u_D(f^{(0)}), \end{aligned}$$

i.e., u_H satisfies the commutative graph Fig. 11.1.

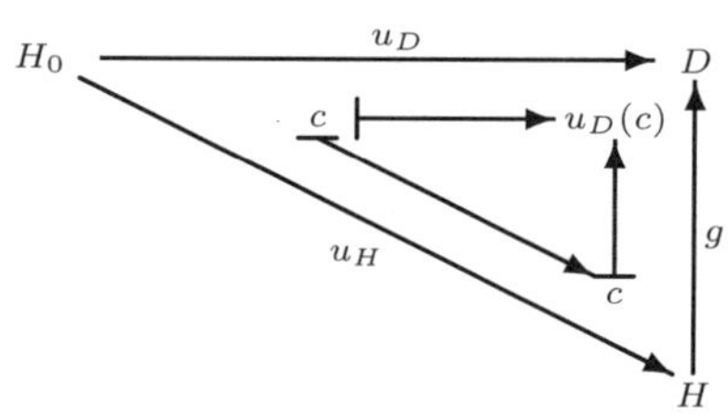

Fig. 11.1. Commutative Graph 1 for u_H.

(2).

$$\begin{aligned} u_H:\ & (F_{G^*} \setminus C_{G^*}) \longrightarrow U_H^* = \{f_H^{(n)}\colon\ H^n \longrightarrow H \mid n \in \mathbb{N}^+\}, \\ & f^{(n)} \longmapsto u_H(f^{(n)}), \end{aligned}$$

such that

$$\begin{aligned} g(u_H(f^{(n)})(h_1,\cdots,h_n)) &= u_D(f^{(n)})(g(h_1),\cdots,g(h_n)) \\ &= f_D^{(n)}(g(h_1),\cdots,g(h_n)), \end{aligned}$$

i.e., u_H satisfies the commutative graph Fig. 11.2.

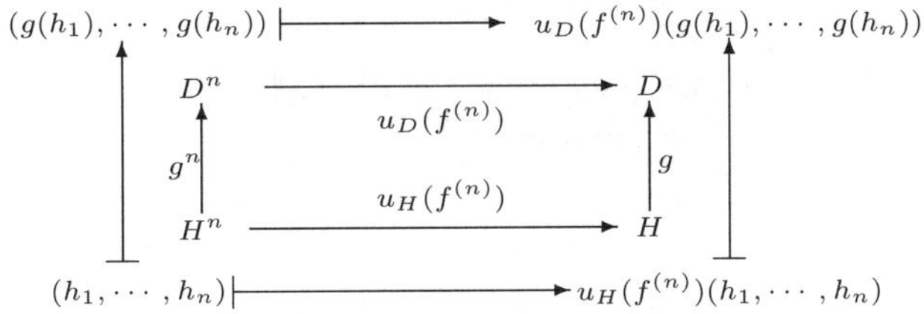

Fig. 11.2. Commutative Graph 2 for u_H.

(3). v_H: $P_{G^*} \longrightarrow V_H$, $p^{(n)} \longmapsto v_H(p^{(n)})$, such that

$$\begin{aligned} v_H(p^{(n)})(h_1, \cdots, h_n) &= p_H^{(n)}(h_1, \cdots, h_n) \\ &= (v_D(p^{(n)}) \circ g^n)(h_1, \cdots, h_n) \\ &= (p_D^{(n)} \circ g^n)(h_1, \cdots, h_n) \\ &= p_D^{(n)}(g(h_1), \cdots, g(h_n)), \end{aligned}$$

i.e., v_H satisfies the commutative graph Fig. 11.3.

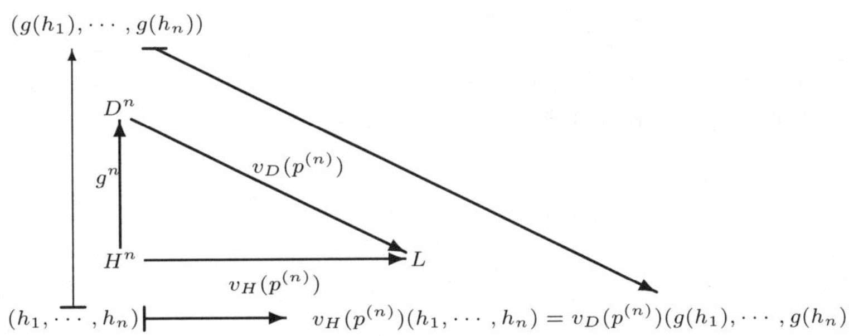

Fig. 11.3. Commutative Graph for v_H.

11.4.2 α-Resolution Principle

Theorem 11.4.2 *Suppose G^* is a generalized-Skölem standard form of a formula G. If an interpretation $\mathscr{I}_D = \langle D, u_D, v_D \rangle$ α-satisfies G^*, then the H-interpretation $\mathscr{I}_H = \langle H, u_H, v_H \rangle$ of G^* corresponding to $\mathscr{I}_D$ also α-satisfies G^*.*

Proof. Suppose

$$G^* = (\forall x_1)\cdots(\forall x_n)(C_1 \wedge \cdots \wedge C_m),$$

where C_i $(i = 1, \cdots, m)$ are generalized-clauses. It is easy to see from the assumption that $v_D(G^*) \geqslant \alpha$. Hence,

$$v_D(C_i(d_1, \cdots, d_n)) \geqslant \alpha, i = 1, \cdots, m$$

hold for any $(d_1, \cdots, d_n) \in D^n$. Moreover, note that

$$v_H(p^{(r)})(h_1, \cdots, h_r) = v_D(p^{(r)})(g(h_1), \cdots, g(h_r))$$

hold for any $(h_1, \cdots, h_r) \in H^r$ and any arbitrary r-ary predicate $p^{(r)}$ occurring in G^*. It follows that

$$\begin{aligned} v_H(C_i(h_1, \cdots, h_n)) &= v_D(C_i(g(h_1), \cdots, g(h_n))) \\ &\geqslant \alpha, \qquad i = 1, \cdots, m \end{aligned}$$

hold for any $(h_1, \cdots, h_n) \in H^n$. Consequently, $v_H(G^*) \geqslant \alpha$. □

Theorem 11.4.3 *Suppose G^* is a generalized-Skölem standard form of a formula G. G^* is α-false if and only if $v_H(G^*) \ngeqslant \alpha$ holds for any H-interpretations $\mathscr{I}_H$ of G^*.*

Proof. The necessity is straightforward. Now we only need to prove the sufficiency. If G^* is α-satisfiable, *i.e.*, there exists an interpretation $\mathscr{I}_D$ such that $v_D(G^*) \geqslant \alpha$, it follows from Theorem 11.4.2 that $v_H(G^*) \geqslant \alpha$ holds for the H-interpretation $\mathscr{I}_H$ of G^* corresponding to $\mathscr{I}_D$. It is a contradiction with the fact that $v_H(G^*) \ngeqslant \alpha$ holds for any H-interpretation $\mathscr{I}_H$ of G^*. □

In the following, by Theorem 11.4.3, we only consider the H-interpretation.

Suppose G^* is a generalized-Skölem standard from of a formula G,and $\mathscr{A}(G^*)$ is the atom set of G^*. It follows from the construction of H that $\mathscr{A}(G^*)$ is enumerable in the H-interpretation. Hence, we can assume that

$$\mathscr{A}(G^*) = \{p_1, \cdots, p_n, \cdots\}.$$

Note that $v_H \in \mathscr{F}_L(\mathscr{A}(G^*))$ for an arbitrary H-interpretation $\mathscr{I}_H$ of G^*. Therefore, v_H can be expressed as

$$v_H = (\alpha_1, \cdots, \alpha_n, \cdots),$$

where $\alpha_n = v_H(p_n)$, $n = 1, 2, \cdots$.

Conversely, for any $A = (\alpha_1, \cdots, \alpha_n, \cdots) \in L^{\mathbb{N}^+}$, an H-interpretation $\mathscr{I}_H$ can be determined by defining v_H such that $v_H(p_n) = \alpha_n$, $n = 1, 2, \cdots$.

Remark 17. Here we need to define u_H at the same time, but note that u_H and v_H can be defined independently. Consequently, we conclude that v_H in any H-interpretation can be one-to-one corresponding to the element in $L^{\mathbb{N}^+}$.

Remark 18. For any $K \in \mathbb{N}^+$, and for arbitrary K-tuple $(\alpha_1, \cdots, \alpha_K) \in L^K$, if there exists an H-interpretation $\mathscr{I}_H$ of G^* such that

$$v_H(p_n) = \alpha_n, n = 1, 2, \cdots, K,$$

where $\mathscr{A}(G^*) = \{p_1, \cdots, p_n, \cdots\}$, then v_H is called an adjoint of $(\alpha_1, \cdots, \alpha_K)$.

Theorem 11.4.4 *Suppose G^* is a generalized-Skölem standard form of a formula G, and $|L| < \aleph_0$, G^* is α-false if and only if there exists $K \in \mathbb{N}^+$ such that $v_H(G^*) \ngeq \alpha$ holds for any adjoint v_H of every element in L^K.*

Proof. (necessity). Suppose the atom set of G^* is $\mathscr{A}(G^*) = \{p_1, \cdots, p_n, \cdots\}$. For any $A = (\alpha_1, \cdots, \alpha_n, \cdots) \in L^{\mathbb{N}^+}$, suppose that v_H in the H-interpretation $\mathscr{I}_H$ corresponds to A, *i.e.*,

$$v_H(p_n) = \alpha_n, n = 1, 2, \cdots.$$

Therefore, there exists a ground instance C^* of a generalized-clause C in G^* such that $v_H(C^*) \ngeq \alpha$. Let

$$C_a^* = \{p \mid p \text{ is an atom in } C^*\}.$$

Since C_a^* is finite, there exists an n_A such that $C_a^* \subseteq \{p_1, \cdots, p_{n_A}\}$. Hence, $v_H(C^*) \ngeq \alpha$ holds for any adjoint v_H of the first n_A section A_{n_A} of A, where $A_{n_A} \triangleq (\alpha_1, \cdots, \alpha_{n_A})$. Consequently, $v_H(G^*) \ngeq \alpha$.

Let $L = \{\beta_1, \cdots, \beta_m\}$. Note that $L^{\mathbb{N}^+}$ can be expressed as the following graph (Fig. 11.4):

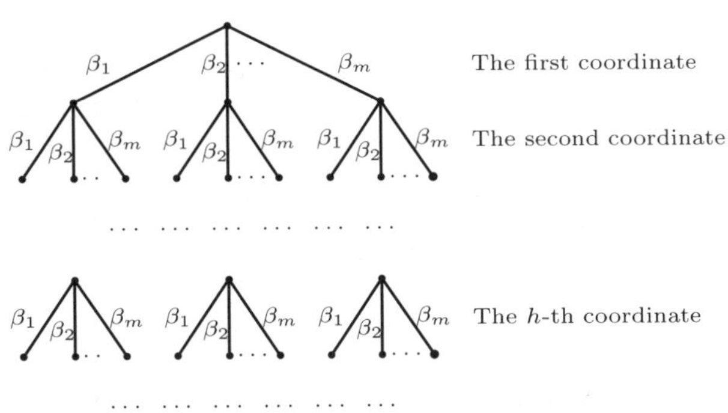

Fig. 11.4. Graph for $L^{\mathbb{N}^+}$

Therefore, every A is uniquely corresponding to a certain branch in the above graph. Conversely, every branch must be uniquely corresponding to a certain A. Hence, there is a bijection from $L^{\mathbb{N}^+}$ to the set of those branches in the above graph. Due to the above conclusion that $v_H(C^*) \ngeq \alpha$ for any adjoint v_H of the first n_A section A_{n_A} of A, and note that the finite (m^h) branches is obtained if the h-th coordinate is given. Therefore, take the first n_A section from every branch A (here we directly denote as A). Let K be the maximal element of all these n_A, and we can note the fact that if $v_H(G^*) \ngeq \alpha$ holds for any adjoint v_H of A_{n_A}, then for any $t \in \mathbb{N}^+$, $v^e_H(G^*) \ngeq \alpha$ also holds for any extension adjoint v^e_H of v_H over A_{n_A+t} (*i.e.*, $v^e_H|A_{n_A} = v_H$). Consequently, there exists K, such that $v_H(G^*) \ngeq \alpha$ holds for any adjoint v_H of every element in L^K (*i.e.*, every A_K).

(sufficiency). Note that for any $A \in L^K$ (*i.e.*, for any v_H in an H-interpretation $\mathscr{I}_H$), $v_H(G^*) \ngeq \alpha$ holds for first K segment of A, *i.e.*, $A_K = (\alpha_1, \cdots, \alpha_K) \in L^K$, therefore $v_H(G^*) \ngeq \alpha$ holds for A. □

Theorem 11.4.5 *Suppose G^* is a generalized-Skölem standard form of a formula G, and $|L| < \aleph_0$, G^* is α-false if and only if there exists a finite ground instance set G^{*0} of G^* such that G^{*0}_c is α-false, where G^{*0}_c is the conjunction of all ground instances of G^{*0}.*

Proof. (necessity). If G^* is α-false, then it follows from Theorem 11.4.4 that there exists K such that for any adjoint v_H of every element in L^K, there exists a ground instance C^* of generalized-clause C in G^* such that $v_H(C^*) \ngeq \alpha$.

Let

$$G^{*0} \triangleq \{C^* \mid C^* \text{ is a ground instance of a } g\text{-clause } C, \text{ and there exists } A_K \in L^K \text{ such that } v_H(C^*) \ngeq \alpha \text{ holds for any adjoint } v_H \text{ of } A_K\}.$$

Since $|L^K| < \aleph_0$ and $A_K (\in L^K)$ is an ordered K tuple, $|G^{*0}| < \aleph_0$. Note that for any interpretation $\mathscr{I}^0$ of G^{*0}_c, there exists an interpretation $\mathscr{I}$ of G^* such that $\mathscr{I}^0 \subseteq \mathscr{I}$. Moreover, since G^* is α-false, $v_H(G^{*0}_c) \ngeq \alpha$ for the v_H in $\mathscr{I}$. Therefore, there exists a ground instance $C^* \in G^{*0}$ of generalized-clause C in G^* such that $v_H(C^*) \ngeq \alpha$. Since $\mathscr{I}^0$ is a part of $\mathscr{I}$, it follows that $v^0_H(C^*) \ngeq \alpha$ holds for the v^0_H in $\mathscr{I}^0$. So

$$v^0_H(G^*_c) \ngeq \alpha.$$

Because $\mathscr{I}^0$ is arbitrary, we conclude that G^{*0}_c is α-false.

(sufficiency). Since for any interpretation $\mathscr{I}$ of G^* contains an interpretation $\mathscr{I}^0$ of G^{*0}_c and G^{*0}_c is α-false, $v_H(G^{*0}_c) \ngeq \alpha$ holds for the parallelism v_H of every interpretation $\mathscr{I}$ of G^*. It follows that there exists a ground clause C^* in G^{*0}_c such that $v_H(C^*) \ngeq \alpha$. Note that C^* is a ground instance of a generalized-clause C in G^*. Therefore, $v_H(G^*) \ngeq \alpha$. Since $\mathscr{I}$ is arbitrary, we conclude that G^* is α-false. □

Definition 11.4.11 *Let $\alpha \in L$, C_1^0 and C_2^0 be two generalized-clauses without common variables. For any generalized-clauses C_1 and C_2, if there exists a substitution ε such that the following conditions hold:*

(1). $C_1^\varepsilon = C_1^0$, $C_2^\varepsilon = C_2^0$.

(2). $C_1 = C_1^* \vee p_1 \vee p_{11} \vee \cdots \vee p_{1k_1}$, $C_2 = C_2^* \vee p_2 \vee p_{21} \vee \cdots \vee p_{2k_2}$, *satisfy*

a)

$$\{p_1, p_{11}, \cdots, p_{1k_1}\} = \{q \mid q \text{ is a g-literal in } C_1, q^\varepsilon = p_1^\varepsilon\},$$

and

$$\{p_2, p_{21}, \cdots, p_{2k_2}\} = \{q \mid q \text{ is a g-literal in } C_2, q^\varepsilon = p_2^\varepsilon\}.$$

So,

$$C_1^0 = C_1^{*\varepsilon} \vee p_1^\varepsilon, C_2^0 = C_2^{*\varepsilon} \vee p_2^\varepsilon.$$

b) There exists a generalized-literal p, $p \wedge p_2 \leqslant \alpha$.

c) p_1^ε *and* p^ε *have the most general unifier* σ.

Then the generalized-clauses

$$C_1^{0*\sigma} \vee C_2^{0*\sigma} \triangleq R_\alpha(C_1^0, C_2^0)$$

is called an α-resolvent of C_1^0 and C_2^0; p_1^ε and p_2^ε are called an α-resolution pair, denoted by $(p_1^\varepsilon, p_2^\varepsilon) - \alpha$, here

(1).

$$C_1^{0\sigma} = C_1^{0*\sigma} \vee p_1^{\varepsilon\sigma} \vee q_{11}^{\varepsilon\sigma} \vee \cdots \vee q_{1t_1}^{\varepsilon\sigma},$$

and

$$\{p_1^\varepsilon, p_{11}^\varepsilon, \cdots, p_{1t_1}^\varepsilon\} = \{q \mid q \text{ is a g-literal in } C_1^0, q^\sigma = p_1^{\varepsilon\sigma}\}.$$

Hence

$$C_1^{0\sigma} = C_1^{0*\sigma} \vee p_1^{\varepsilon\sigma}.$$

(2).

$$C_2^{0\sigma} = C_2^{0*\sigma} \vee p_2^{\varepsilon\sigma} \vee q_{21}^{\varepsilon\sigma} \vee \cdots \vee q_{2t_2}^{\varepsilon\sigma},$$

and

$$\{p_2^\varepsilon, p_{21}^\varepsilon, \cdots, p_{2t_2}^\varepsilon\} = \{q \mid q \text{ is a g-literal in } C_2^0, q^\sigma = p_2^{\varepsilon\sigma}\}.$$

Hence

$$C_2^{0\sigma} = C_2^{0*\sigma} \vee p_2^{\varepsilon\sigma}.$$

Theorem 11.4.6 *If C_1^0 and C_2^0 are instances of generalized-clauses C_1 and C_2 respectively, C^0 is an α-resolvent of C_1^0 and C_2^0, then there exists an α-resolvent C of C_1 and C_2 such that C^0 is an instance of C, i.e., Fig. 11.5 holds.*

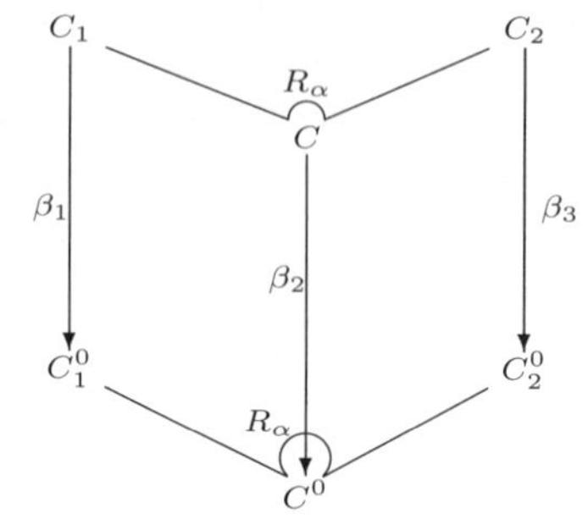

Fig. 11.5. Lift Lemma of α-Resolution Principle.

Proof. Since any two generalized-clauses C_1 and C_2 can always be transformed to two generalized-clauses without common variable by a substitution of rename, we suppose that C_1 and C_2 have no common variables.

Since C_1^0 and C_2^0 are instances of C_1 and C_2 respectively, it follows that there exists a substitution ε such that:

(1). $C_1^\varepsilon = C_1^0$, $C_2^\varepsilon = C_2^0$.
(2). $C_1 = C_1^* \vee p_1 \vee p_{11} \vee \cdots \vee p_{1k_1}$, $C_2 = C_2^* \vee p_2 \vee p_{21} \vee \cdots \vee p_{2k_2}$, satisfy:
 a)

$$\{p_1, p_{11}, \cdots, p_{1k_1}\} = \{q \mid q \text{ is a } g\text{-literal in } C_1, q^\varepsilon = p_1^\varepsilon\},$$

and

$$\{p_2, p_{21}, \cdots, p_{2k_2}\} = \{q \mid q \text{ is a } g\text{-literal in } C_2, q^\varepsilon = p_2^\varepsilon\}.$$

So,

$$C_1^0 = C_1^{*\varepsilon} \vee p_1^\varepsilon, C_2^0 = C_2^{*\varepsilon} \vee p_2^\varepsilon.$$

 b) There exists a generalized-literal p, $p \wedge p_2 \leqslant \alpha$.
 c) p_1^ε and p^ε have the most general unifier σ.

Hence, the generalized-clause $C_1^{0*\sigma} \vee C_2^{0*\sigma} = R_\alpha(C_1^0, C_2^0) = C^0$, and $(p_1^\varepsilon, p_2^\varepsilon)$-$\alpha$, here,

(1).

$$C_1^{0\sigma} = C_1^{0*\sigma} \vee p_1^{\varepsilon\sigma} \vee q_{11}^{\varepsilon\sigma} \vee \cdots \vee q_{1t_1}^{\varepsilon\sigma}$$

and

$$\{p_1^\varepsilon, q_{11}^\varepsilon, \cdots, q_{1t_1}^\varepsilon\} = \{q \mid q \text{ is a } g\text{-literal in } C_1^0, q^\sigma = p_1^{\varepsilon\sigma}\}.$$

Hence,

$$C_1^{0\sigma} = C_1^{0*\sigma} \vee p_1^{\varepsilon\sigma}.$$

(2).
$$C_2^{0\sigma} = C_2^{0*\sigma} \vee p_2^{\varepsilon\sigma} \vee q_{21}^{\varepsilon\sigma} \vee \cdots \vee q_{2t_2}^{\varepsilon\sigma},$$

and

$$\{p_2^\varepsilon, q_{21}^\varepsilon, \cdots, q_{2t_2}^\varepsilon\} = \{q \mid q \text{ is a } g\text{-literal in } C_2^0, q^\sigma = p_2^{\varepsilon\sigma}\}.$$

Hence,
$$C_2^{0\sigma} = C_2^{0*\sigma} \vee p_2^{\varepsilon\sigma}.$$

Note that $p_1^{\varepsilon\sigma} = p^{\varepsilon\sigma}$. It means that p_1 and p have the most general unifier, demoted by λ. Hence,

(1). There exists a substitution β such that

$$\varepsilon\sigma = \lambda\beta.$$

(2). $C_1^\lambda = (C_1)^{\lambda*} \vee p_1^\lambda \vee r_{11}^\lambda \vee \cdots \vee r_{1m_1}^\lambda$, where $\{p_1, r_{11}, \cdots, r_{1m_1}\} = \{q \mid q$ is a generalized-literal in C_1, $q^\lambda = p_1^\lambda\}$. Hence $C_1^\lambda = (C_1)^{\lambda*} \vee p_1^\lambda$. $C_2^\lambda = (C_2)^{\lambda*} \vee p_2^\lambda \vee r_{21}^\lambda \vee \cdots \vee r_{2m_2}^\lambda$, where $\{p_2, r_{21}, \cdots, r_{2m_2}\} = \{q \mid q$ is a generalized-literal in C_2, $q^\lambda = p_2^\lambda\}$. Hence $C_2^\lambda = (C_2)^{\lambda*} \vee p_2^\lambda$.

Moreover, the following properties hold:

(1). p_1 and p have the most general unifier λ.
(2). $p \wedge p_2 \leqslant \alpha$. Hence $R_\alpha(C_1, C_2) = C_1^{\lambda*} \vee C_2^{\lambda*}$. Hence,

$$\begin{aligned} C^0 &= C_1^{0*\sigma} \vee C_2^{0*\sigma} \\ &= C0\sigma^*{}_1 \vee C0\sigma^*{}_2 \\ &= C_1^{*\varepsilon\sigma} \vee C_2^{*\varepsilon\sigma} \\ &= C_1^{*\lambda\beta} \vee C_2^{*\lambda\beta} \\ &= (C_1^{\lambda*} \vee C_2^{\lambda*})^\beta \\ &= (R_\alpha(C_1, C_2))^\beta. \ \square \end{aligned}$$

Theorem 11.4.7 (Weak Completeness) *Suppose $|L| < \aleph_0$, $\bigvee_{a\in L}(a \wedge a') \leqslant \alpha < I$, α is a numerator, and there exists $\beta \in L$ such that*

$$\beta \wedge (\beta \to \beta') \nleqslant \alpha,$$

G^* is a regular generalized-Skölem standard from of G. If G^* is α-false, then there exists an α-resolution deduction from G^* to an α-◎.

Proof. If G^* is α-false, then it follows from Theorem 11.4.5 that there exists a finite ground instance set G^{*0} of G^* such that G^{*0} is α-false. By Theorem 11.3.2, there exists a resolution deduction D^0 from G^{*0} to α-◎. Moreover, we can use Theorem 11.4.6 to lift D^0 to a resolution deduction D from G^* to α-◎. $\square$

References

1. Akdag, H., Pacholczyk, D. (1989): Incertitude et iogique multivalente-Premiere partie: Etude theorique. BUSEFAL, **38**, 75–82
2. Antoniou, G. (1997): On the decomposition of default theories into parts. International Conference of Information Science, **2**, 184–187
3. Antoniou, G. (1998): Non-monotonic Reasoning. Cambridge, MIT Press, MA
4. Baaz, M., Fermüller, C.G. (1995): Resolution-based theorem proving for many-valued logics. J. Symbolic Comput., **19** (4), 353–391
5. Baaz, M., Fermüller, C.G., Salzer, G., Zach, R. (1998): Labeled calculi and finite-valued logics. Studia Logica, **61** (1), 7–33
6. Baaz, M., Fermüller, C.G., Salzer, G. (1999): Automated deduction for many-valued logics. In: Robinson, A., Voronkow, A. (Eds.) Handbook of Automated Reasoning. Elsevier
7. Bachmair, L., Ganzinger, H. (1990): On restrictions of ordered paramodulation with simplification. In: Proc. of CADE-10, LNCS 449, 427–441, Springer
8. Bachmair, L., Ganzinger, H. (1997): A theory of resolution. Technical report of the Max-Planck-Institut für Informatik, MPI-I-97-2-005, April, Germany
9. Bachmair, L., Ganzinger, H. (1998): Ordered chaining calculi for first-order theories of transitive relations. J. ACM, **45** (6), 1007–1049
10. Baker, A.B. (1991): Non-monotonic reasoning in situation calculus. Artificial Intelligence, **49**, 5–23
11. Baldwin, J.F. (1979): Fuzzy logic and fuzzy reasoning. Internat. J. Man-Machine Stud., **11**, 465–480
12. Bandler, W. (1980): Fuzzy power sets and fuzzy implication operators. Fuzzy Sets and Systems, **4**, 13–30
13. Barnes, D.W., Mack, J.M. (1975): An Algebraic Introduction to Mathematical Logic. Springer
14. Bauer, M. (1997): Approximation algorithms and decision making in the dempster-shafer theory of evidence–an impirical study. Internat. J. Uncertainty Reason., **17**, 217–237
15. Beckert, B., Hähnle, R., Manya, F. (1999): Transformations between signed and classical clause logic. In: Proc. of the 1999 ISMVL, 248–255, IEEE Press
16. Belluce, L.P. (1986): Semisimple algebras on infinite valued logic and bold fuzzy set theory. J. Funct. Anal., **38** (6), 1356–1379
17. Belnap, N.D. (1977): A useful four-valued logic. In: Dunn, J.M., Epstein, G. (Eds.) Modern Uses of Multiple-Valued Logic, 5-37, D. Reidel Publishing Company
18. Ben-David, S., Ben-Eliyahu-Zohary, R. (2000): A modal logic for subjective default reasoning, Artificial Intelligence, **116**, 217–236

19. Ben-Eliyahu, R., Dechter, R. (1996): Default reasoning using classical logic, Artificial Intelligence, **84**, 113–150
20. Berger, J.O. (1985): Statistical Decision Theory and Bayesian Analysis. New York: Springer
21. Besnard, P. (1989): An Introduction to Default Logics. Berlin: Springer
22. Besnard, P., Schaub, T. (1994): Possible worlds semantics for default logics. Fundamenta Mathematicae, **21** (1-2), 39-66
23. Bezdek, J.C. (1993): Fuzzy models – what are they, and why? IEEE Trans. Fuzzy Systems, **1** (1), 1–6
24. Bien, Z., Chun, M.G. (1994): An inference network for bidirectional uncertainty reasoning based on equality measure. IEEE Trans. Fuzzy Systems, **2** (2), 177–180
25. Birkhoff, G., Neumann, J. (1936): The logic of quantum mechanics. Ann. of Math., **37** (4)
26. Birkhoff, G. (1967): Lattice Theory. Trans. Amer. Math. Soc., New York
27. Blair, H.A., Subrahamnian, V.S. (1989): Paraconsistent logic programming. Theoret. Comput. Sci., **68**, 135–154
28. Blovkley, D., Godfrey, P. (2001): Doing It Differently. Thomas Telford, London
29. Blyth, T.S., Janowitz, M.F. (1972): Residuation Theory. Pergamon Press
30. Bochvar, D.A. (1939): On a 3-valued logic calculus and its application to the analysis of contradictions. Mat. Sb., **4**, 287–308 (in Russian)
31. Bolc, L., Borowik, P. (1992): Many-Valued Logics. Springer-Verlag
32. Bonissone, P.P., Tong, R.M. (1985): Reasoning with uncertainty in expert systems. Internat. J. Man-Machine Stud., **22**, 241–250
33. Booker, L.B., Hota, N. (1988): Probabilistic reasoning about ship images. Uncertainty in Artificial Intelligence, **2**, 371–379
34. Borns, D.W., Mack, J.M. (1975): An Algebraic Introduction to Mathematical Logic. New York-Heidelberg-Berlin: Springer
35. Boutilier, C. (1984): Model logics for qualitative possibility theory. Internat. J. Approx. Reason., **10** (2), 173–201
36. Brewka, G. (1991): Non-monotonic Reasoning: Logical Foundations of Commonsense. Cambridge University Press, Cambridge
37. Brewka, G. (1991): Cumulative default logic: in defense of non-monotonic inference rules. Artificial Intelligence, **50**, 183–205
38. Brooks, R.A. (1991): Intelligent without representation. Artificial Intelligence, **47**, 139–159
39. Brown, A.L., Gaucas, D.E., Benanav D. (1987): An algebraic foundation for truth maintenance. In: Proc. of the 1987 IJCAI, Milan, Italy, 973–980
40. Buchanan, B.G., Shortliffe, E.H. (1990): Rule-Based Expert Systems: The MYCIN Experiments of the Stanford Heuristic Programming Project. The MIT Press, Cambridge, MA
41. Buede, D.M., Girardi, P. (1997): A target identification comparison of Bayesian and Dempster-Shafer multisensor fusion. IEEE Trans. Systems, Man and Cybernet. Part A: Systems and Humans, **27** (5), 569–77
42. Burris, S., Sankappanavar, H.P. (1981): A Course in Universal Algebra. Springer
43. Busch, D.R. (1993): A sequent axiomatization of three valued logic with two negations. In: Pereira, L.M., Nerode, A. (Eds.) Logic Programming and Non-monotonic Reasoning, MIT Press, Cambridge, MA, 476–494

44. Butnariu, D., Klement, E.P., Zafrany, S. (1995): On triangular norm-based propositional fuzzy logics. Fuzzy Sets and Systems, **69**, 241–255
45. Cadoli, M., Schaerf, M. (1993): A survey of complexity results for non-monotonic logics. J. Log. Program., **17**, 127–160
46. Cao, Z., Kandel, A., Li, L. (1990): A new model of fuzzy reasoning. Fuzzy Sets and Systems, **36**, 311–325
47. Carnielli, W.A. (1987): Systematization of finite many-valued logics through the method of tableaux. J. Symbolic Logic, **52** (2), 473–493
48. Carnielli, W.A. (1991): On sequents and tableaux for many-valued logics. J. Non-Classical Logic, **8** (1), 59–76
49. Caselton, W.T., Luo, W. (1992): Decision making with imprecise probabilities-Dempster-Shafer theory and application. Water Resources Res., **28**, 3071–3083
50. Caselton, W.F., Luo, W. (1994): Inference and decision under near ignorance conditions. In: Duckstein, L., Parent, E. (Eds.) Engineering Risk in Natural Resources Management. NATO Advanced Study Institute Series, 291–304, Dordrecht: Kluwer Academic
51. Castro, J.L. (1994): Fuzzy logics as families of bivalued logics. Fuzzy Sets and Systems, **64**, 321–332
52. Chang, C.C. (1958): Algebraic analysis of many valued logic. Trans. Amer. Math. Soc., **88**, 467–490
53. Chang, C.C. (1959): A new proof of the completeness of the Łukasiewicz axioms. Trans. Amer. Math. Soc., **93**, 74–80
54. Chang, C.C. (1968): Fuzzy topological spaces. J. Math. Anal. Appl., **24**, 182–190
55. Chang, C.C., Keisler, H.J. (1973): Model Theory. North-Holland
56. Chang, C.L., Lee, R.C.T. (1973): Symbolic Logic and Mechanical Theorem Proving. Academic Press, New York
57. Charniak, E. (1991), Bayesian Networks without tears. AI Magazine, **12**(4), 50–63
58. Chechik, M., Easterbrook, S., Petrovykh, V. (2001): Model-checking over multi-valued logics. In: Proc. of the 2001 FME. LNCS, **2021**, 72–98
59. Chen, S.M. (1995): New methodology to fuzzy reasoning for rule-based expert systems. Cybernetics and Systems, **26**, 237–263
60. Chen, H.P., Parng, T.M. (1996): A new approach of multi-stage fuzzy logic inference. Fuzzy Sets and Systems, **78** (1), 51–72
61. Chen, S.M., Hsiao, W.H., Jong, W.T. (1997): Bidirectional uncertainty reasoning based on interval-valued fuzzy sets. Fuzzy Sets and Systems, **91**, 339 –353
62. Chen, S.M., Hsiao, W.H. (2000): Bidirectional uncertainty reasoning for rule-based systems using interval-valued fuzzy sets. Fuzzy Sets and Systems, **113**, 185–203
63. Cholewinski, P., Marek, V.W., Mikitiuk, A., Truszczynski, M. (1999): Computing with default logic. Artificial Intelligence, **112** (2-3), 105–146
64. Cignoli, R., Esteva, E., Godo, L., Torrens, A. (2000): Basic fuzzy logic is the logic of continuous t-norms and their residua. Soft Computing, **4** (2), 106–112
65. Clark, D.A. (1990): Numerical and symbolic approaches to uncertainty management in AI. Artif. Intell. Rev., **4**, 109–146
66. Cohen-Tannoudji, C., Dui, B., Laloe, F. (1992): Quantum Mechanics. John Wiley & Sons

67. Cortes-Rello, E., Golshani, F. (1990): Uncertain reasoning using the dempster-shafer method: an application in forecasting and marketing management. Expert Systems, **7** (1), 9–18
68. Czogala, E., Leski, J. (2001): On equivalence of uncertainty reasoning results using different interpretations of fuzzy if-then rules. Fuzzy Sets and Systems, **117**, 279–296
69. da Costa, N.C.A. (1990): Automated theorem proving in paraconsistent logics: theory and implementation. In: Proc. of the 1990 ICAD. LNAI, **449**, 72–86
70. D'Aquila, R.O., Crespo, C., Mate, J.L., Pazos, J. (2002): An inference engine based on fuzzy logic for uncertain and imprecise expert reasoning. Fuzzy Sets and Systems, **129**, 187 – 202
71. Das, P.S. (1981): Fuzzy groups and level subgroups. J. Math. Anal. Appl., **84**, 264–269
72. Davey, B.A., Priestley, H.A. (1990): Introduction to Lattices and Order. Cambridge University Press
73. Day, A. (1990): The lattice theory of functional dependencies and normal decompositions. Internat. J. Algebra Comput., **2**, 409–431
74. De Glas, M. (1989): Knowledge representation in a fuzzy setting, Technical Report 89/48, LAFORIA, Universite Paris VI, France
75. Delgrande, J.P., Schaub, T., Jackson, W.K. (1994): Alternative approaches to default logic. Artificial Intelligence, **70**, 167–237
76. Dempster, A.P. (1968): A generalization of Bayesian inference. J. Roy. Statist. Soc. Ser. B, **30**, 205–247
77. Dempster, A.P. (1968): Upper and lower probabilities generalized by a random closed interval. Ann. Math. Statist., **39**, 957–966
78. Dempster, A.P. (1969): Upper and lower probability inference for families of hypotheses with monotone density ratios. Ann. Math. Statist., **40**, 953–969
79. Denoeux, T. (1995): A k-nearest neighbour classification rule based on Dempster-Shafer theory. IEEE Trans. Systems, Man and Cybernet., **25**(5), 804–813
80. Denoeux, T. (1999): Reasoning with imprecise belief structures. Internat. J. Approx. Reason., **20**, 79–111
81. Dilworth, R.P. (1938): Abstract residuation over lattice. Bull. Amer. Math. Soc., **44**, 262–268
82. Dilworth, R.P., Ward, M. (1939): Residuated lattices. Trans. Amer. Math. Soc., **45**, 335–354
83. di Nola, A., Sessa, S., Pedrycz, W., Sanchez, E. (1989): Fuzzy Relation Equations and Their Application to Knowledge Engineering. Kluwer, Dordrecht
84. Dix, J. (1992): On cumulativity in default logic and its relation to Poole's approach. In: Neumann, B. (Ed.) 10th European Conference on Artificial Intelligence (ECAI-92), Vienna, Austria, Wiley, New York, 289–293
85. Dix, J., Furbach, U., Nerode, A. (1997): Logic Programming and Nonmonotonic Reasoning. Springer, Berlin
86. Di Zenzo, A. (1986): A new many-valued logic and its applications to uncertainty reasoning. In: Kugler, H. (Ed.) Information Processing 86. North-Holland, Amsterdam, 421–427
87. Di Zenzo, A., Ventre, A.G.S. (1989): On fuzzy implication in De Morgan algebra. Fuzzy Sets and Systems, **33**, 155–164
88. Donini, F.M., Lenzerini, M., Nardi, D., Pirri, F., Schaerf, M. (1990): Nonmonotonic reasoning. Artif. Intell. Rev., **4**, 163–210

89. Douglas, R.B. (1993): A sequent axiomatization of three valued logic with two negations. In: Pereira, L.M., Nerode, A. (Eds.) Logic Programming and Non-monotonic Reasoning. MIT Press, 476–494
90. Doyle, J. (1979): A truth maintenance system. Artificial Intelligence, **12**.
91. Dubois, D., Prade, H. (1980): Fuzzy Sets and System: Theory and Applications. Academic Press, New York
92. Dubois, D., Prade, H. (1985): The generalized modus ponens under sup-min composition–A theoretical study. In: Gupta, M.M. et al. (Eds.) Uncertainty Reasoning in Expert Systems, North–Holland, 217–232
93. Dubois, D., Prade, H. (1986): Possibility Theory. Plenum Press, New York
94. Dubois, D., Lang, J., Prade, H. (1987): Theorem proving under uncertainty–A possibility theory based approach. In: Proc. of the 1987 IJCAI. Mitano, Italy, 984–986
95. Dubois, D., Prade, H. (1988): An introduction to possibilistic and fuzzy logics. In: Smets, P., Mamdani, A., Dubois, D., Prade, H. (Eds.) Non-standard logics for automated reasoning. Academic Press
96. Dubois, D., Prade, H. (1988): Representation and combination of uncertainty with belief functions and possibility measures. Computational Intelligence, Blackwell Publishing, **4**, 244–264
97. Dubois, D., Prade, H., Farreny, H., Martin-Clouaire, R., Testemale, C. (1988): Possibility Theory – An Approach to the Computerized Processing of Uncertainty. Plenum Press, New York
98. Dubois, D., Lang, J., Prade, H. (1989): Automated reasoning using possibility logic semantics, belief revision and variable certainty weights. In: Proc. of the 1989 Workshop on Uncertainty in Artificial Intelligence. Windor, Ontario, 81–87
99. Dubois, D., Prade, H. (1990): Resolution principle in possibilistic logic. Internat. J. Approx. Reason., **4** (1), 1–21
100. Dubois, D., Lang, J., Prade, H. (1991): Fuzzy sets in uncertainty reasoning, Part I: Inference with possibility distributions, Part II: Logical approaches. Fuzzy Sets and Systems, **40**, 143–202, 203–244
101. Dubois, D., Prade, H., Yager, R.R. (1993): Readings in Fuzzy Sets for Intelligent Systems. Morgan and Kaufmann, CA
102. Dubois, D., Lang, J., Prade, H. (1994): Possibilistic logic. In: Gabbay, D.M., Hogger, C.J., Robinson, J.A. (Eds.) Handbook of Logic in Artificial Intelligence and Logic Programming, **3**: Non-monotonic Reasoning and Uncertain Reasoning, Clarendon Press, 439–513
103. Dubois, D., Prade, H. (1995): Fuzzy relation equations and causal reasoning. Fuzzy Sets and Systems, **75**, 119–134
104. Dubois, D., Prade, H. (1996): What are fuzzy rules and how to use them. Fuzzy Sets and Systems, **84**, 169–185
105. Dubois, D., Prade, H. (1996): Combining hypothetical reasoning and plausible inference in possibility logic. Internat. J. Multiple-Valued Logic, **1** (3), 219–239
106. Dubois, D., Prade, H., Yager, R.R. (1997): Fuzzy Information Engineering: a Guided Tour of Applications. New York: Wiley
107. Dubois, D., Prade, H. (1998): An introduction to fuzzy systems. Clinica Chimica Acta, **270**, 3–29
108. Duda, R.O., Haet, P.E., Nileson, N.J. (1976): Subjective Bayesian methods for rule-based inference systems. In: Proc. of AFIPS Conference. **45**, AFIPS Press, 1075–1082

109. Dummett, M. (1959): A propositional calculus with denumerable matrix. J. Symbolic Logic, **24**, 97–106
110. Dundici, D. (1986): Interpretation of AF C^*-algebras in Łukasiewicz Sentential Calculus. J. Funct. Anal., **65** (1), 15–63
111. Düntsch, (1994): Rough relation algebras. Fundamenta Mathematicae, **21**, 321–331
112. El-Azhary, E.S., Edrees, A., Rafea, A. (2002): Diagnostic expert system using non-monotonic reasoning. Expert Systems with Applications, **23** (2), 137–144
113. Eiter, T., Gottlob G. (1993): Propositional circumscription and extended closed world reasoning are complete. Theoret. Comput. Sci., **114**, 231–245
114. Eiter, T., Gottlob G., Leone, N. (1997): Semantics and complexity of abduction from default theories. Artificial Intelligence, **90**, 177–223
115. Emami, M.R., Türksen, I.B., Goldenberg, A.A. (1999): A unified parameterized formulation of reasoning in fuzzy modeling and control. Fuzzy Sets and Systems, **108**, 59–81
116. Engemann, K.J., Miller, H.E., Yager, R.R. (1999): Evaluating risk and transnational decision support systems using interval probabilities. In: Lasker, G. (Ed.) Advances in Support Systems Research, **V**: Decision Support Methodology for Human Systems Management and Its Application to Economics and Commerce, 45–59
117. Engemann, K.J., Miller, H.E., Yager, R.R. (2000): A decision methodology using belief structures with interval probabilities. In: Engemann, K., Lasker, G. (Eds.) Advances in Decision Technology and Intelligent Information Systems, **1**, Institute for Advanced Studies in Systems Research and Cybernetics: Windsor, Canada, 1–5
118. Esteva, F. (1981): On a representation theorem of De Morgan algebras by fuzzy sets. Stochastica, **5** (2), 109–115
119. Esteva, F. (1984): Some representable De Morgan algebras. J. Math. Anal. Appl., **100** (2), 463–469
120. Esteva, F., Garcia, C.P., Godo, L. (1994): Enriched interval bilattices and partial many valued logics: An approach to deal with graded truth and imprecision. Internat. J. Uncertain. Fuzziness Knowledge-Based Systems, **2**, 37–64
121. Esteva, F., Godo, L., Hájek, P., Navara, M. (2000): Residuated fuzzy logic with an involutive negation. Arch. Math. Logic, **39**, 103–124
122. Etherington, D. (1988): Reasoning with Incomplete Information. Pitman, London
123. Fitting, M.C. (1988): Logic programming on a topological bilattice. Fundamenta Mathematicae, **11**, 209–218
124. Fitting, M.C. (1989): Bilattices and the theory of truth. J. Philos. Logic, **18**, 225–256
125. Fitting, M.C. (1991): Many-Valued Modal Logics. Fundamenta Mathematicae, **15** (3-4), 335–350
126. Fitting, M.C. (1992): Many-valued modal logics II. Fundamenta Mathematicae, **17**, 55–73
127. Font, J.M., Rodriquez, A.J., Torres, A. (1984): Wajsberg algebras. Stochastica, **8** (1), 5–31
128. Font, J.M., Moussavi, M. (1993): Note on a six valued extension of three valued logic. J. Appl. Non-Classical Logics, **3**, 173–187
129. Friedman, N., Halpern, J.Y. (2001): Plausibility measures and default reasoning. J. ACM, **48**, 648–685

130. Gabbay, D., Hogger, C.J., Robinson, J.A. (1994): Handbook of Logic in Artificial Intelligence and Logic Programming. **3**: Non-monotonic Reasoning and Uncertain Reasoning. Clarendon Press, Oxford
131. Gaines, B.R. (1976): Foundation of fuzzy reasoning. Internat. J. Man-Machine Stud., **6**, 623–668
132. Gaines, B.R. (1976): Fuzzy reasoning and the logics of uncertainty. Proc. of the 1976 ISMVL, USA, 179–188
133. Gaines, B.R. (1979): Logical foundations for database systems. Internat. J. Man-Machine Stud., **11** (4), 481–500
134. Ganzinger, H., Meyer, C., Hustadt, U., Schmitt, R. (1998): A resolution-based decision procedure for extensions of $K4$. In: Advances in Modal Logic'98 workshop, October, Uppsala, Sweden, also in: Zakharyaschev, M., Segerberg, K., M. de Rijke, Wansing, H. (Eds.) Advances in Modal Logic, Volume 2 Lecture Notes. Stanford, CA: CSLI Publications, 2000.
135. Ganzinger, H., Stokkermans, V.S. (2000): Chaining techniques for automated theorem proving in many-valued logics. In: Proc. of the 2000 ISMVL, Oregon, May, 23 - 25, IEEE Computer Society, Los Alamitos, 337–344
136. Garcia, P., Esteva, F. (1988): Representation of symmetric algebras and its subvarieties. In: Proc. of the 1988 ISMVL. Palma de Mallorca, IEEE Computer Society Press, 277–284
137. Gardenfors, P., Makinson, D. (1994): Non-monotonic inference based on expectations. Artificial Intelligence, **65**, 197–245
138. Gatsonis, C., Hodges, J., Kass, R. Singpurwalla, N. (1992): Bayesian Statistics and Technology: Case Studies. Springer, New York
139. Geffner, H. (1992): Default Reasoning: Causal and Conditional Theories. MIT Press, Cambridge
140. Gelfond, M., Przymusinska, H. (1986): Negation as failure: careful closure procedure. Artificial Intelligence, **30**, 273–287
141. Gelfond, M., Przymusinska, H., Przymusinska, T. (1989): On the relationship between circumscription and negation as failure. Artificial Intelligence, **38**, 49–73
142. Geneserth, M.R., Nileson, N.J. (1987): Logical Foundations of Artificial Intelligence. Morgan-Kaufmann, Los Altos, CA
143. Gerla, G. (1994): Comparing fuzzy and crisp deduction systems. Fuzzy Sets and Systems, **67**, 317–328
144. Gerla, G. (1994): An extension principle for fuzzy logic. Mathematical Logic Quarterly, **40**, 357–380
145. Giarrantano, J., Riley, G. (1994): Expert Systems: Principles and Programming (2Ed). PWS Publishing Company, Boston
146. Giles, R. (1976): Łukasiewicz logic and fuzzy sets theory. Internat. J. Man-Machine Stud., **8** 3–27
147. Ginsberg, M.L. (1986): Counterfactuals. Artificial Intelligence, **30**, 35–79
148. Ginsberg, M.L. (1987): Multi-valued logic. In: Ginsberg, M. (Ed.) Readings in Non-monotonic Reasoning. Morgan-Kaufmann Pub., 251–255
149. Ginsberg, M.L. (1987): Readings in Non-monotonic Reasoning. Morgan Kaufmann, San Mateo
150. Ginsberg, M.L. (1988): Multivalued logics: a uniform approach to inference in artificial intelligence. Comput. Intell., **4** (3), 265–316
151. Glymour, C. (1985): Independence assumpations and Bayesian updating. Artificial Intelligence, **16** (1), 95–99

152. Goguen, J.A. (1967): L-type fuzzy sets. J. Math. Anal. Appl., **18**, 145–174
153. Goguen, J.A. (1969): The logic of inexact concepts. Synthese, **19**, 325–373
154. Goldblatt, R. (1989): Varieties of complex algebras. Ann. Pure Appl. Logic, **44** (3), 153–301
155. Goldszmidt, M., Pearl, J. (1996): Qualitative probabilities for default reasoning. Belief revision, and causal modeling. Artificial Intelligence, **84**, 57–112
156. Goodwin, J.W. (1987): A theory and system for non-monotonic reasoning, Linköping Studies in Science and Technology, Dissertations No. 165, Department of Computer and Information Science, Linköping University.
157. Gorzalczany, M.B. (1987): A method of inference in uncertainty reasoning based on interval-valued fuzzy sets. Fuzzy Sets and Systems, **21**, 1–17
158. Gottlob, G. (1992): Complexity results for non-monotonic logics. J. Logic Comput., **2**(3), 397–425
159. Gottlob, G., Zhang, M.Y. (1994): Cumulative default logic: finite characterization, algorithms, and complexity. Artificial Intelligence, **69**, 329–345
160. Gottwald, S. (1980): Fuzzy propositional logics. Fuzzy Sets and Systems, **3**, 181–192
161. Gottwald, S. (1989): Mehrwertige Logik. Akademie-verlag, Berlin
162. Guan, J.W., Guan, Z., Bell, D.A. (1997): Bayesian probability on boolean algebras and applications to decision theory. Inform. Sci., **97**(3–4), 267–280
163. Gupta, M.M. (1992): Intelligence, uncertainty and information. In: Ayyub, B.M., Gupta, M.M., Kanal, L.N. (Eds.) Analysis and Management of Uncertainty–Theory and Applications. Elsevier, 3–12
164. Hähnle, R. (1990): Towards an efficient tableau proof procedure for multiple-valued logics. In: Proc. of the Workshop on Computer Science Logic, Heidelberg. LNCS. **533**, 248–260
165. Hähnle, R. (1991): Uniform notation of tableaux rules for multiple-valued logics. In: Proc. of the 1991 ISMVL. 238–245
166. Hähnle, R. (1993): Short CNF in finitely-valued logics. In: Komorowski, J., Ras, Z.W. (Eds.) Proc. of the 1993 ISMIS. 49–58
167. Hähnle, R. (1994): Automated Deduction in Multiple-Valued Logics. Oxford University Press, Oxford
168. Hähnle, R. (1994): Short conjunctive normal forms in finitely valued logics. J. Logic Comput., **4** (6), 905–927
169. Hähnle, R. (1996): Exploiting data dependencies in many-valued logics. J. Appl. Non-Classical Logics, **6** (1), 49–69
170. Hähnle, R., Escalada Imaz, G. (1997): Deduction in many-valued logics: a survey. Mathware Soft Comput., **4** (2), 69–97
171. Hähnle, R. (1998): Commodious axiomatization of quantifiers in multiple-valued logic. Studia Logica, **61** (1), 101–121
172. Hájek, P. (1995): Fuzzy logic and arithmetical hierarchy. Fuzzy Sets and Systems, **73**, 359–363
173. Hájek, P., Godo, L., Esteva, F. (1996): A complete many-valued logic with product conjunction. Arch. Math. Logic, **35**, 191–208
174. Hájek, P. (1998): Basic fuzzy logic and BL-algebras. Soft Computing, **2**(3), 124–128
175. Hájek, P. (2000): Meta-Mathematics of Fuzzy Logic. Kluwer Academic Publishers-Dordrecht
176. Halpern, J.Y., Fagin, R. (1992): Two views of belief: belief as generalized probability and belief as evidence. Artificial Intelligence, **54**, 275–317

177. Hansom, O., Mayer, A. (1989): Heuristic search as evidential reasoning. In: Proc. of the 1989 Workshop on Uncertainty in Artificial Intelligence. Mountain View, CA
178. Hazelhurst (1996): Compositional Model Checking of Partially Ordered State Spaces. PhD thesis, Department of Computer Science, University of British Columbia
179. Heckerman, D., Wellman, M. (1995): Bayesian networks. Comm. ACM, **38**, 27–30
180. Henkind, S.J., Harrison, M.C. (1988): Analysis of four uncertainty calculi. IEEE Trans. Man, System and Cybernet., **18** (5), 700–714
181. Hinde, C.J. (1986): Fuzzy Prolog. Internat. J. Man-Machine Stud., **11**, 569–595
182. Höhle, U. (1995): Commutative, residuated L-monoids. In: Höhle, U., Klement, E.P. (Eds.) Non-Classical Logics and Their Applications to Fuzzy Subsets. Kluwer, Boston, 53–106
183. Höhle, U., Klement, E.P. (1995): Non-Classical Logics and Their Applications to Fuzzy Subsets. Kluwer, Boston
184. Höhle, U. (1996): On the fundamental of fuzzy set theory. J. Math. Anal. Appl., **201**, 786–826
185. Horn, A. (1969): Logic and truth values in a linearly ordered Heyting algebra. J. Symbolic Logic, **34** (3), 395–408
186. Horty, J.F. (1994): Some direct theories of non-monotonic inheritance. In: Gabbay, D., Hogger, C.J., Robinson, J.A. (Eds.) Handbook of Logic in Artificial Intelligence and Logic Programming, **3**: Non-monotonic Reasoning and Uncertain Reasoning. Clarendon Press, Oxford, 111–187
187. Hsiao, W.H., Jong, W.T., Chen, S.M., Lee, C.H. (1996): Interval-valued bidirectional uncertainty reasoning techniques for rule-based systems. In: Proc. of the 4th National Conference on Defense Management, Taipei, Taiwan, 915–927
188. Hu, Q.P. (1987): BCI-algebras (in Chinese). Shannxi Science and Technoloy Press, Xi'an, China
189. Hu, C.L., Song, Z.M. (1990): Foundation of Lattice Theory (in Chinese). Henan University Press, Kaifeng, China
190. Hunter, D. (1987): Dempster-Shafer vs. probabilistic logic. In: Proc. of the Third AAAI uncertainty in artificial intelligence workshop, 22–29
191. Iancu, I. (1997): Reasoning system with fuzzy uncertainty. Fuzzy Sets and Systems, **92**, 51–59
192. Ibrahim, Ayyub, B.M. (1992): Uncertainties in risk-based Inspection of complex systems. In: Ayyub, B.M., Gupta, M.M., Kanal, L.N. (Eds.) Analysis and Management of Uncertainty, Theory and Applications. Elsevier, New York, 247–262
193. Ip, H.H., Ng, J.M. (1994): Human face recognition using Dempster-Shafer theory. In: ICIP. 1st International Conference on Image Processing, **2**, 292–295
194. Iseki, K. (1966): An algebra related with a propositional calculus. In: Proc. Japan. Acad., **42**, 26–29
195. Iseki, K., Tanaka, S. (1976): Ideal theory of BCK-algebras. Math. Japanica, **21**, 351–366
196. Iseki, K., Tanaka, S. (1978): An introduction to the theory of BCK-algebras. Math. Japanica, **23**, 1–26

197. Iturrioz, L. (1982): Modal operators on symmetrical Heyting algebras. In: Traczyk, T. (Ed.) Universal Algebra and Applications. Banach Center Publications, PWN-Polish Scientific Publishers, **9**, 289–303
198. Iturrioz, L. (1983): Symmetrical Heyting algebras with operators. Zeit. f. Math. Logik und Grundlagen d. Math., **29**, 33–70
199. Iturrioz, L., Orlowska, E. (1996): A Kripke-style and relational semantics for logic based on Łukasiewicz algebras. In: Conference in honour of J. Łukasiewicz. Dublin
200. Jensen, F.V. (1996): An Introduction to Bayesian Networks. Springer, New York
201. Jiang, H., Deng, F.A. (1991): The relations between fuzzy Lattice implication algebras and bounded BCK-algebras. Fuzzy Systems Math., **5** (1), 88–89 (in Chinese)
202. Jiang, H., Deng, F.A. (1993): The relations between fuzzy Lattice implication algebra and bounded $BCK-$algebra. J. Southwest Jiaotong Univ., **89** (4), 20–27 (in Chinese)
203. Johnson, R.W. (1986): Independence and Bayesian updating methods. In: Kanal, L.N., Lemmer, J.F., (Eds.) Uncertainty in Artificial Intelligence. North-Holland, Amsterdam, 137–151
204. Jordan, P., Wigner, E., Von Neumann, J. (1934): On an algebraic generalization of the quantum mechanical formulation. Ann. Math., **35**
205. Jun, Y.B. (1997): Implication filter of lattice implication algebras. Bull. Korean Math. Soc., **34**, 193–198
206. Jun, Y.B., Xu, Y. (1997): On right regular maps of lattice implication algebras. J. Inst. Math. Comput. Sci. Math. Ser., **10** (1), 51–58
207. Jun, Y.B., Roh, E.H., Xu, Y. (1998): LI-Ideals in lattice implication algebras. Bull. Korean Math. Soc., **35**, 13–24
208. Jun, Y.B., Xu, Y., Qin, K.Y. (1998): Positive implicative and associative filters of lattice implication algebras. Bull. Korean Math. Soc., **35** (1), 53–61
209. Jun, Y.B., Xu, Y. (1998): Quasi-uniformity on lattice implication algebras. Chinese Quart. J. Math., **13** (1), 84–86
210. Jun, Y.B., Xu, Y. (1999): Fuzzy LI-ideals in lattica implication algebras. J. Fuzzy Math., **7** (4), 997–1003
211. Jun, Y.B., Xu, Y., Liu, J. (1999): Quotient lattice implication algebras induced by fuzzy LI-ideals. J. Fuzzy Math., **7** (4), 1013–1020
212. Jun, Y.B., Xu, Y., Qin, K.Y. (1999): Normal fuzzy LI-ideals in lattice implication algebras. J. Fuzzy Math., **7** (4), 1045-1052
213. Jun, Y.B., Xu, Y., Qin, K.Y., Roh, E.H. (2000): Intuitionistic fuzzy LI-ideals in lattice implication algebras. Sci. Math., **3** (2), 185–192
214. Jun, Y.B. (2001): The prime filters theorem of lattice implication algebras. Int. J. Math. Math. Sci., **25** (2), 115–118
215. Jun, Y.B. (2001): Fuzzy positive implicative and fuzzy associative filters of lattice implication algebra. Fuzzy Sets and Systems, **121**, 353–357
216. Kahneman, D., Slovic, P., Tversky, A. (1982): Judgment under Uncertainty: Heuristics and Biases. Cambridge University Press, New York
217. Kalmbach, G. (1983): Orthomodular Lattices. Academic Press, London
218. Kanal, L.N., Lemmer, J.F. (1986): Uncertainty in Artificial Intelligence. Elsevier Science Publishers BV North Holland, Amsterdam
219. Kaufmann, A. (1988): Theory of expertons and fuzzy logic. Fuzzy Sets and Systems, **28**, 295–304

220. Kautz, H. A., Selman, B. (1991): Hard problems for simple default logics. Artificial Intelligence, **49**, 243–279
221. Kifer, M. (1989): On the expressive power of annotated logic program. In: Proc. of the 1989 North American Conference on Logic Programming, Cleveland, OH, 1069–1089
222. Kifer, M., Lozinskii, E. (1991): RI: A logic for reasoning with inconsistency. In: Proc. of the 1991 ISMVL. 253–262
223. Kifer, M., Subrahamnian, V.S. (1992): Theory of generalized annotated logic programming and its application. J. Log. Program., **12**, 335–367
224. Kirin, V.G. (1966): Gentzen's method of the many-valued propositional calculi. Zeit. f. Math. Logik und Grundlagen d. Math., **12**, 317-332
225. Kleene, S.C. (1952): Introduction to Metamathematics. Van Nostrand, New York
226. Klir, G.J., Yuan, B. (1995): Fuzzy Sets and Fuzzy Logic: Theory and Applications. Prentice-Hall, Upper Saddle River, NJ
227. Klir, G.J., Yuan, B. (1995): Fuzzy Sets, Fuzzy Logic, and Fuzzy Systems - Selected Papers by Zadeh L.A.. World Scientific, Singapore
228. Konolige, K. (1988): On the relation between default and autoepistemic logic. Artificial Intelligence, **35**, 343–382
229. Krause, P., Clark, D. (1993): Representing Uncertain Knowledge: An artificial Intelligence Approach. Kluwer
230. Kruse, R., Schwecke, E., Heinsohn, J. (1992): Uncertainty and Vagueness in Knowledge-Based Systems. Springer, Berlin
231. Kruse, R., Gebhardt, J., Klawonn, F. (1994): Foundations of Fuzzy Systems. Wiley, New York
232. Kuhu, P., Nikhil, R.P. (1999): A neural-fuzzy system for inference. Internat. J. Intell. Systems, **14**, 1155–1182
233. Lano, K. (1992): Fuzzy sets and residuated logic. Fuzzy Sets and System, **47** (2), 203–220
234. Lee, R.C.T., Chang, C.L. (1971): Some properties of fuzzy logic. Inform. Control, **19**, 417–431
235. Lee, R.C.T. (1972): Fuzzy logic and the resolution principle. J. ACM, **19**, 109–119
236. Lehmke, S. (1996): Weight structures for approximate reasoning with weighted expressions. In: Proc. of the 1996 ISMVL. Santiage de Compostela, Spain, 178–183
237. Lehmke, S. (1996): On Resolution-Based Theorem Proving in Propositional Fuzzy Logic with 'Bold' Connectives. Ph. D thesis, available at: http://ls1-www.cs.uni-dortmund.de/lehmke/EFL990/Diplomarbeit.pdf
238. Lehmke, S. (1997): On the semantics of Uncertainty logics. In: Fifth European Congress on Intelligent Techniques and Soft Computing, Aachen, Germany, **1**, 137–141
239. Lehmke, S. (1999): A resolution-based axiomatisation of 'bold' propositional fuzzy logic. In: Dubois, D., Klement, E.P., Prade, H. (Eds.) Fuzzy Sets, Logics, and Reasoning about Knowledge. Kluwer Academic Publishers, Dordrecht
240. Lehmann, D. (2001): Non-monotonic logics and semantics. J. Logic Comput., **11**, 229–256
241. Leitsch, A. (1989): On different concept of resolution. Zeit. f. Math. Logik und Grundlagen d. Math., **35**, 71–77

242. Lewis, D.K. (1973): Counterfactuals. Cambridge, Harvard University Press, MA
243. Liau, C.J. (1996): Possibilistic residuated implication logics with applications. In: Proc. of the 1996 IPMU. 935–940
244. Liau, C.J., Lin, I.P. (1997): Possibilistic reasoning – A mini-survey and uniform semantics. Artificial Intelligence, **88**, 163–193
245. Lifschitz, V. (1985): Computing circumscription. In: Proceedings of the Ninth International Joint Conference on Artificial Intelligence (IJCAZ-85), 121–127
246. Lifschitz, V. (1985): Closed-world databases and circumscription. Artificial Intelligence, **27**, 229–235
247. Lifschitz, V. (1988): On the declarative semantics of logic programs with negation. In: Minker, J. (Ed.) Foundation of Deductive Databases and Logic Programming, Morgan Kaufmann, 177–192
248. Lifschitz, V. (1989): Between circumscription and autoepistemic logic. Research Report, Stanford University, Stanford, CA
249. Lindley, D.V. (1987): The probability approach to the treatment of uncertainty in artificial intelligence and expert systems. Statist. Sci., **2**, 3–44
250. Lingras, P., Wong, S.K.M. (1990): Two perspectives of the Dempster-Shafer theory of belief functions. Internat. J. Man-Machine Stud., **33**, 467–487
251. Linke, T., Schaub, T. (2000): Alternative foundations for Reiters default logic. Artificial Intelligence, **124**, 31–86
252. Liu, J., Xu, Y. (1996): On syntactical problem of lattice-valued propositional logic system $LP(X)$. In: Proc. of the Celebration of Southwest Jiaotong Univ.'s Centenary–Special Issue for Graduates Students. Southwest Jiaotong Univ. Press, 305–308 (in Chinese)
253. Liu, J., Xu, Y. (1996): On certain filters in lattice implication algebras. Chinese Quart. J. Math., **11** (4), 106–110
254. Liu, J., Xu, Y. (1996): On the Properties (P) of Lattice Implication Algebras. J. Lanzhou Univ. Nat. Sci.: Fuzzy Mathematics and Systems (Supplement), **32**, 344–348 (in Chinese)
255. Liu, J. (1996): Study on the Structure of Lattice Implication Algebras. MA Thesis, Southwest Jiaotong Univ., China (in Chinese)
256. Liu, J., Xu, Y. (1997): On filters and structure of lattice implication algebras. Chinese Sci. Bull., **42** (18), 1517–1520 (in Chinese)
257. Liu, J., Xu, Y. (1998): On prime filters and decomposition theorem of lattice implication algebras. J. Fuzzy Math., **6** (4), 1001–1008
258. Liu, J., Xu, Y. (1998): On lattice implication algebras and MV-algebras. BUSEFAL, **77**
259. Liu, J., Xu, Y. (1999): On the representation of lattice implication algebras. J. Fuzzy Math., **7** (1), 251–258
260. Liu, J. (1999): Study on lattice-valued logic system and lattice-valued resolution principle. Ph.D. Thesis, Southwest Jiaotong Univ., Chengdu, China (in Chinese)
261. Liu, J., Song, Z.M., Qin, K.Y. (2000): A resolution procedure based on a fuzzy logic. In: Proc. of the 2000 ICFS. San Antonio, Texas
262. Liu, X.H., Xiao, H. (1985): Operator fuzzy logic and fuzzy resolution. In: Proc. of the 1985 ISMVL. Kingstion, Canada, 68–75
263. Liu, X.H. (1989): Fuzzy Logic and Fuzzy Reasoning. Jilin University Press, Changchun, China (in Chinese)

264. Liu, X.H., Xiao, H. (1989): Operator fuzzy logic and λ-resolution. Chinese J. Computer, **2**, 81–91 (in Chinese)
265. Liu, X.H. (1994): Resolution-Based Automated Reasoning. Academic Press of China, Beijing, China (in Chinese)
266. Liu, L.Z., Wang, G.J. (1998): Fuzzy implication algebras and MV-algebras. J. Fuzzy Systems Math., **12** (1), 20–25 (in Chinese)
267. Liu, W.R., Hong, J. (2000): Reinvestigating Dempster's idea on evidence combination. Knowledge Inform. Systems, **2**, 223–241
268. Loveland, D.W. (1978): Automated Theorem Proving: A Logical Basis. North-Holland
269. Łukasiewicz, J. (1920): On 3-valued logic. Ruch Filozoficzny, **5**, 169–170
270. Łukasiewicz, J. (1970): Selected Works. Amsterdam, North-Holland
271. Łukaszewicz, W. (1990): Non-Monotonic Reasoning: Formalization of Commonsense Reasoning. Ellis-Horwood, Chichester
272. Lu, J.J., Henschen, L.J. (1992): The completeness of gp-resolution for annotated logics. Inform. Process. Lett., **44**, 135–140
273. Lu, J.J., Murray, N.V., Rosenthal, E. (1993): Signed formulas and annotated logics. In: Proc. of the 1993 ISMVL. 48–53
274. Lu, J.J., Murray, N.V., Rosenthal, E. (1993): Signed formulas and fuzzy operator logics. In: Komorowski, J., Ras, Z.W. (Eds.) Proc. of the 1993 ISMIS. 75–84
275. Lu, J.J., Murray, N.V., Rosenthal, E. (1998): A framework for reasoning in multiple-valued logics. J. Automated Reason., **21** (1), 39–67
276. Luo, X., Zhang, C. (1998): LSLNCF: A hybrid uncertain reasoning model based on probability. Internat. J. Uncertain. Fuzziness Knowledge-Based Systems, **6** (4), 401–422
277. Makinson, D. (1994): General patterns in non-monotonic reasoning. In: Gabbay, D., Hogger, C., Robinson, J.A. (Eds.) Handbook of Logic in Artificial Intelligence and Logic Programming, **3**, 35–110. Oxford University Press
278. Ma, J.(1999): Study on Lattice Implication Algbra and Lattice-Valued Propositional Logic $LP(X)$. MA Thesis, Southwest Jiaotong Univ., China (in Chinese)
279. Ma, J.(2002): Lattice-Valued Logical System and Automated Reasoning Based on Lattice Implication Algebra. Ph.D. Thesis, Southwest Jiaotong Univ., China (in Chinese)
280. Manna, Z., Waldinger, R. (1986): Special relations in automated deduction. J. ACM, **33** (1), 1–59
281. Marco, C., Schaerf, M. (1993): A survey of complexity results for non-monotonic logics. J. Log. Program., **17**(2-4), 127–160
282. Marek, W., Trusczynski, M. (1989): Relating autoepistemic and default logics. Research Report, University of Kentucky, Lexington, KY
283. Marek, W., Trusczynski, M. (1993): Non-monotonic Logic: Context Dependent Reasoning. Springer, Berlin
284. Marek, W., Trusczynski, M. (1994): Default Logic, Springer
285. McCarthy, J. (1980): Circumscription-a form of non-monotonic reasoning. Artificial Intelligence, **13**, 27–39
286. McCarthy, J. (1983): Some expert systems need common sense, In: Pagels, H. (Ed.) Computer Culture: The Scientific, Intellectual and Social Impact of the Computer. **426**, Annals of the New York Academy of Sciences

287. McCarthy, J. (1986): Applications of circumscription to formalizing commonsense knowledge. Artificial Intelligence, **28**, 89–116
288. McCarthy, J. (1990): Two puzzles involving knowledge, In: McCarthy, J. (Ed.) Formalizing Common Sense: Papers by John McCarthy. Ablex Publishing Corporation
289. McDermott, D., Doyle, J. (1980): Non-monotonic logic I. Artificial Intelligence, **13** (1), 41–72
290. McDermott, D., Doyle, J. (1982): Non-monotonic Logic II: Mon-monotonic Modal Theories. J. ACM, **29** (1), 33–57
291. McDermott, D. (1989): A general framework for reason maintenance. Report 691, Computer Science Department, Yale University, New Haven, CT
292. McDermott, D. (1991): A general framework for reason maintenance. Artificial Intelligence, **50**, 289–329
293. McNaughton, R. (1951): A theorem about infinite-valued sentential logic. J. Symbolic Logic, **16** (1), 1–13
294. Meng, B.L. (1998): The prime filters of lattice implication algebras. J. Northwest Univ., **28** (3), 189–192 (in Chinese)
295. Meng, J., Jun, Y.B. (1994): BCK-Algebra. Kyung Moon SA Co., Korea
296. Menger, K. (1942): Statistical metrics. Proc. Natl. Acad. Sci. USA, **8**, 535–537
297. Michalski, R.S. (1977): Variable-Valued Logic and its Applications to Pattern Recognition and Machine Learning. In: Rine, D.C. (Ed.) Computer Science and Multiple-Valued Logic: Theory and Applications. North-Holland, Amsterdam, 506–534
298. Minari, P. (1980): Completeness theorems for some intermediate predicate calculus. Studia Logica, **42** (4), 431–441
299. Minker, J. (1982): On indefinite databases and the closed world assumption. In: Proc. of the Sixth Conference on Automated Deduction (CADE-82), 292–308
300. Mizumoto, M., Zimmermann, H.J. (1982): Comparison of fuzzy reasoning methods. Fuzzy Sets and Systems, **8**, 253–283
301. Mizumoto, M. (1985): Extended fuzzy reasoning. In: Gupta, M.M., Kandel, A., Bandler, W., Kiszka, J.B. (Eds.) Approximate Reasoning in Expert Systems. North-Holland, Amsterdam, 67–76
302. Moinard, Y. (1994): Reasoning by cases without contraposition in default logic. In: Proc. of the ECAI-94, Amsterdam, 381–385
303. Montero, J., Cappelle, B., Kerre, E.E. (1995): The usefulness of complete lattices in reliability theory. In: Onisawa, T., Kacprzyck, J. (Eds.) Reliability and Safety Analyses under Fuzziness. Springer, Heidelberg, 95–110
304. Moore, R. (1985): Semantical considerations on non-monotonic logic. Artificial Intelligence **25** (1), 75–94
305. Morgan, C. (1976): A resolution principle for a class of many-valued logics. Logique et Analyse, **19** (74–76), 311–339
306. Morgan, C.G. (1976): Resolution for many-valued logics. Logique et Anal., **19** (74-76), 311–339
307. Morgan, C., Pelletier, F.J. (1997): Some notes concern fuzzy logic. Linguistics and Philosophy, **1** (1), 79–97
308. Morgenstern, L., Guerreiro, R. (1993): Epistemic logics for multiple agent non-monotonic reasoning I. Proceedings of the Second Symposium on Logical Formalizations of Commonsense Reasoning (CS-93). Austin, TX, 147–156

309. Morgenstern, L. (1998): Inheritance comes of age: Applying non-monotonic techniques to problems in industry. Artificial Intelligence, **103**, 237–271
310. Mukaidono, M. (1975): On some properties of fuzzy logic. Syst. Comput. Control, **6** (2), 36–43
311. Mukaidono, M. (1982): Fuzzy inference of resolution style. In: Yager, R.R. (Ed.) Fuzzy Sets and Possibility Theory. Pergamon Press, New York, 224–231
312. D. Mundici. (1982): Compactness, interpolation and Firedman's third problem. Ann. Math. Logic, **22**, 197–211
313. Mundici, D. (1984): Abstract model theory nets of C^*-algebras: noncommutative interpolation and preservation properties. Lecture Notes in Math, **1103**, 351–377
314. Mundici, D. (1986): Interpretation of AF C^*-algabras in Łukasicwize sentential calculus. J. Funct. Anal., **65**, 15–63
315. Mundici, D. (1986): Mapping Albelian I-group with strong unit one-one into MV-algebras. J. Algebra, **98**, 76–81
316. Mundici, D. (1987): Satisfiability in many-valued sentential logic is NP-complete. Theoret. Comput. Sci., **52**, 145–153
317. Mundici, D. (1989): The C^*-algebras of three-valued logic. In: Ferro, R*et al.* (Eds.) Logic Colloquium'88. Elsevier
318. Murray, N.V. (1982): Completely non-clausal theorem proving. Artificial Intelligence, **18**, 67–85
319. Murray, N.V., Rosenthal, E. (1991): Resolution and path dissolution in multiple-valued logics. In: Proc. of the 1991 ISMIS, Carlotte. In: LNAI, **542**, 570–579
320. Murray, N.V., Rosenthal, E. (1991): Improving tableau proof in multiple-valued logic. In: Proc. of the 1991 ISMVL. Victoria, BC, Canada, 230–237
321. Nachtnebel, H.P., Duckstein, L. (1994): Incorporating risk and imprecision into water related decision making: an Austrian case study for instream water requirements. In: Nachtnebel, H.P. (Ed.) Decision Support Systems in Water Resources Management, 431–451. UNESCO Press, Paris
322. Neapolitan, R.E. (1990): Probabilistic Reasoning in Expert Systems: Theory and Algorithms. John Wiley & Sons: New York, NY
323. Ng, K.C., Abramson, B. (1990): Uncertainty management in expert systems. IEEE Intelligent Systems and Their Applications, **5** (2), 29–48
324. Nguyen, C.H., Tran, D.K., Huynh, V.N., Nguyen, H.C. (1999): Hedge algebras, linguistic-valued logic and their application to fuzzy reasoning, Internat. J. Uncertain. Fuzziness Knowledge-Based Systems, **7** (4), 347–361
325. Niittymäki, J., Turunen, E. (2003): Traffic signal control on similarity logic reasoning. Fuzzy Sets and Systems, **133**(1), 109–131
326. Nilsson, N.J. (1980): Principle of Artificial Intelligence. Tioga
327. Nilsson, N.J. (1986): Probabilistic logic. Artificial Intelligence, **28**, 71–87
328. Novak, V. (1982): First-order fuzzy logic. Studia Logica, **46** (1), 87–109
329. Novak, V., Pedrycz, W. (1988): Fuzzy sets and T–norm in the light of fuzzy logic. Internat. J. Man-Machine Stud., **29**, 113–127
330. Novak, V. (1989): Fuzzy Sets and Their Applications. Adam Hilger, Bristol-Philadelphia
331. Novak, V. (1990): On the syntactico-semantical completeness of first-order fuzzy logic, Part I: Syntax and semantics. Kybernetika, **26** (1), 47–66
332. Novak, V. (1990): On the syntactico-semantical completeness of first-order fuzzy logic, Par II: Main results. Kybernetika, **26** (2), 134–152

333. Novak, V. (1992): Fuzzy logic as a basis of approximate reasoning. In: Zadeh, L.A., Kacprayk, J. (Eds.) Fuzzy Approach to Reasoning and Decision Making. Wiley, New York
334. Novak, V. (1994): Fuzzy control from the point of view of fuzzy logic. Fuzzy Sets and System, **66**, 159–173
335. Novak, V. (1995): A new proof of completeness of fuzzy logic and some conclusion for approximate reasoning. In: Proc. of Internat. Conf. FUZZ-IEEE/IFES'95, Yokohama, 1461–1468
336. Novak, V. (1998): Open theories, consistency and related results in fuzzy logic. Internat. J. Approx. Reason., **18**, 191-200
337. Novak, V., Perfilieva, I., Mojckojr, J. (1999): Mathematical Principles of Fuzzy Logic. Kluwer
338. O'hearn, P., Stachniak, Z. (1989): Note on theorem proving strategies for resolution counterparts of non-classical logics. In: Proc. of the 1989 ISSAC, Poreland, Oregon, 364–372
339. Ohlbach, H. (1993): Translation methods for non-classical logics: An overview. Bull. of the IGPL, **1** (1), 69–89
340. Oliver, R.M., Smith, J.Q. (1990): Influence Diagrams, Belief Nets, and Decision Analysis. John Wiley & Sons, Chichester, UK
341. Orlowska, E. (1978): The resolution principle for ω^+-valued logic. Fundamenta Mathematicae, **2** (1), 1–15
342. Orlowska, E. (1979): Resolution systems and their applications I, II. Fundamenta Mathematicae, **3**, 235–268
343. Orlowska, E. (1980): Resolution system and their application II. Fundamenta Mathematicae, **III** (3), 333–362
344. Orlowska, E., Wierzchon, S. (1985): Mechanical reasoning in fuzzy logics. Logique et Analyse, **28** (110-111), 193–207
345. Orlowska, E. (1985): Mechanical proof methods for Post logics. Logique et Analyse, **28** (110), 173–192
346. Orlowska, E. (1999): Many-valuedness and uncertainty. Internat. J. Multiple-Valued Logic, **4** (3), 207–227
347. Oshins, E., Ford, K.M., Rodriguez, R.V., Anger, F.D. (1992): Classical, fuzzy and quantum logics: relations and implications. Advances in Artificial Intelligence Research, **2**, 37-57
348. Pacholczyk, D. (1989): Incertitude et logique multivalente-Interpretation et completude. BUSEFAL, **40**, 80–88
349. Papadimitriou, C.H., Sideri, M. (1994): Default theories that always have extensions. Artificial Intelligence, **69**, 347–357.
350. Pate'-Cornell, M. (1996): Uncertainties in risk analysis: Six levels of treatment. Reliability Engineering and System Safety, **54**, 95–111
351. Pavelka, J. (1979): On fuzzy logic I: Many-valued rules of inference, II: Enriched residuated lattices and semantics of propositional calculi, III: Semantical completeness of some many-valued propositional calculi. Zeitschr. F. Math. Logik und Grundlagend. Math., **25**, 45–52, 119–134, 447–464
352. Pearl, J. (1986): A constraint-propagation approach to probabilistic reasoning. In: Kanal, L.N., Lemmer, J.F. (Eds.) Uncertainty in Artificial Intelligence, North-Holland, Amsterdam, 357–369
353. Pearl, J. (1987): Bayesian decision methods. In: Shapiro, S.C. (Ed.) Encyclopedia of Artificial Intelligence, 48–56. John Wiley

354. Pearl, J. (1988): Probabilistic Reasoning in Intelligent Systems: Networks of Plausible Inference. 2nd ed., Morgan Kaufmann Publishers, San Mateo, CA
355. Pearl, J. (1990): Probabilistic semantics for non-monotonic reasoning: A survey. In: Shafer, G., Pearl, J. (Eds.) Readings in Uncertain Reasoning. Morgan Kaufmann, San Mateo, 699–710
356. Peddle, D.R. (1995): Knowledge formulation for supervised evidential classification, Photogrammetric Engineering & Remote Sensing, **61** (4), 409–417
357. Pedgham, L. (1989): Negative reasoning using inheritance. In: Proceedings of the Eleventh International Joint Conference on Artificial Intelligence (IJCAI'89), Detroit, Michigan USA, 1086-1092
358. Pei, D.W. (2000): Study on Fuzzy Logic and Fuzzy Reasoning Based on Formal System L^*. Ph.D. Thesis, Sichuan University, China (in Chinese)
359. Pei, Z. (2002): Study on Automated Reasoning Theory and Method Based on Neurol Network. Ph.D. Thesis, Southwest Jiaotong Univ., China (in Chinese)
360. Pelletier, F.J., Elio, R. (1997): What should default reasoning be, by default?. Computational Intelligence, **13** (2), 165–187
361. Polymeris, A. (1994): Conjunctive normal forms in three valued propositional ligics. J. Logic Comput., **4**, 157–173
362. Poole, D. (1988): A logical framework for default reasoning. Artificial Intelligence, **36**, 27–47
363. Post, E. (1921): Introduction to a general theory of elementary propositions. American J. Mathematics, **43**, 163–185. In: Van Heijenoort, J. (Ed.) From Frege to Godel: A source book in Mathematical Logic 1879–1931. Cambridge, Mass
364. Preparata, F.P., Ye, R.T. (1972): Continuously valued logic. J. Comput. and Sys. Sci., **6**, 397–418
365. Priestley, H. (1970): Representation of distributive lattices by means of ordered Stone spaces. Bull. London Math. Soc., **2**, 186–190
366. Priestley, H. (1984): Ordered sets and duality for distributive lattices. Ann. Discr. Math., **23**, 39–60
367. Przymusinski, T. (1991): Three-valued non-monotonic formalisms and semantics of logic programs. Artificial Intelligence, **49**, 309–343
368. Qin, K.Y., Xu, Y. (1993): Fuzzy power sets based on lattice implication algebras. Fuzzy Systems Math., **7** (2), 17–24 (in Chinese)
369. Qin, K.Y., Song, Z.M., Xu, Y. (1993): On product and factorization of Lattice implication algebras. Chinese Quart. J. Math., **8** (4), 53–57
370. Qin, K.Y., Xu, Y. (1993): Łukasiewicz implication algebra on $\{0, 1\}$. In: Proc. of 5th IFSA. Seoul, **7**, 410–412
371. Qin, K.Y. (1994): Study on Lattice Implication Algebras and Lattice-Valued Logic System. MA Thesis, Southwest Jiaotong Univ., China (in Chinese)
372. Qin, K.Y., Xu, Y. (1994): A lattice-valued propositional logic system. Fuzzy Systems Math., **8** (suppl.), 322–326 (in Chinese)
373. Qin, K.Y., Xu, Y. (1994): Lattice-valued proposition logic (II). J. Southwest Jiaotong Univ., **2** (1), 22–27 (in Chinese)
374. Qin, K.Y., Xu, Y. (1994): Fuzzy propositional logic system $FP(X)$(I). J. Fuzzy Math., **2** (2), 351–357
375. Qin, K.Y. (1996): Study on Lattice-valued Logic System Based on Lattice Implication Algebras and Their Applications. Ph.D. Thesis, Southwest Jiaotong Univ., China (in Chinese)

376. Qin, K.Y., Xu, Y. (1996): Semantics of lattice-valued first-order logic FM. In: Proc. of the Celebration of Southwest Jiaotong Univ.'s Centenary–Special Issue for academic papers. Press of Southwest Jiaotong Univ., Chengdu, China (in Chinese)
377. Qin, K.Y., Xu, Y. (1997): Lattice-valued propositional logic system $L(X)$(I). Fuzzy Systems Math., **11** (4), 5–11 (in Chinese)
378. Qin, K.Y., Xu, Y. (1998): On some properties of fuzzy filters of lattice implication algebra. In: Liu, Y.M. (Ed.) Fuzzy Set Theory and its Application. Press of Hebei University, Baoding, China, 179–182 (in Chinese)
379. Qin, K.Y., Xu, Y., Song, Z.M. (1998): Lattice valued propositional logic system $L(X)$(II). Fuzzy Systems Math., **12** (1), 10–19 (in Chinese)
380. Qin, K.Y., Xu, Y., Song, Z.M. (1998): Some kinds of approximate reasoning based on system $L(X)$. Fuzzy Systems Math., **12** (2), 55–60 (in Chinese)
381. Qin, K.Y., Xu, Y., Song, Z.M. (1998): Compactness of logic system FM and Löwenheim-Skolem property. Chinese Sci. Bull., **43**(4), 371–375 (in Chinese)
382. Qin, K.Y., Xu, Y. (1998): The category of fuzzy lattice implication algebras. Math. Appl., **11** (3), 13–16 (in Chinese)
383. Qin, K.Y., Xu, Y. (2000): On model properties of lattice implication algebras. J. Southwest Jiaotong Univ., **35** (5), 546–550 (in Chinese)
384. Qiu, Y.H., Zhang, W.Q. (1992): An Introduction of Automated Reasoning. Press of Electronic and Technology University, Chengdu, China (in Chinese)
385. Quinlan, J.R. (1983): Inferno: a cautious approach to uncertain inference. The Computer Journal, **26**, 255–269
386. Raha, S., Ray, K.S. (1996): Uncertainty reasoning with time. Fuzzy Sets and Systems, **107** (1), 59–79
387. Raha, S., Ray, K.S. (1997): Reasoning with vague default. Fuzzy Sets and Systems, **91** (3), 327–338
388. Raha, S., Ray, K.S. (1999): Reasoning with vague truth. Fuzzy Sets and Systems, **105** (3), 385–399
389. Raiffa, H. (1968): Decision Analysis, Introductory Lectures on Choices under Uncertainty. Reading, Addison-Wesley, MA
390. Rajasekar, A., Lobo, J., Minker, J. (1989): Weak generalized closed world assumption. J. Automated Reason., **5**, 293–307
391. Ramamurti, S. (1994): Principles of Quantum Mechanics, 2nd Edition, Plenum Pub. Corp., New York, California, USA
392. Rasiowa, H., Sikorski, R. (1958): On the isomorphism of Lindenbaum algebras with fields of sets. Colloq. Math., **5**, 143–158
393. Rasiowa, H. (1974): An Algebraic Approach to Non-Classical Logics. Stud. in Logic, 78. North Holland, Amsterdam
394. Rasiowa, H. (1976): Algorithmic logic: multi-valued extensions. Stuidia Logica, **38** (4), 317–335
395. Rasiowa, H. (1987): Logic of approximate reasoning. In: Proc. of the 1987 Workshop on Computer Science Logic, Karlsruhe. In: LNCS, **329**
396. Rayner, M. (1991): On the applicability of non-monotonic logic to formal reasoning in continuous time. Artificial Intelligence, **49**, 345–360
397. Reggia, J.A., Nau, D.S., Wang, P. (1983): Diagnostic expert systems based on a set covering model. Internat. J. Man-Machine Studies, **19**, 437–460
398. Reiter, R. (1978): On reasoning by default, In: Proceedings of 2nd Symp. on Theoretical Issues in Natural Language Processing-2 (TINLP-2), University of Illinois at Urbana-Champaign, 210–218

399. Reiter, R. (1980): A logic for default reasoning. Artificial Intelligence, **13**, 81–132
400. Rescher, N. (1969): Many-Valued Logic. McGraw-Hill, New York
401. Robinson, J.P. (1965): A machine-oriented logic based on the resolution principle. J. ACM, **12**, 23–41
402. Roehrig, S.F. (1996): Incompletely specified probabilistic networks, J. Management Information Systems, **12**
403. Ron, E.H., Kim, S.Y., Xu, Y., Jun, Y.B. (2001): Some operations on lattice implication algebras. Int. J. Math. Math. Sci., **27** (1), 45–52
404. Roos, N.N. (1992): A logic for reasoning with inconsistent knowledge. Artificial Intelligence, **57**, 69–103
405. Roos, N.N. (1998): Reasoning by cases in Default Logic. Artificial Intelligence, **99**, 165–183
406. Rose, A., Rosser, J.B. (1958): Fragments of many-valued statement calculi. Trans. Amer. Math. Soc., **87**, 1–3
407. Rosenfel, A. (1971): Fuzzy groups. J. Math. Anal. Appl., **35**, 512–517
408. Rosser, J.B., Turquette, A.R. (1952): Many-Valued Logic. Amsterdam, North-Holland
409. Rousseau, G. (1967): Sequents in many valued logic I. Fundamenta Mathematicae, **60**, 23–33
410. Rousseau, G. (1970): Sequents in many valued logic II. Fundamenta Mathematicae, LXVII, 125–131
411. Routley, R., Meyer, R., Plumwood, V., Brady, R. (1982): Relevant Logics and Their Rivals 1. Ridgeview Publ. Comp., Atascadero, California
412. Russell, A., Froese, T., Caselton, W. (1988): A Dempster-Shafer based construction expert system. In: Proc. of the Third International Conference on Computing in Civil Engineering. Vancouver, Canada, 469–478
413. Salzer, G. (1996): Optimal axiomatizations for multiple-valued operators and quantifiers based on semilattices. In: McRobbie, M., Slaney, J. (Eds.) Proc. of the 1996 CADE. In: LNCS, **1104**, 688–702
414. Santos, E. (1977): Regular fuzzy expressions. In: Gupta, M.M., Saridis, G.N., Gaines, B.R. (Eds.) Fuzzy Automata and Decision Processes. New York, North–Holland, 169–176
415. Saunders, M.L. (1971): Category of the Working Mathematician. Springer
416. Schlechta, K. (1997): Non-monotonic Logics: Basic Concepts, Results and Techniques. Springer, Berlin: LNAI Series 1187
417. Schmitt, P.H. (1986): Computational aspects of three-valued logic. In: Siekmann, J.H. (Ed.) Proc. of the Eighth International Conference on Automated Deduction. Springer, Berlin, 190–198
418. Schocken, S. Kleindorfer, P.R. (1989): Artificial intelligence dialects of the Bayesian belief revision language. IEEE Trans. on Systems, Man, and Cybernetics, **19** (5), 1106–1121
419. Schweizer, B., Sklar, A. (1961): Associative functions and statistical trangle inequalities, Publ. Math. Debrecen, **8**, 169–186
420. Schweizer, B., Sklar, A. (1963): Associative functions and abstract semigroups. Publ. Math. Debreecen, **10**, 69–81
421. Schweizer, B., Sklar, A. (1983): Probabilistic Metric Spaces. North-Holland, Amsterdam

422. Schwind, C.B. (1990): A tableaux-based theorem prover for a decidable subset of default logic. In: Proc. of the Tenth Conference on Automated Deduction, 528–542
423. Selman, B., Levesque, H.J. (1990): Abductive and default reasoning: a computational core. In: Proc. of the Eighth National Conference on Artificial Intelligence (AAAZ-90), 343–348
424. Sessa, S., Turunen, E. (2001): Local BL-algebras. Multi-Valued Logic, **6**, 229–249
425. Shafer, G. (1976): A Mathematical Theory of Evidence. Princeton University Press, New Jersey
426. Shafer, G. (1990): Perspectives on the theory and practice of belief functions. Working paper no. 218. University of Kansas, School of Business
427. Shanahan, M. (1997): Solving the Frame Problem. MIT Press, Cambridge, MA
428. Shen, Z.L., Ding, L., Mukaidono, M. (1988): A theoretical framework of fuzzy prolog machine. In: Gupta, M.M., Yamakawam, T. (Eds.) Fuzzy Computing-Theory, Hardware and Applications. North-Holland, 89–100
429. Shen, Z.L., Ding, L., Mukaidono, M. (1988): Designing fuzzy expert system by fuzzy Prolog. In: Proc. of the 1988 IWFSA. Iizuka, 207–208
430. Shi, Z.Z. (1998): Advanced Artificial Intelligence. Academic Press of China, Beijing, China (in Chinese)
431. Shortliffe, E.H., Buchanan, B.G. (1975): A model of inexact reasoning in medicine. Mathematical Biosciences, **23**, 351–379
432. Shortliffe, E.H. (1976): Computer-based medical consultion, MYCIN. Elsevier
433. Sikorski, R. (1964): Boolean Algebras (2ed). Academic Press, New York
434. Skala, H.J. (1978): On many-valued logics, fuzzy sets, fuzzy logics and their applications. Fuzzy Sets and Systems, **1**, 129–149
435. Stalnaker, R. (1994): Non-monotonic consequence relations. Fundamenta Informaticae, **21**, 7–21
436. Smith, J.Q. (1990): Statistical principles on graphs. In: Oliver, R.M., Smith, J.Q. (Eds.) Influence Diagrams, Belief Nets, and Decision Analysis. John Wiley & Sons, New York
437. Smith, P. (1988): Review of [47]. Mathematical Reviews 88g: 03034
438. Smets, P. (1988): Belief functions (with discussion). In: Smets, P., Mamdani, A., Dubois, D., Prade, H. (Eds.). Non-standard logics for automated reasoning, 253-286
439. Sofronie-Stokkermans, V. (1997): Fibered Structures and Applications to Automated Theorem Proving in Certain Classes of Finitely-Valued Logics and to Modeling Interacting Systems. PhD thesis, RISC-Linz, J. Kepler Univ. Linz
440. Sofronie-Stokkermans, V. (1998): On translation of finitely-valued logics to classical first-order logic. In: Prade, H. (Ed.) Proc. of the 1998 ECAI. John Wiley, 410–411
441. Sofronie-Stokkermans, V. (1999): Representation theorems and automated theorem proving in non-classical logics. In: Proc. of the 1999 ISMVL, Freiburg, IEEE Computer Society Press, 242–247
442. Sofronie-Stokkermans, V. (2000): Chaining techniques for automated theorem proving in finitely-valued logics. In: Proc. of the 2000 ISMVL. Portland, Oregon, 337–344

443. Sofronie-Stokkermans, V. (2000): Priestley duality for SH_n-algebras and applications to the study of Kripke-style models for SH_n-logics to appear in multiple-valued logics. Internat. J. Multiple-Valued Logics, **5** (4), 281–305
444. Sofronie-Stokkermans, V. (2000): Duality and canonical extensions of bounded distributive lattices with operators and applications to the semantics of non-classical logics, Part I. Studia Logica, **64** (1), 93–132
445. Sofronie-Stokkermans, V. (2000): Duality and canonical extensions of bounded distributive lattices with operators and applications to the semantics of non-classical logics, Part II. Studia Logica, **64** (2), 151–172
446. Sofronie-Stokkermans, V., Ganzinger, H. (2000): Chaining techniques for automated theorem proving in finitely-valued logics. In: Proc. of the 2000 ISMVL. Portland, Oregon, 337–344
447. Sofronie-Stokkermans, V. (2000): Resolution-based theorem proving for SH_n-logics. In: Caferra, R., Salzer, G. (Eds.) Automated Deduction in Classical and Non-Classical Logics. In: LNCS, **1761** (subseries LNAI), 268–282
448. Sofronie-Stokkermans, V. (2001): Representation theorems and the semantics of (semi)lattice based logics. In: Proc. of the 2001 ISMVL. Warsaw, Poland, 125–134
449. Song, S.J., Wu, C. (2002): Backward three-I algorithm in fuzzy reasoning. Science in China: Series E, **32** (2), 230–246 (in Chinese)
450. Song, Z.M., Xu, Y. (1996): A representation of lattice H implication algebras. J. Lanzhou University (Natural Science): Fuzzy Mathematics and Systems Supplement, **32**, 349–351 (in Chinese)
451. Song, Z.M., Xu, Y. (1997): On congruence relations of lattice implication algebras. Math. Appl., **10** (3), 121–124 (in Chinese)
452. Song, Z.M., Guo, T.B. (1997): Prime space of lattice implication algebras. Chinese Quart. J. Math., **12** (4), 46–50 (in Chinese)
453. Song, Z.M. (1998): Study on Uncertain Reasoning in Intelligent Control Based on Lattice Implication Algebras. Ph.D Thesis, Southwest Jiaotong Univ., Chengdu, China (in Chinese)
454. Song, Z.M. (2000): Implication filter spaces. J. Fuzzy Math., **8** (1), 263–266
455. Sovinov, A.A. (1993): Fuzzy propositional logic. Fuzzy Sets and Systems, **60** (1), 9–17
456. Spiegelhalter, D.J. (1989): A unified approach to imprecision and sensitivity of beliefs in expert systems. In: Kanal, L.N., Lemmer, J., Levitt, T.S. (Eds.) Artificial Intelligence and Statistics. North-Holland, Amsterdam, 47–68
457. Stachniak, Z. (1988): The resolution rule: an algebraic perspective. In: Proc. of Algebraic Logic and Universal Algebra in Computer Science Conf., Springer. In: LNCS, **425**, 227–242
458. Stachniak, Z. (1990): Resolution rule: algebraic perspective. In: Proc. of the European Workshop on Logics in AI. Amsterdam, LNCS, **425**, 227–242
459. Stachniak, Z. (1996): Resolution Proof Systems: An Algebraic Theory. Kluwer
460. Stefik, M. (1995): Introduction to Knowledge Systems. Morgan Kaufmann Publishers, San Francisco, CA
461. Steve, G. (1995): Probabilistic inferential engines in expert systems: how the strength of rules should be expressed. In: Bouchon, B., Yager, R.R. (Eds.) Uncertainty in Knowledge-based Systems. LNCS 286, Springer, Berlin, 179–188
462. Stillman, J. (1992): The complexity of propositional default logics. In: Proc. of the AAAI-92, SanJose, CA, 794–799

463. Subasic, P., Hirota, K. (1998): Similarity rules and gradual rules for analogical and interpolative reasoning with imprecise data. Fuzzy Sets and Systems, **96**, 53–75
464. Subrahamnian, V.S. (1992): Paraconsistent disjunctive databases. Theoretical Computer Science, **93**, 115–141
465. Surma, S.J. (1979): On the origin and subsequent applications of the concept of the Lindenbaum algebra. In: Cohen, L.J., Los, M., Pfeiffer, H., Podewski, K.-P. (Eds.) Proceedings of the 6th International Congress of Logic, Methodology and Philosophy of Science, Hannover August 22-29, Amsterdam, **VI**, 719–734
466. Surma, S.J. (1984): An algorithm for axiomatizing every finite logic. In: Lusk, E., Overbeek, R. (Eds.) Computer Science and Multiple-valued Logics, 143–149. North–Holland, Amsterdam
467. Suzuki, H., Arimoto, S. (1994): Embedment of a fuzzy logic system into a Boolean lattice for satisfying a complementary law. Information Science, **78**, 257–268
468. Svozil, K. (1998): Quantum Logic. Springer-Verlag, Singapore
469. Takagi, T., Imura, A., Ushida, H., Yamaguchi, T. (1996): Multi-layered reasoning by means of conceptual fuzzy sets. Internat. J. Intelligent Systems, **11**, 97–111
470. Takano, M. (1987): Another proof of the strong completeness of the intuitionistic fuzzy logic. Tsukuba J. Math., **11** (1), 101–105
471. Takeuti, G., Titani, S. (1984): Intuitionistic fuzzy logic and intuitionistic fuzzy set theory. J. Symbolic Logic, **49** (3), 851–886
472. Takeuti, G., Titani, S. (1986): Global intuitionistic fuzzy set theory. In: Di Nola, A., Ventre, A.G.S. (Eds.) The Mathematics of Fuzzy Systems. Verlag TUV Rheinland Koln, Germany, 291–302
473. Tan, S., Pearl, J. (1995): Specificity and inference in default reasoning. In: Proc. of 14th International Joint Conference on Artificial Intelligence (IJCAI 95), August, Montréal, Québec, Canada
474. Tawflk, A., Nenfeld, E. (1998): Model-based diagnosis: a probabilistic extension. In: Hunter, A., Parsons, S. (Eds.) Applications of Uncertainty Formalisms. Springer, Berlin
475. Thirunarayan, K. (1993): A theory of non-monotonic inheritance based on annotated logic. Artificial Intelligence, **60**, 23–50
476. Todt, G. (1983): Fuzzy logic and modal logic. In: Ballmer, T.T., Pinkal, M. (Eds.) Approaching Vagueness. North-Holland, Amsterdam, 213–260
477. Togai, M.A. (1985): Fuzzy inverse relation based on Gödelian logic and its applications. Fuzzy Sets and Systems, **17**, 211–219
478. Tonn, B.E., Schavhauser, A.J. (1994): An Approach to Understanding, Representing, and Managing Uncertainty in Integrated Resource Planning. ORNL/CON-399. Oak Ridge, TN: Oak Ridge National Laboratory
479. Trillas, E., Valverde, L. (1985): On mode and implication in approximate reasoning. In: Gupta, M.M. (Eds.) Approximate Reasoning in Expert Systems, Elsevier, Amsterdam
480. Tsukamoto, T. (1979): An approach to fuzzy reasoning method. In: Gupta, M.M., Ragade, R.K., Yager, R.R. (Eds.) Advances in Fuzzy Set Theory and Applications. North-Holland, Amsterdam, 137–149
481. Türksen, I.B. (1999): Type I and Type II fuzzy system modeling, Fuzzy Sets and Systems, **106** (1), 11–34

482. Türksen, I.B. (1999): Interval valued fuzzy sets based on normal forms, Fuzzy Sets and Systems, **20**, 191–210
483. Turunen, E. (1991): A note on Pavelka's fuzzy logic. Zeitschr. f. Math. Logik und Grundlagen d. Math., **37**, 39–40
484. Turunen, E. (1991): Residuated lattice in fuzzy logical system. In: Terano, T., Sugeno, M., Mukaidono, M., and Shigematsu, K. (Eds.) Fuzzy Engineering toward Human Friendly System, Ohmusha, Japan, **1** 60–69
485. Turunen, E. (1992): Algebraic structures in fuzzy logic. Fuzzy Sets and Systems, **52**, 181–188
486. Turunen, E. (1992): Algebraic structures in fuzzy logic. Fuzzy Sets and Systems, **52**, 195–234
487. Turunen, E. (1995): Well-defined Fuzzy Sentential Logic. Mathematical Logic Quarterly, **41**, 236–248
488. Turunen, E. (1996): Injective MV-algebras, fuzzy predicate logic and completeness. In: Proc. of Information Processing and Management of Uncertainty in Knowledge-based Systems, **2**, Granada, Spain, 1015–1019
489. Turunen, E. (1999): BL-algebras of basic fuzzy logic. Mathware Soft Comput., **6**, 49–61
490. Turunen, E., Sessa, S. (2000): Local BL-algebras. Internat. Internat. J. Multiple-Valued Logic, **6**, 1–21
491. Valverde, L., Trillas, E. (1985), On modus ponens in fuzzy logic. In: Proc. of the 1985 ISMVL, Kington, 294–301
492. Wang, G.J. (1983): Generalized topological molecular lattices. Science in China: Series A, **12**, 1063–1072 (in Chinese)
493. Wang, G.J. (1992): Theory of topological molecular lattices. Fuzzy Sets and Systems, **47**, 351–376
494. Wang, G.J. (1997): A formal deductive system of fuzzy propositional calculus. Chinese Sci. Bull., **42**, 1041–1045
495. Wang, G.J. (1997): On the logic foundations of fuzzy reasoning. Lecture Notes in Fuzzy mathematics and Computer Science, Creighton University, **4**, Omaha, USA, 1–48
496. Wang, G.J. (1997): On the logic foundations of FMP and FMT. J. Fuzzy Math., **5**, 229–250
497. Wang, G.J. (1997): Logic system based on a certain type algebra (I). J. Shannxi Normal University, **25** (1), 1–8 (in Chinese)
498. Wang, G.J. (1997): Logic system based on a certain type algebra (II). J. Shannxi Normal University, **25** (3), 1–8 (in Chinese)
499. Wang, G.J. (1999): Full implication three-I algorithm in fuzzy reasoning. Science in China: Series E, **29** (1), 43–53 (in Chinese)
500. Wang, G.J. (2000): Non-Classical Mathematical Logic and Approximate Reasoning. Academic Press of China, Beijing, China (in Chinese)
501. Wang, J., Yang, J.B., Sen, P. (1995): Multi-person and multi-attribute design evaluations using evidential reasoning based on subjective safety and cost analyses. Reliability Engineering and System Safety, **52** (2), 113–129 (in Chinese)
502. Wang, J., Yang, J.B., Sen, P. (1995): Safety analysis and synthesis using fuzzy set modeling and evidential reasoning. Reliability Engineering and System Safety, **47** (3), 103–118 (in Chinese)
503. Wang, J. (1997): A subjective methodology for safety analysis of safety requirements specifications. IEEE Transactions on Fuzzy Systems, **5** (3), 418–430

504. Wang, S.Q., Lu, J.B. (1981): Basic ultra-product theorem in lattice-valued model theory. Chinese Sci. Bull., **27** (2), 71–74 (in Chinese)
505. Wang, S.Q. (1981): Two applications of constant construction in lattice-valued model theory. Chinese Sci. Bull., **27** (3), 129–130 (in Chinese)
506. Wang, S.Q. (1982): Theorem of lattice-valued model theory. Acta Mathematica Sinica, **2**, 202–207 (in Chinese)
507. Wang, W. (2001): The left idempotent elements of lattice implication algebras. Fuzzy Systems Math., **15** (4), 20–23 (in Chinese)
508. Wang, W., Xu, Y., Song, Z.M. (2001): The logical properties of filters of lattice implication algebras. Chinese Quart. J. Math., **16** (3), 8–13
509. Wang, W. (2003): Study of Reasoning Method Based on α-Resolution Principle in Lattice-Valued Propositional Logic $L(X)$. Ph.D Thesis, Southwest Jiaotong Univ., China (in Chinese)
510. Wang, X., You, J., Yuan, L.Y. (1996): Non-monotonic reasoning by monotonic inferences with priority constraints. In: Proceedings of the 2nd International Workshop on non-monotonic extension of logic programming, Lecture Notes in Artificial Intelligence 1216, Springer-Verlag, Berlin, 91–109.
511. Wang, X.F. (2001): Study on Lattice Implication Algebra and Filter. MA Thesis, Southwest Jiaotong Univ., China (in Chinese)
512. Wang, X.F., Xu, Y., Song, Z.M. (2001): Some properties of filters in lattice implication algebras. J. Southwest Jiaotong Univ., **36** (5), 536–539 (in Chinese)
513. Wang, X.H., Liu, X.H. (1982): Generalized resolution. J. Comput., **2**, 81–92 (in Chinese)
514. Wang, Y.Q. (1998): Principle and Methods of Artificial Intelligence. Xi'an Jiaotong University Press, Xi'an, China (in Chinese)
515. Wasserman, L.A. (1990): Belief function and statistical inference. Canadian J. Statistics, **18**, 197–204
516. Wasserman, L.A. (1990): Prior envelopes based on belief functions. Annals of Statistics, **18**, 454–464
517. Waterman, D.A. (1986): Expert Systems. Addison-Wesley, MA
518. Weigert, T.J., Tsai, J.P., Liu, X.H. (1993): Fuzzy operator logic and fuzzy resolution. J. Automated Reason., **10**, 59–78
519. Welman, M.P., Heckerman, D. (1995): Real-world applications of Bayesian networks. Comm. ACM, **8**, 24–30
520. Wille, R. (1982): Restructuring lattice theory: an approach based on hierarchies of concepts. In: Rival, I. (Ed.) Ordered Sets. Reidel, Dordrecht, Boston, 445–470
521. Willmott, R. (1980): Two fuzzier implication operators in the theory of fuzzy power sets. Fuzzy Sets and Systems, **4**, 31–36
522. Wojcicki R. (1988): Theory of Logical Calculi: Basic Theory of Consequence Operations. Kluwer Academic Publisher, Dordrecht
523. Wos, L. (1988): Automated Reasoning: 33 Basic Research Problems. Prentice Hall, Englewood Cliffs, New Jersey
524. Wu, W.M. (1988): A multi-valued logic system with respect to T-norms. In: Gupta, M.M., Yamakama, T. (Eds.) Fuzzy Computing. North-Holland, Amsterdam, 101–118
525. Wu, W.M. (1990): Fuzzy implication algebra. Fuzzy Systems Math., **4** (1), 56–64 (in Chinese)

526. Wu, W.M. (1994): Fuzzy Reasoning Principle and Methods. Guizhou Science and Technology Press, Guiyang, China (in Chinese)
527. Voorbraak, F (1991): On the justification of Dempster's rule of combination. Artificial Intelligence, **48**, 171–197
528. Xu, Y. (1992): Complemented lattice. J. Southwest Jiaotong Univ., **83**, 37–41 (in Chinese)
529. Xu, Y., Qin, K.Y. (1992): On lattice properties of lattice implication algebra. In: Wang, Y.Q. (Ed.) Collection works on Applied Mathematics. Press of Chengdu Science and Technology University, China, 54–58 (in Chinese)
530. Xu, Y., Qin, K.Y. (1992): The (S) condition of lattice implication algebras. In: Wang, Y.Q. (Eds.) Operations and Decision. Press of Chengdu Science and Technology University, 2039–2044 (in Chinese)
531. Xu, Y. (1992): Homomorphisms in lattice implication algebras. In: Proc. of 5th Many-Valued Logical Congress of China, Nanjing, China, 206–211 (in Chinese)
532. Xu, Y., Qin, K.Y. (1992): Lattice properties of lattice implication algebras. In: Wang, Y.Q. (Ed.) Collected works on Applied Mathematics. Press of Chengdu Science and Technology University, Chengdu, China (in Chinese)
533. Xu, Y., Qin, K.Y. (1992): Lattice H implication algebras and lattice implication algebra classes. J. Hebei Mining and Civil Engineering Institute, **3**, 139–143 (in Chinese)
534. Xu, Y. (1993): Lattice implication algebras. J. Southwest Jiaotong Univ., Sum. **89** (1), 20–27 (in Chinese)
535. Xu, Y., Qin, K.Y. (1993): On filters of lattice implication algebras. J. Fuzzy Math., **1** (2), 251–260
536. Xu, Y., Qin, K.Y. (1993): On the category of lattice implication algebras. In: Proc. of the First Asian Fuzzy Systems Symposium, Singapore, **11**, 1098–1101
537. Xu, Y., Qin, K.Y. (1993): Lattice-valued propositional logic (I). J. Southwest Jiaotong Univ., **1** (2), 123–128 (in Chinese)
538. Xu, Y., Qin, K.Y. (1994): On the characteristic of proper lattice implication algebras. Chinese Quart. J. Math., **10** (4), 14–17
539. Xu, Y., Qiao, Q.X., Chen, C.P., Qin K.Y. (1994): Uncertainty Inference. Southwest Jiaotong Univ. Press, Chengdu, China (in Chinese)
540. Xu, Y., Qin, K.Y. (1995): Fuzzy lattice implication algebras. J. Southwest Jiaotong Univ., **30** (2), 121–127 (in Chinese)
541. Xu, Y., Qin, K.Y. (1995): Fuzzy-valued propositional logic system $FP^*(X)$(II). In: Proc. of China Association for Science and Technology Second Academic Annual meeting of Youths (Sichuan). Press of Southwest Jiaotong Univ., Chengdu, China, **8**, 253–257 (in Chinese)
542. Xu, Y., Qin, K.Y. (1996): On the lattice-valued logic systems. In: Proc. of Celebration of Southwest Jiaotong Univ.'s Centenary–Special Issue for Applied Science. Press of Southwest Jiaotong Univ., Chengdu, China, 72–80 (in Chinese)
543. Xu, Y., Qin, K.Y., Liu, J. (1996): On the lattice implication algebras and the lattice valued logic system. In: Proc. of Theory and Application of Neutral Network, Chengdu, China, 38–47 (in Chinese)
544. Xu, Y., Qin, K.Y. (1997): On semantic of first-order lattice-valued logic systems FM. System Analysis, Modeling and Simulation, Germany
545. Xu, Y., Qin, K.Y., Song, Z.M. (1997): Syntax of lattice-valued first-order logic FM. Chinese Sci. Bull., **42** (10), 1052–1055 (in Chinese)

546. Xu, Y. (1997): On the relations between lattice implication algebras and BCK-algebras. Fuzzy Systems Math., **11** (1), 10–15 (in Chinese)
547. Xu, Y. (1997): Relations between lattice implication algebras and BCK-algebras. Fuzzy Systems Math., **11** (1), 10–15 (in Chinese)
548. Xu, Y., Qin, K.Y. (1998): The category of fuzzy lattice implication algebras. Math. Appl., **11** (3), 13–16
549. Xu, Y., Qin, K.Y., Liu, J., Song, Z.M. (1999): L-valued propositional logic L_{vpl}. Inform. Sci., **114**, 205–235
550. Xu, Y., Ruan, D., Liu, J. (1999): Uncertainty automated reasoning in intelligent learning of soft knowledge. In: Proc. of the 1999 ICST Workshop. Gent, Belgium, 29–45
551. Xu, Y., Liu, J., Song, Z.M., Qin, K.Y. (2000): On semantics of L-valued first-order logic L_{vfl}. Internat. J. Gen. Systems, **29** (1), 53–79
552. Xu, Y., Liu, J., Song, Z.M., Qin, K.Y. (2000): L-valued first-order logic L_{vfl} (I): Semantics of L_{vfl}. Internat. J. Gen. Systems, **29** (1), 53–79
553. Xu, Y., Qin, K.Y. (2000): A simplified axioms system for lattice implication algebras. J. Huanggang Norm. Univ., **20**, 6–8 (in Chinese)
554. Xu, Y., Ruan, D., Kerre, E.E., Liu, J. (2000): α-resolution principle based on lattice-valued propositional logic $LP(X)$. Inform. Sci., **130**, 195–223
555. Xu, Y., Ruan, D., Liu, J. (2000): Approximate reasoning based on lattice-valued propositional logic L_{vpl}, Fuzzy If-Then rules in computational intelligence. In: Ruan, D., Kerre, E.E. (Eds.) Fuzzy Sets Theory and Applications. Kluwer Academic Publishers, 81–105
556. Xu, Y., Eun, H.R., Qin, K.Y. (2001): On quotient lattice implication algebras induced by fuzzy filters. J. Fuzzy Math., **9** (3), 611–617
557. Xu, Y., Song, Z.M., Qin, K.Y., Liu, J. (2001): Syntax of L-valued first-order logic L_{vfl}. Internat. J. Multiple-Valued Logic, **7**, 213–257
558. Xu, Y., Kerre, E.E., Ruan, D., Song, Z.M. (2001): Fuzzy reasoning based on extension principle. Internat. J. Intelligent Systems, **16** (4), 469–495
559. Yager, R.R. (1985): Inference in a multiple-valued logic system. Internat. J. Man-Machine Stud., **23**, 27–24
560. Yager, R.R., (1987): On the Dempster-Shafer framework. Artificial Sciences, **41**, 93–137
561. Yager, R.R., Ovchinnikov, S., Tong, R.M., Nguyen, H.T. (1987): Fuzzy Sets and Applications – Selected Papers by Zadeh L.A.. Wiley, New York
562. Yager, R.R., (1987): Using approximate reasoning to represent default knowledge. Artificial Intelligence, **31**, 99-112
563. Yager, R.R., Filev, D.P. (1994): Essential of Fuzzy Modeling and Control. Wiley, New York
564. Yager, R.R., Fedrizzi, M., Kacprzyk, J. (1994): Advances in the Dempster-Shafer Theory of Evidence. Wiley, New York
565. Yager, R.R. (1996): On the interpretation of fuzzy if-then rules. Appl. Intelligence, **6**, 141–151
566. Yager, R.R., Kreinovich, V. (1999): Decision making under interval probabilities. Internat. J. Uncertainty Reason., **22**, 195–215
567. Yager, R.R. (2000): Uncertainty reasoning and conflict resolution. Internat. J. Uncertainty Reason., **25**, 15–42
568. Yahya, A., Henschen, L.J. (1985): Deduction non-Horn databases. J. Automated Reason., **1**, 141–160

569. Yang, J.B. and Singh, M.G. (1994): An evidential reasoning approach for multiple attribute decision making with uncertainty. IEEE Trans. System, Man and Cybernetic, **24** (1), 1–18
570. Yang, J.B. and Sen, P. (1994): A general multi-level evaluation process for hybrid MADM with uncertainty. IEEE Trans. System, Man and Cybernetic., **24** (10), 1458–1473
571. Yang, J.B., Sen, P. (1997): Multiple attribute design evaluation of large engineering products using the evidential reasoning approach. J. Engineering Design, **8** (3), 211–230
572. Yang, J.B. (2001): Rule and utility based evidential reasoning approach for multi-attribute decision analysis under uncertainties. Europe J. Operational Research, **131**, 2001, 31–61
573. Yang, J.B., Wang, J., Xu, D.L., Dale, B.G., Kieran, O., Ruxton, T. (2001):, Multiple criteria decision analysis applied to safety and cost synthesis. The Journal of the Safety & Reliability Society, **21** (2), 42–63
574. Yen, J. (1989): GERTIS: A Dempster-Shafer approach to diagnosing hierarchical hypotheses. Commun. ACM, **32** (5), 573–585
575. Yeung, D.S., Tsang, E.C.C. (1997): Weighted fuzzy production rules. Fuzzy Sets and Systems, **88** (3), 299–313
576. Yeung, D.S., Tsang, E.C.C. (1997): A comparative study on similarity-based fuzzy reasoning method. IEEE Trans. System, Man and Cybernetic. Part-B, **27** (2), 216–227
577. Ying, H.S. (1988): On standard models of fuzzy modal logics. Fuzzy Sets and Systems, **26**, 357–363
578. Ying, M.S. (1992): Fuzzy reasoning under approximate match. Chinese Sci. Bull., **37**, 1244–1245 (in Chinese)
579. Ying, M.S. (1992): The foundamental theorem of ultraproduct in Pavelka's Logic. Zeit. f. Math. Logik und Grundlagen d. Math., **38**, 197–201
580. Ying, M.S. (1992): Compactness, the Löwenheim-Skölem property and the direct product of lattices of truth values. Zeit. f. Math. Logik und Grundlagen d. Math., **38**, 521–524
581. Ying, M.S. (1993): Fuzzy topology based complete residuated lattice-valued logic (I). Fuzzy Sets and Systems, **56**, 337–373
582. Ying, M.S. (1994): A logic for approximate reasoning. J. Symbolic Logic, **59**, 830–838
583. Ying, M.S. (2000): Declarative semantics of programming in residuated lattice-valued logic. Science in China: Series E, **43**, 481–494 (in Chinese)
584. Ying, M.S. (2001): Fuzzy topology based on residuated lattice-valued logic. Acta Math. Sinica: English Series, **17**, 89–102
585. You, J., Wang, X., Yuan, L.Y. (2001): Non-monotonic reasoning as prioritized argumentation, IEEE TKDE. Nov/Dec, **13** (6), 968–979
586. Yuan, B., Wu, W. (1990): Fuzzy ideals on a distributive lattice. Fuzzy Sets and Systems, **35**, 231–240
587. Yuan, B., Pan, Y., Wu, W. (1995): On normal from based interval-valued fuzzy sets and their applications to uncertainty reasoning. Internat. J. Gen. Systems, **23**, 241–254
588. Zadeh, L.A. (1965): Fuzzy sets. Inform. Control., **8**, 338–353
589. Zadeh, L.A. (1975): Fuzzy logic and approximate reasoning. Syntheses, **30**, 407–428

590. Zadeh, L.A. (1975): The concept of linguistic variable and its application to approximate reasoning Part I. Inform. Sci., **8**, 199–251
591. Zadeh, L.A. (1975): The concept of linguistic variable and its application to approximate reasoning Part II. Inform. Sci., **8**, 301–357
592. Zadeh, L.A. (1975): The concept of linguistic variable and its application to approximate reasoning Part III. Inform. Sci., **9**, 43–80
593. Zadeh, L.A. (1978): A fuzzy sets as a basis for a theory of possibility. Fuzzy Sets and Systems, **1**, 3–28
594. Zadeh, L.A. (1979): A theory of approximate reasoning. Machine Intelligence, **9**, 149–194
595. Zenzo, S.D. (1986): A new many-valued logic and its applications to approximate reasoning. Information Processing, **86**, 421–427
596. Zhang, A., Marek, W. (1990): On the classification and existence of structures in default logic. Fundamenta Informaticae, **13** (4), 485–499
597. Zhang, M.Y. (1993): On extensions of general default theories. Science in China (English edition), Ser. A, **36** (10), 1273–1280
598. Zhang, S.C., Yan, X.W., Wang, C.M., Nie, W.L. (1996): Uncertain Inference Technology. Guangxi Normal University Press, Nanning, China (in Chinese)
599. Zhang, W.X., Liang, Y. (1996): Uncertainty Inference. Xi'an Jiaotong University Press, Xi'an, China (in Chinese)
600. Zhang, Y.J. (1995): A note on lattice implication algebras. J. Liaoning Engineering University, **15** (2), 535–536 (in Chinese)
601. Zhao, X.S., Wang, X.S. (1992): On the ultra-product of fuzzy subsets and its application. In: Proc. of Fuzzy Mathematics and Systems of China. Press of Hunan Science and Technology, Changsha, China (in Chinese)
602. Zhao, G.F., Xu, Y., Qin, K.Y. (2002): On the fuzzy LI-ideals and fuzzy filters of lattice implication algebras. J. Fuzzy Math., **10** (1), 137–146
603. Zhu, W.J., Xiao, X.A. (1985): Propositional calculus system of medium logic (I). J. Nature, **8** (4), 315 (in Chinese)
604. Zhu, W.J., Xiao, X.A. (1985): Propositional calculus system of medium logic (II). J. Nature, **8** (5), 394 (in Chinese)
605. Zhu, W.J., Xiao, X.A. (1985): Propositional calculus system of medium logic (III). J. Nature, **8** (6), 473 (in Chinese)
606. Zhu, W.J., Xiao, X.A. (1985): Predicate calculus system of medium logic (I). J. Nature, **8** (7), 540 (in Chinese)
607. Zhu, W.J., Xiao, X.A. (1985): Predicate calculus system of medium logic (II). J. Nature, **8** (8), 601 (in Chinese)
608. Zhu, Y.Q. (1997): On lattice implication algebras. J. Huanggang Norm. Univ., **17** (4), 11–14 (in Chinese)
609. Zhu, Y.Q., Tu, W.B. (2001): A note on lattice implication algebras. Bull. Korean Math. Soc., **38** (1), 191–195
610. Zou, K.Q., Xu, Y. (1989): Fuzzy Systems and Expert Systems. Southwest Jiaotong Univ. (in Chinese)
611. Zimmermann, H.-J. (1991): Fuzzy Set Theory and its Application. Norwell, Kluwer, MA
612. Zimmermann, H.-J. (2000): An application-oriented view of modeling uncertainty. European J. Operational Research, **122**, 190–199

Index

Druck: Strauss Offsetdruck, Mörlenbach
Verarbeitung: Schäffer, Grünstadt